# Integrative Plant Anatomy

# Integrative
# Plant Anatomy

WILLIAM C. DICKISON

HARCOURT
ACADEMIC
PRESS

San Diego   San Francisco   New York   Boston
London   Toronto   Sydney   Tokyo

This book is printed on acid-free paper. ∞

*Cover credit: Charles Krebs/The Stock Market*

Academic Press
A Harcourt Science and Technology Company
525 B Street, Suite 1900, San Diego, California 92101-4495, USA
http://www.academicpress.com

Academic Press
Harcourt Place, 32 Jamestown Road, London, NW1 7BY, UK
http://www.hbuk.co.uk/ap/

Harcourt/Academic Press
200 Wheeler Road, Burlington, Massachusetts 01803, USA
http://www.harcourt-ap.com

Library of Congress Catalog Card Number: 99-68568

International Standard Book Number: 0-12-215170-4

1001950748

PRINTED IN THE UNITED STATES OF AMERICA
99 00 01 02 03 EB 9 8 7 6 5 4 3 2 1

As I was writing the final chapter of this book, I realized that I was living the final chapter of my life. It is with this thought that I dedicate this book to my grandchildren

Andrew Campbell
Erica Grace
William Campbell

in remembrance of their grandfather.

# ◼◼ CONTENTS

## 3 Tissue Organization in Stems, Leaves, and Roots   121

## 4 Origin and Structure of the Secondary Plant Body   161

## II  EVOLUTIONARY, PHYSIOLOGICAL, AND ECOLOGICAL PLANT ANATOMY   203

## 5 Evolution and Systematics   205

## 6 Macromorphology   237

## 12 Fibers, Fiber Products, and Forage Fiber  399

## 13 Forensic Science and Animal Food Habits  421

## 14 Archaeology, Anthropology, and Climatology  437

## 15 Properties and Utilization of Wood    453

## 16 The Arts and Antiques    473

# PREFACE

During the course of my career as a structural botanist, I have occasionally been asked: What value is plant anatomy and how does it relate to other fields of study? *Integrative Plant Anatomy* was written in response to such questions and to present the subject of plant anatomy as a multifaceted, interdisciplinary, and relevant field of inquiry. Plant anatomy, or the developmental and comparative study of plant cells, tissues, and organs, is a botanical discipline with a long tradition. Many individuals have emphasized the fact that anatomy is both a descriptive and an experimental science. In other words, anatomists employ critical and extensive observation, resulting in the compilation, codification, and analysis of descriptive data, but they also use the analytical methods of the experimental sciences. Each methodology has different advantages and limitations, and each requires distinct skills on the part of the investigator. Much anatomical research and most practical courses in plant anatomy historically have included topics centered on the systematic, phylogenetic, developmental, and functional aspects of the subject. Although there are now many outstanding textbooks and comprehensive reference works that emphasize one or more of these approaches, inadequate attention has been given to the overall relationship of anatomy to other areas of botany, as well as to the nonbiological sciences, the arts, and numerous other fields of human endeavor. Clearly, a solid foundation in basic plant anatomy is required in order for individuals to fully participate in the diversity of modern interdisciplinary studies. The intent of this book is to bridge these gaps by consolidating widely scattered information into a series of chapters that highlight the principal contributions of plant anatomy to the solution of a number of current and future problems. The selection and detail of the topics is meant to provide a balanced and informative synthesis both for students of botany and for people without any special knowledge of plants but who might want to become acquainted with this interesting branch of plant biology. Most books treat the subject of plant anatomy in basically the same way, discussing cells, tissues, tissue systems, and organs in sequence. I attempt to present the discipline of plant anatomy in a different light and show how structural botany is integrated with other modern fields. At the very least, I hope that this volume will result in a renewed awareness of the many interrelated applications of this classical subject and why individuals conversant with the details of structural botany are necessary and useful.

The first four chapters provide an overview of higher plant organization and structure. This can serve as the basis for further study of the subject, or provide students of botany with an introduction to the field. Later chapters discuss these anatomical concepts and approaches at greater length in relation to more specialized topics. This book centers on seed plants, especially angiosperms. It presents the basic concepts and terminology of the study of plant anatomy, but it also emphasizes applied and economic aspects of the discipline. Examples or case histories are provided in which a knowledge of plant structure has been or could be an important tool in resolving a problem or contributing to our general understanding of some other discipline. Because the literature is voluminous, the additional readings at the end of each chapter include a sampling of recent references as well as classical citations.

In 1992, the late Professor Dr. Klaus Napp-Zinn of the University of Cologne and I agreed to collaborate on a book whose central theme would be the integration of plant anatomy with other fields. Professor Napp-Zinn died unexpectedly during the early stages of our collaboration. This volume represents the book that we had in mind, although its content reflects entirely my own judgment and work. I wish to thank my departed friend not only for his numerous contributions to the field of plant anatomy but also for his enthusiasm for this project.

William C. Dickison
November, 1999

# ACKNOWLEDGMENTS

I thank the many individuals, too numerous to mention, who have knowingly or unknowingly assisted me through their teachings and writings. Many individuals and publishers generously granted me permission to use illustrations previously printed in their publications. Where a borrowed illustration or table is used, its source is indicated. I am particularly grateful to W. Barthlott, D. Dobbins, J. A. Heitmann, Jr., A. Jones, P. G. Mahlberg, T. Perdue, J. Seago, R. Sexton, M. Simpson, T. Terrazas, and E. Wheeler who supplied me with photographs. I acknowledge my indebtedness to Michael Donoghue (Harvard University) and Elisabeth Wheeler (North Carolina State University) for allowing me to use valuable illustrations from the Bailey-Wetmore Wood Collection of Harvard University. I will always be grateful to Yvonne Strickland and Martha Hall for typing the manuscript and Susan Whitfield for preparing the illustrations. John Sperry, Edward Schneider, and Elisabeth Wheeler kindly provided invaluable professional assistance in reading, correcting, and improving specific portions of the manuscript. Thank you to Patricia Gensel for proofing the pages. I further wish to express my appreciation to David Phanco of Harcourt/Academic Press for having faith in this project and providing enthusiastic support.

Finally, and most of all, I acknowledge my wife, Marlene, for her constant love, unfailing encouragement, and sustaining interest in the progress of this book.

# I

# ANATOMICAL FOUNDATION OF THE PLANT BODY: AN OVERVIEW

# 1
# PLANT GROWTH, DEVELOPMENT, AND CELLULAR ORGANIZATION

Despite enormous differences in plant size, all higher plant bodies consist of a central axis divided into two regions, the mostly above ground **shoot** and the below ground **root.** The shoot is further differentiated into the lateral photosynthetic appendages, the **leaves,** and the axial component on which the leaves are attached, the **stem.** Reproductive organs, sporangia and/or sporophylls, form at reproductive maturity. Each organ in turn is composed of **tissues,** collections of one or more cell types that share a common physiological and structural function and often a common origin. As the shoot grows, it forms new increments of stem and new leaves in an acropetal direction, that is, from the base of the plant upward.

## PLANT GROWTH

Growth is a genetically programmed developmental process involving cell division, selective cell enlargement, and maturation. Roots, like shoots, also become extended apically, increasing in length. This manner of growth is traceable to a small region of cells at the tip of the stem and the root that is characterized by the formation of new, and eventually more differentiated cells, by repeated subdivision and cell expansion. These regions of potentially unlimited growth and active cell division are termed **meristems.** They represent a defining feature of plants. All growth and organ formation of the plant is initiated and largely controlled by the meristems and their subordinate regions.

The concept of meristem refers to the collection of actively dividing cells, termed **initials,** plus their most recent **derivatives.** Functionally, vegetative meristems are **indeterminate,** that is, not limited in their capacity to continue

**3**

the development of the axis. The actual elongation of the axis occurs by means of cellular enlargement, a process that can result in a 10- to 1000-fold increase in cell volume and that is associated with an increase in internal turgor pressure and irreversible changes in the cell wall. Vascular plants are characterized by a so-called open growth pattern in which new cells, tissues, and organs are regularly initiated throughout the plant's life. This type of growth relies upon the existence of small aggregations of cells that occur at specific sites in the plant body, retaining their genetic potential of active cell division.

In addition to initiating new tissues and organs and controlling the pattern of cell and tissue differentiation, meristems communicate signals to and from the remainder of the plant body, while at the same time maintaining themselves as organized, formative regions. It has long been assumed that variations in the behavior of meristems, that is, the frequency and orientation of division planes, are major factors in determining the shape of plant organs. Recent studies, however, have challenged a number of long-standing assumptions about the significance and function of cell division in organ formation and the role of localized meristems in morphogenesis. Nevertheless, there is agreement that the remarkably complex process of overall plant growth is dependent upon the presence of permanently "embryonic" meristematic tissue.

## Meristem Function

It is possible to identify meristems by their topographic positions and by their capacity for cell division. Meristems restricted to the extreme tips of stems, branches, and roots are called **apical meristems** and their activity results in **primary growth** and an associated increase in axis length. The concept of the **promeristem** has been utilized to refer to the core of centrally positioned initial cells and their immediate derivatives that have yet to show visible evidence of differentiation and a progression to a more specialized form or function. The tissues that arise directly from these apically located regions are called **primary tissues.** They contribute to growth in length as well as to limited growth in the width of organs. The growing tip of the shoot consists of a pool of meristematic cells whose derivatives result not only in organ length but also in the formation of regularly spaced, undifferentiated subapical mounds of cells on the flanks of the apical dome. These cell groups are called **primordia.** They develop into new leaves, branches, or floral parts in regular sequence following a course of development that is species specific. The apical meristem of roots, on the other hand, is subapical and not only contributes derivatives to the root tissue that account for an increase in root length but also produces cells distally to form the outer root cap. The exact boundary, however, between the core of actively dividing cells and the proliferative subapical zone is often difficult to clearly demarcate.

The differential expansion and multiplication of cells or groups of cells is important in these growth processes. Leaf primordia that are more removed from the apex become progressively larger as they grow outward. The upper or **adaxial surface** of the leaf primordium is directed toward the stem apex, whereas the lower or **abaxial surface** faces away from the side. The resulting **dorsiventral symmetry** is associated with two sides that are morphologically and anatomically different. In the upward angle or axil (formed by the leaf and petiole with the stem) is a small mound of cells that represents the fore-runner of a bud. This is the **axillary bud primordium.** It contains the **axillary meristem,** which when stimulated will give rise to a branch shoot indistin-guishable from the primary shoot.

By definition, the **primary body** of the plant and its constituent **primary tissues** include all the tissues that arise more or less directly from the activity of the apical meristems and cell divisions in subapical regions. On the other hand, **lateral meristems,** termed **cambia,** are those regions of dividing cells positioned as a continuous thin sheet on the periphery of the axis, and whose activity is re-sponsible for axis thickness. Lateral meristems participate in **secondary growth** and the formation of **secondary tissues** that will in turn form the **secondary plant body.** The major lateral meristems are the **vascular cambium** that forms

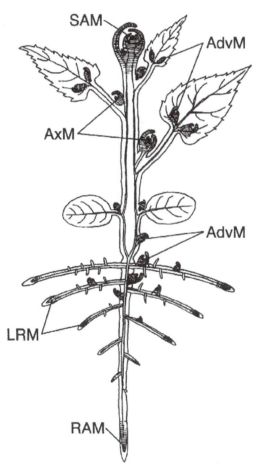

**FIGURE 1.1**  Vegetative shoot and meristems. An idealized dicotyledonous plant is shown with adventitious buds forming on leaves, roots, and hypocotyl. Abbreviations: AdvM, adventitious meristem; AxM, axillary meristem; LRM, lateral root meristem; RAM, root apical meristem; SAM, shoot apical meri-stem. Reprinted with permission from Kerstetter and Hake (1997), *Plant Cell* **9**, 1001–1010. Copyright © American Society of Plant Physiologists.

secondary conducting tissues and the **cork cambium** that deposits an outer zone of periderm. A variety of monocots have **intercalary meristems** located at the base of internodal regions. Intercalary meristems produce cells upward into the internode, where they differentiate progressively. There are also specialized meristems such as the so-called **thickening meristem** of some arborescent monocotyledons and various anomalous types of cambial activity.

In woody dicotyledons and all gymnosperms, the region of meristematic tissue positioned between the primary food- and water-conducting cells becomes transformed into the **vascular cambium.** In a narrow sense, the vascular cambium consists of one layer of cells. Each cambial cell divides in a tangential plane to form either a new xylem or phloem mother cell and another meristematic cell. In this way, the vascular cambium is responsible for the increase in diameter of gymnospermous and dicotyledonous roots and stems. The increase in diameter of most monocotyledonous stems and roots results from the enlargement of cells produced in the apical meristem because a vascular cambium is absent in the majority of those plants.

## Cell Differentiation

Meristems are composed of two categories of cells, the dividing initials that are capable of proliferation and self-maintenance and the immediate undifferentiated derivatives of cell division that represent the progenitors of the various kinds of cells and tissues possessed by an organ. Under the appropriate influence, these recent derivatives become **differentiated** and changed into various types of mature cells and tissues. From an anatomical point of view, cell differentiation is related to changes in cell size and shape, modifications of the wall, changes in staining characteristics of the nucleus or cytoplasm, as well as the degree of vacuolation and the ultimate loss of the protoplast in some cases. At the subcellular level, major alterations in organelle composition and structure can occur. As cells differentiate, they become specialized to perform a particular function. Generally speaking, the more removed from the embryonic state a cell becomes, the more specialized it is. The mature condition may be irreversible, as in the case of fibers and tracheary elements that have a layered secondary wall and are devoid of a protoplast when functional. In the case of living, nucleated parenchymatous cells, however, the mature condition is potentially reversible, so that a cell retains its capacity for growth and enlargement if subjected to specific influences. If a parenchymatous cell becomes transformed into another cell type without undergoing an intervening cell division, it is said to undergo **transdifferentiation.** When a parenchymatous cell begins to divide and then changes into another cell type, it undergoes **dedifferentiation.**

The transformation of cultured *Zinnia elegans* leaf mesophyll cells into nonliving tracheary elements is an excellent example of transdifferentiation occurring at the cellular level in higher plants. This species is currently being used as an experimental system to study development. Isolated leaf mesophyll cells are induced to differentiate into tracheary elements *in vitro* by stimulation with the appropriate ratio and amounts of the plant hormones auxin and cytokinin. Auxin also plays a role in the differentiation of wound-induced tracheary elements from parenchymatous elements. The *de novo* origin of adventitious meristems from differentiated tissues provides an example of dedifferentiation.

During the process of differentiation, individual cells or groups of cells undergo **determination,** during which there is a commitment to a particular course of development. As a result, undifferentiated cells assume new identities and eventually become transformed into specialized cells that transport water or food or carry out some other vital function. A cell is considered determined if it has undergone a self-perpetuating change of cytological character states that distinguishes it from other cells and irreversibly commits it to a specialized course of development. Following this definition, a leaf mesophyll cell is differentiated but not strictly determined because its pathway of differentiation is reversible. Water-conducting tracheary cells, on the other hand, become fully determined during their differentiation because they irreversibly enter a differentiation pathway that leads to a dead element.

As tissues mature, certain cells form specific patterns of distribution. The concept of **pattern** refers to the spatial arrangement of elements, in this case cells. The process of patterning involves identifying which undifferentiated cells become transformed into particular mature cell types. It does not include the cytological steps of their differentiation. Cell and tissue patterning is under genetic control and results from complex cell division rates and planes, cell-to-cell communication, and the position of cell types relative to one another.

Determination is a complex and poorly understand process. The structural and physiological characteristics of cells and tissues are the result of a response to a developmental signal (**induction**) and then a commitment to a specific developmental fate. The determined state is expressed in the form of leaf cells, vascular conducting cells, fibers, and so on. This sequence of events is thought to be strongly related to the final position of a cell within the plant and to the position of cells relative to one another. This hypothesis is

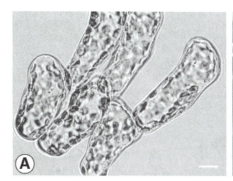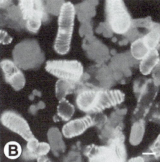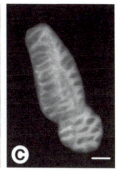

**FIGURE 1.2**  Morphological changes during tracheary element differentiation in cultured *Zinnia elegans* cells. (A) Isolated leaf mesophyll cells as introduced into culture and observed with phase contrast light microscopy. Scale bar = 10 μm. (B) Cultured cells 96 hr after isolation as viewed with fluorescence microscopy. Differentiated tracheary elements are noted by the yellow autofluorescence from their lignified secondary cell walls. Undifferentiated cells are noted by red autofluorescence from their chloroplasts. Scale bar = 20 μm. (C) Differentiated tracheary cells stained with Calcofluor white to reveal the cellulose-containing secondary cell wall. Scale bar = 10 μm. Courtesy of A. T. Groover and A. S. Jones. (See Color Plate.)

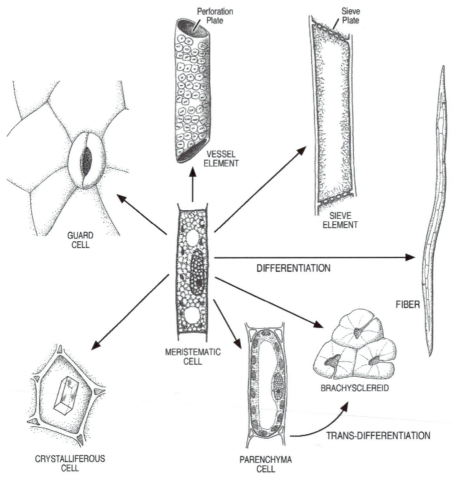

**FIGURE 1.3** Representative end products of plant cell differentiation from a meristematic cell. The parenchyma cell retains a living protoplast and is able to transdifferentiate into another cell type, such as a sclereid, that becomes devoid of living contents at maturity.

supported by gene expression patterns and tissue development in plants capable of a specialized type of photosynthesis known as $C_4$ photosynthesis.

$C_4$ plants have a distinctive leaf structure, termed **Kranz anatomy,** that involves two major photosynthetic cell types: mesophyll cells that initially absorb the $CO_2$ entering the leaf through the stomata, and bundle sheath cells that immediately surround the veins and collect $CO_2$ from the mesophyll. Sugars are manufactured within the latter cells and are then transported into the conducting phloem. Evidence indicates that these two closely spaced and interacting cell types interpret positional information that is distributed around each vein to express cell-specific genes correctly during the differentiation of the Kranz pattern. The fact that position-dependent information is essential in establishing cell fate also has been documented in experiments performed on developing roots. If an epidermal initial cell is removed, the underlying cortical cell alters its course of differentiation and becomes transformed into an epidermal cell. If a cortical cell initial is removed, an underlying pericycle cell replaces the missing cell. It has been hypothesized that daughter cells within the root apex produce positional signals that control cell fate.

The morphogenetic signals that control development are largely unknown. They lie at the heart of some of the most difficult, unsolved problems of morphogenesis. How do cells receive positional information from their neighbors and how does this information initiate the events leading to cell differentiation? One possibility involves a functional connection between the plasma membrane and the cell wall substrate. Surface adhesions may form interactive networks that are involved in the perception and movement of positional information. Such networks then may function directly in sensing the environment, polarity, tropic movement, and differentiation. It is important to emphasize that plant cells are not mobile and that their position is fixed at the time of cell division. In plants, therefore, there are none of the complexities of cell mobility and adhesion that result in different morphogenetic phenomena in animals. However, plant cells do undergo considerable enlargement during the differentiation process, and this may produce cell displacement, intercellular spaces, and particular cell arrangements. As a consequence of the lack of plant cell mobility, cell position appears to be more important than cell lineage in influencing the direction of plant cell maturation.

A complete understanding of the patterns and underlying causes of cellular determination and differentiation, as well as the origin, development, and structural modifications of divergent organs, remains a major challenge to plant scientists. In recent years, the concept of **heterochrony** (changes in developmental timing) has been advocated as a major influence on the evolutionary change in plant and organ form. However, in order to understand how tissues arise (**histogenesis**) and organs develop (**organogenesis**), it will be necessary in subsequent chapters to clarify the remarkable structural characteristics of vegetative meristems and how they function.


## TERMS USED IN THE DESCRIPTION OF PLANT CELLS

The minute, careful, and at times even tedious observations and descriptions that form the bases of plant anatomical study require an understanding of basic terminology denoting cellular orientation. The terms "transverse," "axial," "radial," and "tangential" are used to describe cell and tissue surfaces in sectioned tissues with reference to the long axis of the stem, branch, or root. A **transverse section** or wall (also known as a transection or cross section) pertains to the side perpendicular to the longitudinal axis. A **radial** (or axial) **section** or cell wall is a longitudinal surface along a radius of the axis. A **tangential section** or cell wall refers to a longitudinal surface at right angles to an axis radius. Cell orientation in relation to the long axis may not be clear or may be descriptively unimportant for individual cells, as, for example, when the cells of a tissue become disassociated into a **maceration** of separate cells. Here the terms "end wall" and "longitudinal wall" or "lateral wall" are most often used in relation to the long axis of the cell.

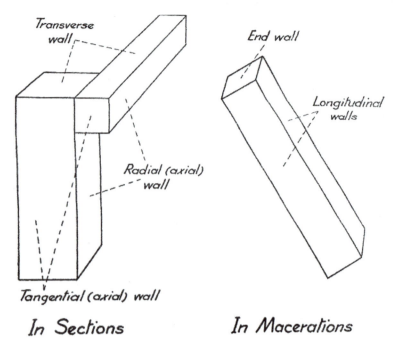

**FIGURE 1.4**   Terminology for the orientation of cell walls. From Committee on Nomenclature, International Association of Wood Anatomists (1957).

Separate terminology is used to describe the orientation of planes during cell division. A cell division that occurs so that the new cell wall is oriented at right angles to the surface is termed an **anticlinal division.** A **periclinal division,** on the other hand, forms a new wall that is parallel to the surface or to an organ, or that is parallel to the circumference of the organ in which it occurs.

## EMBRYOLOGY

During the sexual life of a higher plant, the primary plant body is initiated as a single-celled **zygote** following fertilization. The first division of the zygote is usually a very precise asymmetrical division that gives rise to a smaller **apical (or terminal) cell** and a larger **basal cell.** Although the first division of the zygote is often transverse to the future embryonic axis, different dicotyledonous taxa have zygotes in which the initial division is longitudinal, oblique, or even variable. The apical cell undergoes subsequent cell divisions and gives rise to most of the embryo proper, whereas the basal cell undergoes a limited series of transverse divisions to form the typically filamentous **suspensor.** The metabolically active suspensor attaches the embryo proper to surrounding nutritive tissue and functions in the absorption and transport of various nutrients to the developing embryo. It is fully differentiated early in development. The distal portion of the suspensor in contact with the embryo proper is known as the **hypophysis.** In size and appearance, the mature suspensor differs greatly among flowering plants, ranging from a single cell to a massive column of several hundred cells. Some plants form large **haustoria** that invade surrounding tissues and apparently facilitate the transfer of nutrients to the developing embryo.

Although many flowering plants have predictable patterns of cell division during the early stages of embryogenesis, a number of species exhibit irregular patterns of growth even in these earliest stages. In either case, the initial cell divisions of the zygote result in a four-celled, eight-celled, and sixteen-celled (often globose) parenchymatous **proembryo.** Out of the intricately regulated patterns of cell division that characterize early embryology will ultimately originate all the diverse cells and tissues of a completely developed embryo. As embryonic development proceeds in a dicotyledonous plant, a pair of **cotyledons** or seedling leaves are initiated at the embryonic shoot apical meristem, and the embryo becomes heart shaped. The shoot apex forms a small, dome-shaped cluster of cells between the expanding cotyledons and may form leaves prior to seed germination. In some plants, the shoot meristem remains essentially inactive until the first stage of germination.

A number of critical events occur during the transition from the globular to the heart-shaped stages of embryogenesis and the early development of cotyledons. The embryo proper establishes bilateral symmetry with the initiation of cotyledons. Future major **tissue systems** subsequently become established in the undifferentiated state as transitional meristematic zones or regions. The future surface, or epidermis, of the plant becomes recognizable as an incompletely differentiated outer layer of cells (called the **protoderm**) following periclinal divisions within the globular embryo. The single-layered protoderm subsequently differentiates into the outermost layer of cells, the **epidermis.** The immature ground or fundamental tissue is called **ground meristem** and forms the future **cortex** and **pith** regions. The undifferentiated primary vascular or conducting tissue is derived from **procambium (provascular) tissue.** Even though they may continue to divide, these transitional **primary meristematic tissues** differ from the more actively dividing initial cells by undergoing elongation and showing various wall modifications. Although all permanent primary tissues arise from these three primary meristems—protoderm, procambium, and ground meristem—little is known regarding the time and pattern of differentiation of the vascular cells in the embryo or early seedling. In some cases, no differentiating or mature vascular elements are evident in the mature embryo. In other taxa, mature or differentiating tracheary cells have been described in the fully mature embryo. Eventually the triploid nutritive **endosperm** tissue within the seed, present as free nuclei during early stages of embryo growth, begins to become cellular. A period of temporary dormancy may or may not designate the end of embryo development. Embryo development and subsequent plant growth are continuous processes in some plants. Both the suspensor and cellular endosperm degenerate as the mature embryo begins to grow.

The mature dicotyledonous embryo is distinguished by the organization of the shoot and root apical meristems at either end of its short axis. The low apical dome of the shoot occurs between the two cotyledons and is characterized by the early establishment of cytohistological zonation. The apical meristem of the young root, or the **radicle,** becomes defined by the covering

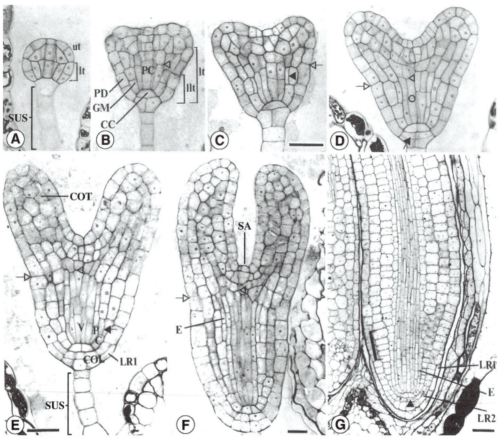

**FIGURE 1.5** Median longitudinal sections showing embryogenesis of the dicotyledon *Arabidopsis thaliana*. (A) Early globular stage. (B) Triangular stage. (C) Early heart stage. (D) Mid heart stage. (E) Late heart stage. (F) Mid torpedo stage. (G) Mature embryo stage. Symbols: Open arrowhead, apical wall of four-tier procambium; Open arrow, protoderm indention; Closed arrow, arrowheads and open circle, cell walls marking divisions. Abbreviations: ut, upper tier; lt, lower tier; llt, lower region of lt; PD, protoderm; PC, procambium; GM, ground meristem; CC, central wall; VP, vascular primordium; C, cortex; COL, columella; LR1, lateral root cap layer; E, endodermis; LR2, second lateral root cap layer; SUS, suspensor; COT, cotyledon; SA, shoot apex. Bar = 25 µm. Reprinted with permission of Scheres *et al.* (1994), *Development* **120**, 2475–2487. Copyright © the Company of Biologists Limited.

root cap. In more mature plants, apical meristems are quite varied in form, size, and organization and are composed of rather small, thin-walled cells that have prominent nuclei and are nonvacuolated. In such non-seed-producing vascular plants as ferns, horsetails (*Equisetum*), and spike mosses (*Selaginella*), a single initial cell is present in the apices of the rhizome, aerial stems, and roots. The rootless whisk fern (*Psilotum*) also has a single apical cell at the stem apex. The apical cell is often tetrahedral, appearing as an inverted pyramid. Derivative cells are formed on three sides of the apical initial. Axis development originates in and is maintained by this single apical cell, which divides in a regular and predictable manner and contributes all other cells of the primary body. The shoot apex in higher seed plants, in con-

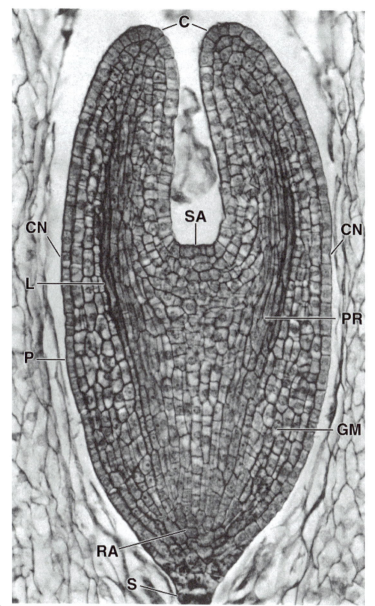

**FIGURE 1.6**   Median longitudinal section of the embryo of *Nerium oleander* (Apocynaceae) showing undifferentiated primary tissues, × 340. Abbreviations: C, cotyledons; CN, cotyledonary node; GM, ground meristem; L, laticifer; P, protoderm; PR, procambium; RA, root apex; S, upper part of suspensor; SA, shoot apex. Courtesy of P. G. Mahlberg.

trast, consists of multiple surface layers organized into zones variously demarcated on the basis of cell size, degree of vacuolation, wall thickness, staining characteristics, and frequency and plane of cell division. The demonstration that some apical meristems possess a small central quiescent zone of cells that are relatively inactive mitotically has further challenged traditional histological views of apex organization and function.

Upon germination, the active shoot apical meristem grows and is genetically programmed to initiate new leaves and their associated **nodes** and **internodes** (i.e., new shoots) in regular sequence. Apical meristems in the axils of leaves produce axillary shoots. These in turn may form their own lateral shoots. As a result of such development, the plant bears a three-dimensional system of branches upon the main or primary stem. This represents the vegetative stage of growth. Under the proper conditions, the angiospermous plant enters a phase of reproductive growth in which some vegetative apical meristems undergo changes in the pattern of cell division and transition into an indeterminate or determinate reproductive floral apex, which produces an inflorescence or flower.

## THE PLANT CELL

The plant kingdom includes taxa of varied size and structure. Despite the enormous variation, however, all plants are multicellular and **eukaryotic,** that is, each cell possesses a membrane-bound nucleus with chromosomes. In a generalized cell, the term **protoplast** denotes a mass of semiliquid, colloidal **cytoplasm** and an associated dense **nucleus** that contains the genetic material. The outer boundary of the cytoplasm is defined by a very flexible **cell** or **plasma membrane** that effectively separates and isolates the internal cell components from the surrounding external environment. It is through this membrane that materials must pass in order to enter and leave the cell.

The cytoplasm displays an internal compartmentalization of membranes that encompasses an array of cellular **organelles,** within or on which various metabolic processes occur. Most organelles are composed of membranes that contain diverse enzyme complements that lead to metabolic channeling and act as barriers against free diffusion into and out of the organelle. Compartmentalization allows for specialization and a division of labor for efficient metabolic activity. The nonmembranous area of the cytoplasm is referred to as the **hyaloplasm,** or **cytosol,** and contains an aqueous mixture of ribosomes and enzymes, as well as the intermediary and end products of metabolism. The cytoplasm surrounds a large, fluid-filled space, the **vacuole,** and is separated from it by a delicate membrane. Plant cells are surrounded by an outer, more or less rigid **cell wall** that influences cell differentiation and growth, cell-to-cell communication, and the exchange of materials between cells. Exactly how the term "cell" should be used has sometimes been a matter of some disagreement among plant biologists. Most traditional plant scientists use the term "cell" to refer to the protoplast and the surrounding wall. Others restrict the word "cell" to mean only the protoplast. These scientists view the cell wall as an elaborate extracellular matrix that encloses the cell.

### The Cell Wall

One of the most distinctive structural features of plant cells is the surrounding wall, characteristically composed of the carbohydrate cellulose deposited in fibrillar form. The cell wall is widely considered to be a nonliving entity or secretion of the protoplast that further protects and supports the protoplast while strengthening the cell and tissue to which it belongs. Cell walls provide rigidity and define cell shape. They must be strong enough to maintain turgor pressure, yet they must also possess properties capable of permitting cell expansion. The wall is a dynamic structure that grows and has the ability to alter both its shape and composition. The rigid cell wall represents a major design innovation and enabled plants to grow as tall, erect organisms with a maximum of exposed surface area. Cell walls represent the bulk of plant biomass.

The wall of a plant cell is not chemically and structurally homogeneous. Rather, the ontogenetic development of the wall usually results in the production of more or less clearly defined layers. Different cells possess wall layers of varying composition, thickness, and structure. Cell walls play important roles in plant development, cell-to-cell communication, physiology, and environmental adaptation and stress responses. These varied functions are reflected in diverse wall structures. Historically, much attention has centered upon the study of plant cell walls, and both morphological and physiological concepts of cell wall organization have been advocated. Structural botanists currently recognize a series of major wall layers, often three, that can be distinguished by their chemistry, ultrastructure, and refractive properties when viewed under polarized light. Each of the three successive layers has significance for the mechanical properties of the cell and tissue.

## The Cell Plate and Middle Lamella

The cell wall first becomes visible as a **cell plate** that arises during late telophase of mitosis. The cell plate is a thin layer of largely pectic materials laid down centrifugally across the **phragmoplast,** the microtubular structure that forms midway between the two nuclei and disassembling spindle during **cytokinensis.** The cell plate grows in the region of the phragmoplast where the ends of the microtubules overlap and eventually is in continuity with the already existing wall. Cell plate formation consists of the creation of a plate-like membranous network that is derived from the fusion of Golgi-derived vesicles in the equatorial plane. This is followed by cell wall assembly within this network as the margins of the network unite with the parent cell membrane and cellulose synthesis begins. As more pectic substances are synthesized by the dictyosomes (Golgi apparati) and transported to the cell plate in vesicles, the cell plate is transformed into the **middle lamella.** The middle lamella acts as an intercellular glue to bind the walls of daughter cells together. It is regarded as the first true cell wall layer. This layer consists largely of highly hydrated, pectinaceous substances and can be identified as an extremely thin layer between two adjacent cells. The middle lamella is **optically isotropic,** which means that it is composed of substances having the same optical properties along all axes. Under the light microscope, the middle lamella is typically at the limits of resolution.

## The Primary Wall

The **primary wall** is the first readily visible layer of the cell wall, and its formation accompanies extension growth. It develops on either side of the

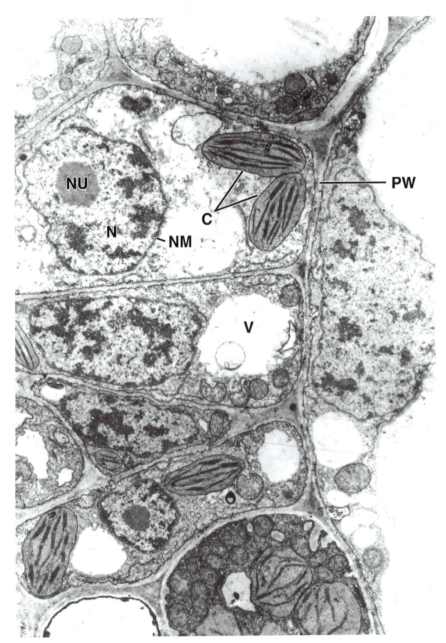

**FIGURE 1.7** EM of pea (*Pisum*) leaf mesophyll cells showing primary wall and cellular organelles. Abbreviations: C, chloroplast; N, nucleus; NM, nuclear membrane; NU, nucleolus; PW, primary wall; V, vacuole. Courtesy of T. Perdue.

middle lamella when two cells are adjacent and largely determines cell shape and size during plant growth and development. It is composed of a continuous interconnected, fortifying system of aggregated, threadlike cellulosic **microfibrils** that result from the simultaneous polymerization and crystallization of cellulose molecules. **Cellulose** is a long polymeric carbohydrate composed of unbranched chains and a degree of polymerization of up to 10,000 glucose units. Cellulose molecules consist of 1→4–linked β-D-glucan subunits joined into crystalline microfibrils of indeterminate length and varying width and degree of order. Individual microfibrils are constructed of highly organized crystalline regions and less crystalline "amorphous areas."

The polymerization of sugar molecules into cellulose involves activity of a **cellulose-synthesizing complex** containing cellulose-synthesizing enzymes that are clustered together within the plasma membrane. During synthesis, a **globular component** is bound to the terminus of a microfibril. The globular component is positioned primarily on the outer surface of the plasma membrane, whereas a protein cluster of contiguous particle subunits, called a **rosette,** is located on the innermost surface of the membrane. The rosette and globular complex constitute what is believed to be the functional **terminal complex** in higher plant cells. It has been proposed that the terminal complex moves in the plane of the membrane during the assembly of sugar monomers into cellulose microfibrils. While passing from the interior to the exterior surface of the membrane, the microfibrils become embedded in the cell wall.

Microfibrils are highly stable and their crystalline structure plays a major role in the structural characteristics of the cell wall. Differences in the orientation and other mechanical properties of cellulosic fibrils within cell walls are primarily responsible for the high specific strength and resistance to tensile

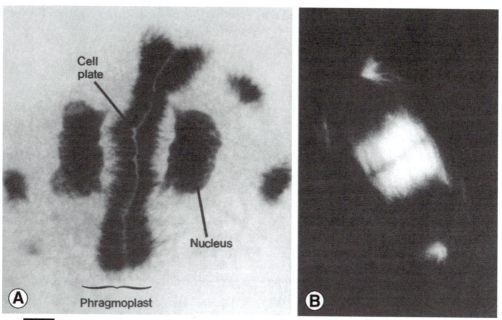

**FIGURE 1.8**  The phragmoplast of dividing plant cells. (A) The microtubules of the phragmoplast are visualized by an immunogold procedure using an antibody specific for tubulin. The developing cell plate is indicated. (B) The phragmoplast microtubules are visualized by an immunofluorescence procedure using a fluorescein-labeled antitubulin antibody. Both the immunofluorescence and immunogold procedures demonstrate that the phragmoplast consists of two overlapping sets of microtubules. Reprinted with permission from Fosket (1994). "Plant Growth and Development.  A Molecular Approach," Academic Press.

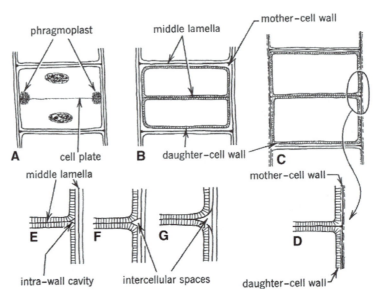

**FIGURE I.9**   Concepts regarding the adjustments between new and old cell walls after cell division. (A) Cell plate has been formed. (B) Two primary walls cemented by intercellular substance occupy the position of cell plate; primary daughter-cell walls have been laid down on the inside of primary mother cell wall. (C, D) Daughter cells have expanded vertically and the mother-cell wall has been stretched and ruptured opposite the new wall. Thus, old and new intercellular lamellae have joined. (E, G) Establishment of continuity between old and new middle lamellae through formation of intercellular space. (E) Appearance of cavity between daughter- and mother-cell walls. (F) Dissolution of mother-cell wall next to cavity. (G) Completion of change of the intra-wall cavity into an intercellular space. Reprinted with permission from Esau (1965), "Plant Anatomy," 2nd ed., John Wiley & Sons.

forces that cells exhibit at the cellular level. In the primary wall, the microfibrils can have a more or less random or dispersed orientation, or they can be predominantly transverse in arrangment, embedded in a less ordered matrix rich in **hemicellulosic** materials and pectic polysaccharides. The primary wall is usually poor in cellulose, being composed of less than 25 to 30% fortifying cellulose. The other hydrophilous matrix constituents are hemicelluloses (30%), pectins (35%), and glycoproteins (1–5%), on a dry weight basis. Water is also an important constituent of this wall layer, about 75% by mass.

Hemicelluloses are a heterogeneous group of celluloselike polysaccharides that are variable in composition and readily decomposable into several differ-

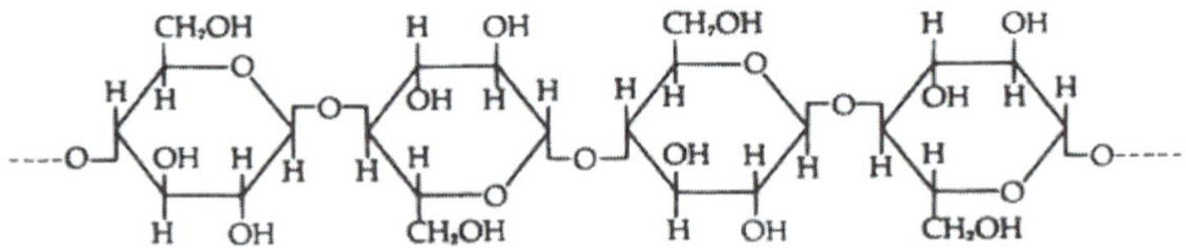

**FIGURE I.10**   Small segment of cellulose molecule showing chemical structure. The OH groups that project from both sides of the chain form hydrogen bonds with neighboring OH groups, resulting in bundles of cross-linked parallel chains.

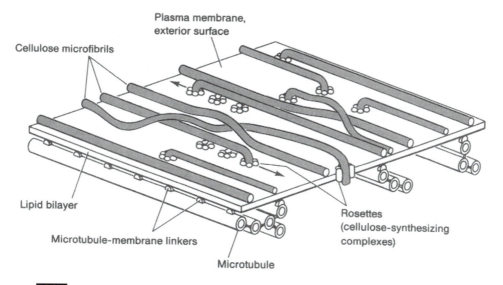

Plasma membrane, exterior surface

Cellulose microfibrils

Lipid bilayer

Microtubule-membrane linkers

Microtubule

Rosettes (cellulose-synthesizing complexes)

**FIGURE 1.11**   Model illustrating how cortical microtubules might determine the orientation of cellulose microfibril deposition in the cell wall. Reprinted with permission from Fosket (1994), "Plant Growth and Development. A Molecular Approach," Academic Press.

ent simple sugars. They are also amorphous, polymeric carbohydrates, with a slightly branched structure and a degree of polymerization of about 200 to 500 sugar units per molecule. Hemicelluloses are synthesized in the Golgi apparatus where they are packaged into vesicles and subsequently deposited in the wall. The structural similarity between cellulose and hemicellulose results in strong hydrogen bond interactions between these two polymers. **Xyloglucan** is the predominant hemicellulose in the primary wall of most plants. It is believed to cross link and anchor cellulose microfibrils and establish a strong three-dimensional network. Xyloglucan is a branched polymer consisting of a 1→4–linked β-D-glucan backbone with short side chains containing xylose, galactose, and often, although not always, a terminal fucose. Cell wall **pectins** are a diverse collection of polysaccharides that, like hemicelluloses, are secreted by the Golgi apparatus and are a major component of the wall matrix. Pectins are acidic polysaccharides in which the cellulose and hemicellulose framework is interwoven. As cell walls grow, vesicles containing these wall materials or precursor molecules move across the plasma membrane and become deposited within the growing wall.

In addition to carbohydrates, the cell walls of many species have a structural and enzymic protein component. One extensively studied primary cell wall protein is named **extensin**. The most abundant cell wall proteins are rich in the amino acids hydroxyproline and glycine, which are found in highly repetitive sequences. When the hydroxyproline content in the wall is low, glycine levels often are high. Cytoplasmic microtubules are arranged in the same orientation as the developing cell wall microfibrils, pointing to a probable role in transporting materials to the site of wall synthesis or in orienting the deposition of wall substances.

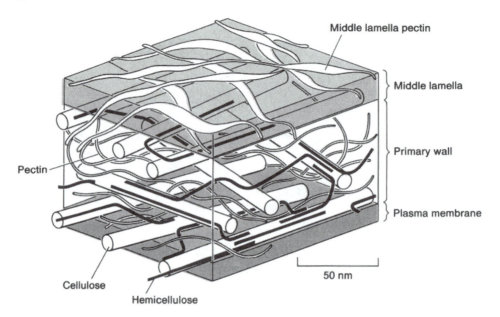

Middle lamella pectin

Middle lamella

Primary wall

Plasma membrane

Pectin

Cellulose

Hemicellulose

50 nm

**FIGURE 1.12** Simplified model for the structure of the cell wall showing the interactions of the three classes of polysaccharides. Hemicellulose xyloglucans adhere tightly to the surface of the cellulose microfibrils and cross-link them. The cellulose microfibrils probably are completely coated with hemicellulose chains. The pectins are considered to form a separate network of fibrous molecules that interdigitate with the cellulose and hemicellulose network, except in the region of the middle lamella, which is composed primarily of pectin. Reprinted with permission from Fosket (1994), "Plant Growth and Development. A Molecular Approach," Academic Press.

The primary wall is optically **anisotropic,** meaning that its wall materials have unequal optical properties along different axes. As a result, primary walls exhibit **birefringence** (appear bright) when viewed under polarized light. Primary walls are constructed so that they are capable of expanding and increasing in surface area as the cell grows. As a result, the terminology used to describe plant cell walls has often been a contentious issue. According to one view, the primary wall is best characterized from a structural or textural viewpoint, with cellulose fibrils arranged in a dispersed texture. Longitudinal growth of the wall is not considered important. The later-forming secondary wall, in contrast, is ultrastructurally layered. According to a second, more physiological view of the primary wall, any wall should be regarded as primary as long as it grows in area by elongation or general extension. Following this definition, regions of the secondary wall must be called "primary" as long as the cell is expanding. Some authors prefer the term "growing walls" for all expanding portions of the cell wall.

The rate and degree of cell elongation and expansion are correlated with the rate and degree of organ growth. The question of how plant cells and their surrounding walls expand, however, has intrigued plant scientists for nearly two centuries. It is still not completely understood. Cell growth or expansion is a regulated and highly cell-specific irreversible process that occurs through water uptake and turgor pressure accompanied by cell wall synthesis and expansion. Two major interpretations of wall expansion have been advocated. One view suggests that wall expansion primarily results from the secretion and synthesis of wall materials. Studies indicating that cell wall expansion is coordinated with wall synthesis, so that walls rarely become thinner as cells enlarge, appear to support this view. Wall deposition is highest in the zone of maximal cell expansion in young stems and roots. Cell expansion is usually greatest in the direction perpendicular to the previous division plane. Plant cells frequently elongate 10 times their length following their origin from meristematic initials. In some cases, cells can elongate 100-fold. This can be seen in root apices where cells are arranged in longitudinal files, and divisions perpendicular to the long axis of the root contribute additional cells to each file. The resulting daughter cells expand in the long radius of the root axis, causing root elongation. The orientation of cell division and expansion is governed to a large degree by the system of parallel microtubules in the cytoplasm that regulate expansion by guiding the deposition of cellulose microfibrils. Cell wall microfibrils are deposited in the same direction as the underlying microtubules. As a result, the microfibrils girdle around the cell. Major cell expansion occurs in the direction transverse to the orientation of microtubules and cellulose microfibrils.

An alternative view of wall expansion envisions a biochemical loosening of the wall followed by a turgor-driven expansion of the wall polymer network. This idea relates the elongation of cell walls (wall stretchability) to alterations of chemical bonds between cellulosic molecules by acid-activated enzymes in association with the hormone auxin. Clearly the process of cell expansion involves many interrelated processes. Following the cessation of growth at the time of maturation, cell walls typically become more rigid and less susceptible to expansion. This process of wall rigidification is thought to occur as a result of structural modifications in its molecular components.

Cells with only primary walls are usually metabolically active and conspicuously vacuolated. They are able to undergo various reversible changes. For some cell types, the primary wall may become very thick, but more often it is thin. Most primary walls are approximately 0.1 µm in thickness. Especially thick primary walls are seen in the endosperm tissue of some seeds (*Diospyros, Aesculus, Phoenix*), where food storage has significantly increased wall thickness. The primary wall may not be of uniform thickness, but instead may contain thin areas known as **primary pit fields** through which numerous, clustered plasmodesmatal complexes can pass to connect the protoplasts of adjacent cells. The middle lamella and primary walls of two adjacent cells combine to constitute what is conveniently termed the **compound middle lamella,** visible with the light microscope. The compound middle lamella is sometimes composed primarily of a complex organic molecule called **lignin,** along with some pectins, cellulose, hemicellulose, and other minor constituents. In mature tissues, it is common to observe **intercellular spaces** between cells, where the primary walls of adjacent cells have separated or never joined. The sum total of the contiguous plant cytoplasm within the plant body that is connected by plasmodesmata is referred to as the **symplast,** whereas the region of the plant occupied by cell walls and intercellular spaces outside the plasma membranes is the **apoplast.**

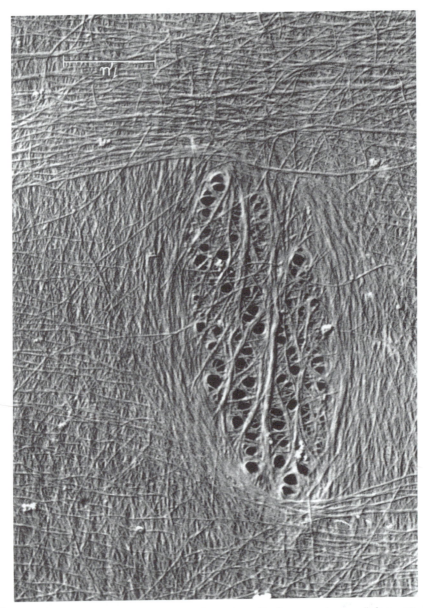

**FIGURE 1.13** Electron micrograph of primary wall of a parenchyma cell from oat, *Avena* (Gramineae), coleoptile. The parallel-oriented microfibrils at top occurred in one of the angles of the cell. Other microfibrils show a random orientation. Plasmodesmatal pores are clustered in a primary pit field, × 26,000. Reprinted with permission from Böhmer (1958), *Planta* **50,** 461–497.

Associated with the formation of intercellular spaces in a number of widely unrelated vascular plants are structures known as **intercellular pectin protuberances.** These pectinaceous intercellular wall projections have diverse morphologies, occurring as scales, warts, strands, or filaments on the outer cell surface. They usually develop during tissue expansion at the time of intercellular space formation. The exact function of these structures, if any, is not known.

## Plasmodesmata

The primary walls of plant cells are traversed by microscopic strands of cytoplasm called **plasmodesmata** that form a unique mode of communication between neighboring cells. Plasmodesmata consist of membranes and proteins in the form of structurally complex membrane-lined pores that form uninterrupted cytoplasmic bridges between adjacent cells. Plasmodesmata play an important role in establishing and regulating short-distance cell-to-cell communication. Collectively they form an integrated cytoplasmic system throughout the plant body, from the early stages of embryo development to plant maturity. Plasmodesmata form during the last stage of cell division known as cytokinesis. These first-formed intercellular connections are called **primary plasmodesmata.** As the initial wall material is laid down, strands of endoplasmic reticulum become extended across the path of the developing cell plate. The subsequent fusion of Golgi-derived vesicles in the vicinity of the developing plasmodesmata provides matrix materials to the developing cell wall and acts as

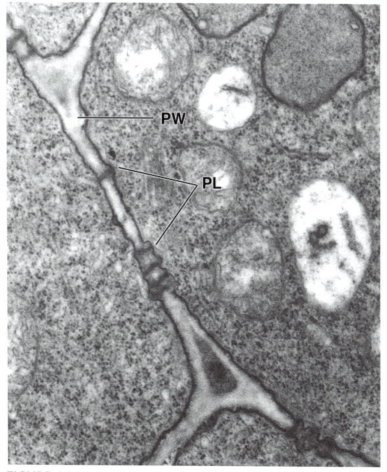

**FIGURE 1.14**   Transverse section of primary walls (PW) of *Pisum* root cortical cells showing plasmodesmata (PL). Courtesy of T. Perdue.

a source of new membranes. Plasmodesmatal channels can be unbranched or branched. The dynamic nature of cell-to-cell contact is further evidenced by the formation of **secondary plasmodesmata** across existing primary walls. At the sites of secondary plasmodesmatal formation, the adjacent walls develop thin areas. It is here that endoplasmic reticulum and Golgi-derived vesicles aggregate. Higher plants have the ability to regulate both the number of plasmodesmatal connections between cells and to determine which cell groups will be symplastically connected. It is important to note that existing connections can be temporarily or permanently removed or sealed off during this process.

Estimates place the diameter of the pores at about 40 to 60 nm, with exceptionally large diameters up to 1000 nm (= 1μm) in special cases. The central region of a pore is characteristically occupied by the **desmotubule** (or axial component). The desmotubule appears to be continuous with the **endoplasmic reticulum** of the adjacent cells. The region between the plasma membrane and the desmotubule forms the "channel" through which macromolecular traffic passes. The structurally advanced plasmodesmata of higher plants regulate the movement of macromolecules, including proteins and informational molecules. This selective molecular traffic and intercellular communication is essential for normal plant development, including cell differentiation, tissue formation, organogenesis, and other physiological processes.

## The Secondary Wall

Some cells, particularly those with strengthening and supporting functions, continue to add wall material inside the primary wall during cell expansion,

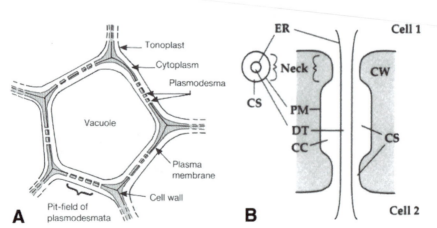

**FIGURE 1.15**   Plasmodesmata of higher plants. (A) Cell-to-cell contact is established within special locations where numerous plasmodesmata are clustered into pit fields. (B) Structural model emphasizing the membranes that delimit plasmodesmata. The plasma membrane (PM), adjacent to the cell wall (CW), forms an outer boundary and is continuous between two cells. The DT, a tube of appressed endoplasmic reticulum (ER), is located in the center of the plasmodesmata. Between the DT and plasma membrane is the cytoplasmic sleeve (CS), which may include a central cavity (CC) region. A, reprinted with permission from Lucas and Wolf (1993), *Trends in Cell Biology* **3**, 308–315, published by Elsevier Science. B, reprinted with permission from McLean *et al.* (1997), *Plant Cell* **9**, 1043–1054. Copyright © American Society of Plant Physiologists.

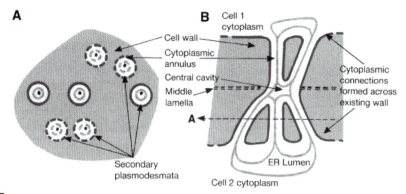

**FIGURE 1.16** Structural details of primary and secondary plasmodesmata of higher plants. (A) Transverse section through the position indicated by the broken line in B illustrating the increase in cell-to-cell contact established by new cytoplasmic bridges. The solid images are located within the plane of the longitudinal section illustrated in B (central solid image represents the initial primary plasmodesma), while the broken images are positioned in front of and behind this plane. (B) Schematic representation of a secondary plasmodesma formed by the modification of an initial primary plasmodesma. The appressed endoplasmic reticulum (ER) of the primary plasmodesma has been aligned with the equivalent structure. New cytoplasmic connections appear to form from the center of the wall, in the vicinity of the middle lamella, and project out in both directions to make contact with the plasma membrane and endoplasmic reticulum of the neighboring cells. From Lucas and Wolf (1993), *Trends in Cell Biology* **3**, 308–315, permission from Elsevier Science.

before the cell has reached its final size. This additional wall material is called the **secondary wall** and is represented by the further deposition of laminated cellulose upon the primary wall. The cellulose microfibrils are laid down by successive deposition of layer upon layer in a process known as **apposition.** The transformation of cultured leaf mesophyll cells into tracheary elements reveals that the development of secondary wall patterns is under genetic control. The early role of microtubules in wall deposition also has been documented. Cellulose may provide a framework upon which other molecules become incorporated into the wall pattern, in the view of some scientists. The wall may have some ability to undergo self-assembly, providing that cellulose molecules are initially deposited in the proper pattern.

The proportion of cellulose in secondary walls is typically higher than in primary walls. In contrast to primary walls, secondary walls lack glycoproteins. The secondary wall can be comparatively thin, as in the large "springwood" cells of some dicotyledons. However, in most fibrous cells, the secondary wall becomes very thick and, in a few cases, nearly occludes the cell lumen. Secondary walls also are strongly anisotropic. In cells like the gymnospermous secondary xylem water-conducting tracheid, they are microscopically layered, consisting of a relatively narrow outer layer ($S_1$), a middle layer of variable width ($S_2$), and narrow inner layer ($S_3$). The three layers are characterized by a different fibril angle of orientation of the abundant and extremely long cellulose microfibrils. Variations in total thickness of the secondary wall are primarily due to fluctuations in the middle or central $S_2$ layer. Multiple-layered secondary walls are characteristic of some fibrous cells (e.g., those in the primary stem tissues of many monocotyledons).

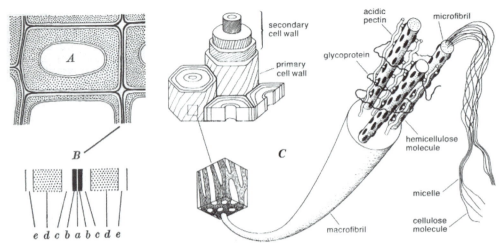

**FIGURE 1.17**   Cell wall structure. (A) Transverse section of a fiber and parts of seven neighboring fibers showing sequence of wall layers. (B) Section of adjacent walls more highly magnified; a, isotropic intercellular middle lamella; b, primary wall; c, outer layer of secondary wall with a mostly flat helical texture; d, central layer of secondary wall with a steep helical texture; e, inner layer of secondary wall with a flat helical texture. (C) Cell with primary and secondary walls. Cellulose molecules are united to form microfibrils, which in turn compose macrofibrils. A, B, reprinted with permission from Bailey (1938), *Industrial and Engineering Chemistry* **30**, 40–47. Copyright ©1938. American Chemical Society. C, reprinted with permission from Niklas (1989), *American Scientist* **77**, 344–349. Copyright © Sigma Xi Scientific Society.

Fibril angle refers to the angle between the longitudinal axis of the cell and the direction of the microfibrils in the cell wall. In contrast to the random pattern of microfibril deposition in the primary wall, microfibrils in the secondary wall are arranged in an ordered, parallel manner. As more secondary wall is deposited, microfibrils become increasingly more aligned in the long axis of the cell. In commerce, microfibril angle usually refers to the angle of microfibrils in the $S_2$ layer of softwood tracheids. Not only does microfibril angle vary within different layers of the same wall, but also it can decrease from approximately 40° to 10° moving radially from pith to cambium and upward in tree, as it does in the case of pine. The precise parallel texture of the microfibrils results from the tendency of the cellulose chains to form lateral interconnections. The crossed structure of the microfibrils prevents the formation of splits and gaps in the framework of the wall.

Cellulosic microfibrillar components make up over 50% of secondary walls, a higher percentage than the highly hydrated isotropic substances such as hemicelluloses and pectins. These wall materials are embedded in a matrix that normally contains the amorphous, highly cross-linked, partly aromatic, polymerized molecule lignin. Lignins form the second most abundant group of plant biopolymers after cellulose. Lignin is deposited within the cellulose framework by the process of **intussusception.** It is structurally composed of *p*-hydroxyphenyl, guaiacyl, and/or syringyl monomers. Monomer composition varies across species and plant groups, as well as between specific cell types within a stem. For example, guaiacyl subunits are predominant in the

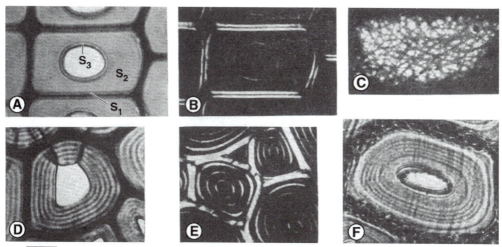

**FIGURE 1.18**   Cell wall structure. (A) Transverse section of a latewood tracheid of *Pinus* showing S₁, S₂, S₃ layers of secondary wall, ×670. (B) Same as A but photographed with polarized light between crossed nicols, ×670. (C) Primary wall of parenchyma cell showing reticulate appearance of primary pit fields, ×730. (D) Transverse section of wood fiber showing alternating cellulosic (light) and noncellulosic (dark) layers of secondary wall, ×1000. (E) Transverse section of fibers showing alternating birefringent and dark layers of secondary wall in polarized light, ×770. (F) Unswollen transverse section of lignified wood fiber, showing concentric layering of secondary wall, ×1330. Reprinted with permission from Bailey (1938), *Industrial and Engineering Chemistry* **30**, 40–47. Copyright © 1938 American Chemical Society.

tracheary elements of *Arabidopsis* stems, whereas adjacent sclerenchymatous cells contain syringyl subunits. Secondary walls of xylem cells contain approximately 25% lignin based on dry weight.

The incorporation of the three-dimensional phenolic polymers called lignins into the space between the cell wall microfibrils of vascular plants is termed **lignification.** Although it is not known exactly how lignin precursors are transported to the cell wall during the complex process of lignin deposition, this event represented a major evolutionary landmark in the evolution of early land plants. They provided the mechanical strength and decay resistance that was crucial during the transition of plants from an aquatic to a terrestrial habitat. Lignin also decreases the permeability and degradability of walls and is important in determining the mechanical behavior of predominantly nonliving tissues such as wood. Molecules of lignin frequently permeate the existing primary wall as well as the middle lamella during differentiation. In conifer water-conducting cells, the middle lamella is the most heavily lignified part of the wall. Cells with heavily lignified secondary walls undergo irreversible changes that render them incapable of further growth or volume increase. Most often the protoplasts of these cells autolyze during the final stages of maturation so that the wall layers surround an empty cavity or lumen.

## Warty Layer

As tracheary cells and fibers mature and undergo cytoplasmic breakdown, the membrane fragments that result from this process are deposited as a continu-

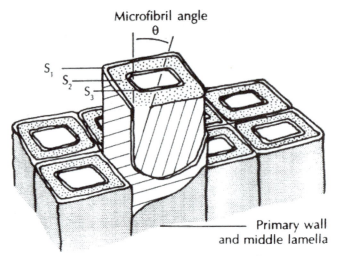

**FIGURE 1.19**  Orientation of the microfibrils in the secondary wall of a conifer tracheid showing microfibril angle. Reprinted with permission from Walker and Butterfield (1996), *N. Z. For.* **40**, 34–40.

ous, distinct layer over the inner surface of the secondary cell wall and the chambers of bordered pits. This material is referred to as the **warty layer** in reference to the small, wartlike protuberances that are visible with the electron microscope. The warts are generally 0.1 to 0.5 μm in diameter. The warty layer is known from both hardwood and conifer wood cells. The warts represent localized outgrowths of the $S_3$ wall layer and proteinaceous materials derived from cytoplasmic debris.

## Wall Pits

The secondary wall is not easily permeable to solutions and is therefore deposited in characteristic patterns that result in numerous interruptions along its surface. Any interruption in the secondary wall is termed a **pit**. Often pits are superimposed over the primary pit fields of the primary wall. As this occurs, the plasmodesmata are typically severed. The pit of one cell usually lies exactly opposite the pit of an adjacent cell wall, forming a functional intercellular **pit pair** that is of fundamental importance in facilitating the lateral movement of water and dissolved materials between cells. The two pits composing the pair between water-conducting cells are separated by a thin, permeable, and unlignified primary wall called the **pit membrane**. The pit membrane between a water-conducting cell, and the neighboring parenchyma cell is typically thickened.

Several types of pits can be distinguished, and these are sometimes characteristic of a particular cell type. A **simple pit** is one in which the opening, or **pit aperture**, is wider or of the same width as the base of the pit cavity. In a **bordered pit**, the pit membrane between the two cells consists of randomly or radially arranged microfibrils and is overarched by the secondary wall that forms a border. The aperture is located in the center of the border. In the tracheids of conifers, *Ginkgo*, and the Gnetales, the delicately bordered pit membrane possesses a central, thickened portion called the **torus**. With few

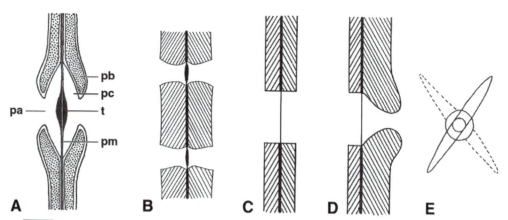

**FIGURE 1.20** Pit structure. (A) Diagrammatic bordered pit pair in sectional view. (B) Bordered pit pairs of thick-walled cells of summerwood of pine, showing slight overhang of secondary walls, thick membrane, and poorly developed valvelike characters. (C) Simple pit pair between two cells. (D) Half-bordered pit pair between water-conducting tracheid and parenchyma cell; (E) Face view of bordered pit pair with extended and crossed pit apertures beyond the limits of the reduced pit border. Abbreviations: pa, pit aperture; pb, pit border; pc, pit chamber; pm, pit membrane; t, torus. A, reprinted with permission from Kerr and Bailey (1934), *J. Arnold Arb.* **15**, 327–349. B–D, reprinted with permission from Bailey (1915), *American Railway Engineering Association Bulletin* **174**, 835–853.

exceptions, a well-defined torus is typically absent in flowering plants. Because the border is often circular in outline in face view and the bordered pits of adjacent cells are exactly opposite one another (occurring in pairs), the functional unit is a **circular bordered pit pair.** Some bordered pits are elongate in outline with an extended aperture. These are called **scalariform pits.** Tracheids from woods of pinaceous affinity have **crassulae** thickenings ("Bars of Sanio") on the radial wall above and below the bordered pits. These are interpreted as thicker portions of the middle lamella and primary walls. The formation of pit pairs between adjacent cells is the culmination of closely coordinated developmental pathways during the formation of the secondary wall, and their conspicuous presence on the lateral walls of water-conducting tracheary elements allows for efficient lateral water movement without impairing the strength and rigidity of the cell. Chapter 7 contains a more complete account of the structure and functioning of bordered pits.

At the point of contact between a wood ray cell with simple pits and a neighboring tracheary cell with bordered pits, **half-bordered pit pairs** are evident in which a bordered pit of one cell joins a simple pit of the neighbor. **Vestured pits** represent a distinctive and specialized type of pit structure in which the pit cavity, and sometimes the aperture as well, is wholly or partly lined with small, often branched projections (or vestures), extending from the secondary wall. According to one hypothesis, the dense network of vestures filling the pit chambers functions to minimize the effects of pit membrane displacement (aspiration) in angiosperms, in response to the high-pressure drops between adjoining vessel elements caused by air embolism. It has been suggested that this condition reduces the risk of membrane rupture during such events. If this explanation is correct, the occurrence of vestured pits represents an important structural adaptation in woody plants.

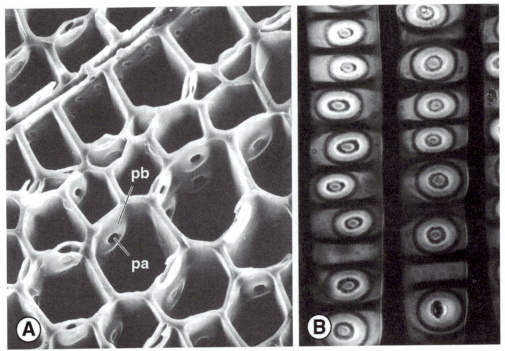

**FIGURE 1.21**   Bordered pits of Pinaceae. (A) SEM of wood of eastern spruce (*Picea*) showing tracheids with circular bordered pit pairs on the lateral walls. Pit apertures (pa) and pit borders (pb) are visible, ×1300. (B) Light micrograph of tracheids from *Pinus strobus* showing prominent circular bordered pits on the lateral walls. Note outlines of the borders, pit membranes, and tori. Crassulae occur above and below each pit. A, Courtesy of the International Association of Wood Anatomists. B, Courtesy of the Bailey-Wetmore Wood Collection, Harvard University.

## CELL MEMBRANES

The membranes that surround the cell protoplast and that subdivide the cytoplasm by delimiting various organelles not only act as boundaries but also provide sites and a surface area for the numerous biochemical processes that continuously unfold within the cell. An unbroken plasma membrane is essential for every living cell. In growing cells, the plasma membrane must be able to enlarge and change shape. The membranes that surround the protoplast and define the membrane system of the cytoplasm possess a number of important properties. These include flexibility, fluidity, and an ability to self-seal and show selective permeability to molecular movement. Although determining the fundamental structure of cell membranes has been a contentious issue, there is now widespread agreement that the chemical composition and physiological properties of membranes is best explained by a **fluid mosaic model.** Investigators now agree that all cell membranes consist of two layers of phospholipids. The individual lipid molecules are polar, with their forked hydrophobic tails directed inward and in contact with each other, while their outward positioned hydrophilic heads are in contact with the surrounding aqueous medium. The membrane is about 10 nm thick. Buried in various sites within the lipid layers are protein molecules of different sizes. Some proteins

**FIGURE 1.22**  Lateral segment of vessel element of *Asimina triloba* (Annonaceae) showing multiseriate intervessel pitting of the alternate type. ×300.

extend completely through the lipid bilayer and protrude on both faces (**transmembrane proteins**), other proteins have one or more hydrophobic parts embedded in the membrane and a hydrophilic part protruding on the outer or inner surfaces (**integral** or **intrinsic proteins**), whereas still other proteins may be present only on the surfaces of the membrane without penetrating the interior (**peripheral** or **extrinsic proteins**). Because of the fluidity of the membrane, some intrinsic proteins are free to move laterally within the two lipid layers.

Membranes serve to regulate the passage of different substances in and out of the cell and its various organelles. Not only do the sugars, amino acids, and various other molecules and ions that are essential to life move across membranes, but metabolic waste products also must pass outward to avoid accumulating inside. How membranes function as selectively permeable barriers to molecular traffic depends on the proteins they contain. Certain transmembrane proteins act as specific carriers for the movement of substances, sometimes acting as highly specific gates, channels, or molecular pumps through which ions and molecules pass. In this way, membrane proteins are able to control not only what passes through the membrane of a particular cell or organelle but also the rate at which it moves.

Not only do membranes serve important transport processes, but also the chemical signaling between and within cells in multicellular organisms is frequently mediated by cell surface receptor membrane components. Many of the proteins exposed on the outside surface of membranes possess sugar components. These glycoproteins act as surface receptors and contain short, branched sugar chains with about 4 to 15 sugar monomers. The numerous types of cell surface receptors function to receive information regarding cell or molecule recognition and to communicate it to the inside.

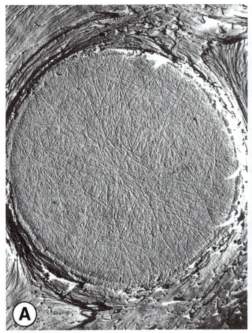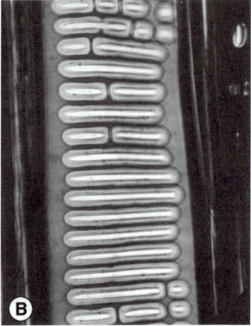

**FIGURE 1.23** Intervessel pitting in dicotyledons. (A) SEM surface view of pit membrane from an intervessel pit pair of *Quercus* (Fagaceae). Note absence of thickened central torus. × ca. 12,000. (B) Scalariform and transitional intervascular pitting showing a mixture of scalariform and opposite pitting in *Magnolia acuminata* (Magnoliaceae). × 400. A, courtesy of E. A. Wheeler. B, courtesy of the Bailey-Wetmore Wood Collection, Harvard University.

## Cytoplasmic Organelles

The chemical reactions within a plant make up its **metabolism.** The systems that carry out metabolism are located in the cytoplasm, particularly in a number of major **organelles** that regulate cellular processes. Plant cells contain a variety of these discrete subcellular organelles, each with specialized structure and function. Many organelles are membranous structures, possessing varied biosynthetic capacities within which a wide array of cell functions are compartmentalized.

The energy-related activities of aerobic respiration and photosynthesis are packaged in organelles known as **mitochondria** and **chloroplasts.** Mitochondria serve as the main centers of respiration and oxidative energy production. Their shape ranges from rodlike to almost spherical, with dimensions about the size of bacteria (i.e., approximately 1 μm in diameter and 1–5 μm in length). The ultrastructure of a mitochondrion consists of an envelope of two membranes: a continuous outer membrane and an inner membrane that invaginates into the interior of the mitochondrial body to form membranous infolded **cristae.** Chains of respiratory enzymes are embedded along the inner face of the cristae. The central fluid-filled matrix of the mitochondrion appears to be structureless.

Green plants are autotrophic; that is, they have the ability to manufacture their own oxidizable organic food molecules. Photosynthesis is the complex process whereby plants utilize radiant energy to convert carbon dioxide and

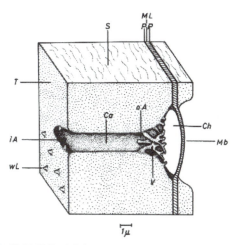

**FIGURE 1.24** Schematic representation of a vestured pit from *Goniorrhachis marginata* (Leguminosae). Abbreviations: Ca, pit canal; Ch, pit chamber; iA, inner aperture; Mb, pit membrane; ML, middle lamella; oA, outer aperture; P, primary wall; S, secondary wall; T, tertiary wall; wL, warty layer, V, vestures. Reprinted with permission from Schmid and Machado (1964), *Planta* **60**, 612–626.

water into the high bond energy of carbohydrates. The process of photosynthesis is carried out in the chloroplasts. Chloroplasts are typically larger than mitochondria, between 5 and 10 µm in diameter, but like mitochondria they are surrounded by an envelope of two membranes. An internal system of

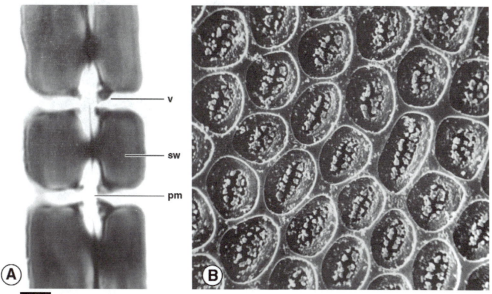

**FIGURE 1.25** Vestured pits on vessel walls of Combretaceae. (A) Light microscope sectional view of vestured pits in *Terminalia chebula*. ×ca. 3000. (B) SEM of vestured intervessel pits with pit floors removed in *Anogeissus acuminata*. Abbreviations: pm, pit membrane, sw, secondary wall; v, vesture. A, courtesy of the Bailey-Wetmore Wood Collection, Harvard University. B, courtesy of The International Association of Wood Anatomists.

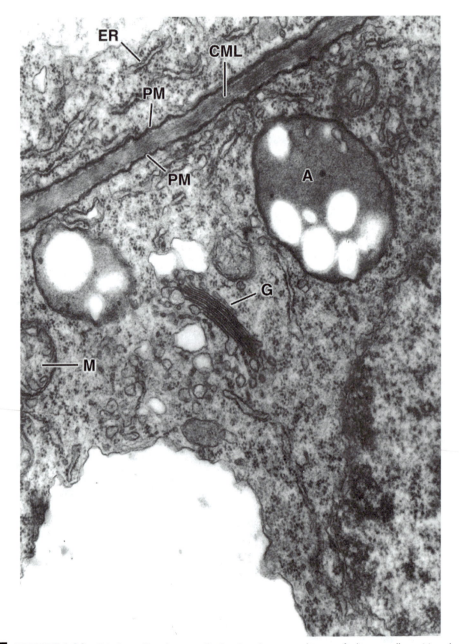

**FIGURE 1.26**   EM of pea (*Pisum*) root cells showing plasma membranes of adjacent cells positioned against the compound middle lamella. Abbreviations: A, amyloplast; CML, compound middle lamella; ER, endoplasmic reticulum with associated ribosomes; G, Golgi; M, mitochondrion; PM, plasma membrane. Courtesy of T. Perdue.

chlorophyll-containing membranes is in the form of flattened, disklike, stacked lamellae called **thylakoids.** Each membrane stack is called a **granum.** Each chloroplast contains a large number of grana. Grana are cylindrical in outline and form an increased surface area for photosynthetic pigments and reactions. Grana are interconnected by additional lamellae called **stroma** lamellae. But unlike mitochondria, the internal membrane system of a chloroplast is not connected to a peripheral membrane. The internal nonmembranous matrix of the plastid is known as **stroma.** Excess photosynthate often accumulates within the chloroplast as starch grains. Both mitochondria and chloroplasts are involved in electron transport processes, both contain small amounts of genetic information and reproduce by division, and in many other ways both resemble primitive prokaryotic organisms.

Chloroplasts belong to a broader category of pigment-containing bodies, the plastids, that are collectively referred to as the **plastidome.** Yellow, orange, and red plastids are grouped under the heading of **chromoplasts.** Chromoplasts can be disk shaped, angular, or irregularly shaped and are characterized by an internal membrane system that is not organized into grana. Chromoplasts lack photosynthetic enzymes. Examples of chromoplasts are seen in the petals of various flowers; in fleshy fruits such as tomatoes, pumpkins, and peppers; and in the roots of carrots. **Leucoplasts** are nonpigmented plastids that frequently contain starch and are typically located in underground plant parts, for example, in potato tubers or *Iris* rhizomes as well as in other bulbs, corms, and roots. In some cases, leucoplasts are involved in fat and lipid synthesis. A leucoplast type of plastid that forms and stores oil is an **elaioplast.** Another special type of plastid is the **amyloplast.** It is typically unpigmented and lacks photosynthetic enzymes. It accumulates and stores large amounts of sugar and starch in the form of large starch grains. One mature amyloplast may contain over a hundred starch grains. Structurally amyloplasts contain few internal membranes. Plastids are believed to develop from smaller, less well defined precursors, the **proplastids,** which are able to increase in number by division.

The cytoplasm is frequently rich in elongated sheets of membrane called the endoplasmic reticulum (ER) that are involved in the regulation of cellular exports. The ER provides a large surface area for cellular reactions. It gives rise to the nuclear membrane during cell division and is an important center of membrane biosynthesis in the cell. The ER membranes form a system of tubes or sheets (cisternae) that extend throughout the cytoplasm. On the inside, the membrane walls of ER cisternae are flat and smooth. Portions of the external membrane surface, however, can be studded with electron-dense particles called **ribosomes.** Ribosomes are nonmembranous particles composed of RNA and protein. They represent the sites of protein production. Those portions of the ER membrane that are covered with ribosomes are called **rough ER,** whereas membrane devoid of ribosomes is termed **smooth ER.** Protein chains manufactured on the external surface of the endoplasmic reticulum may be combined with carbohydrates and exported through the interior ER canal as glycoprotein. Smooth ER is involved in lipid synthesis and membrane biosynthesis.

The **dictyosome (Golgi apparatus)** is a complex and highly variable organelle. It looks like a system of stacked, often netlike, membranes. The dictyosome is generally more distinct and clearly separated from other intracellular membranes in plant cells than in animal cells. Its stacked, flattened membranes are called **cisternae.** This organelle secretes and is surrounded by numerous small, spherical, membrane-bound **vesicles** containing proteins, lipids, and other substances. The Golgi apparatus of plant cells serves two

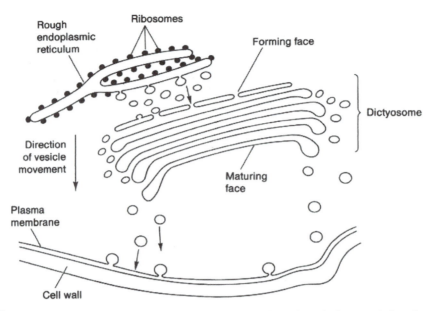

**FIGURE I.27**  Flow of membranes and other materials from the endoplasmic reticulum through the Golgi to the plasma membrane. Proteins synthesized on the endoplasmic reticulum, such as the wall protein extensin, and noncellulosic carbohydrates synthesized in the Golgi are transported to the plasma membrane. This is the source both of new cell wall components and of plasma membrane. Reprinted with permission from Fosket (1994), "Plant Growth and Development. A Molecular Approach," Academic Press.

major functions: it assembles and processes the oligosaccharide side chains of glycoproteins, and it synthesizes the complex polysaccharides of the cell wall matrix, the hemicelluloses, and pectins. The dictyosome is a membranous system that possesses a distinct polarity. It consists of an endoplasmic, convex, **forming face** (cis cisternae) and an exoplasmic, concave side called the **maturing face** (trans cisternae). Secretory proteins undergo processing as they move through the dictyosomes, including the partial trimming or addition of molecules. Because protein secretion is less pronounced in plant cells than in animal cells, plant dictyosomes are engaged primarily in the synthesis of cell wall polysaccharides, including those involved in cell plate formation during cytokinensis. In further contrast to animal cells, in which the dictyosomal membranes break up during cell division, the dictyosome membrane stacks remain intact throughout mitosis in plant cells. The membrane-bound vesicles secreted by the dictyosome move to the plasma membrane, where vesicles fuse with the plasma membrane and then the contents extrude from the cell and fuse with the developing cell wall. The endoplasmic reticulum and the dictyosomes form a highly regulated secretory pathway. Proteins and other organic materials are synthesized, transported in bulk, processed, packaged, and exported by the endoplasmic reticulum and ribosomal complex via the dictyosome. The pathways that connect the endoplasmic reticulum and dictyosome result in the two organelles functioning as a dynamic membrane system specialized for the biosynthesis and sorting of membranes.

Cells possess an internal proteinaceous "cytoskeleton" or framework that reinforces specific areas of the cell, influences cell shape, and actively participates in the movement of dictyosome-derived and other vesicles. The cytoskeleton is composed of filamentous structural elements called **microtubules** and **microfilaments**. Microtubules take the form of straight, unbranched, hollow cylinders or tubes. They are varied in length and are approximately 200 to 300 nm in diameter. Microtubules are composed of aggregate subunits of the protein **tubulin** that polymerizes and depolymerizes during their rapid assembly and disassembly. In addition to their involvement in directional movement within the cell during cell division and elsewhere and as a structural component of flagella, they are involved in the channeling of vesicles filled with wall materials to the growing cell plate as well as in the orientation of cell wall molecules. Microtubules are especially abundant directly beneath the plasma membrane where they are oriented transverse to the long axis of the cell wall. Microfilaments are thin filaments with a diameter of about 6 nm. They are composed of the globular protein **actin**. Along with microtubules, microfilaments control certain aspects of cell growth, differentiation, and movement.

The active metabolism of the cell entails the production and consumption of many different compounds at various times and in varying amounts. **Microbodies** are small, roughly spherical cell compartments (0.5–1.0 μm in diameter) that are invested by a selectively permeable membrane and possess diverse metabolic capacities. The matrix of the microbody usually is amorphous, although in some cells there is a dense crystalloid core. The most common form of microbody is the **peroxisome**. It plays a key role in the metabolism of hydrogen peroxide, $H_2O_2$. Peroxisomes serve to isolate this toxic compound, and through a series of oxidation–reduction reactions, they immediately metabolize it, converting it to water and oxygen. Peroxisomes contain a large number of multienzyme complexes that mediate this process. In leaf mesophyll cells, peroxisomes also function to metabolize certain by-products following the successive reactions of photorespiration. A final type of microbody, the **glyoxysome**, is essential for the conversion of fat into carbohydrates, which is necessary during the germination of some oil-rich seeds. The transient appearance of glyoxysomes in the cells of castor bean seedlings, for example, results in the mobilization and utilization of storage lipids.

## Vacuome

The cytoplasm of all living plant cells typically surrounds a fluid-filled space, the vacuole, and its associated **transvacuolar strands**. The **vacuome**, or totality of the diverse vacuoles within a cell, is sometimes referred to as the vacuolar system. Vacuoles are a characteristic feature of plant cells and a large vacuole can account for as much as 99% of the cell's volume. A vacuole is bounded by a single membrane termed the **tonoplast**. Active transport processes at the tonoplast generate internal turgor pressure and, in some instances, concentrate inorganic ions of calcium and sodium within the vacuolar fluid. Vacuoles occur in varied sizes and shapes and perform different functions, often within the same cell. In addition to producing turgor pressure and associated cell growth, vacuoles carry out a storage function, accumulating the end products of secondary metabolism, as well as starch and proteins that can be transported across the tonoplast and metabolized at a later period. As cells differentiate or as the climatic seasons change, the shape of the vacuome also may change as smaller vacuoles merge. Vacuoles typically arise from preexisting vacuoles through fusion or dispersion. They also may originate from other membrane systems in the cell. In some cells endocytic **secondary vacuoles** form when the plasma membrane invaginates and the edges of the invaginated membrane fuse. The fold becomes pinched off to the cell interior and becomes a membrane-bound compartment.

Among the diverse assemblage of membrane-bound sacs in cells are the **lysosomes.** Lysosomes contain enzymes called acid hydrolases that function in the segregation and digestion of cellular components from the surrounding cytoplasm. Many different hydrolases have been identified. They are capable of breaking down the major organic constituents of cells, including proteins, polysaccharides, lipids, and nucleic acids. Lysosomes are therefore involved in the turnover of large molecules and organelles as well as in the clearance of dead cells and foreign material. Lysosomes are approximately spherical bodies that are surrounded by a single membrane and exhibit little internal structure. Lysosomes are not as readily identifiable in plant cells as in animal cells, although biochemical and cytochemical evidences indicate that bodies with lysosomelike activity are present in many plant cells during the early developmental stages of vacuole formation.

## Starch

Like cellulose, starch is a polymer composed of glucose subunits, but it is joined by α rather than β glycosidic linkages. Starch is normally a mixture of two types of polysaccharides: amylose and amylopectin. Most starches consist of about 25% amylose and 75% amylopectin, but the ratio may be reversed in some cases. Starch grains may be spherical, pear shaped, or even angular and range in size from 1 to 150 μm. The overall shape and size of the grains depends, in part, on the number of grains formed within a plastid and how each grain develops. The first part of the starch grain to be produced is termed the **hilum.** Successive layers of starch encircle the hilum. Depending upon how the grain was formed, the position of the hilum results in a grain being either concentric or eccentric. The outline of the hilum can be pointed, straight fissured, or stellate.

The multiple grains of chloroplasts are usually transient because starch generally is quickly decomposed after formation. The large grain or grains within a leucoplast, on the other hand, remains for an extended period. Because starch grains vary in structure as well as in size, their form can be characteristic of particular species. Starch synthesis occurs along various pathways, and apparently all the pathways may occur within a single species. The synthesis of starch and its degradation to sugar by various pathways are influenced by a number of factors both internal and external, and these factors may determine whether starch synthesis or breakdown is predominant at any particular time.

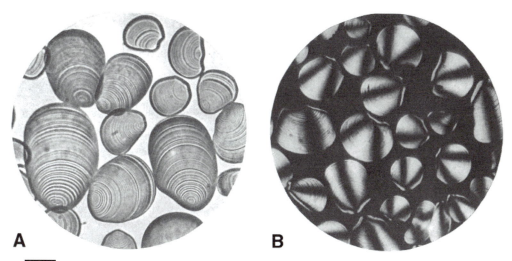

**FIGURE 1.28** Starch grains of the monocotyledon *Canna roscoeana* (Cannaceae). (A) Note laminations. (B) Photographed under polarized light. Reprinted with permission from Reichert (1913). Copyright © Carnegie Institution of Washington.

Polarized light is typically used to detect the presence of starch bodies. The behavior of starch grains toward polarized light results in their appearing bright under the condition of "crossed" nicols. In polarized light, starch grains show an interference figure in the form of a "cross," which is a result of the double refraction of light through cylindrical or spherical lamellar structures, such as starch grains. The point of intersection of the two parts of the cross usually corresponds to the position of the hilum.

### Nucleus

In eukaryotic cells, the hereditary material (DNA) responsible for information storage and transfer is packaged in highly organized chromosomes located within the **nucleus.** The nucleus is delimited by a double-layered **nuclear envelope** that is a specialized component of the endoplasmic reticulum. The gap between the two membranes is called the **perinuclear space,** ranging from 10 to 60 nm in width. At the beginning of cell division, the nuclear envelope breaks down and the resulting membrane fragments are indistinguishable from strands of the endoplasmic reticulum. During the telophase of mitosis, the nuclear membrane is reformed from the membrane of the ER. The outer nuclear membrane is continuous with the ER, providing an open channel of communication between the perinuclear space and the ER cisternae. The nuclear envelopes are interrupted by thousands of structurally complex and uniformly distributed channels or pores, through which proteins and ribonucleoprotein molecules are actively transported in and out of the nucleus. Individual **nuclear pores** are lined with a proteinaceous complex and regulate the passage of important molecules across the membranes. Within the nucleus is another spherical body, the **nucleolus,** which is without an investing membrane. It is involved in ribosome biogenesis and information transfer functions. There can be one, few, or hundreds of nucleoli within a nucleus.

## NONPROTOPLASMIC MINERAL DEPOSITION

Among the nonprotoplasmic inclusions of plant cells are starch granules and various mineral depositions such as silica, gypsum, calcium carbonate, and calcium oxalate. The most common are those composed of calcium salts, usually calcium oxalate, secreted as monohydric or trihydric **crystals,** and occurring in special cells that function in the metabolism of calcium. Infrequently crystals occur as depositions of calcium malate or potassium calcium sulfate. Although the physiological function of crystals, if any, has not been fully clarified, calcium crystal formation appears to play an important role in calcium metabolism and the regulation of ionic equilibrium. It has been suggested that calcium oxalate crystals provide a reserve of calcium in the cell in addition to providing a means of removing toxic accumulations of oxalate. They can be widely distributed throughout the plant body and are found in members of all but a very few families of flowering plants. Crystals are usually found within the cell vacuole but can be located within the cytoplasm and cell wall. Although crystals are most commonly found within cells, they can occasionally be deposited on the outer wall surfaces. Extracellular crystals have been reported on the outer walls of collenchyma cells and on the walls of phloem fibers of four genera of Taxaceae: *Amentotaxus, Taxus, Pseudotaxus,* and *Torreya.* Crystals also occur in the walls of sclereids of a number of genera. The crystal-studded sclereids of *Nymphaea* and *Schisandra* are especially well known. In some cases the crystal-containing cells are essentially the same size and shape as neighboring cells, but some of these cells are highly differentiated and conspicuously enlarged and are referred to as **crystalliferous idioblasts.** Because crystals are highly anisotropic, examination of tissues with polarized light is the best method for confirming their presence.

Crystals can be quite variable in occurrence. In some species, crystals are characteristically abundant; in other species, they are consistently present but not abundant. In still other species, they are present in some individuals but absent in others. Within the plant body, crystals can be restricted to a single cell or tissue type, or they can be distributed in more than one cell and tissue type. As an example, crystals can be found only in the wood ray cells of some woody plants or they can be widely distributed throughout the wood, leaf, and reproductive tissues. In leaves, crystals most commonly develop within the bundle sheath cells or bundle sheath extensions.

Crystals are quite varied in form and several classification schemes have been devised to account for the many morphological intergradations. **Raphides** are long, thin, needlelike, generally monohydric crystals with pointed ends that are usually aggregated into sheathlike bundles that are considered a single unit. The individual crystals are surrounded by a membrane, and the entire bundle is usually covered by a mass of mucilage. Raphides are often deposited within enlarged idioblastic cells, known as raphide sacs, that are conspicuously distinct from neighboring cells. Raphides are fairly common in many monocot families, but they also are abundant in the dicotyledonous families Balsaminaceae, Rubiaceae, and Dilleniaceae. Unusual raphides with barbs and grooves have been reported to occur in *Xanthosoma sagittifolium* of the family Araceae.

**Acicular** crystals, although not always distinguished from raphides, are small needlelike forms that do not occur in bundles. This crystal type can be seen in Acanthaceae, Lauraceae, and Myristicaceae. **Druses,** sometimes called cluster crystals, are compound crystals that are loosely united into a sharp pointed spherical mass in which the many component crystals protrude from the surface to give the whole structure a star-shaped appearance. This crystal type displays many variations in form, including the pointed-star crystal. Druses are common in several families, including Leguminosae, Juglandaceae, Polygonaceae, and Tiliaceae. **Prismatic crystals** are solitary rhombohedrons or octohedrons of varied size that are widespread but commonly found in secondary xylem and phloem parenchyma cells. This is the most common type of crystal in wood.

**Crystal sand** consists of a granular mass of many fine particulate accumulations of microcrystals scattered throughout the cell, in which they have a sandy appearance. Examples of crystal sand can be seen in the stem cells of *Acuba* (Cornaceae), *Amaranthus* (Amaranthaceae), and *Sambucus* (Caprifoliaceae). **Styloids** are large, long, solitary crystals at least four times as long as they are broad, with pointed or square ends. They are less commonly encountered than the preceding types. Styloid crystals can be seen in Liliaceae and in the leaves of *Iris* (Iridaceae). Unusually large styloids have been recorded in the secondary xylem of a few woody species. In the genus *Henriettea* of the Melastomataceae, for example, extremely large styloids occur measuring 24 to 80 µm in diameter and 240 to 560 µm in length.

The occurrence of calcium oxalate crystals in plants has been shown to be related to the level of calcium concentration in the surrounding medium. In one recent study, calcium concentration was shown to influence the development of leaf tissues and cells in connection with crystal formation. When young plants of *Phaseolus vulgaris* (bean) are grown in nutrient solution with low calcium concentrations, more palisade leaf parenchyma cells and less extended bundle sheath cells are formed on the adaxial side of minor leaf veins. As the concentration of calcium is raised in the growing medium, more bundle sheath cells form, and there is an increase in the number of calcium oxalate prismatic crystals in the sheathing elements.

Although less common than calcium oxalate, calcium carbonate (lime) is found in the cells of some plants as an encrustation on the walls or in the form of an unusual deposit called a **cystolith.** Cystoliths are typically located in enlarged surface cells of leaves called **lithocysts,** where the calcium carbonate is deposited over a cellulosic extension that hangs from the wall. Cystoliths are found in several unrelated dicotyledonous families but are especially abundant and variable in members of the order Urticales (Ulmaceae, Cannabaceae, Moraceae, Urticaceae). Spherical, bacilliform, and fusiform (curved or straight) types of cystoliths are most common; stellate or vermiform shapes occur rarely. Cystoliths may be systematically useful at various taxonomic levels, as evidenced by the general correspondence between cystolith shape and tribal circumscription within the Urticaceae. Different species within a

**FIGURE 1.29**   SEMS of calcium oxalate crystals. (A) Single raphide crystal from leaf raphide bundle of *Psychotria punctata* (Rubiaceae). (B) Leaf raphide crystal bundle of *Psychotria punctata*. (C) Styloid crystal from leaf spongy mesophyll of *Peperomia* (Piperaceae). (D) Druse crystal from leaf of *Opuntia* (Cactaceae). (E) Crystal sand from petiole of *Nicotiana glauca* (Solanaceae). (F) Prismatic crystals from leaf of *Begonia* (Begoniaceae). (G) Aggregate crystal complex from leaf spongy mesophyll of *Peperomia astrid* (Piperaceae). All line scales equal 5 μm. From V. R. Franceschi and H. T. Horner (1980), Calcium oxalate crystals in plants, *Botanical Review* **46**(4), 361–427. Used with permission of the New York Botanical Garden, copyright 1980.

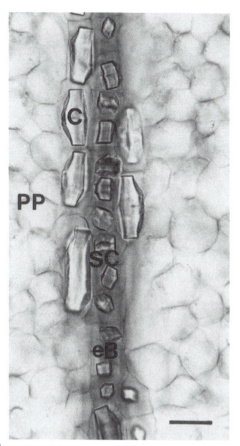

**FIGURE 1.30**   Cleared leaflet of *Phaseolus* (Leguminosae) showing crystals in leaf vein bundle sheath cells. Abbreviations: C, calcium oxalate crystal; eB, extended bundle sheath; PP, palisade parenchyma; SC, "short cell" of bundle sheath extension. From Zinder-Frank (1995), *Botanica Acta* **108**, 144–148. Used with permission of Georg Thieme Verlag, Stuttgart, New York.

genus may have cystoliths with markings of distinctive shape. Cystoliths are of very rare occurrence in wood, although one notable exception is the family Opiliaceae, where they are conspicuous in the ray cells.

Mention also needs to be made of myrosin, a glucoside of mustard oil that is sometimes present in crystalline form in specialized parenchymalike cells called **myrosin cells.** Myrosin cells are typically elongate cells with dark contents, associated with the phloem tissue. Myrosin is characteristic of the order Capparales (Capparaceae, Brassicaceae, and others).

Deposits of silica ($SiO_2$) are often abundant in the wood cells of dicotyledons and in the stems and leaf cells of some monocotyledons. However, silica is clearly characteristic of some families and not of others. Among those species of woody dicotyledons that have been studied, approximately 80% of the wood silica was present in the xylem ray cells. Silica may form as an incrustation on the wall or as inclusions called **silica grains** or **silica bodies** in the cell lumen. In many cases but not always, the silica body conforms to the

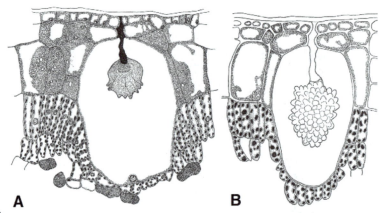

**A**                                        **B**

**FIGURE 1.31**  Cystoliths in leaves of *Ficus elastica* (Moraceae). (A) Mature lithocyst. The calcium carbonate has been dissolved away showing the cellulose matrix of the cystolith. (B) Mature lithocyst showing the grapelike deposit of calcium carbonate upon the cystolith stalk. Both ×927. Reprinted with permission from Ajello (1941), *Am. J. Bot.* **28**, 589–594.

shape of the cell in which it occurs. The size of silica bodies tends to be quite variable. Silica may resemble calcium oxalate crystals, but it is amorphous and noncrystalline and so does not show birefringence when viewed under polarized light. Because silica is resistant to breakdown, it can be used to identify fossil or subfossil plant materials that have undergone certain types of fossilization. In addition to its presence or absence, the shape of individual silica bodies can be diagnostically important as an aid in classification. Solitary silica bodies are often located within small, almost isodiametrically shaped cells called **stegmata,** which may have thickened inner and anticlinal walls and which are sometimes lignified or suberized. In all palms, stegmata are arranged in longitudinal rows adjacent to the vascular or nonvascular fibrous bundles of the stem and leaf. These continuous or discontinuous files of silica-containing cells are never positioned in the epidermis of Palmae.

## SUMMARY

The plant body of a typical angiosperm consists of two major organ systems, the root and the shoot. The shoot is composed of stems, leaves, and flowers. Each major organ in turn is composed of tissues that represent collections of one or more specialized cell types. All growth and organ formation of the plant body is initiated and controlled by regions of active cell division termed meristems. Vegetative meristems are indeterminate in their activity. Meristems retricted to the extreme tips of stems, branches, and roots are called apical meristems, and their activity results in primary growth and an associated increase in axis length. Lateral meristems, termed cambia, are those regions of dividing cells positioned as a continuous thin sheet on the outside of the axis. Their activity is responsible for axis thickness. Tissues formed from the cambia are called secondary tissues and constitute the secondary plant body.

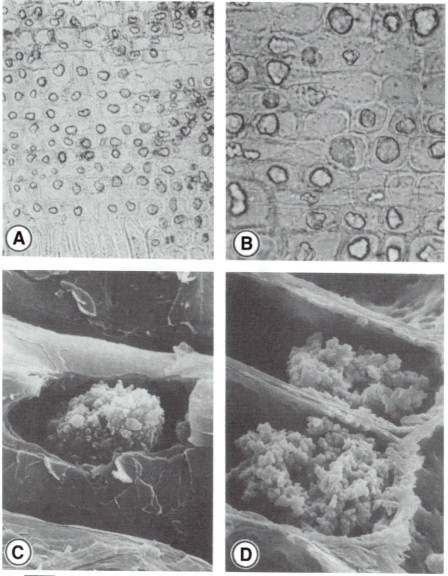

**FIGURE 1.32** Silica grains in wood of *Hibbertia* (Dilleniaceae). (A) *H. trachyphylla,* radial section showing abundant silica grains in ray parenchyma cells. × 125. (B) *H. altigena,* radial section showing solitary silica grains in wood ray cells. ×230. (C) *H. deplancheana,* SEM illustrating granular surface of silica grain. × 2600. (D) *H. tontoutensis,* SEM illustrating very granular silica grain. × 4300. Reprinted with permission from Dickison (1984). *IAWA Bull. n.s.* **5,** 341–343. The International Association of Wood Anatomists.

Meristems are composed of two categories of cells, the dividing initials and the immediate undifferentiated derivatives of cell division.

The derivatives subsequently become differentiated and change into different types of mature cells and tissues. Cell lineage studies have demonstrated that plant cell differentiation is not dependent on cell lineage. Rather, plant cell differentiation is position dependent, and cell fate is only determined gradually. Following fertilization, the plant body begins development as a single-celled zygote that will undergo embryogenesis. After a series of early cell divisions, a globose proembryo is formed. During the later stages of embryo development, the embryonic axis of the plant body is formed, along with the root and shoot apical meristems from which the rest of the plant body will develop following seed germination. As embryonic development proceeds, future major tissue systems become established as transitional meristematic zones or regions. The primary meristems produce cells that differentiate to form the three main tissue systems of the plant: vascular, ground, and dermal tissues.

One of the most distinctive structural features of plant cells is the surrounding cell wall, typically composed of the carbohydrate cellulose. Cellulose molecules consist of $1 \rightarrow 4$ linked β-D-glucan subunits joined into crystalline microfibrils of indeterminate length and varying width and degrees of order. Different cells possess wall layers of varying composition, thickness, and structure. Cell walls play important roles in plant development, cell-to-cell communication, physiology, and environmental adaptation and stress response. These varied functions are reflected in the cell's diverse wall structures. The cell wall arises as a cell plate during cytokinensis. It is transformed into the middle lamella that binds the walls of adjacent cells together and is regarded as the first true cell wall layer. The primary wall is the first readily visible wall layer, and its formation accompanies extension growth. In addition to cellulose, primary walls are composed of hemicelluloses, pectins, glycoproteins, and water. Primary walls are traversed by microscopic strands of cytoplasm called plasmodesmata that form a unique mode of communication between neighboring cells. Some cells (particularly those with strengthening, supporting, and water conduction functions) continue to add wall material inside the primary wall and form a secondary wall. Secondary walls are microscopically layered and normally contain the polymerized molecule lignin. The incorporation of lignins into the space between cell wall microfibrils is termed lignification and provides mechanical strength and decay resistance to cells. Cells with heavily lignified secondary walls undergo irreversible changes that render them incapable of further growth. The secondary wall is not easily permeable to solutions and is therefore interrupted by numerous openings or pits along its surface that occur in a variety of structural types.

Like all eukaryotic cells, plant cells possess a variety of membranous organelles possessing varied biosynthetic capacities within which a wide array of cell functions are compartmentalized. The cytoplasm of all living plant cells typically surrounds a fluid-filled space, the vacuole. Vacuoles occur in many sizes and shapes and perform different functions, often within the same cell. Among the nonprotoplasmic components of plant cells are starch granules and various mineral depositions; the most frequent of these are silica, gypsum, calcium carbonate, and calcium oxalate. The most common are those composed of calcium salt, especially calcium oxalate, usually secreted as crystals of diverse form.

## ADDITIONAL READING

1. Ajello, L. (1941). Cytology and interrelations of cystolith formation in *Ficus elastica*. *Am. J. Bot.* **28**, 589–594.

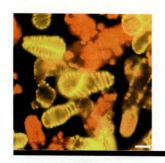

# Color Plates

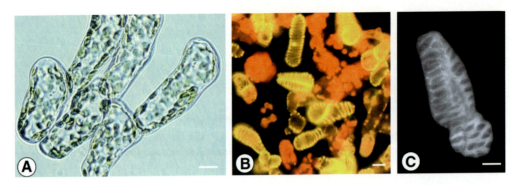

**FIGURE 1.2** Morphological changes during tracheary element differentiation in cultured *Zinnia elegans* cells. (A) Isolated leaf mesophyll cells as introduced into culture and observed with phase contrast light microscopy. Scale bar = 10 μm. (B) Cultured cells 96 hr after isolation as viewed with fluorescence microscopy. Differentiated tracheary elements are noted by the yellow autofluorescence from their lignified secondary cell walls. Undifferentiated cells are noted by red autofluorescence from their chloroplasts. Scale bar = 20 μm. (C) Differentiated tracheary cells stained with Calcofluor white to reveal the cellulose-containing secondary cell wall. Scale bar = 10 μm. Courtesy of A. T. Groover and A. S. Jones.

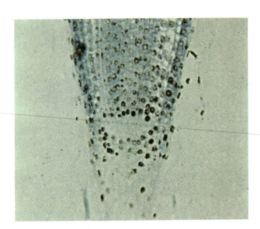

**FIGURE 2.9** Autoradiograph of longitudinal section of root tip of *Sinapsis alba* fed with tritium-labeled thymidine for 72 hours. The dark spots show location of cells that incorporated the tracer into newly synthesized DNA. The region devoid of spots is the quiescent center. Reprinted from Clowes (1965), *Endeavour* **24**, 8–12, with permission from Elsevier Science.

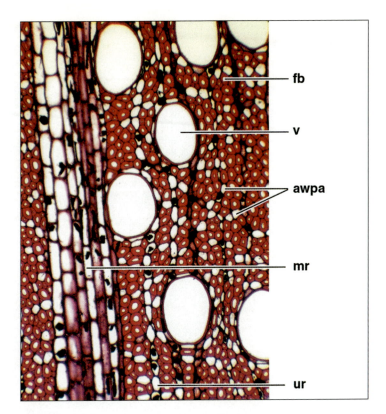

**FIGURE 4.21** Transverse section of wood of *Dillenia pentagyna* (Dilleniaceae) showing cells of the axial and radial systems. Abbreviations: awpa, axial wood parenchyma (diffuse and diffuse in aggregates); fb, xylary fiber; mr, multiseriate ray; ur, uniseriate ray; v, vessel element.

**FIGURE 16.6** Green-stained wood in intarsia masterpieces. (A) Photograph of a hand-colored print showing a variety of naturally colored wood blocks from a German book published in 1773. A block of green-stained *Populus* wood is located in the center. This figure is used courtesy of Smithsonian Institution Libraries, Cooper-Hewitt Branch, New York. (B, C) Greenish blue fruiting bodies of *Chlorociboria* on the surface of wood from *Populus* sp. collected from the forest. (D, E) Green stain within wood from dead *Populus* sp. trees recently collected from the forest. (F, G) Light micrographs of transverse sections made from green-stained wood in the Gubbio *studiolo*. A combination of dark green and yellowish-orange pigments are located in ray parenchyma cells and in some fibers and vessels. Extensive coloration is found within the ray parenchyma cells. Reprinted with permission of Blanchette *et al.* (1992), *Holzforschung* **46**, 225–232.

2.  Albersheim, P. (1975). The walls of growing plant cells. *Sci. Am.* **232**, 81–95.
3.  Bailey, I. W. (1915). The effect of the structure of wood upon its permeability. I. The tracheids of coniferous timbers. *Am. Railway Engineering Assoc. Bull.* **174**, 835–853.
4.  Bailey, I. W. (1930). The cambium and its derivative tissues. V. A reconnaissance of the vacuome in living cells. *Z. Zellf. Microskop. Anat.* **10**, 651–682.
5.  Bailey, I. W. (1938). Cell wall structure of higher plants. *Ind. Engineering Chem.* **30**, 40–47.
6.  Bailey, I. W., and Kerr, T. (1935). The visible structure of the secondary wall and its significance in physical and chemical investigations of tracheary cells and fibers. *J. Arn. Arb.* **16**, 273–300.
7.  Bailey, I. W., and Vestal, M. R. (1937). The orientation of cellulose in the secondary wall of tracheary cells. *J. Arn. Arb.* **18**, 185–195.
8.  Boudet, A. M. (1998). A new view of lignification. *Trends Plant Sci.* **3**, 67–71.
9.  Brown, M. R., Jr. (1985). Cellulose microfibril assembly and orientation: Recent developments. *In* "The Cell Surface in Plant Growth and Development" (K. Robert, A. W. B., Johnston, C. W. Lloyd, P. Shaw, and H. W. Woolhouse, Eds.), pp. 13–32. *J. Cell Sci.* Suppl. 2. Company of Biologists, Cambridge.
10. Carr, D. J. (1976a). Plasmodesmata in growth and development. *In* "Intercellular Communication in Plants: Studies on Plasmodesmata" (B. E. S. Gunning and A. W. Robards, Eds.), pp. 243–289. Springer-Verlag, Berlin.
11. Carr, D. J. (1976b). Historical perspectives on plasmodesmata. *In* "Intercellular Communication in Plants: Studies on Plasmodesmata". (B. E. S. Gunning and A. W. Robards, Eds.), pp. 291–295. Springer-Verlag, Berlin.
12. Chafe, S. C., and Wardrop, A. B. (1970). Microfibril orientation in plant cell walls. *Planta* **92**, 13–24.
13. Cook, M. E., Graham, L. E., Botha, C. E. J., and Lavin, C. A. (1997). Comparative ultrastructure of plasmodesmata of *Chara* and selected bryophytes: Toward an elucidation of the evolutionary origin of plant plasmodesmata. *Am. J. Bot.* **84**, 1169–1178.
14. Cosgrove, D. J. (1993). Wall extensibility: Its nature, measurement and relationship to plant cell growth. *New Phytol.* **124**, 1–23.
15. Cutter, E. G. (1969). "Plant Anatomy: Experiment and Interpretation. Part 1. Cells and Tissues". Addison-Wesley, Reading, Massachusetts.
16. de Duve, C. (1984). "A Guided Tour of the Living Cell." Sci. Am. Library. W. H. Freeman, New York.
17. Dickison, W. C. (1984). On the occurrence of silica grains in woods of *Hibbertia* (Dilleniaceae). *IAWA Bull. n. s.* **5**, 341–343.
18. Douglas, C. J. (1996). Phenylpropanoid metabolism and lignin biosynthesis: From weeds to trees. *Trends Plant Sci.* **1**, 171–178.
19. Fosket, D. E. (1994). "Plant Growth and Development. A Molecular Approach." Academic Press, San Diego.
20. Franceschi, V. R., and Horner, H. T. (1980). Calcium oxalate crystals in plants. *Bot. Rev.* **46**, 361–427.
21. Frank, E., and Jensen, W. A. (1970). On the formation of the pattern of crystal idioblasts in *Canavalia ensiformis* DC. IV. The fine structure of crystal cells. *Planta* **95**, 202–217.
22. Frey-Wyssling, A. (1976). "The Plant Cell Wall," 3rd ed. Handbuch der Pflanzenanatomie. Bd. 3, T. 4. Gebrüder Borntraeger, Berlin.
23. Fry, S. C. (1988). "The Growing Plant Cell Wall." John Wiley & Sons, New York.
24. Fukuda, H. (1994). Redifferentiation of single mesophyll cells into tracheary elements. *Inter. J. Plant Sci.* **155**, 262–271.
25. Gunning, B. E. S. (1976). Introduction to plasmodesmata. *In* "Intercellular Communication in Plants: Studies on Plasmodesmata" (B. E. S. Gunning and A. W. Robards, Eds.), pp. 1–13. Springer-Verlag, Berlin.
26. Gunning, B. E. S., and Steer, M. W. (1986). "Plant Cell Biology: An Ultrastructural Approach." M. W. Steer, Dublin.
27. Hill, W. E. (Ed) (1990). "The Ribosome: Structure, Function, and Evolution." American Society of Microbiology, Washington, D.C.
28. Horner, H. T., and Wagner, B. L. (1995). Calcium oxalate formation in higher plants. *In* "Calcium Oxalate in Biological Systems" (S. R. Khan, Ed.), pp. 53–72. CRC Press, Boca Raton.

29. Hyams, J. S., and Lloyd, C. W. (Eds.) (1994). "Microtubules." Wiley-Liss, New York.
30. Jansen, S., Smets, E., and Baas, P. (1998). Vestures in woody plants: A review. *IAWA J.* **19**, 347–383.
31. Jones, M. G. K. (1976). The origin and development of plasmodesmata. *In* "Intercellular Communication in Plants: Studies on Plasmodesmata" (B. E. S. Gunning and A. W. Robards, Eds.), pp. 81–105. Springer-Verlag, Berlin.
32. Kaplan, D. R., and Cooke, T. J. (1997). Fundamental concepts in the embryogenesis of dicotyledons: A morphological interpretation of embryo mutants. *Plant Cell* **9**, 1903–1919.
33. Kaplan, D. R., and Hagemann, W. (1991). The relationship of cell and organism in vascular plants. *BioScience* **41**, 693–703.
34. Kerr, T., and Bailey, I. W. (1934). The cambium and its derivative tissues. X. Structure, optical properties and chemical composition of the so-called middle lamella. *J. Arn. Arb.* **15**, 327–349.
35. Kerstetter, R. A., and Hake, S. (1997). Shoot meristem formation in vegetative development. *Plant Cell* **9**, 1001–1010.
36. Kragler, F., Lucas, W. J., and Mcnzer, J. (1998). Plasmodesmata: Dynamics, domains and patterning. *Ann. Bot.* **81**, 1–10.
37. Kuo-Huang, L.-L., and Zindler-Frank, E. (1998). Structure of crystal cells and influences of leaf development on crystal cell development and vice versa in *Phaseolus vulgaris* (Leguminosae). *Bot. Acta* **111**, 337–345.
38. Laux, T., and Jürgens, G. (1997). Embryogenesis: A new start in life. *Plant Cell* **9**, 989–1000.
39. Lewis, N. G., and Paice, M. G. (Eds.) (1989). "Plant Cell Wall Polymers: Biogenesis and Biodegradation." Am. Chem. Soc., Washington, D.C.
40. Liese, W. (1965). The fine structure of bordered pits in softwoods. *In* "Cellular Ultrastructure of Woody Plants" (W. A. Côté, Jr., Ed.), pp. 271–290. Syracuse Univ. Press, Syracuse, New York.
41. Lin, J., and Hu, Y. (1998). Taxonomic significance of extracellular crystals on the phloem fibers of Taxaceae. *Flora* **193**, 173–178.
42. Lloyd, C. W. (Ed.) 1991. "The Cytoskeletal Basis of Plant Growth and Form." Academic Press, San Diego.
43. Lucas, W. J., Ding, B., and van der Schoot, C. (1993). Plasmodesmata and the supracellular nature of plants. *New Phytol.* **125**, 435–476.
44. Lydon, R. F. (1990). "Plant Development: The Cellular Basis." Unwin Hyman, London.
45. Marin, B. (Ed.) (1987). "Plant Vacuoles: Their Importance in Solute Compartmentation in Cells and Their Applications in Plant Biotechnology." Plenum Press, New York.
46. McLean, B. G., Hempel, F. D., and Zambryski, P. C. (1997). Plant intercellular communication via plasmodesmata. *Plant Cell* **9**, 1043–1054.
47. Meeuse, A. D. J. (1941). Plasmodesmata. *Bot. Rev.* **7**, 249–262.
48. Middendorf, E. A. (1968). Plants with blowguns. *Turtox News* **46**, 162–164.
49. Mühlethaler, K. (1965). Growth theories and the development of the cell wall. *In* "Cellular Ultrastructure of Woody Plants" (W. A. Côté, Jr., Ed.). pp. 51–60. Syracuse Univ. Press, Syracuse, New York.
50. Mühlethaler, K. (1967). Ultrastructure and formation of plant cell walls. *Annu. Rev. Pl. Physiol.* **18**, 1–24.
51. Netolitzky, F. (1929). "Die Kieselkörper. Die Kalksalze als Zellinhaltskörper." *In* "Handbuch der Pflanzenanatomie".(K. Linsbauer, Ed.), Vol. 3, No. 12.
52. Niklas, K. J. (1989). The cellular mechanics of plants. *Am. Sci.* **77**, 344–349.
53. Overall, R. L., and Blackman, L. M. (1996). A model of the macromolecular structure of plasmodesmata. *Trends Plant Sci.* **1**, 307–311.
54. Pavelka, M. (1987). "Functional Morphology of the Golgi Apparatus." Springer-Verlag, New York.
55. Pobeguin, T. (1943). Les oxalates de calcium chez quelques Angiospermes. *Ann. Sci. Nat. Bot.* Ser. 11 **4**, 1–95.
56. Preston, R. D. (1974). Plant cell walls. *In* "Dynamic Aspects of Plant Ultrastructure"(A. W. Robards, Ed.), pp. 256–309. McGraw-Hill, London.
57. Price, J. L. (1970). Ultrastructure of druse crystal idioblasts in leaves of *Cercidium floridum*. *Am. J. Bot.* **57**, 1004–1009.

58. Reichert, E. T. (1913). "The differentiation and specificity of starches in relation to genera, species, etc.; stereochemistry applied to protoplasmic processes and products, and as a strictly scientific basis for the classification of plants and animals." Carnegie Institution of Washington Publ. 173, pts. 1 & 2. Washington, D.C.

59. Robards, A. W. (1975). Plasmodesmata. *Annu. Rev. Pl. Physiol.* **26**, 13–29.

60. Robards, A. W. (1976). Plasmodesmata in higher plants. *In* "Intercellular Communication in Plants: Studies on Plasmodesmata" (B. E. S. Gunning and A. W. Robards, Eds.), pp. 15–57. Springer-Verlag, Berlin.

61. Scheres, B., Wolkenfelt, H., Willemsen, V., Terlouw, M., Lawson, E., Dean, C., and Weisbeek, P. (1994). Embryonic origin of the *Arabidopsis* primary root and root meristem initials. *Development* **120**, 2475–2487.

62. Schmid, R., and Machado, R. D. (1964). Zur Entstehung und Feinstruktur Skulpturierter Hoftüpfel bei Leguminosen. *Planta* **60**, 612–626.

63. Scott, F. M. (1941). Distribution of calcium oxalate crystals in *Ricinus communis* in relation to tissue differentiation and presence of other ergastic substances. *Bot. Gaz.* **103**, 225–246.

64. Steeves, T. A., and Sussex, I. M. (1989). "Patterns in Plant Development." 2nd ed. Cambridge Univ. Press, Cambridge, New York.

65. Sugiyama, M., and Konmamine, A. (1990). Transdifferentiation of quiescent parenchymatous cells into tracheary elements. *Cell Diff. Dev.* **31**, 77–87.

66. ter Welle, B. J. H. (1976). Silica grains in woody plants of the neotropics, especially Surinam. *Leiden Bot. Ser.* **3**, 107–142.

67. Tolbert, N. E. (1971). Microbodies—peroxisomes and glyoxysomes. *Annu. Rev. Pl. Physiol.* **22**, 45–74.

68. Torrey, J. G., Fosket, D. E., and Hepler, P. K. (1971). Xylem formation: A paradigm of cytodifferentiation in higher plants. *Am. Sci.* **59**, 338–352.

69. Trockenbrodt, M. (1995). Calcium oxalate crystals in the bark of *Quercus robur*, *Ulmus glabra*, *Populus tremula*, and *Betula pendula*. *Ann. Bot.* **75**, 281–284.

70. Tsoumis, G. (1965). Light and electron microscopic evidence on the structure of the membrane of bordered pits in the tracheids of conifers. *In* "Cellular Ultrastructure of Woody Plants" (W. A. Côté, Jr., Ed.), pp. 305–317. Syracuse Univ. Press, Syracuse, New York.

71. Walker, J. C. F., and Butterfield, B. G. (1996). The importance of microfibril angle for the processing industries. *N. Z. For.* **40**, 34–40.

72. Wardrop, A. B., Ingle, H. D., and Davies, G. W. (1963). Nature of vestured pits in angiosperms. *Nature* **197**, 202–203.

73. Yeagle, P. (1993). "The Membranes of Cells." 2nd ed. Academic Press, San Diego.

74. Zinder-Frank, E. (1995). Calcium, calcium oxalate, and leaf differentiation in the common bean (*Phaseolus vulgaris* L.). *Bot. Acta* **108**, 144–148.

75. Zinder-Frank, E., Wichmann, E., and Korneli, M. (1988). Cells with crystals of calcium oxalate in the leaves of *Phaseolus vulgaris*—A comparison with those in *Canavalia ensiformis*. *Bot. Acta* **101**, 246–253.

76. Zirkle, C. (1937). The plant vacuole. *Bot. Rev.* **3**, 1–30.

# 2

# ORIGIN AND STRUCTURE OF THE PRIMARY PLANT BODY

The general concept that higher plants possess small, organized regions of active cell division, cell enlargement, and cell differentiation termed **meristems** was not formulated until the mid 19th century. As the concept of the meristem gradually developed, however, it became clear that all plant growth and morphogenesis is largely controlled by these localized regions of the stem and root. It is now well established that in many plants the entire plant body has its origin from growth of the shoot and root apical meristems. Attempts to clarify the cellular organization of the shoot and root apical meristem have received considerable attention in studies of plant development and can provide important clues for understanding meristem function and morphogenesis. Current research focuses on genetic control of meristem activity. The cells, tissues, and organs that result from this growth make up the **primary plant body,** and although secondary growth may take place to alter, conceal, or change the primary tissues, the fully differentiated primary body represents a complete morphological, structural, and functional unit.

## STRUCTURE AND FUNCTION OF THE ANGIOSPERM VEGETATIVE SHOOT APEX

Despite numerous studies of the angiosperm shoot apex, no one model satisfactorily explains shoot apex organization and function. The shoot apical meristem is typically a dome-shaped or nearly flattened collection or reservoir of permanently embryonic cells at the terminus of the stem. It has the capacity for unlimited growth. No laterally positioned structures form on the shoot apex above the level of the youngest **leaf primordia.** This region of perpetual juvenility varies in size, shape, and aspects of internal structure from species to species, as well as with age and other factors within the same plant. Submerged stems, for example, tend to have larger, more elongated apices than aerial ones, and the transition from the juvenile to adult plant is generally marked by a corresponding change to a larger shoot apical meristem. Some authors restrict the term "shoot apical meristem" to the distal-most collection of dividing cells, including their immediate derivatives. The more inclusive term "shoot apex," by comparison, includes the apical meristem proper as well as the subapical zone, where leaf primordia are initiated.

In view of the unique properties of the apical meristem and its obvious importance for plant development, it is not surprising that scientists have concentrated on this region of the plant body. Their studies have been conducted with the objective of revealing the potentialities for shoot apex growth and development under particular conditions, and of understanding the contribution of one part of the apex to another and the relationship of the meristematic cells to the older, more mature tissues of the plant. Included among these important questions are the following. Is the apex autonomous or is it influenced by the mature shoot beneath? What causes the initiation of leaf primorida and what determines where they will form?

Two general methodologies have been employed to help answer these questions. Microsurgical techniques have placed punctures or incisions in specific sites on the apex in order to observe changes in growth and interactions between parts. Apical meristems also have been surgically isolated from the rest of the plant and placed in sterile culture to determine whether they constitute self-regulating units. Additional studies have attempted to clarify their requirements for nutrients and other growth factors and to study the effects of various substances on their development. Both surgical and cultural techniques have demonstrated the capacity of even small portions of the terminal region of the shoot apical meristem to undergo normal growth and development in the absence of more mature tissues. These studies have shown, for the most part, that the apex is a largely autonomous region. Given an adequate supply of nutrients, this region of the plant is capable of continuous growth. The self-regulating ability of the shoot apical meristem can be observed following an injury to any portion of its structure. Upon destruction of any part of the meristem, the remaining cells divide and reconstitute a meristem of normal size, organization, and function.

## SHOOT APEX ORGANIZATION

Various theories of angiosperm shoot apex organization have been proposed. The data they attempt to explain come from direct histological observation, analysis of patterns and frequency of cell division, observed changes in cell DNA concentration, and examination of cell lineage patterns of genetically altered apical cells. There are currently two views of angiosperm shoot apex organization. One describes the apex as a stratified region of clonally distinct cells that is composed of one or more superficial layers that extend across the meristem to enclose a subsurface region. The outer cell layers constitute the **tunica** and undergo surface growth. The tunica is composed of cells that have a common orientation and that undergo a preponderance of **anticlinal** divisions in which the wall separating two daughter cells is formed at right angles to the surface. The number of layers comprising the tunica may vary during different stages of plant development and from species to species. The tunica is commonly two or three cells deep with layers, from the outside in, referred to as $L_1$, $L_2$, and $L_3$. As would be expected, the young surface layer, and the mature epidermis that it differentiates into, is derived from the outer $L_1$ layer. The inner layers of the tunica give rise to the corpus and eventually the ground tissue and cells that differentiate into vascular tissue. The subsurface mass of cells, called the **corpus,** is unlayered in appearance and undergoes growth in volume following cell divisions that occur in all planes (**anticlinally** and **periclinally**). By controlling the rates and planes of cell division in relation to the position of cells within the apex, the tunica and corpus remain histologically

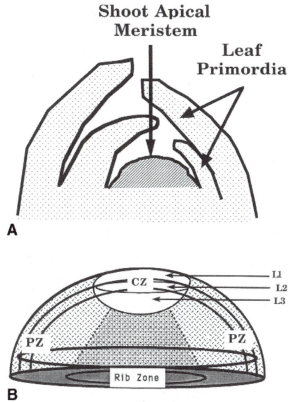

**A**

**B**

**FIGURE 2.1**  Diagram of shoot apical meristem. (A) Comparison of shoot apex and shoot apical meristem. The entire figure represents the shoot apex, whereas the shoot apical meristem is represented by the dome-shaped, shaded area. (B) Zones and layers within the shoot apical meristem. The central zone (CZ) is a small, oval, distally located group of cells that serves as the source of cells for other parts of the meristem. The peripheral zone (PZ) is located to the side and beneath the central zone. Organ initiation takes place in the peripheral zone. The rib zone forms the boundary between the meristem and the rest of the plant and is the source of the interior tissues of the stem. The zones are drawn with distinct lines, but zonal boundaries are less distinct. $L_1$, $L_2$, $L_3$ correspond to the genetically defined three cell layers from the exterior of the meristem to the interior, respectively. Reprinted with permission from Medford (1992), *Plant Cell* **4**, 1029–1039. Copyright © American Society of Plant Physiologists.

segregated, and the apex maintains a layered organization. Interestingly, if cells are displaced from the $L_1$ layer to the $L_2$ layer, they change fate according to their new position.

A second and more recent interpretation of the shoot apex recognizes cytologically and histologically definable zones and is referred to as **cytohistological zonation.** Cytohistological zonation has come to be regarded as a fundamental feature of vegetative shoot apicies and can be seen during all stages of plant development, from embryo to adult individual. Zone boundaries in actively dividing meristems are distinguished by differences in cell size, shape, degree of vacuolation, stainability, and plane of cell division. Positional information and intercellular communication regulate the identity of cells in the shoot meristem.

The **initial zone,** also called the central zone, represents a conspicuous group of enlarged initial cells located at the summit of the stem apex. These cells undergo infrequent cell division, possess prominent nuclei, and are often highly vacuolated. The initial zone functions as the source of all other cells of the apex and, by extension, of the entire primary shoot. The flanking **peripheral zone** is derived from the central zone and encircles the apical region. Cells of the peripheral zone are smaller and mitotically more active and have denser cytoplasms. They form the sites of origin of the cortex and leaf primorida that arise sequentially in precisely ordered ontogenetic patterns. An analysis of *Arabidopsis* mutants such as *wuschel* (*wus*) indicates that the central and peripheral zones are established early in embryogenesis. The third zone is designated the **transition** (or **rib**) **zone** and is located at the base of the apical meristem. This zone is thought to arise directly form the central zone and, as the name suggests, represents an intermediate region between the initial cells and partially differentiated derivative cells below. Histologically, cells in the

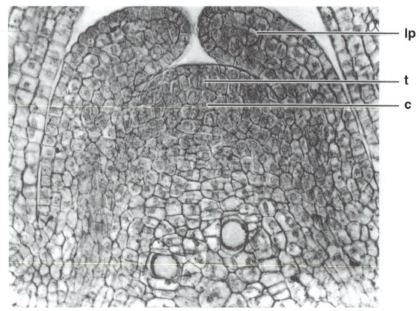

**FIGURE 2.2**   Median longitudinal section of shoot apex of *Liriodendron tulipifera* (Magnoliaceae) showing tunica (t), corpus (c), and leaf primordia (lp).

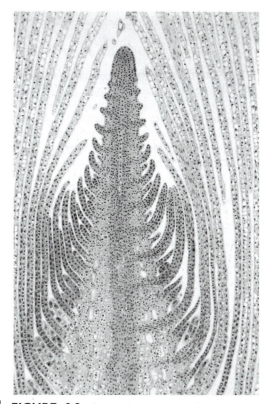

**FIGURE 2.3**  Longitudinal section of shoot apex of the aquatic plant *Elodea canadensis* (Hydrocharitaceae).

transition zone are arranged in longitudinal files as a result of cell division occurring predominantly at right angles to the stem axis. The products of the transition zone form the central pith of the stem where cells undergo significant elongation, thus expanding the internodal region. This region also may serve the important function of providing a pathway for transmission of hormone mediated signals to the rest of the plant. Plants that lack a rib zone possess shortened internodes and congested leaves that are produced in rosettes. Despite the wide acceptance of the concept of apical zonation, the precise boundaries of the zones are often ill defined and the contribution of each zone to histogenesis is still in dispute. Descriptive accounts of shoot apex organization have provided limited insight about the fate of individual cells and when a cell or group of cells at a specific location becomes destined to produce a particular tissue or structure. Furthermore, it is unclear how the different interpretations of apical organization can be reconciled.

## CHIMERAS

One approach to analyzing the fates of apical cells and their derivatives is to study cell lineages that have been generated by producing genetically distinguishable cell clones. The relative contribution of the various sectors of the shoot apex to the formation of tissues and organs can be followed with some precision by the analysis of somatic cell lineage patterns that arise from genetic mosaics. One type of genetic mosaic is a **chimera**. A chimera is a combination of tissues of different genetic constitution in the same part of the plant. A **polyploid chimera,** for example, can be induced by applying a dilute solution of the alkaloid colchicine directly to the shoot apex. Following absorption, colchicine disrupts mitotic spindle formation so that the divided chromosomes fail to separate. As a result, these cells have twice the usual number of chromosomes and because they are larger, they are readily recognized histologically and analyzed at the microscopic level. Colchicine-induced chimeras can result in even higher ploidy levels. Due to the differences in cell size and chromosome number, the descendants of the various apical regions can be followed as cell lineages in virtually all tissues of the leaf and stem. In a famous series of experiments utilizing cytochimeras, seeds of *Datura* were subjected to colchicine, resulting in changes in the ploidy level of some embryonic apical cells of the resulting plants. Analysis showed the presence of three independent layers in the shoot apex. Each entire layer of the apex can have cells of the same ploidy level. The changes in chromosome number are stable, and each region of the apex transmits its ploidy condition to all parts of the plant derived from it, such as leaves and floral apices. By using mosaics of this type, from which cell lineage analyses can be undertaken, it is possible to develop detailed pictures of the developmental fate of cells in the shoot meristem.

Ionizing radiation produces another kind of chimera that has contributed to our understanding of cell lineage patterns. **Variegated chimeras** fall into this category. Following treatment, certain cells of the apex have normal chloro-

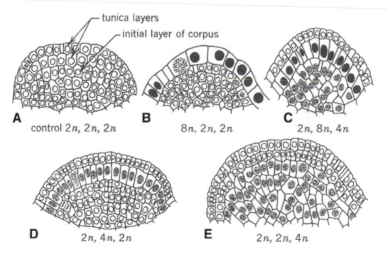

**FIGURE 2.4**   Longitudinal sections of shoot apices of *Datura* (Solanaceae), showing periclinal chimeras induced by treatment with colchicine. (A) Untreated plant with all cells diploid. (B) Polyploidy in outer layer of the tunica. (C) Polyploidy in the inner layer of tunica and corpus. (D) Polyploidy in inner layer of tunica. (E) Polyploidy in the corpus. Reprinted with permission from Esau (1965), "Plant Anatomy," 2nd ed., John Wiley & Sons. Adapted from Satina, Blakeslee and Avery, *Am. J. Bot.* (1940), **27**, 895–905.

plasts, whereas other mutated cells have plastids devoid of chlorophyll. The different initials form visible variegated patterns in the mature tissue derived from them. These studies have provided evidence that variability in cell lineage patterns is common and that cell lineage itself can play little, if any, role in plant development.

## STRUCTURE AND FUNCTION OF THE ROOT APEX

The root system of a plant begins its development from the **radicle** of the embryo. The radicle grows out of the seed after the seed has absorbed sufficient amounts of water; it then continues to grow as the primary root of the new plant. The tip of the root is covered by a collection of parenchymatous cells that constitute the **root cap.** In addition to shielding the meristem, root cap cells function in growth regulation (such as gravity perception) and in the production and secretion of abundant mucilage. The root cap originates from the activity of the root apical meristem and consists in some roots of centrally positioned, longitudinally aligned **columella cells** and outer **peripheral cells.** As root cap cells are formed, they are pushed through the cap and are eventually sloughed off in great numbers by friction as the root grows through the soil. Columella cells are distinguished by their elongate shape and by containing dense aggregations of starchy amyloplasts that sediment to lower sides within the cell in response to gravity. The peripheral cells of the root cap secrete enormous quantities of a mucilage called **mucigel.** This slimy substance benefits the plant by protecting and lubricating the growing root apex and aiding in water and nutrient absorption.

Just proximal to the root cap are the mitotically active cells of the root **apical meristem.** These constitute the earliest detectable progenitors of the more mature cells. This zone is sometimes termed the **region of cell division** (meristematic zone). Behind the point where cells divide less actively is a narrow region where the rate of extension or expansion growth increases. This is the **region of cell elongation** or radial enlargement (elongation zone). The **region of maturation** (differentiation zone), which comprises a longitudinal sequence of cell and tissue differentiation, also is evident near the apical meristem. In some root apices, the differentiating cells are typically positioned in files that can be visibly traced back to a specific number of initials at the tip. The younger, less differentiated cells are present near the root apex, whereas progressively larger and more mature cells are located at greater distances from the apex. The zone of maturation is recognizable by the presence of root hairs on the surface. As a result of their radial symmetry, roots are a favorite tool for the study of the origins of patterning.

Some investigators have traced root tip cell lineages in stained longitudinal sections by studying the frequencies of what are called T divisions. In a T cell division, a meristematic cell divides transversely followed by a longitudinal division in one of the daughter cells. The two cells formed by the longitudinal division then act as the progenitors of longitudinal rows of cells into the root axis and root cap. In the cells giving rise to the root axis, the upper daughter cell divides longitudinally to form an inverted T. In the cells giving

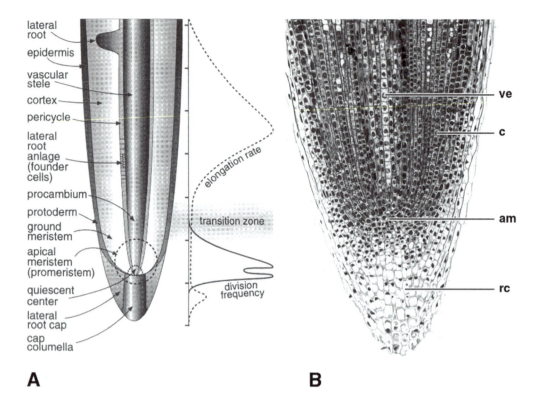

**FIGURE 2.5**   General angiosperm root apex organization. (A) Pertinent anatomical features of a root apex are labeled at left. Relative elongation rates (dashed line) and division frequencies (solid line) for epidermal cells are plotted at right. Ticks on the vertical scale are in 1-μm intervals for maize and at 50-μm intervals for *Arabidopsis* (Cruciferae). (B) Longitudinal section of root apex of *Hyacinthus orientalis* (Liliaceae). Abbreviations: am, apical meristem; c, cortex; rc, root cap; ve, differentiating vessel. A, reprinted with permission from Jacobs (1997), *Plant Cell* **9**, 1021–1029. Copyright © American Society of Plant Physiologists.

rise to the root cap, the lower daughter cell divides longitudinally, forming an upright T.

Roots differ from stems in many aspects of anatomy and function. For example, no primordia are formed on or near the root apex, making the angiosperm root in some respects an anatomically less complex organ. As has been the case with shoot apices, a number of conflicting explanations have been given for the organization and function of the root apical meristem. From a descriptive standpoint, it has long been thought that there exist distinct tiers of ultimate meristematic initials (**histogens**) that function as a source of patterning. Each initial is considered to be programmed to give rise to a particular mature tissue that converges in a differentiating file upon the apex. Species can vary in the number of initials that are present and in the pattern of root meristem growth. Four sets of initials occur in some roots. One gives rise to the protoderm and a portion of the root cap, a second forms the central portion of the root cap, a third gives rise to the ground meristem and

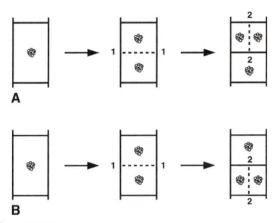

**FIGURE 2.6**   Diagram illustrating T divisions within the root apex. (A) Inverted T divisions 1–1 and 2–2, which give rise to all tissues of the root except the root cap. (B) Upright T divisions 1–1 and 2–2, which give rise to the root cap.

undifferentiated endodermal regions, and a fourth produces the pericycle and procambium. Experiments have shown that cells of the outer cortical region of the root also can participate in the formation of portions of the root cap. The histogen concept of root apical meristem organization assumes that there is a perfect correlation between cell lineage and mature cell and tissue type. Evidence suggests, however, that positional information plays a more significant role in regulating the developmental fate of root cells.

Although the distinction is not always clear cut, root meristems are sometimes described as "closed" if the meristem is divided into separate regions that generate the outer root cap cells and inner tissues. In these cases there is a discrete region of the meristem producing either root cap cells alone or cap and epidermal cells. Many monocotyledons—for example, grasses—possess closed meristems in which the meristem originates cap cells separately from the epidermis, that differentiates from the outer layer of the differentiating cortical region. An "open meristem," on the other hand, occurs when the meristem is not separated into distinct zones. The cells between the stelar pole and root cap divide to form an unstable boundary between the cap and the rest of the root. In median sections of open meristems, therefore, it is not easy to relate regions to the pole of the apex to the tissues of the root, as in the closed type.

The use of sophisticated radioactive labeling and staining techniques by the British plant scientist F.A.L. Clowes led to the discovery of a remarkable region of infrequently dividing cells in the root tip called the quiescent center. The concept of a **quiescent center** refers to a hemispherically shaped aggregation of mitotically and metabolically inactive cells that are positioned just behind the root cap. These cells divide at rates 10 to 20 times slower than adjacent cells and do not participate in forming mature root tissues. Cells lying within this region show a low rate of DNA synthesis and a corresponding slow cell cycle, in which cells are arrested in the $G_1$ phase of mitosis or are noncycling. The existence of such a region has now been confirmed in every

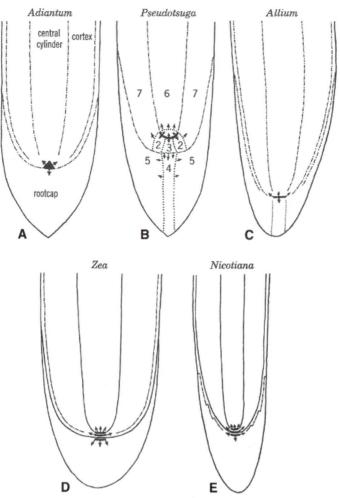

**FIGURE 2.7**   Organization of distal region of the apical meristems of roots. (A) A single apical cell (black triangle) is the source of all parts of root and rootcap. (B) Initial zone (black arc) initiates mother cell zones of the various root parts as follows: I (below 6, not marked) of central cylinder (6); 2, of cortex (7); 3, of column of rootcap (4). Longitudinal divisions on periphery of column give cells to peripheral part of rootcap (5). C–E are based on the classical histogen concept. (C) Distal region with poorly individualized initials is source of central cylinder, cortex, and column. (D) Three tiers of initials in the initial zone. The first is related to central cylinder; the second, to cortex; the third, to rootcap. The epidermis differentiates from the outermost layer of the cortex. (E) Three tiers of initials, the first related to central cylinder; the second, to cortex; the third, to rootcap. The epidermis originates from the rootcap by periclinal divisions. Reprinted with permission from Esau (1965), "Plant Anatomy," 2nd ed., John Wiley & Sons.

angiosperm and gymnosperm root apex thus far examined, including both primary and lateral roots.

This region can easily be identified by feeding roots with radioactive precursors of DNA and by using a technique called autoradiography in which the labeled cells can be detected on photographic film and shown to surround a zone in which DNA synthesis is infrequent. Under this view, the root

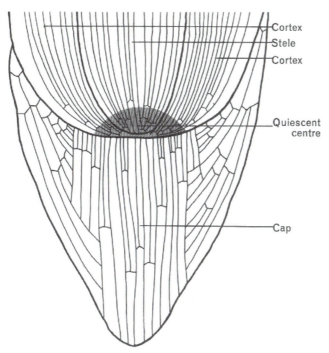

**FIGURE 2.8**   Pattern of cell files in the root tip of *Zea mays*. Note the quiescent center and the major T divisions that give rise to the different cell lineages. Reprinted from Clowes (1965), *Endeavour* **24**, 8–12, with permission form Elsevier Science.

promeristem refers to the highly meristematic cells on the distal and proximal surfaces of the quiescent zone that give rise to the mature root cap, epidermis, cortex, and stele. Initials on the distal side of the quiescent center give rise to the columella portion of the root cap, whereas the initials forming the main body of the root are located on the basal side of the quiescent center.

The quiescent center becomes established during early embryogenesis or following seed germination when the radicle has undergone some elongation; it is highly variable in size and activity thereafter. The boundaries of the quiescent zone are clearly defined by the root cap at the distal side but are ill defined on the proximal side. The size of the zone is generally proportional to the size of the root apex. Estimates of the number of cells composing the quiescent center range from as few as 2 to 4 in *Alectra vogelii*, *Striga gesnerioides*, and *Arabidopsis* to over 1000 cells in very large apices such as *Zea mays*. In roots with small quiescent centers, the initials and quiescent cells are in direct contact with one another. When quiescent centers are large, this is usually not the case. Dramatic ontogenetic changes can occur in this region, as evidenced by the fact that it can become expanded during root growth after cortical and central cylinder initials and their immediate derivatives become quiescent. Although the function of the quiescent center remains unanswered, various suggestions have been offered, including: (1) it may serve as a pool for replacing the more peripheral meristematic cells, (2) it may be necessary for the root to maintain its geometric integrity, and (3) it may function as a source or sink for metabolites and growth regulators, specifically cytokinins. Recent

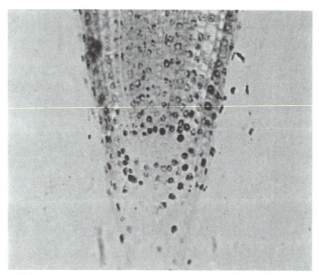

**FIGURE 2.9**  Autoradiograph of longitudinal section of root tip of *Sinapsis alba* fed with tritium-labeled thymidine for 72 hours. The dark spots show location of cells that incorporated the tracer into newly synthesized DNA. The region devoid of spots is the quiescent center. Reprinted from Clowes (1965), *Endeavour* **24**, 8–12, with permission from Elsevier Science. (See Color Plate.)

evidence indicates that cells in the quiescent center normally function in maintaining the surrounding initials in an undifferentiated state. According to this explanation, cells remain as initials because the quiescent center acts to inhibit the differentiation process. In the absence of a quiescent center, cell division in the surrounding initials is aberrant.

Subjecting root apices to ionizing radiation results in damaged cells that no longer divide normally. Roots can survive damaging irradiation, however, because the quiescent center cells still can initiate division and so can produce a new set of derivatives. Quiescent cells also can be induced to divide and repopulate the meristem following root decapitation or other experimental treatments. The fact that root tip cells display differential radiosensitivity, in which the quiescent center is less susceptible than the rest of the apex, has an apparent counterpart in animal tissues in which rapidly dividing cells (such as bone marrow stem cells) are very sensitive to radiation damage. Because radiation is a major component of some cancer therapies, understanding the effects of irradiation on cell organization and function has medical import.

## DIFFERENTIATION OF PRIMARY TISSUES

Similar to the developing embryo, three categories of so-called primary meristematic tissues can be recognized in the undifferentiated region immediately behind the apical meristem. These undifferentiated tissues are derived from the various cell groups or zones of the apex and, without exception, mature into the primary tissues of the shoot. This generalization has been regarded as a unifying principle of developmental plant anatomy, and a complete understanding of the patterns of cell differentiation is essential for anyone working in this area.

The outermost, single layer of cells at the apex is the **protoderm.** It will mature into the epidermis of the primary plant body. The **procambium,** also called provascular tissue, is identifiable as strands of elongated cells that connect to the plant's mature vascular tissue and differentiate into the primary vascular tissues, namely the primary xylem and primary phloem. Some of the procambial strands extend into the developing leaf primordia. In still older leaves, most of the procambium has matured into the primary xylem and primary phloem of the leaf veins. The remainder of the cells within an apex constitute the **ground meristem.** The ground meristem is located between the protoderm

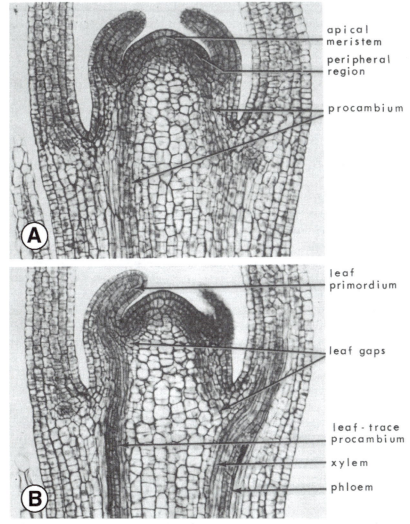

**FIGURE 2.10**  Longitudinal sections of shoot tips of flax (*Linum*), median section in A, nonmedian in B. (A) Procambium merges with the peripheral region of the apical meristem where leaf primordia are initiated. (B) Procambium is shown diverging toward leaf primordia. Both ×175. Reprinted by permission of the publisher from PLANTS, VIRUSES, AND INSECTS by K. Esau, Cambridge, Mass.: Harvard University Press, Copyright © 1961 by the President and Fellows of Harvard College.

and the procambial strands and fills the center of the young axis. It matures into the cortex and pith of the plant axis in the ontogenetic sequence from apex to tissue of the mature shoot.

## LEAF INITIATION AND DEVELOPMENT

Leaves of angiosperms are initiated as groups of cells on the flanks of the shoot apical meristem. Primordial cells can originate as discrete mounds of undifferentiated tissue on the sides of the apex, or they can arise as a raised structure that completely encircles the meristem. Although the formation of leaves has been regarded as one of the most important functions of the shoot apex, the direct developmental relationship between the shoot meristem proper and leaf initiation has never been clearly established. For example, leaves or leaflike structures have been observed to arise on meristemless shoots or on shoots with nonfunctional meristems.

Leaves originate in remarkably precise and stable patterns that can be expressed mathematically. The arrangement of leaves on the stem is called **phyllotaxy**. A number of different phyllotactic patterns exist among flowering plants (e.g., spiral, paired, whorled). Included among the major unresolved questions of plant development are those that ask why leaves form at predictable sites on the apex and what controls their development. Experimental manipulation of the apical meristem has established that leaf morphogenesis generally does not depend on the presence of existing primorida or other more mature tissues, such as procambial strands. One popular proposal suggests that each successive primordium arises in the next largest or available space on the apex that becomes available. This concept does not totally explain, however, why a primordium forms in a particular place. The suggestion that a recently formed primordium inhibits the development of a new primordium in its immediate vicinity by chemical signaling or by creation of stresses also has been widely discussed, although the evidence supporting this hypothesis is not totally satisfactory. In this model, a primordium could arise only at the farthest site from an existing primordium.

Organogenesis commences with coordinated changes in the rate and pattern of cell division and with expansion within localized cells in the peripheral zone of the meristem. One recent biophysical proposal suggests that leaf initiation is associated with the reorientation of cellulose microfibrils on the meristem flanks to form new circular alignments around each incipient primordium. This is thought to be accomplished through changes in the cellular cytoskeleton. The newly organized microfibrils presumably hinder the lateral enlargement of cells, with the result that the primordium protrudes out from the meristem surface. Primordial initials become committed to form determinate lateral organs extremely early, prior to their emergence as small protrusions on the apex.

Careful studies of leaf ontogeny have emphasized that the enormous diversity of leaf forms among angiosperms arises principally from highly coordinated patterns of differential growth occurring in localized regions of the primordium at different stages of ontogeny. These growth regions show quantitative variations in sets of generating factors involving size, position, and

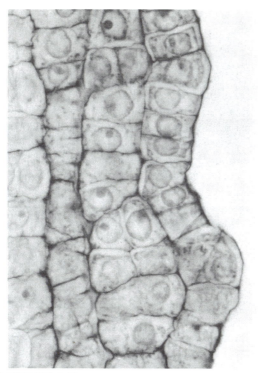

**FIGURE 2.11** Longitudinal section of shoot apex of *Anacharis* (Hydrocharitaceae) showing two early stages in the initiation of leaf primordia. × 900. Courtesy of P. G. Mahlberg.

degree of activity. Shifts in the amount of growth expressed in different regions of the developing leaf lead to variation in the adult form. Regardless of variation in adult leaf morphology, however, all simple leaves appear to follow a similar general pattern of early development. At a very young state, following a change in the plane of cell divisions in protodermal and subdermal layers, the primordial aggregation of cells grows upward from the shoot apex. In most cases a leaf primordium originates from the three cell layers of the shoot apex. In contrast to lower vascular plants such as ferns, developing leaf tissues of angiosperms are not derived from the activity of a primordium apical cell or a group of apical cells. Most of the increase in length of the primordium is the result of intercalary growth. Lateral growth in the basal intercalary meristem forms a sheathing leaf base. Unlike most stems and roots, growth in the length of the leaf ceases early in ontogeny.

In the majority of dicotyledons, the early steps in leaf development are followed by the later transformation of the undifferentiated peglike primordium into a dorsiventrally flattened organ. This has long been thought to be primarily accomplished by the gradual lateral expansion of the winglike lamina by **marginal** and **submarginal meristem** activity. In petiolate leaves, this is followed by extension of the petiole. Traditionally, growth of the blade has been interpreted to involve anticlinal cell divisions in the outer row of marginal protodermal cells, giving rise to the two epidermal layers—one abaxial, the other adaxial. This is accompanied by periclinal and anticlinal divisions in the cells immediately be-neath the protoderm that contribute to the production of the multilayered meso-phyll. This widely held assumption has been brought into serious question in recent years by the clonal analysis of genetically marked cell lineages during leaf morphogenesis. These studies provide evidence that neither the orientation of cell division at the margin of the blade nor the frequency of cell division in this region supports the existence of a marginal meristem. Rather, the complicated processes of blade expansion and differentiation appear to result primarily from cell division and expansion occurring throughout the developing blade.

In the case of compound leaves, basal intercalary meristem activity is combined with differential growth ("fractionation") on the flanks, to form marginal lobing. The great German botanist Wilhelm Troll described three patterns of leaflet morphogenesis that are characteristic of compound leaves. These are (1) basipetal, the oldest leaflets near the primordium tip forming and developing first; (2) acropetal, the oldest leaflets near the primordium base forming first; and (3) divergent, leaflets initiated first in one direction and then in both acropetal and basipetal directions. A petiole is typically interca-lated between the leaf base and lower-most pair of leaflets.

The period of early primordium growth in monocotyledons is followed by the establishment of a **basal leaf meristem.** This meristem functions for only a short period but is accompanied by the general enlargement of cells to pro-duce the linear leaves that are characteristic of this group. In contrast to dicotyledons, the maturation of monocot leaf cells occurs in a basipetal direc-tion, that is, from the leaf apex toward the base. Enlargement of mesophyll cells occurs in both vertical and horizontal planes.

## CLASSIFICATION OF TISSUES

Higher plant cells take many forms. The shape and structure of individual cells are determined primarily by the manner in which they are grouped into tissues and by the nature of the function they perform. At maturity each of these diverse forms of cells has a distinctive shape, particular wall character-istics, and specific physiological properties. In some cells this includes the ontogenetic elimination or modification of the entire protoplast or specific organelles. The wide variety in form and function of plant cells presents problems for the classification of cell types, tissues, and tissue systems. Classification is more difficult because there are sometimes no clearly defined morphological boundaries between cell types, for example, between parenchyma and collenchyma, parenchyma and sclerenchyma, fibers and scle-reids, and even, in certain cases, fibers and tracheary cells. There also is a con-tinuum from wood fibers with distinctly bordered pits (fiber tracheids) to those with simple pits (libriform fibers). At the present time there is no con-sensus regarding a scheme of classification that can satisfactorily account for the totality of cytological and histological variation encountered within the higher plant body. Some cells develop highly specialized and limited structures and functions, whereas others appear to carry out multiple functions and may even resume growth, cell division, and differentiation. Some plant tissues appear to be relatively homogeneous (so-called **"simple" tissue**), whereas others are composed of a variety of cell types (so-called **"complex" tissue**). As a result, previous attempts to classify plant cells and tissues on the basis of mature structure, principal cell or tissue function, or cell and tissue origin have been largely unsuccessful. In view of these difficulties, classifications of plant cells and tissues are, for the most part, artificial and should be viewed only as a convenience.

The German botanist Julius von Sachs (1832–1897) devised a scheme of classification for mature vegetative tissues that was based primarily upon the recognition of major tissue aggregations of general topographical and physiological similarity. These tissue aggregations were called "tissue systems" and were viewed as being the result of evolutionary specialization within the higher plant body. Three principal tissue systems can be recognized. These are: (1) **dermal** (includes the primary epidermis and secondarily produced cork layers); (2) **fascicular** (contains the conducting xylem and phloem of primary and secondary origin); and (3) **fundamental** (includes the primary pith, cortex, and mesophyll). Although objections to this scheme can be offered, it continues to provide a useful framework for the student of plant anatomy. The three mature tissue systems of the primary body can be related ontogenetically to the differentiating tissues produced by the apical meristem, namely protoderm, procambium, and ground meristem, respectively.

## Dermal Tissue System

The **epidermis** comprises the mature, typically uniseriate surface layer of the entire primary plant body. It is derived from the protoderm and by definition encompasses a variety of cell types, including ordinary epidermal cells, guard cells, subsidiary cells, trichomes or emergences, and various idioblasts. In a limited number of plant families, the foliar protoderm undergoes periclinal division somewhat late in development to form a **multiple epidermis** of two or more layers in thickness. A multiple epidermis is visible in plants belonging to a number of dicotyledonous families, such as Begoniaceae, Bombacaceae, Malvaceae, Moraceae, and Piperaceae, as well as many monocotyledons. The multiple epidermis is thought to function as a water storage tissue, at least in some plants. During secondary growth of the stem and root, the epidermis is most often sloughed off and replaced by periderm.

Epidermal cells are variable in shape as well as wall characteristics and retain an active protoplast. Walls of epidermal cells show a diversity of structure and chemical composition in keeping with their function as the protective boundary layer between the environment and internal plant tissues. The plant surface has been subdivided into four recognizable categories of epidermal features. These are: (1) cellular arrangement or pattern; (2) shape of the epidermal cells ("primary sculpture"); (3) relief of the outer epidermal cell walls caused primarily by **cuticular striations** and visible wall thickenings ("secondary sculpture"); and (4) **epicuticular wax** secretions ("tertiary sculpture"). The shape of epidermal cells varies with respect to their outline in surface view, configuration of anticlinal walls (generally straight, curved, or undulated), relief of the cell boundary (channeled or raised), and degree of curvature of the outer periclinal

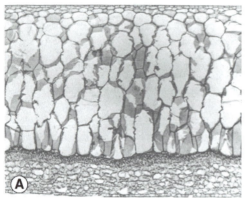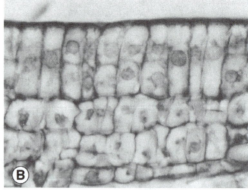

**FIGURE 2.12**   Multiple epidermis structure and development. (A) Transverse section of leaf of *Peperomia* (Piperaceae) showing adaxial multiple epidermis. (B) Early stage in the ontogeny of the multiple epidermis in a young leaf of *Ficus* (Moraceae). × 900. B, courtesy of P. G. Mahlberg.

walls. In some taxa, epidermal cells become **papillate** or **mucilaginous** in appearance, in a few cases they become sclerotic.

## Cuticle and Epicuticular Waxes

The outer walls of cells covering the aerial organs are characteristically overlaid by a structurally complex waxy layer, the **cuticle,** that is largely impervious to liquids and gases. It represents one of the major adaptations of plants to life on land. Based on the stage of development, some individuals have employed the terms "primary" and "secondary" cuticle. The **primary cuticle** forms while epidermal cells are in the process of expansion. This stage is followed by the development of **secondary cuticle** after epidermal cells have reached their full size. The cuticle, or "cuticular membrane" as it is sometimes referred to, is chiefly composed of **cutin,** a fatty substance that becomes oxidized and polymerized on the outer cell surface by a process known as **cuticularization.** By contrast, the process of impregnation of the cell wall with cutin is called **cutinization.** In addition to cutin, epidermal cells can have lignin, silica, waxes, or a mixture of other materials in or on the walls.

The broadly defined plant cuticle is a complex region whose many organizational details are not well understood. Because the cuticular surface is the boundary layer between the environment and plant body, it is not surprising that this region shows a great deal of structural variability. Various interpretations of the cuticle and its relationship to the epidermal cells have been proposed, and the terminology that has been used to describe the plant surface has not been standardized. Some workers, for example, separate the cuticle (strictly termed the "cuticular membrane") into an outermost layer termed the "cuticle proper" and an inner "cuticular layer." The cuticular layer often is the major component of the cuticle. Different plants possess cuticles of different thickness, ultrastructure, and chemical composition. At the electron microscopic level the cuticle can appear more or less amorphous in appearance or lamellate. Under light microscopy, the cuticle is seen as a transparent, color-

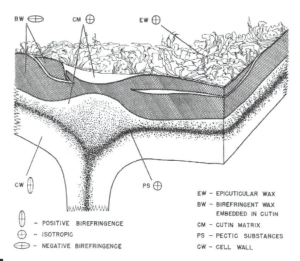

BW ⊖   CM ⊕   EW ⊕

CW ⊖   PS ⊕

⊖ — POSITIVE BIREFRINGENCE
⊕ — ISOTROPIC
⊖ — NEGATIVE BIREFRINGENCE

EW – EPICUTICULAR WAX
BW – BIREFRINGENT WAX
        EMBEDDED IN CUTIN
CM – CUTIN MATRIX
PS – PECTIC SUBSTANCES
CW – CELL WALL

**FIGURE 2.13** Diagrammatic representation of the structure of the pear (*Pyrus communis*) leaf upper cuticle, not necessarily drawn to scale. Lower cuticle may differ in certain structural details in quantity of oriented wax molecules. Reprinted with permission from Norris and Bukovac (1968), *Am. J. Bot.* **55**, 975–983.

less, or pale gray layer. The cuticle typically is separated from and attached to the underlying epidermal cell walls by a layer of pectinaceous material.

The formation of both epicuticular waxes and cutin begins with fatty acid synthesis. To form cutin, fatty acids are hydroxylated, esterified, and linked into a complex polymer. To form waxes, fatty acids are elongated and modified to form the wide variety of compounds found in waxes. Some waxes are of considerable commercial importance. Carnauba wax, for example, is derived from the wax palm *Copernicia cerifera*; thick wax forms on the undersurface of the leaves, which are cut and processed. Bayberry wax is derived from the fruit of *Myrica*. In this case, the granular wax is removed and made into candles. Because epicuticular waxes are made in the epidermal cells, they must pass through the epidermal cell membrane, cell wall, and cuticle to reach the surface. Exactly how the wax arises in the epidermis and how it is transported to the surface is unclear. It is generally believed that epicuticular wax deposition occurs either by diffusion of wax through the cuticle (diffusion hypothesis), by passage through small complex anastomosing pores (pore hypothesis), or by some combination of these two processes.

Waxes are deposited within the cuticle either intracuticularly or epicuticularly as a thin film or layer on the surface of leaves and other structures. Surface waxes restrict transpirational water loss and provide an additional barrier to the entry of air pollution into leaves and other plant parts. Epicuticular wax secretions show a wide range of ultrastructural and chemical variation and produce a notably glaucous appearance to the organ surface. The amount of wax deposition on leaves varies widely and differs across species. Leaf age and environmental conditions also affect deposition. Waxes consist mainly of solid lipophilic substances secreted by nonspecialized epidermal cells. Scanning electron microscopy (SEM) has revealed the fine detail of wax aggregations and a variety of self-assembling shapes. These SEM data have led to their classification and their arrangement into at least 23 distinct types.

Some waxes are deposited as continuous thin films or crusts. The most frequent shapes are described as platelets, rodlets, filaments, ribbons, and dendritic structures. The tubule is one of the most uniform, complex, and characteristic forms of these epicuticular wax aggregates. This form of wax structure was initially described as tiny needles and subsequently as cylindrical rods. They are now known to be hollow tubules. A few dicotyledons and monocotyledons, especially succulents and xerophytic plants, have complex **stomatal wax chimneys** surrounding the stomata. These cylindrical structures can form longitudinally aggregated rodlets, (as in *Heliconia collinsiana*) or form excessive, localized wax production in the epidermal cells surrounding the stomata (as in *Euphorbia tirucalli*).

Various functions can be attributed to the plant surface under different environmental conditions. All forms of epicuticular wax aggregations create a rough hydrophobic surface that repels water. Studies have also shown a relationship between the wettability of leaf surfaces and their degree of susceptibility to contamination by particulate depositions (dust, spores, etc.). Epicuticular wax crystals create surface roughness, reduced particle adhesion, and water repellency. In water-repellent leaves, contamination particles are removed by adhering to water droplets that roll off the surface following rain, heavy fog, or dew. The geometry and chemistry of wax tubules, for example, are reported to be especially well suited to lowering surface wettability. Although cutin has long been known to reduce water loss in some specialized plants, certain surface features may actually aid in increasing the rate of water loss by causing turbulences in the airflow over the surface, thus resulting in increased transpiration and gas exchange. An extremely thick deposition of cutin also can increase the mechanical stability of the surface. In some plants the surface microsculpture also can have a major influence in controlling surface temperatures, by reflecting or altering incoming radiation. The role of the cuticle as a plant defense mechanism is discussed in Chapter 10.

## Specialized Epidermal Cells

A diverse collection of epidermal cell types can occur on the young stems and foliar organs of different angiosperms. For example, large, thin-walled, and highly vacuolated **bulliform** (balloonlike) **cells** are present in the leaves of many monocotyledons, where they are thought to function in the rolling up or unrolling of leaves following the loss or uptake of water. The cells of various taxa form complex stalked, irregularly shaped mineral depositions of calcium carbonate (termed **cystoliths**), which are deposited over an internal cellulosic framework. These can completely fill a cell. Cystoliths occur in the parenchymatous cells of various parts of the plant, but they are most frequently found in the epidermis, in hairs, or in greatly enlarged cells (termed **lithocysts**). Surface cells with distinctly papillate outer walls are found in diverse genera. Although a perfectly smooth or nonsculptured epidermal surface occurs in many plants, a large number of species show a well-developed **indumentum** composed of hairs or trichomes. Diverse types of **trichomes,** or hairs, commonly arise from the surface of many plants. Included in this diversity are various unicellular and multicellular types, as well as **grandular hairs** that produce essential oils, resulting in

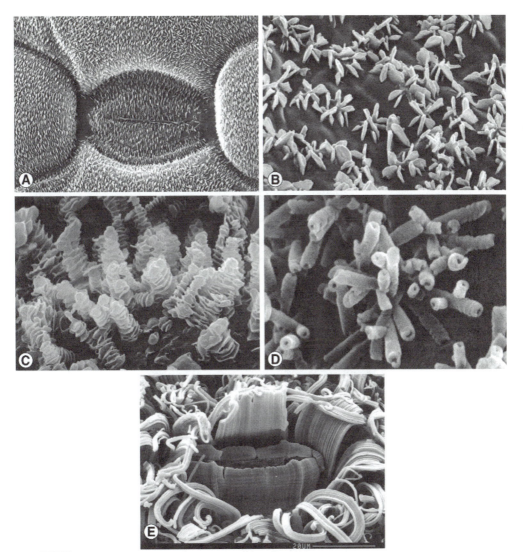

**FIGURE 2.14**   SEM views of epicuticular wax crystals on leaf surfaces. (A) Parallel oriented platelets (*Convallaria* type). Platelets are evenly spread in parallel lines. Around stomata the lines form a pattern resembling electromagnetic field lines. (B) Rosettes (*Fabales* type). Platelets in a stellate arrangement. The number and individual morphology of the platelets varies within certain limits. (C) Transversely ridged rodlets (*Aristolochia* type). Rodlets with perpendicular ridges. Diameter of rodlets and distance of ridges to each other may vary within certain limits. Transversely ridged rodlets can be divided into two groups, either containing palmitone or other dominating compounds (e.g., alkanes) that are supposed to determine crystal morphology. (D) Nonacosanol tubules (*Berberis* type). Hollow crystals with an outer diameter approximately 1.2 μm. Branching is perpendicular. Often arranged in clusters. (E) Longitudinally aggregated rodlets (*Strelitzia* type). Large, usually curved rodlets (up to 50 μm long), which are fused along their axis. They form characteristic "chimneys" around stomata. Reprinted with permission of W. Barthlott. Copyright © W. Barthlott.

a distinctive odor or secretion. Trichomes commonly arise from asymmetric divisions within a single protodermal mother cell, although in some instances the first division of the mother cell is symmetric and more than one initial is involved. At maturity hair cells may be without protoplasm, but they are more often living. The distributions of trichomes on plants display distinct patterns of nonrandom spacing that arise along controlled developmental programs.

The epidermis of all aerial organs such as leaves, stems, floral parts and fruits are typically provided with specialized openings called **stomata.** The pores are surrounded by a pair of specialized and often reniform epidermal cells, the **guard cells,** which open or close the stomatal opening by changing the turgor pressure, thereby regulating the rate of transpiration and gaseous exchange between the atmosphere and the internal air spaces. Guard cells possess starch-accumulating chloroplasts, whereas ordinary epidermal cells generally lack plastids or have only rudimentary plastids. Following an increase in internal pressure, the structure of the pectin-rich guard cell walls allows for expansion only along the direction of the curved longitudinal axis. This results in the opening of the pore. Two or more morphologically distinct **subsidiary cells** may surround the guard cells. The pore, guard cells, and subsidiary cells collectively are sometimes referred to as the **stomatal apparatus** or **stomatal complex.** Subsidiary cells store large amounts of water and ions, and because no functional plasmodesmata occur between mature guard cells, subsidiary cells, and ordinary epidermal cells, these materials must move through the apoplast.

Stomata form on both very young leaves and later in development. The cell division planes and the patterns of stomatal development are very precise, and both the development and mature appearance of the stomatal complex can be important for determing evolutionary relationships. Like trichomes, stomata have an ordered distribution or pattern in the epidermis. In monocotyledons and some conifers, stomata are typically aligned in linear cell files that are parallel to the long axis of the leaf and that are separated by areas devoid of stomates. Some plants possess stomata that are clustered together in discrete groups. In all plants, the guard cells of neighboring stomata are not in contact with each other. This results in stomata being separated on the surface by at least one ordinary epidermal cell. Different theories of stomatal

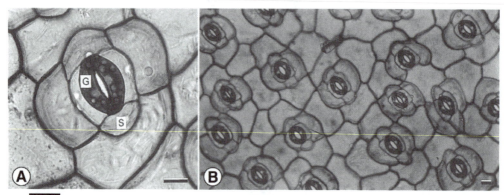

**FIGURE 2.15** Stomatal arrangement in *Commelina communis* (Commelinaceae). (A) High magnification of a single stomatal complex on the abaxial leaf surface, showing the two paired guard cells (G) and six surrounding subsidiary cells (S). (B) Low magnification of stomatal complexes demonstrating their regular, two-dimensional arrangement. Scale bar represents 20 μm. Reprinted with permission from Hall and Langdale (1996), *New Phytol.* **132,** 533–553.

patterning have been advocated. Some emphasize cell lineage, whereas others emphasize cell inhibition within a field of influence, or cell cycle features in which only cells in a particular stage in the cell cycle are developmentally responsive.

The ontogenies of stomata follow a number of different, often taxon-specific, developmental pathways. All stomata begin their development following an initial asymmetric division of a protodermal precursor cell to form two daughter cells that are unequal in both size and ultimate fate. The smaller cell is the stomatal initial or **meristemoid**. In a representative monocot such as *Zea mays*, the smaller of the two daughter cells becomes a guard cell mother cell that undergoes a subsequent equal division to form a pair of guard cells. A general developmental pattern in dicotyledons is characterized by the formation of a small triangular cell that continues to undergo a variable number of precise divisions to form the guard cells. Depending upon the manner of origin of subsidiary cells, angiosperm stomata can be categorized into three main types. **Mesogenous stomata** have subsidiary cells that arise from the same initial as the guard cells. **Perigenous stomata** have subsidiary cells that do not have a common origin with the guard cells, but are formed from cells lying around the guard cell initial.

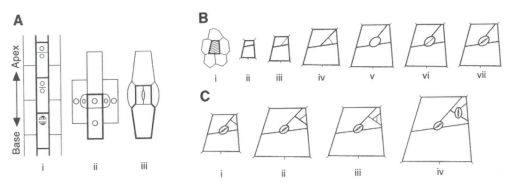

**FIGURE 2.16**   Stomatal initials and patterning in grasses and *Arabidopsis*. (A) Stomatal patterning in grasses. i, Each stomatal initial forms closer to the leaf apex. New transverse walls are offset from wall in adjacent files; ii, subsidiary cells in asymmetic divisions in adjacent cell files; iii, the guard mother cell divides symmetrically to form two guard cells that surround a pore. The lower neighbor cell originated in the asymmetric division in i; the darker line outlines the two cells produced by that division. (B) Formation of a stomatal complex from a primary meristemoid in *Arabidopsis*. The oriented production of neighbor cells can create stomatal patterning, regardless of the original placement of the primary meristemoid. i and ii, A primary meristemoid forms through the asymmetric division of a protodermal cell (shaded) that functions as a meristemoid mother cell. iii and iv, Additional asymmetric divisions are oriented so that the primary meristemoid becomes placed approximately in the center of the future stomatal complex. v, The triangular meristemoid converts to an oval guard mother cell. vi and vii, The guard mother cell divides symmetrically to form two guard cells, which then differentiate. (C) Formation of a stomatal complex from a satellite meristemoid in *Arabidopsis*. The initial placement of the satellite meristemoid establishes stomatal patterning regardless of the number of subsequent oriented divisions. i, Satellite meristemoids form in an asymmetric division of a neighbor cell. The placement of this type of meristemoid is regulated so that it forms away from an existing stomate. ii and iii, Satellite meristemoids can also divide asymmetrically to produce additional neighbor cells, iv. Reprinted with permission from Larkin *et al.* (1997), *Plant Cell* **9**, 1109–1120. Copyright © American Society of Plant Physiologists.

**Mesoperigenous stomata** have subsidiary cells of mixed origin. Because cell division planes and patterns of stomatal development are very regular, both the ontogeny and the mature appearance of stomata can be systematically useful. Different developmental pathways, however, can produce mature stomata of similar appearance.

## Fundamental Tissue System

The fundamental or ground tissue system, which includes the cortex and pith regions of the stem and root and the mesophyll of leaves, contains a collection of cell and tissue types that are most often derived from the ground meristem and that possess a general topographic and physiological similarity. The principal cell types of this system are **parenchyma, collenchyma,** and **sclerenchyma.**

## Parenchyma

The term "parenchyma" refers to a cell type, or a tissue, that exhibits relatively few distinctive structural characteristics and forms the metabolic system of the plant. Parenchyma may have different origins and is sometimes regarded as the basic ground tissue from which all other more obviously specialized cell types have developed during evolution. In some woody plants, axial wood parenchyma appears to represent transformed tracheary cells. Parenchyma cells can occur as scattered, diffuse elements, or they may constitute a rather homogeneous major tissue. In practice, any living, vacuolated cell that is not readily assignable to another cell category is considered to be parenchyma. Parenchyma cells are generally polyhedral, although extreme morphological variation does occur, such as stellate or armed forms that appear branched. In these branched cell types, adjacent cells are interconnected by means of the branches.

The shape of a cell is the direct result of the pressure exerted upon it by neighboring cells and the internal turgor pressure created during development. Within a homogeneous parenchymatous region like the pith, polyhedral

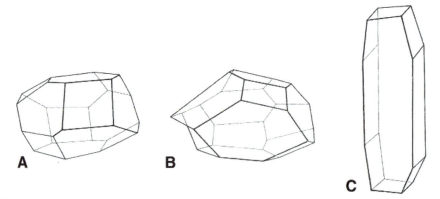

**A**                    **B**

**C**

**FIGURE 2.17**   Three-dimensional shapes of parenchyma cells. (A, B) Pith cells of *Ailanthus glandulosa* (Simarubaceae). (C) Cortical cell from stem of *Elodea* (Hydrocharitaceae). Reprinted with permission of Hulbary (1944), *Am. J. Bot.* **31,** 561–580.

des (geometrically an orthic tetrakaidecahedron)
leal shape is approached because of the uniform
Elsewhere in the plant the internal pressures are
natous cells characteristically have thin primary
y walls and pitted and considerably lignified sec-
op. Parenchyma cells are separated by abundant
ysiologically active in the synthesis, transport, or
ts. They retain the ability to divide even when
y differentiate into other cell types. Despite their
id parenchymatous tissues can provide consider-
sistance to local buckling.
any regions of the plant and forms the bulk of the
tissues as the pith and cortex, the fleshy tissue of
ynthetic cells leaves, and the rays of secondary
e parenchyma occurs in various regions of the
ays. Within the primary plant body, parenchyma
ound meristem, procambium, and protoderm. In
yma cells originate from the activity of both the

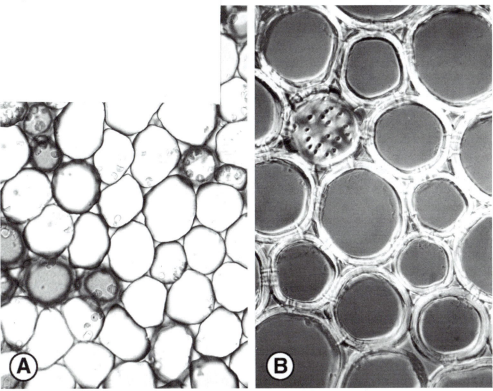

**FIGURE 2.18**   Parenchyma cells. (A) Thin-walled parenchyma from root cortex of *Ranunculus*
(Ranunculaceae). Note starch grains and numerous intercellular spaces. (B) Parenchyma cells from pith
of *Clematis vitalba* (Ranunculaceae) having thick, secondary walls. Note surface view of cell wall in upper
left showing simple pits.

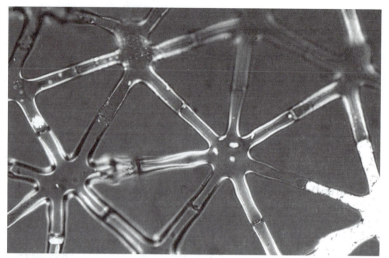

**FIGURE 2.19**  Stellate parenchyma with extended arms in the stem of *Juncus effusus* (Juncaceae).

vascular cambium and cork cambium. Because parenchyma cells retain the potential for division, they participate in wound healing, repair, and regeneration. Obviously, parenchyma is not a homogeneous cell type. It is only our current lack of understanding that causes such functionally and structurally different cells to be grouped together.

## Collenchyma

Collenchyma consists of living and usually axially elongated cells, up to 2 mm long. They have irregularly thickened primary walls of high tensile strength, that is, resistance to lengthwise stress. Collenchyma is encountered in diverse flowering plants, although it is especially well known in the Labiatae (mints) and Umbelliferae (umbells). Collenchyma typically does not occur in roots. It differentiates exclusively from the ground meristem and is found only in the primary body, where it is usually formed in elongating tissues. Some biologists do not recognize collenchyma as a distinct cell type, but view it as a form of thick-walled parenchyma. The cell walls of collenchyma are rich in pectin, with a high water content. The thickened portion of the wall consists of alternating layers of cellulosic and noncellulosic pectic materials. Because fresh walls of collenchyma contain a large amount of water, they typically have a glistening or gleaming appearance. These cells are only rarely lignified, although lignified collenchyma has been reported in members of Labiatae, Piperaceae, and Umbelliferae.

Collenchyma cells are frequently positioned immediately beneath the epidermis and are arranged as either a complete cylinder or in discrete strands (termed **strand collenchyma**); they serve a supportive function in growing herbaceous stems and expanding leaves. In plants that undergo secondary growth, collenchyma gives only temporary support to the axis. In an often cited experiment, it was shown that the mechanical shaking of young, differ-

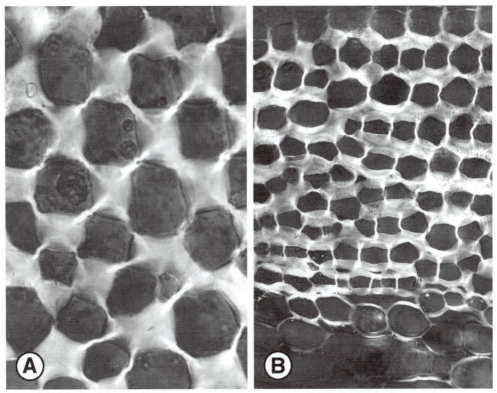

**FIGURE 2.20**   Collenchyma cells. (A) Angular collenchyma from stem of *Urtica dioica* (Urticaceae). (B) Transverse section of stem of *Sambucus nigra* (Caprifoliaceae) showing lamellar collenchyma in outer cortex.

entiating stems of *Datura stramonium* for 9 hr/day for 40 consecutive days stimulated the deposition of wall thickenings in collenchyma cells. Similar experiments with celery also increased the amount of collenchyma, as compared with control plants. Collenchyma retains its capacity for supporting young expanding tissues and the organs of the primary body because of its plastic primary wall.

Different types of collenchyma can be recognized, based on the manner of deposition of the primary cell wall thickenings. The French botanist A. Duchaigne simplified earlier classification schemes by recognizing three types of collenchyma, although in practice the separation of these cell types is not always distinct.

1. **Angular collenchyma.** This is the common classical type of collenchyma in which the cell corners are differentially thickened as viewed in transverse section. In longitudinal view, the vertical wall thickenings are conspicuous at the junction of three cells. This cell type is seen in herbaceous stems and petioles. A variation of this condition is sometimes called **lacunate collenchyma.** In this type, intercellular spaces are present and the wall thickenings are most pronounced adjacent to intercellular spaces. This condition occurs in the stems of various Asteraceae.

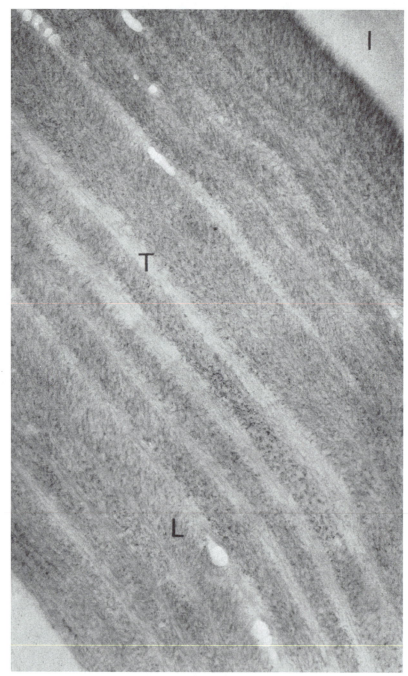

**FIGURE 2.21**    Electron micrograph of transverse section of collenchyma cell wall in *Petasites fragrans* (Asteraceae) showing a lamellation of pectic substances. Those lamellae (L) having the higher pectin content (I) appear to be those in which the orientation of the cellulose microfibrils is longitudinal. Layers low in pectin (T) have transversely oriented microfibrils. × 68,000. Reprinted with permission from Chafe (1970), *Planta* **90,** 12–21.

2. **Tangential (or plate) collenchyma.** This type also is known as **lamellate collenchyma** and is identified by thickening of the inner and outer tangential walls. Tangential collenchyma is readily observed in young stems of *Sambucus nigra*.

3. **Annular collenchyma.** Annular collenchyma is distinguished by having uniformly thickened walls. This is a relatively common type and good examples can be seen in the stems of Labiatae and Umbelliferae and in the petioles of Araliaceae and Magnoliaceae.

## Sclerenchyma

Sclerenchyma comprises a collection of cell types with thickened secondary walls that are usually lignified when fully mature and that generally do not retain a living protoplast at maturity. Sclerenchyma cells most often appear as dead elements following differentiation. Sclerenchyma usually is lignified, although lignin concentration can vary between plant parts and stages of development. In some instances sclerenchyma remains essentially unlignified.

Sclerenchyma serves a supportive or protective function throughout the plant body and is found as isolated elements or cell aggregations and tissues. Fibrous sclerenchymatous elements are flexible enough to bend when subjected to tension and are sufficiently strong to withstand the forces of compression. As a result, supporting tissues are distributed in stems and leaves to cope with the alternating stresses of compression and tension. The arrangement of thick-walled and thin-walled cells within the vertical axis confers mechanical stability to the plant. As a general rule, thin-walled cells are more centrally located, whereas thick-walled elements such as sclerenchyma are positioned near the perimeter to resist the deformation caused by bending. The walls of sclerenchymatous fibers typically possess the thickest walls in the plant body. Because sclerenchyma cells are nonliving and lignified, they are incapable of further growth and division. Sclerenchyma arises within the primary body from the differentiation of protoderm, ground meristem, or procambium, as well as from a solitary primordial initial or from the secondary transformation, or sclerosis, of a parenchyma cell. In the secondary body, sclerenchyma develops as a derivative of the vascular cambium or cork cambium.

As just noted, the distribution of fibrous supporting cells within the primary plant body reflects the heavy stresses to which plant organs are subjected. In addition to the compression forces resulting from the weight of the plant body, stems are subjected to lateral stresses from bending. These forces produce considerable tension in the outer regions of the axis. In those portions of the stem, branches and leaves that are called upon to resist heavy stress during compression, highly specialized supporting tissues are generally laid down as a continuous subdermal layer, or as isolated strands near the perimeter of the stem. In one sense, the lignified fibrous strands can be thought of as construction I-beams that meet the biomechanical needs of the plant. In monocotyledonous stems, each vascular bundle is usually surrounded by large numbers of thick-walled fibrous elements. Such bundles are sometimes compared to the rods used in making reinforced concrete that also resist the strains arising from lateral bending. In flattened leaf blades the upper surface is usually subjected to tension, whereas the lower surface feels compression strain. The concentration of thick-walled sclerenchyma or collenchyma cells is greatest where the greatest strain occurs, near lamina surfaces. Supporting tissues are typically associated with the midrib and lateral veins, where they occur just under the two outer surfaces.

## Sclereids

Two categories of sclerenchyma cells, **sclereids** and **fibers,** are arbitrarily recognized on the basis of form. Sclereids, or the "spicular cells" and "stone cells" of earlier literature, although varied in form, are typically more isodiametric than fibers. Their thick, highly lignified walls often show conspicuous concentric laminations interrupted by abundant pitting. Sclereids take many forms. Typically they are branched, star shaped, or columnar in form. These cells are widely distributed among diverse angiosperm groups, although they are much less common in monocotyledons. A systematic survey reveals that foliar sclereids are found in approximately 500 genera belonging to 121 families of flowering plants. They can occur in the stem, leaf, fruit, and seed coat as solitary elements, clusters, or layers. Rod-shaped or columnar sclereids constitute the outer layer of the seeds of many families, for example, Leguminosae.

Major classical types of foliar sclereids arbitrarily based on form include: (1) **astrosclereids,** branched sclereids; (2) **brachysclereids** (stone cells), short roughly isodiametric sclereids that resemble parenchyma cells; (3) **filiform sclereids,** elongated and slender, resembling fibers; (4) **macrosclereids,** somewhat elongated cells with uneven deposition of secondary walls; (5) **osteosclereids** (bone shaped), with columnar middles and enlargements at both ends; and (6) **trichosclereids,** branched, with thin hairlike branches extending into intercellular spaces. It also is possible to recognize four types of foliar sclereids based on their pattern of distribution. These are: (1) **diffuse sclereids,** which

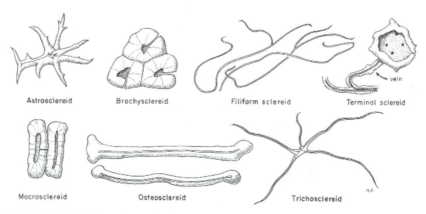

Astrosclereid          Brachysclereid          Filiform sclereid          Terminal sclereid

Macrosclereid          Osteosclereid          Trichosclereid

**FIGURE 2.22**    Sclereid types. From "Vascular Plant Systematics" by Radford *et al.,* copyright © 1974 by Albert E. Radford, William C. Dickison, James R. Massey, and C. Ritchie Bell. Reprinted by permission of Addison-Wesley Educational Publishers.

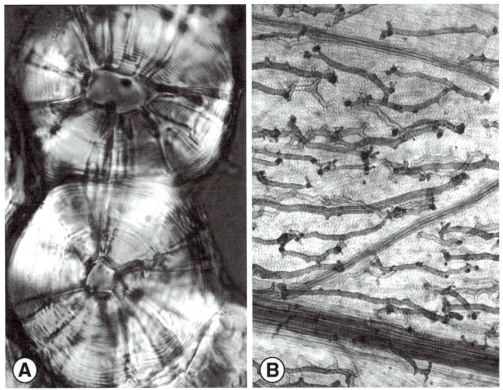

**FIGURE 2.23**   Sclereids. (A) Brachysclereids from pith of *Hoya carnosa* (Asclepiadaceae). Note extremely thick, laminated walls and elongate, branched pit canals. (B) Cleared leaf of the Australian species *Hibbertia salicifolia* (Dilleniaceae) showing abundant elongate sclereids extending throughout the mesophyll.

are dispersed in the leaf mesophyll; (2) **terminal sclereids,** which are confined to the ends of small veins; (3) **mixed patterns** containing both terminal and diffuse sclereids; and (4) **epidermal sclereids.**

Certain families are characterized by having sclereids of one or many types consistently present in the majority of genera. For example, sclereids of varied shapes and sizes are present in such families as Theaceae, Proteaceae, and Vochysiaceae. An extraordinarily diverse sclereid composition is found in the Southeast Asian genus *Cyrtandra* (Gesneriaceae), in which as many as 10 categories of types and patterns can be recognized and correlated with tentative groupings of species made from morphological considerations.

In a limited number of species of both gymnosperms and angiosperms, the walls of sclereids are characterized by the presence of crystals. These cells are called **crystalliferous sclereids.** The crystals, which are usually prismatic in form, have been reported to be calcium oxalate in nature. They are deposited at an early stage of cell development. Crystal-containing sclereids have been described in *Araucaria, Nymphaea, Nuphar, Illicium, Schisandra, Kadsura,* and *Welwitschia.*

Little is known of sclereid function, although in some plants and tissues sclereids appear to function in mechanical support and protection, such as minimizing or deterring herbivory. The terminal sclereids of some taxa, such as *Hakea suaveolens,* act as vein extensions that may conduct water to the leaf epidermal and mesophyll layers. In a few selected plants, such as *Olea europaea,* the basal region of filiform foliar sclereids appears to be a light-transferring source within the dark background of the leaf mesophyll. This observation has led to the interesting idea that certain sclereids might function as synthetic optical fibers, increasing light levels within the leaf, and thus enhancing photosynthetic rates.

We also know relatively little about the origin, induction, development, and sequence of maturation of sclereids. Often they originate late in ontogeny from the secondary sclerosis of a parenchyma cell, followed by intrusive growth that penetrates the spaces between neighboring cells. Within the leaves of some taxa, sclereids develop in a basipetal sequence, coinciding with the development of the foliar venation. In other plants, sclereids originate in the vicinity of the midrib or in both the midrib and leaf margins. Although it has been clearly shown that the formation of sclereids can be influenced by repeated wounding of tissues and that sclereids appear near such wounds, the developmental signals for sclereid differentiation are still not known.

## Fibers

A fiber is an elongated, tapering sclerenchyma cell with a more or less thick secondary wall. Like sclereids, fibers may be found in various parts of the plant. Fibers are slender, elongated cells, owing to extreme apical elonga-

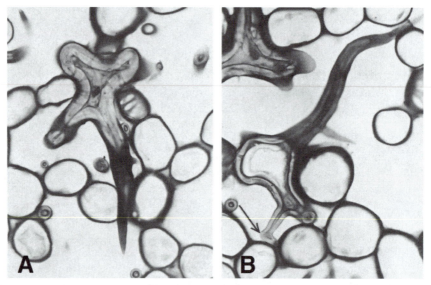

**FIGURE 2.24**  Branched sclereids from transection of stem of *Trochodendron aralioides* (Trochodendraceae). (A, B) Sclereids showing results of intrusive growth of arms and penetration between neighboring cells. Arrow indicates diversion from straight path. Reprinted with permission from Esau (1965). "Cellular Ultrastructrue of Woody Plants," W.A. Côté, Jr., Ed. Syracuse University Press, Syracuse, New York. Copyright © permission Syracuse University Press.

tion during differentiation, and are provided with a multiple-layered secondary wall. The mature fiber wall may be so thick that the cell lumen can be almost or entirely occluded. Most fibers are nonliving cells at maturity, although living fibrous elements are found in the wood of some dicotyledons. Different fiber types look somewhat similar, but are loosely classified on the basis of their position in plant. Fibers are particularly common in the phloem (**phloem fibers**) and xylem (**xylem** or **wood fibers**) tissues. In monocotyledons, fibers often enclose vascular bundles, appear as strands associated with vascular bundles, or occur as independent strands. Fibers are topographically divided into **extraxylary fibers**, those positioned in the cortex or phloem and sometimes referred to as "bast fibers," and **xylary fibers**, those occurring in the wood and originating from the vascular cambium. In extreme cases, extraxylary fibers reach exceedingly long lengths. The longest fibers ever recorded are from ramie (*Boehmeria nivea*) of the family Urticaceae, with a length of 550 mm (about one-half meter). Because of their high tensile strength, bast fibers are of great economic importance. The leaf fibers of some monocotyledons also are commercially important in the manufacture of rope and other items.

Wood fibers exhibit a series of transitional stages in evolutionary specialization that progress from tracheidlike elements with prominent bordered pits (functioning in both longitudinal conduction and support) to more slender

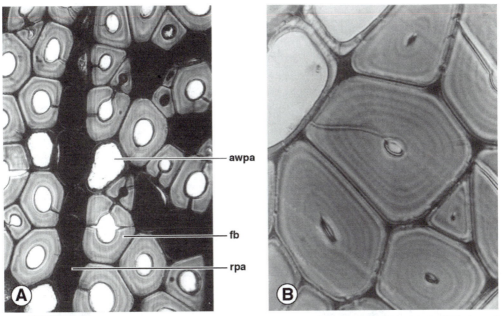

**FIGURE 2.25** Wood fibers. (A) Transverse section of wood of *Dillenia pentagyna* (Dilleniaceae) showing thick-walled xylary fibers with bordered pits. (B) Transverse section of wood of *Tetramerista* (Tetrameristaceae) showing very thick-walled xylary fibers with elongate pit canals. Note laminations within the secondary wall. Abbreviations: awpa, axial wood parenchyma; fb, wood fiber; rpa, ray parenchyma. B, courtesy of the Bailey-Wetmore Wood Collection, Harvard University.

wood fibers (termed **libriform fibers**), that have lost their pit borders and assume an intensified mechanical function. The fibers of a few specialized taxa retain their protoplasts subsequent to the formation of a thick, lignified secondary wall and hence are able to actively store starch. Some fibrous elements develop thin partitions or septa of secondary wall material across the lumen after the secondary wall is laid down and form a category known as **septate fibers.** In these elements the protoplast often divides after the formation of the original secondary wall.

It is known that fiber differentiation and lignification in vascular tissues is influenced by growth-regulating substances, most notably the hormones auxin (indole-3-acetic acid), gibberellin (produced in the leaves), and cytokinin (formed in the roots). The production of more and thicker walled fibers is stimulated by higher auxin levels. Comparatively long fibers with thin secondary walls develop in plants that have been experimentally treated with gibberellin as compared with untreated controls. Together, auxin and gibberellin regulate lignin deposition in fiber walls.

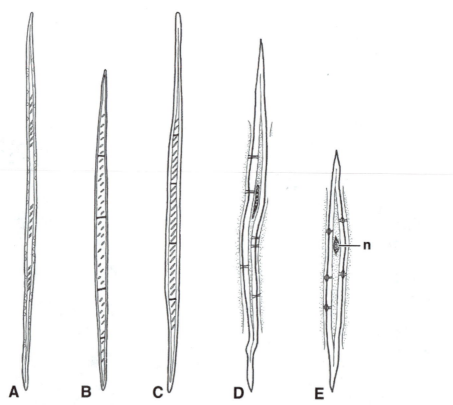

A          B          C          D          E

**FIGURE 2.26**   Xylary fibers. (A) Libriform fiber of white oak, *Quercus alba* (Fagaceae). (B, C) Septate fibers from *Tectona grandis* (Verbenaceae). (D) Living, nucleated libriform fiber of *Suaeda monoica* (Chenopodiaceae). (E) Living, nucleated fiber tracheid of *Teucrium divaricatum* (Labiatae) showing nucleus (n). A, B, reprinted with permission of the publisher from THE ANATOMY OF WOODY PLANTS by E. C. Jeffrey: The University of Chicago Press, Copyright © 1917. D, E, reprinted with permission from Fahn (1962), *IAWA Bulletin,* International Association of Wood Anatomists.

## Laticiferous Tissue System

A restricted number of widely unrelated flowering plant families are characterized by the presence of a distinctive and specialized cell type known as a **laticifer** that contains a viscous, watery fluid called **latex,** which injured cells exude. Laticifers are not readily classifiable into a particular tissue system but have a unique structure and seemingly diverse functions. The exact function of latex has not been clearly identified, but it has been suggested that the latex system is specialized for either secretory or excretory activities while providing the plant with a means of defense against predatory insects. In some species, latex is believed to protect plants by either rendering the mouthparts of chewing insects unmovable or deterring visitors with a variety of repellant compounds. Laticifers are extremely long cells that ramify at maturity throughout various tissues of the plant, such as the cortex, pith, foliar mesophyll, and especially the vascular tissues, phloem, and xylem rays. Laticifers often form a network just external to the phloem, with laticifer branches extending into the cortex. Latex is an emulsion or suspension of many solid particles and has a clear or variously colored, often milky appearance. Latex is variable in composition. It may contain rubber, starch, proteins, alkaloids, waxes, resins, or pigments.

Two phylogenetically unrelated forms of laticifers are known. The **nonarticulated laticifer** is initiated as a single cell in the plant embryo and never develops cross walls. It differs from other cells in its capacity for unlimited elongation and branching during growth, as well as in its synthesis of rubber and terpenoid compounds. The number of laticifer initials can vary between 4 and 64 or more, and these occur in a ring around the periphery of the young vascular tissue in the embryonic cotyledonary node. No additional laticifers are formed as the plant grows. Very young laticifer initials undergo rapid elongation and nuclear division. The mitotic activity of the nucleus within the initial and derived daughter nuclei results in a multinucleated, elongated, thin-walled cell that undergoes a gradual degeneration of the cytoplasm and cytoplasmic organelles. Increasing vacuolation forms a large central vacuole, with the remaining cytoplasm positioned as a thin peripheral layer. In some plants, the large central vacuole contains latex particles, although the origin of these particles is not always clear. In *Papaver sommiferum,* the particles appear to reside in the cytoplasm.

Nonarticulated laticifers are classified into **branched** or **unbranched** types depending upon whether the elongating, growing cells develop lateral extensions. As development occurs in branched laticifers, each unicellular initial begins to grow intrusively, often giving rise to branches that penetrate into surrounding tissues. This form of laticifer development characterizes such families as Apocynaceae, Asclepiadaceae, Moraceae, and Euphorbiaceae. As a result of extreme evolutionary convergence, certain African species of *Euphorbia* are almost indistinguishable vegetatively from New World Cactaceae. The presence of milky latex dripping from detached leaf bases or after injury to epidermal layers of *Euphorbia,* however, easily distinguishes the two groups.

The second form of laticifer is the **articulated type** of latex tube that is often closely associated with the phloem. This cell is found in members of such

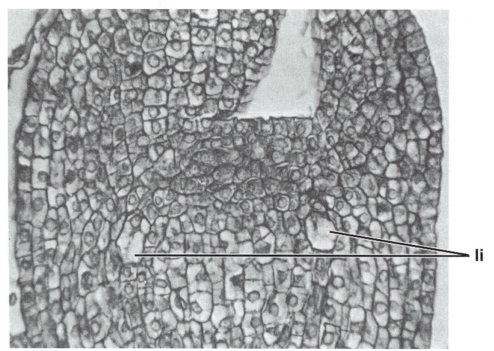

**FIGURE 2.27**    Longitudinal section of embryo of *Euphorbia marginata* (Euphorbiaceae) showing two laticifer initials (li) in the cotyledonary node. × 480. Courtesy of P. G. Mahlberg.

families as the Asteraceae, Euphorbiaceae, and Liliaceae. The laticifer is composed of a series of superimposed cells in which the dissolution of end walls has produced a continuous tube or vessel. New elements are added to the tube when the cross walls between adjacent cells break down, resulting in the union of protoplasts. As is the case with nonarticulated laticifers, the articulated laticifer can be branched or unbranched at maturity. The activity of the vascular cambium may contribute additional laticifers of this type to the plant axis.

Rubber is a commercially important product derived from laticifers. Rubber occurs as microscopic particles within the aqueous fluid of the latex producing cells. In the tropical Para rubber tree (*Hevea brasiliensis*) the latex tubes are located in the secondary phloem tissue of the bark. Cuts in the bark allow the latex to flow and be collected for a limited period of time. Slashes are made on the trunk in spirals or at an angle across the laticifers. Tapping of trees must be done carefully so the cut is deep enough to sever the laticifers yet not cause permanent damage to the inner bark and tree.

## Vascular Tissue Systems

The fundamental feature that distinguishes vascular plants (Tracheophytes) from nonvascular plants is the occurrence of specialized conductive tissues and accompanying mechanisms that make possible the long distance transport of water and inorganic and organic solutes. These tissues are the **xylem,** which conducts large quantities of both water and certain organic and inorganic solutes from the roots to the crown, and the **phloem,** which serves as a con-

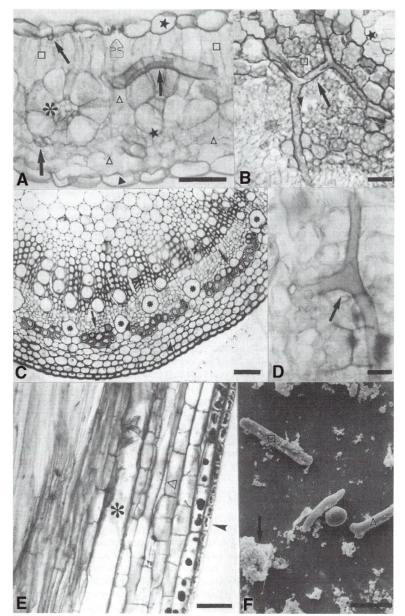

**FIGURE 2.28**   Nonarticulated laticifer system in *Chamaesyce thymifolia* (Euphorbiaceae). (A) Transverse section of leaf showing distribution of latex tubes (arrows) in the lamina. Palisade parenchyma (open square); spongy parenchyma (open triangle); vascular bundle (asterisk); adaxial epidermis (star); abaxial epidermis (black triangle); stomate (open arrow). Bar equals 50 µm. (B) Paradermal section of leaf blade showing a latex tube forming an H profile (arrow). Epidermis (star); palisade parenchyma (open square). Bar equals 10 µm. (C) Transverse section of stem showing distribution of laticifers (asterisks) close to the primary phloem (arrowheads). Gelatinous fibers are indicated by arrows. Bar equals 100 µm. (D) Transverse section of leaf showing a fork or Y-shaped branch in laticifer. Bar equals 10 µm. (E) Longitudinal section of stem. A laticifer is located near the phloem (asterisk) and another is immersed in the cortex (star). Epidermis (arrowhead), parenchyma cells of cortex (open triangle). Bar equals 100 µm. (F) SEM of starch grains from latex. Globular (open circle); rod (open square); spindle (black triangle); and osteoid shaped (open triangle). Latex particles (arrow). Bar equals 100 µm. Reprinted with permission from DaCunha *et al.* (1998), *Acta Bot. Neerl.* **47**, 209–218.

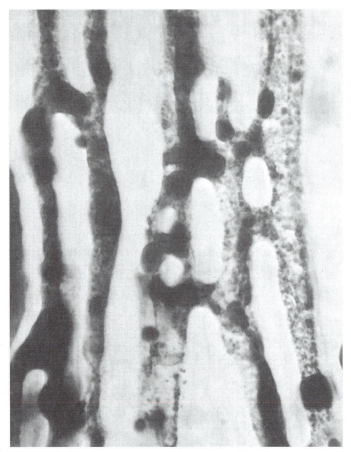

**FIGURE 2.29**   Tangential section of anastomosed laticifer from stem of *Hevea* (Euphorbiaceae). × 260. Courtesy of P. G. Mahlberg.

duit for organic food solutes from one part of the plant to another. These two tissues occur in close proximity and ramify throughout the plant body from the rootlets to foliar and floral veinlets. The development of these complex and distinctive tissues during evolution enabled land plants to assume diverse forms and successfully radiate into varied and often stressful habitats. In woody plants, secondary xylem and secondary phloem are formed later in development by the activity of the vascular cambium, whose functioning results in secondary growth and the formation of the secondary plant body. At maturity, the xylem and phloem form a structurally and functionally integrated unit that is more or less continuous throughout the plant body, providing a low-resistance pathway for conduction.

Structural botanists traditionally regarded the entire primary vascular system as the fundamental morphologic unit of internal organization in land plants. The root and stem were envisioned as basically similar in gross organization because each has a central vascular core enclosed by fundamental tissue. The column of primary vascular tissue within the plant axis, along with any associated ground tissue (such as the pith) is called the **stele.** The variability of stelar structure among vascular plants is now well documented and numerous terms have been introduced to describe stelar types. The merits of the stelar idea as a unifying concept, as well as to theories of stelar evolution, continue to be intensively debated.

## Xylem

Both xylem and phloem tissues are composed of a diverse collection of cell types that have become specialized in structure and function to perform various transport activities. Four cell types are most frequently present in the xylem. In addition to parenchyma and fibers, the two principal conductive cells types of the xylem are **tracheids** and **vessel elements,** collectively termed **tracheary elements.** Tracheary elements are generally elongated cells and are characterized by having rigid, lignified, often extensively pitted secondary walls, and by their elimination of protoplasmic contents at maturity. A most distinctive and constant feature of tracheary elements is the presence of bordered pits in their lateral walls to facilitate lateral conduction and support. As described earlier, in a circular bordered pit a portion of the secondary wall surrounding the pit opening projects away from the primary wall and forms a saucer-shaped region or border that projects into the cavity of the cell. The bordered pit provides a large surface area of pit membrane for water flow, without seriously compromising strength. Immature tracheary elements have symplastic continuity with adjacent xylem parenchyma cells through numerous plasmodesmatal connections in the lateral wall pits. Concomitant with tracheary cell death, plasmodesmata within the pits are severed by the deposition of additional wall material across the channel openings.

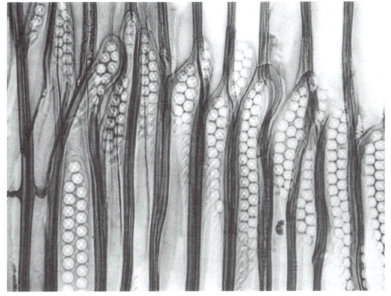

**FIGURE 2.30**  Details of the ends of tracheids of *Agathis australis* (Araucariaceae) showing bordered pits in region of tracheid overlap. × 400. Courtesy of E. A. Wheeler.

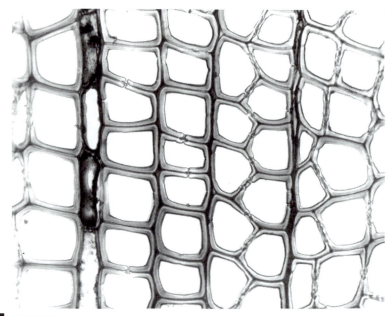

**FIGURE 2.31**   Transverse section of wood of the vesselless dicotyledon *Drimys winteri* (Winteraceae) showing secondary xylem tracheids with angular outlines and bordered pit pairs. Courtesy of the Bailey-Wetmore Wood Collection, Harvard University.

## Tracheids

Tracheids are phylogenetically more primitive than vessel elements and are the principal water-conducting cell in gymnosperms and primitive vascular plants. They also occur to a lesser extent in angiosperms. Vessel elements are characteristic of flowering plants and are now known to be more common in ferns than previously believed. Tracheids are distinguished from vessel elements in that they are very elongated in form and **imperforate;** that is, water must pass from cell to cell through a series of pit membranes of bordered pit pairs present along the extensively overlapping and tapered end walls of two adjacent cells. The xylem of gymnosperms is composed of tracheids that can be as much as 100 times longer than they are wide. Vessel elements, on the other hand, are **perforate** cells. The entire end wall or portions of the end wall are reduced to remnants or, more commonly, lost entirely during the later stages of ontogenetic maturation.

## Vessel Elements

The entire sequence of vessel element differentiation includes the following sequence of distinct events: cell origination, cell enlargement, secondary wall deposition and lignification, and end wall and protoplast lysis by programed cell death. Loss of the vessel element end wall by hydrolysis of noncellulosic matrix components results in the formation of pores, or of an opening, the **perforation plate,** through which xylem sap moves freely. The resulting vessel elements are continuously joined end to end at the perforation plates, generally with less overlapping ends than in tracheids, to effectively form an extended hollow tube, or **vessel,** that enhances the rapid and efficient movement of water. The ascent of sap in the xylem is a movement through dead cells, without the involvement of living, conducting cell contents. An important implication of the distinction between tracheids and vessel elements is that it is often impossible to distinguish these two cell types in transectional view where the ends of the cells cannot be viewed. The classical methods of detecting the presence or absence of pit membranes on tracheary element end walls have been (1) light microscopic examination and (2) the introduction of India ink particles into the conducting stream. We then note whether these pass freely from conducting cell to cell. The intact pit membranes of tracheids prohibits the movement of ink particles. The SEM has proven useful for making this decision with confidence. The SEM studies on tracheary cells in some ferns and flowering plants have shown the presence of small pit membrane pores of various sizes in the end walls. Such membrane remnants and pores were not visible previously with only the light microscope. The presence of porosities in the end walls of tracheary elements has led investigators to designate such cells as vessel elements.

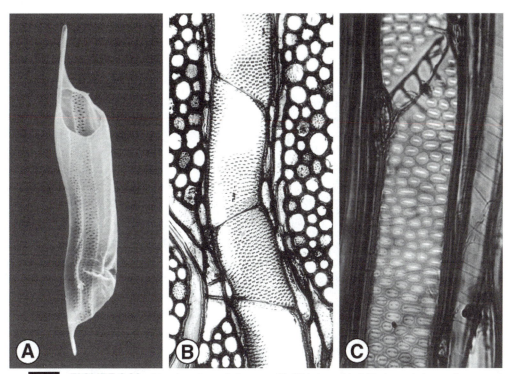

**FIGURE 2.32**   Vessels and vessel elements. (A) SEM view of individual vessel element from wood maceration of *Acer rubrum* (Aceraceae). (B) Axially extending vessel from wood of *Cordia nitida* (Boraginaceae). Individual vessel elements have simple perforation plates and alternate intervessel pitting. × 400. (C) Lateral view of vessel element wall from *Corylus californica* (Betulaceae) with scalariform perforate plate and alternate intervascular pitting. × 400. Courtesy of E. A. Wheeler.

A **simple perforation plate** is an area of end wall that typically exhibits large, single, circular, or elliptical openings that often leave only a narrow rim of primary wall remaining. A **scalariform perforation plate,** in contrast, is a compound plate consisting of elongated, parallel openings separated by one or many (over 100) branched or unbranched bars. Among vessel elements with scalariform perforations, the individual perforations are delimited by the outline of the original wall pit. Examples of woody families characterized by scalariform perforation plates are Betulaceae, Hamamelidaceae, Hydrangeaceae, Magnoliaceae, and Theaceae. Among living angiosperms, the simple plate is the more common type. Most individual vessel elements possess either exclusively simple or exclusively scalariform perforation plates, although some elements can develop a simple plate at one end wall and a scalariform plate at the other. Only rarely do simple to scalariform combination perforation plates occur between the end walls of two adjacent vessel members, a feature that has only been described in a few members of the Verbenaceae and Asteraceae.

That vessel elements evolved from tracheids is certain. Advantageously, the sequence involves characters that are measurable, and therefore can be dealt with statistically. Therefore, it is clear that the most primitive vessel elements are long cells that most resemble tracheids. Vessel elements under 350 µm in length are regarded as short, whereas those over 800 µm are long. Because in dicotyledons tracheids also evolved into thick-walled wood fibers of reduced cross-sectional area, the major evolutionary trends derived from the basic cell type, the tracheid, culminated in a functional dichotomy that resulted in an enhanced rate of longitudinal conduction on the one hand and on the other in intensified mechanical support.

## Primary Xylem Tracheary Elements

Tracheary elements of the primary xylem are characterized by wall deposition in the form of widely spaced, stretchable, or extensible annular rings or loosely coiled helical thickenings of varying pitch. These wall thickenings have gener-

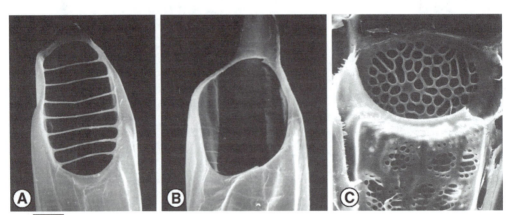

**FIGURE 2.33**  SEMs of vessel element perforation plate types. (A) Scalariform perforation of *Liriodendron tulipifera* (Magnoliaceae). (B) Oblique, simple perforation of *Acer rubrum* (Aceraceae). (C) Reticulate perforation of *Vitex chryscocarpa* (Lecythidaceae). All ca. × 500. A, B, courtesy of E. A. Wheeler.

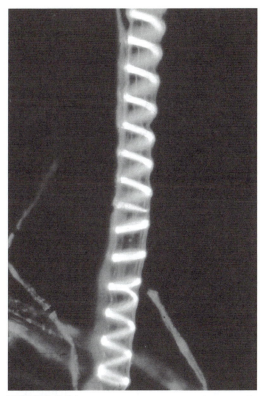

**FIGURE 2.34**   Primary xylem tracheary element with annular and helical secondary wall thickenings.

ally been termed secondary walls. This pattern of wall deposition permits the differential expansion or elongation essential to cell morphogenesis. In later stages of development, the rings are completely separated from each other. These elements are succeeded by ones having more compactly coiled thickenings and eventually by others with more extensive secondary wall deposits that have a scalariform, scalariform and reticulate, or pitted appearance. The more extensively pitted tracheary cells are largely formed after elongation is completed and give enhanced rigidity to the fully mature organ structure.

With increasing specialization, helically thickened primary xylem tracheary elements can become perforate vessel elements. In elements having compactly coiled helical thickenings, scalariform perforation plates can arise through modifications in the form and orientation of the thickenings in the overlapping end walls. In such vessel elements, perforations are formed by the dissolution of the primary wall between the more or less transversely oriented bars. In some cases, the perforation plates of scalariformly perforated elements are further reduced and modified and form a single oval perforation.

## Phloem

Phloem, like the xylem, is a conducting tissue that consists of different cell types. The basic conducting elements are termed **sieve elements,** so named

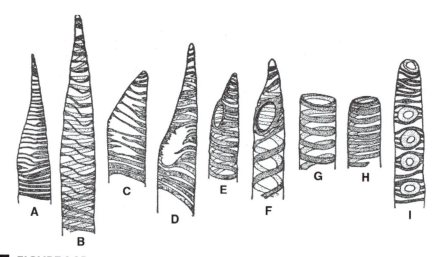

**FIGURE 2.35**  Terminal segments of primary xylem elements. (A, B) Helically thickened vessel elements with scalariform perforation plates. (C, D) Helically thickened vessel elements illustrating transitions between scalariform and porous perforation plates. (E, F) Helically thickened vessel elements with simple perforation plates. (G, H) Helically thickened vessel elements with truncated ends. (I) Helically thickened tracheary element of *Gnetum*, showing circular bordered pits. Reprinted with permission from Bailey (1944), *Am. J. Bot.* **31**, 421–428.

because of the presence of sievelike wall perforations. Wall characteristics and cytological peculiarities are important in characterizing these elements. **Phloem parenchyma cells** that function primarily as long-lived storage cells are universally associated with the sieve elements and together they constitute a physiological unit. Fibers and sclereids also are sometimes located within the phloem tissue. Two major categories of sieve elements are recognized. **Sieve tube elements,** also called sieve tube members, represent a highly specialized cell type that evolved in flowering plants as an adaptation for the more efficient movement of organic solutes at higher rates. **Sieve cells** are a more primitive cell type, confined to gymnosperms and lower vascular plants. Unlike tracheary elements, sieve elements typically possess thin, primary walls and are metabolically active when functional. The lateral walls of sieve elements sometimes become thick and glisten in freshly sectioned tissue. These walls are termed **nacreous walls,** and their role has not yet been determined. Sieve elements function for a limited period in most plants, usually 1 year or less. Exceptions to this rule are found in some monocotyledons such as palms and tree lilies that must have long-lived, functioning sieve elements. The primary phloem in many dicotyledons and monocotyledons is frequently interconnected by anastomosing phloic strands. **Phloem anastomoses** can be simple, branched, or complex connections and have been reported in leaves, stems, and fruits.

### Sieve Tube Elements

Sieve tube elements are slightly elongate cells whose walls bear differentiated regions occupied by numerous narrow **sieve pores** filled with protoplasmic connecting strands that link them to adjacent sieve elements. Each of the numerous recessed areas on the wall containing grouped sieve pores is called a **sieve area.** In sieve cells the wall pores are the same diameter (usually less than 1 μm) on both end walls and lateral walls. Among sieve tube elements, the sieve areas are vestigial on lateral walls but become well defined with wide pores on the end walls. The specialized end wall of a sieve tube element is termed a **sieve plate** and is composed of one or more sieve areas with large pores. These are often wider than 5 μm, and they extend through the entire wall between joining sieve elements. Evolutionary specialization seems to have led to an increase in the size of sieve plate pores.

Through enlarged pores in the opposing walls of sieve plates, the protoplasts of two contiguous and functional sieve elements are interconnected by means of continuous cytoplasmic bridges that resemble enlarged plasmodesmata. Each plasmodesmatally derived interconnecting strand is usually seen to be encased in **callose,** a carbohydrate (β-1,3-glucan) that is chemically distinct from the cellulose and pectin forming the basic primary wall framework of the sieve plate. The plasmodesmatal strands that extend through the sieve plate pores are connected to endoplasmic reticulum cisternae. The question of whether callose is present in the sieve plate of sieve tube elements that have not been killed and prepared for microscopy remains unsettled. Callose clearly accumulates in response to injury of the conducting cell, and some callose also may be present in the normally functioning element.

A sieve plate composed of a single sieve area is a **simple sieve plate,** whereas an end wall containing two or more sieve areas is a **compound sieve plate.** Simple sieve plates are correlated with transversely oriented end walls, and compound sieve plates are usually located in variously inclined end walls. The lateral walls of sieve tube elements possess relatively undifferentiated **lateral sieve areas,** also with connecting strands. Numerous sieve tube elements are joined vertically to form a long end to end series, collectively called a **sieve tube.**

Sieve elements follow an unusual developmental pathway. An especially unique aspect of sieve tube element maturation is the selective loss or disorganization of many of its cellular components. As a young undifferentiated cell, the sieve tube element contains a full complement of the cellular organelles typical of plant cells. However, selective autophagy occurs during the process of maturation, such that the dictyosomes, microtubules,

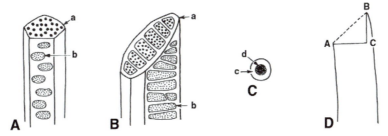

**FIGURE 2.36**   Parts of sieve elements. (A) Sieve tube element with simple transverse sieve plate (a) and lateral sieve areas (b). (B) Sieve tube element with compound inclined sieve plate (a) and lateral sieve areas (b). (C) Transection of a sieve area pore with callose (c) enclosing the connecting strand (d). (D) Diagram of part of a sieve element indicating length of sieve plate at AB and diameter of cell at sieve plate at AC. Reprinted from Esau and Cheadle (1959), *Proc. Natl. Acad. Sci.* (USA) **45**, 156–162.

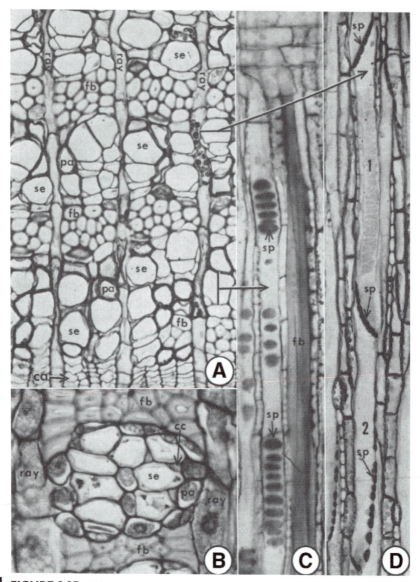

**FIGURE 2.37**    Phloem structure. Light microscope views. (A) Transverse section of stem of *Salix pentandra* (Salicaceae). Sieve elements are wide cells with unstained contents. × 270. (B) Transverse section of *Magnolia tripetala* (Magnoliaceae). × 375. (C) Radial longitudinal section of *Salix pentandra*. Sectioning plane shown by vertical bar in A. One complete sieve element exposes two compound sieve plates in face view. × 270. (D) Tangential longitudinal section of *Salix pentandra*. Sectioning plane shown by horizontal bar in A. Two superimposed sieve elements are number 1 and 2. × 225. Dark thickenings on sieve plate (sp) are masses of callose. Abbreviations: ca, cambium; cc, companion cell; fb, fiber; pa, parenchyma cell; se, sieve element; sp, sieve plate. Reprinted with permission from Esau (1966), *Am. Sci.* **54**, 141–157. Copyright © Sigma Xi.

ribosomes, vacuolar membrane (tonoplast), and most notably the nucleus, undergo disintegration. As a result, at maturity the enucleate sieve tube element has an outer limiting plasma membrane surrounding a thin layer of cyto-

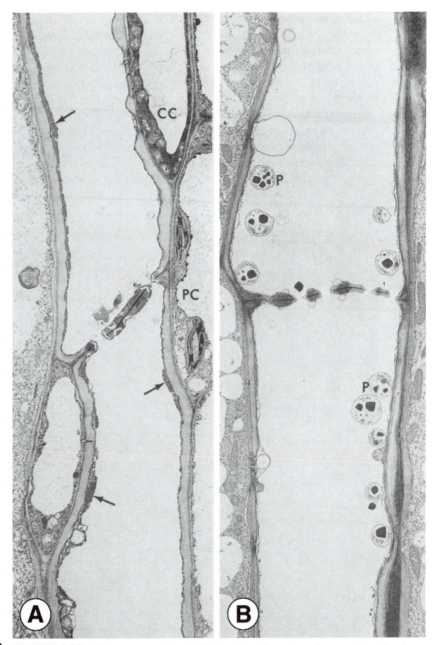

**FIGURE 2.38**   Phloem anatomy. Longitudinal sections of portions of mature sieve tube elements, showing parietal distribution of cytoplasmic components and sieve plates with unoccluded pores. (A) *Cucurbita maxima.* Unlabeled arrows point to P-protein. CC, companion cell; PC, parenchyma cell. × 4180. (B) *Zea mays.* Typical of monocotyledonous sieve tube elements, those of maize contain P-type plastids (P). × 8500. Reproduced with permission of Evert (1984). Comparative structure of phloem. In "Contemporary Problems in Plant Anatomy," R. A. White and W. C. Dickison, Eds., Academic Press.

plasm distributed in a parietal manner. It contains a variable number of such prominent surviving organelles as mitochondria, plastids, and a modified endoplasmic reticulum. It also is of functional significance that the plasma membrane maintains its integrity and differentially permeable properties throughout the life of the sieve tube element. During maturation, plasmodesmata penetrate the enlarged and open sieve plate pores to form the continuous system of protoplasm between contiguous sieve tube elements.

Another unique feature of the sieve tube elements of dicotyledons and some monocotyledons is the presence of proteinaceous substances of as yet unknown function called **phloem protein bodies** (P-protein). In the older literature this substance was referred to as "slime." Different morphological forms of phloem protein were originally identified by numerals ($P_1$, $P_2$) or by descriptive terms like "tubular" or "fibrillar." The exact nature of P-protein and its distribution within the cell have been the subject of lengthy debates.

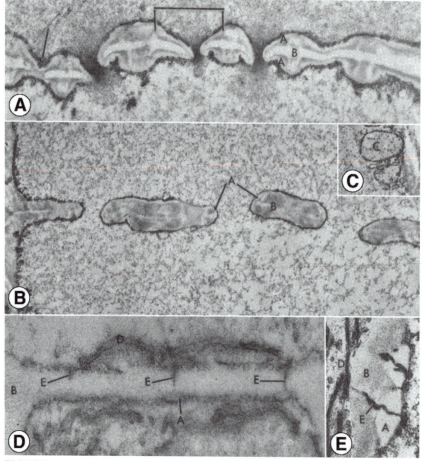

**FIGURE 2.39** Electron micrographs of longitudinal sections through sieve elements of *Cucurbita pepo* (Cucurbitaceae) showing wall structures. (A) Sieve plate with recently opened pores. Arrow indicates a still closed pore site with fused callose and the bracket delimits two halves of one callose platelet. (B) Mature sieve plate. (C) Mitochondria from one of the cells in B. (D) Lateral sieve area between immature sieve elements. (E) Part of wall between mature sieve element (right) and parenchyma cell (left). Abbreviations: A, callose; B, cellulosic part of sieve plate; C, mitochondrion; D, endoplasmic reticulum; E, plasmodesma. A, × 14,000; B, × 10,000; D, × 41,000; E, × 28,500. Reprinted with permission from Esau (1964), "Formation of Wood in Forest Trees," Academic Press.

In microscopic preparations, it sometimes coalesces to form a plug that is near, over, or within the sieve plate pores. As a general rule, P-protein inclusions increase in size during the course of sieve element differentiation and eventually become dispersed throughout the cytoplasm, although this process is not typical of all species. Sieve tube elements are extremely sensitive to injury and manipulation, and information about the unaltered appearance of phloem inclusions at different stages of cell differentiation has been difficult to obtain. Well-preserved sieve tube elements show little apparent effects of injury-induced solute surging. They show P-protein distributed in what has been described as either a loose network of filaments occupying the entire area of the cell, or in a parietal position in some species. In such specimens there is no evidence to suggest that P-protein normally exists in the form of transcellular strands or as a component of such strands, although this has been a matter of dispute.

As noted earlier, another deposit, callose, is associated with the sieve tube element where it lines the pores of the sieve plate and the lateral sieve areas. Callose is a carbohydrate that yields glucose when hydrolyzed. The callose that lines the sieve plate pores is widely regarded to be deposited in response to mechanical injury of the cell or to be an artifact of tissue preparation. As the cell matures, however, massive accumulations of callose appear to form normally over the entire surface of the sieve plate, rendering the element nonfunctional. This callose may be redissolved if the sieve tubes are reactivated.

## Companion Cells

Another distinctive feature of the sieve tube element of flowering plants is its lateral connection to a small, densely cytoplasmic cell or cells, called a **companion cell**. The companion cell is a small parenchymatous cell that is ontogenetically derived by unequal division from the same initial cell as the sieve tube element. Following differentiation it retains a nucleus. The phloem, therefore, has developed a condition in which the sieve tube element is apparently functionally dependent upon associated nucleate cells, especially the companion cells. Undifferentiated companion cells may subdivide transversely so that a single sieve tube element can be associated with more than one companion cell. Not only does the phloem of more primitive vascular plants lack sieve tube elements, but there is typically no difference between companion cells and other phloem parenchyma cells. Typical companion cells also can be absent from the region of the ultimate veinlets of angiosperm leaves.

Companion cell precursors are ontogenetically derived from the same mother cell as the sieve tube element it accompanies and they communicate with the sieve tube element by means of numerous plasmodesmata pore connections across their common walls. Unlike sieve tube elements, companion cells possess a large nucleus at maturity, as well as a rich assemblage of ribosomes and other organelles, such as mitochondria, endoplasmic reticulum, and plastids. The vacuole is typically small or fragmented. It is likely that the companion cell supplies the sieve tube element with energy to drive the translocation process. The fact that the sieve tube element and companion cell are intimately associated and functionally related is supported by the observation that both cease to function and die simultaneously.

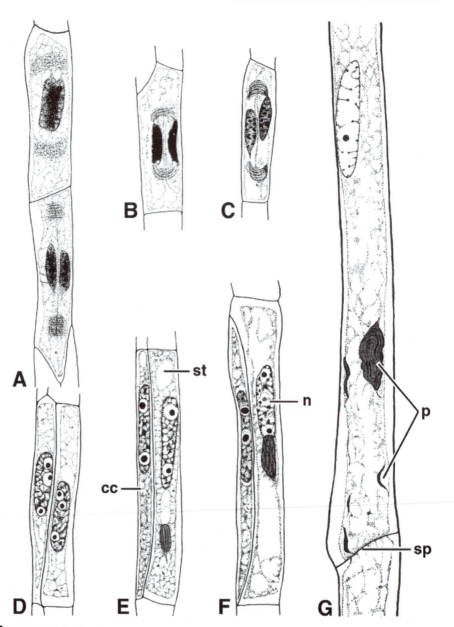

**FIGURE 2.40**   Ontogeny of primary sieve tubes of *Nicotiana tabacum*. (A–C) Dividing phloem mother cells. (D) Young sieve tube, to the right, and companion cell, to the left. (E, F) Two further stages in the development of sieve tubes and companion cells; each sieve tube contains one protein body. (G) Portion of a sieve tube with one large and several small protein bodies and a degenerating nucleus. × 992. Abbreviations: cc, companion cell; n, nucleus; p, protein body; sp, sieve plate; st, sieve tube element. Reprinted with permission from Esau (1938), *Hilgardia* 11, 343–424. Copyright © Regents, University of California.

## Sieve Cells

Gymnospermous plants possess the more primitive form of sieve element called a sieve cell. Typical sieve cells are slender, elongated cells with tapering ends. All have numerous small sieve pores on their lateral walls and none possess distinct sieve plates on their end walls. Sieve cells also lack the P-protein bodies that are characteristic of angiosperm sieve tubes. Throughout the

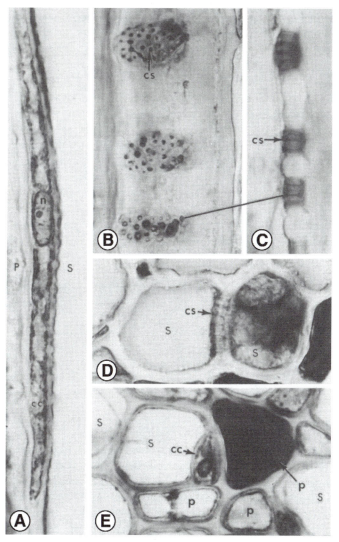

**FIGURE 2.41**  Details of *Vitis* (Vitaceae) phloem in longitudinal (A–C) and transverse (D, E) sections. (A) Companion cell (cc) with pitted wall toward sieve element(s). (B, C) Sieve areas of a sieve plate in surface view (B) and in section (C). The light rings around the connecting strands (cs) in B are callose. In C the dark stain indicates the connecting strands (cs) and the associated callose. Abbreviations: cc, companion cell; cs, connecting strand; n, nucleus; p, parenchyma cell (with tannin shown in black in E); s, sieve element. (A, × 980; B, × 1070; C, × 1340, D, × 800; E, × 890). Reprinted by permission of the publisher from PLANTS, VIRUSES, AND INSECTS by K. Esau, Cambridge, Mass.: Harvard University Press, Copyright (c) 1961 by the President and Fellows of Harvard College.

gymnosperms, sieve cells possess a complex network of tubular endoplasmic reticulum that is continuous from cell to cell through the sieve area pores. Each sieve pore extends only halfway through the common wall of adjacent sieve cells, with symplastic continuity established within the sieve pore of the neighboring cell. Individual sieve cells lack ontogenetically related companion cells, although they are typically associated with cells called **albuminous cells** (also termed **Strasburger cells**). Albuminous cells are not derived from the same mother cell as the associated sieve cell, but the two cells share numerous cytoplasmic connections. Sieve cells are not joined vertically into a longitudinally extended sieve tube.

Comparative study has shown that an evolutionary shortening of conducting cells occurred in both xylem and phloem, accompanied by an increase in diameter and decrease in length and angle of end walls. In the cells of both tissues, the reduction in length of the end wall is correlated statistically with a decrease in the number of openings or perforations in the vessel element, and with a reduction in the number of sieve areas in the sieve plate of the phloem sieve tube member. A long sieve element with highly inclined end walls bearing numerous, relatively undifferentiated sieve areas is considered to be primitive. This type of sieve element predominates in gymnosperms and lower vascular plants and is regarded as the conducting element from which the sieve tube element arose. Conversely, a short element with a transverse end wall orientation, each one composed of a single sieve area (simple sieve plate), is regarded as phylogenetically advanced and is restricted to flowering plants. Less advanced types of sieve elements show little difference in the differentiation of sieve areas between the end walls and the side walls. In the presumably more advanced cell type, the sieve areas on the end walls are composed of conspicuously more highly differentiated and larger pores than those of the side walls. Because there is a range in specialization of sieve areas among living plants, in some taxa the sieve elements are transitional between a less advanced sieve cell and a more advanced sieve tube member.

## DIFFERENTIATION OF PRIMARY VASCULAR TISSUES

Understanding the differentiation of primary vascular tissue is of utmost importance in studies of plant anatomy and is of special interest in investigations of morphogenesis and elsewhere. Procambium arises in the embryo; as the plant grows it continues to extend toward the apices of the developing shoot and root, as well as into the developing leaves. Within the apical regions, procambium development occurs more or less acropetally and continuously from the preexisting procambial strands. In some monocots, procambium formation has been described as occurring in a basipetal manner, that is, from the base of the leaf primordium downward to connect with the procambial cells of the stem. The axial procambial strands in monocotyledons also have been shown to differentiate independently of the appendages they will finally vascularize.

Procambial cells are distinguished at the beginning of tissue differentiation by their retention of densely staining cytoplasm, as contrasted to the vacuolation of cells in the developing cortex and pith. In addition, cell divisions at right angles to the long axis of the stem are infrequent in the procambium. In sectional view, therefore, the procambium appears as a zone of narrow, elongated elements when compared to the surrounding parenchymatous ground meristem. The arrangement of the procambium varies from an almost complete cylinder to discrete strands.

As procambial cells become fully established in the shoot apex, some cells begin to become transformed into sieve elements and tracheary cells. It is now well documented that the vascular tissues and fibrous cells are induced and controlled by the interaction of continuous longitudinal streams of inductive signals, primarily the hormones auxin, gibberellin, and cytokinin, but other endogenous compounds as well. Vascular bundles appear to develop along specific pathways of polar auxin transport. There is considerable evidence that the lateral flow of auxin controls the development of secondary wall thickening among neighboring tracheary elements. The differentiation of primary xylem and phloem also is influenced by different levels of auxin. Phloem differentiation is stimulated by low auxin levels, whereas xylem differentiation, which does not normally occur in the absence of phloem, is promoted at higher auxin concentrations. These observations explain why primary xylem characteristically differentiates in association with phloem, because the hormonal levels required for xylem development also stimulates phloem development.

Differentiation of vascular tissues within the procambial zone occurs along both longitudinal and radial planes, that is, up and down as well as across the procambial strand. To fully comprehend the pattern of differentiation of primary xylem and phloem, it must be kept in mind that the maturation process occurs simultaneously along both planes. In general, phloic elements are the first vascular cells to differentiate and form along the outer edge of the procambial strand, a short distance behind the apical meristem. Primary phloem differentiation then proceeds acropetally along a longitudinal course with newly formed elements in continuous contact with older phloem cells, thereby establishing the vascular bundle of the stem. In seed plants, the first formed xylem elements begin to differentiate later than the phloem and are positioned on the inner margin of the procambial strand.

## PRIMARY XYLEM DIFFERENTIATION

Primary xylem can be temporally and positionally divided into two categories of elements. **Protoxylem** is the initial primary xylem to differentiate and is identified by its position within the organ. **Metaxylem** is that primary xylem component that differentiates later. It is not uncommon for the protoxylem to consist of only one or two elements. In stems, the first recognizable tracheary cells are usually located in the vicinity of the young leaf base without any connection with the mature xylem below, with differentiation proceeding both

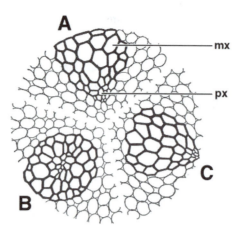

**FIGURE 2.42**   Relationship of protoxylem and metaxylem and types of primary xylem differentiation. (A) Endarch primary xylem. (B) Mesarch primary xylem. (C) Exarch primary xylem. Reprinted with permission of the publisher from THE ANATOMY OF WOODY PLANTS by E.C. Jeffrey: The University of Chicago Press, Copyright © 1917.

acropetally into the developing leaf and basipetally to the level at which contact is made with already mature xylem. The pattern of longitudinal primary xylem differentiation in stems, accordingly, is discontinuous.

The terms **exarch, endarch,** and **mesarch** are used to indicate the direction of radial differentiation of the primary xylem. Exarch primary xylem exhibits a pattern of maturation that progresses centripetally; that is, the earlier formed elements, or protoxylem, are farthest from the center of the axis. This pattern is found in the roots of seed plants, as well as in the structurally primitive stems of some lower vascular plants. In the ontogenetic sequence of endarch primary xylem, in contrast, maturation proceeds in a centrifugal direction with the older protoxylem elements positioned closest to the center of the axis. This is the pattern that is found in the more specialized stems and leaves of seed plants. In mesarch primary xylem, a condition prevalent in some fern rhizomes and flower parts, the direction of differentiation is both centripetal and centrifugal.

Protoxylem generally matures before or during organ elongation and is frequently distinguished by possessing elements having only annular or helical wall thickenings, at least in stems. Metaxylem completes differentiation following the major increase in organ length and tends to be distinguished by a mixture of scalariform, scalariform and reticulate, or pitted secondary wall thickenings. Clearly, however, there is no sharp boundary between protoxylem and metaxylem and intergradations between the two are evident. The reinforcing and patterned wall thickenings prevent the cell from collapsing inward under conditions of negative pressure related to the demands of excessive water loss. The physical pulling and tearing of protoxylem and early metaxylem elements in some plants results in the formation of a space (or lacuna) within the primary xylem that, once formed, also transports water. Tracheary elements of the first formed part of the primary xylem eventually lose their end walls. The occurrence of

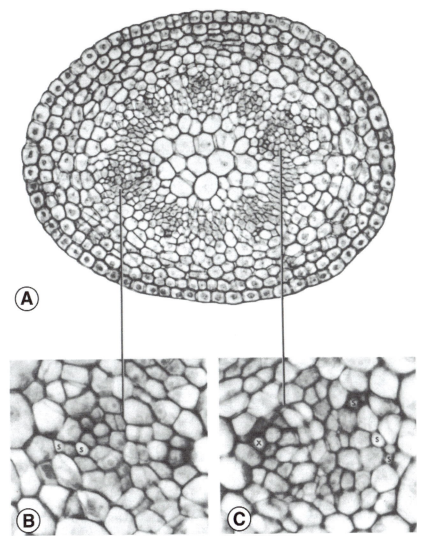

**FIGURE 2.43** Vascularization in flax (*Linum*) stem. (A) Transection of stem 260 μm below the apex. Leaf traces are arranged in a circle. Some consist of procambium, some have the first mature sieve elements (B), and one has the first mature xylem elements (C). Details: s, sieve element; x, xylem element. (A, × 256; B, C, × 740). Reprinted by permission of the publisher from PLANTS, VIRUSES, AND INSECTS by K. Esau, Cambridge, Mass.: Harvard University Press, Copyright © 1961 by the President and Fellows of Harvard College.

true vessel elements in the protoxylem and early metaxylem is an indication of extreme evolutionary structural specialization. Unlike secondary xylem elements, pit outlines do not restrict the size of the primary xylem element perforations. In helically thickened primary xylem elements, a larger area is available for perforation development between gyres of the helical bands.

The criteria used to define protoxylem and metaxylem have varied from author to author. At one time protoxylem and metaxylem were distinguished

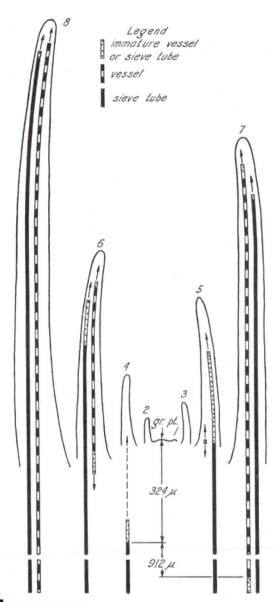

**FIGURE 2.44**   Diagram of the apical region of *Nicotiana glauca,* showing the relative height of the leaves and the position of the first sieve tube and first vessel element in each leaf. The phyllotaxy has been disregarded, but the leaves are placed at their proper levels below the growing point (gr. pt.). The leaves, beginning with the youngest, measured the following number of microns in length: 5, 82, 132, 220, 476, 660, 1032, 1424. ×103. From Esau (1938), *Hilgardia* **11**, 343–424. Copyright © Regents, University of California.

primarily on the basis of the position of elements, the size (i.e., diameter) of cells, the degree of lignification, secondary wall thickening features, and in relation to the degree of elongation that occurs in the organ where they are found. Most contemporary plant anatomists use a positional and temporal characterization. That is to say, protoxylem cells are the first recognizable xylem cells of a system. Metaxylem cells occur later.

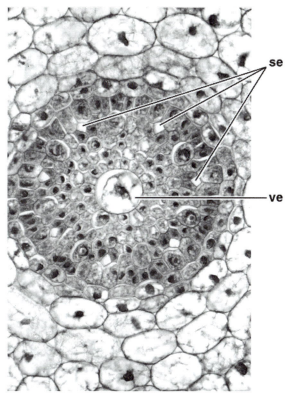

**FIGURE 2.45** Transverse section of central cylinder of young root of barley (*Hordeum vulgare*) showing mature sieve elements (se) and immature centrally positioned vessel element with nucleus (ve).


## CYTODIFFERENTIATION OF TRACHEARY CELLS

The cytodifferentiation of tracheary cells involves a dramatic sequence of cellular changes. Following the enlargement of individual procambial cells, the differentiating primary xylem tracheary element deposits patterned, often discontinuous secondary cell wall thickenings in what have been described as annular, **helical** (spiral), or **reticular** arrangements. Dramatic changes in gene expression and cytoskeleton drive the modeling of the wall materials. Just prior to and during this process, the microtubules that form the internal cell cytoskeleton change from a random or longitudinal orientation to a more ordered transverse arrangement and, along with a proliferated Golgi, presumably establish and guide the distinctly patterned nature of secondary wall deposition. In the differentiated cell, the microtubules follow a more or less parallel arrangement within the ring thickenings, perpendicular to the longitudinal axis of the cell and reflecting the location of secondary thickenings. The wall thickenings are thus considered the local sites of accumulating secondary wall material. With further differentiation, the unlignified intercellular layer (including the primary wall) undergoes hydrolysis. Some workers have suggested that the removal of noncellulosic matrix substances from the wall allows the cellulosic fibrils to undergo extensive stretch-

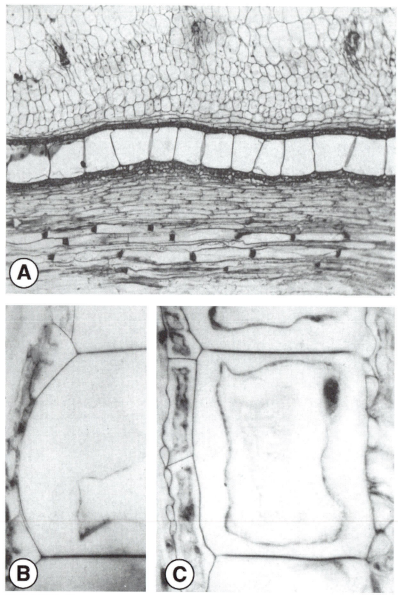

**FIGURE 2.46**   Differentiating vessel elements of *Cucurbita*. (A) Longitudinal section of part of a vascular bundle showing phloem below and above the xylem, a row of wide vessel elements with intact end walls. × 90. (B) Portion of a vessel element with thickened end walls. × 540. (C) Entire vessel element with protoplast and thickened transverse walls. × 540. From Esau (1940), *Hilgardia* **13**, 175–226. Copyright © Regents, University of California.

ing within the rapidly elongating cell. Accompanying these structural changes are biochemical alterations, such as increases in the activity of enzymes controlling reactions associated with lignin formation and the process of wall lignification. Although the thickened regions of the wall become lignified, the compound middle lamella that underlies the bands reportedly remains unlignified.

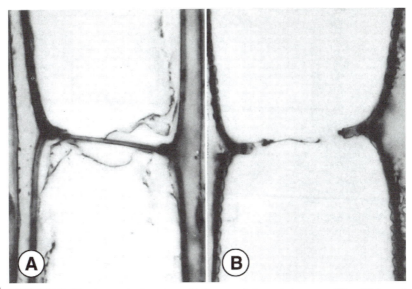

**FIGURE 2.47** Longitudinal sections of fully expanded vessel elements of *Zea*. (A) Secondary longitudinal walls and intact primary end wall. × 540. (B) Disintegrating material in place of end wall. × 540. From Esau (1940), *Hilgardia* **13**, 229–244. Copyright © Regents, University of California.

In isolated and cultured leaf mesophyll cells, secondary wall synthesis typically is completed in a matter of hours. The vacuolar and plasma membranes are intact during wall formation, but the cytoplasm becomes less dense in appearance. Following the completion of secondary wall deposition, the tracheary cell enters the final stage of cytoplasmic disruption. Cell death, or autolysis, is a highly coordinated series of events that begins with a regulated,

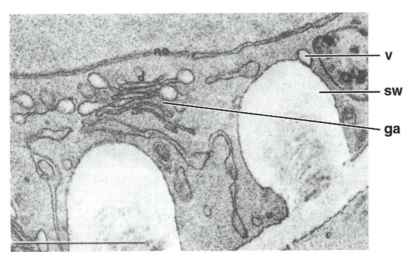

**FIGURE 2.48** Ontogeny of primary xylem tracheary elements. Transverse section of differentiating tracheary element showing the presumed migration and fusion of Golgi-derived vesicles to the developing secondary wall thickening. Abbreviations: ga, Golgi-apparatus; sw, secondary wall; v, vesicle. Reprinted with permission from Esau, Cheadle, and Gill (1966), *Am. J. Bot.* **53**, 756–764.

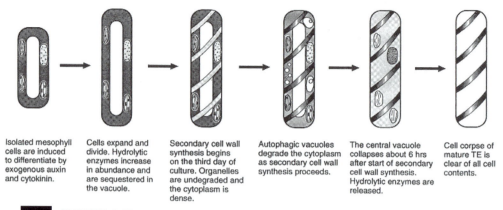

| Isolated mesophyll cells are induced to differentiate by exogenous auxin and cytokinin. | Cells expand and divide. Hydrolytic enzymes increase in abundance and are sequestered in the vacuole. | Secondary cell wall synthesis begins on the third day of culture. Organelles are undegraded and the cytoplasm is dense. | Autophagic vacuoles degrade the cytoplasm as secondary cell wall synthesis proceeds. | The central vacuole collapses about 6 hrs after start of secondary cell wall synthesis. Hydrolytic enzymes are released. | Cell corpse of mature TE is clear of all cell contents. |

**FIGURE 2.49**   Summary of events occurring during tracheary element programmed cell death. Used with permission from Groover et al. (1977), *Protoplasma* **196**, 197–211.

channel-mediated influx of calcium that is triggered by an extracellular signal. Calcium influx results in the cessation of cytoplasmic streaming and collapse of the large central vacuole. This is followed by the complete degradation of cellular contents by hydrolytic enzymes released from the vacuole.

## PRIMARY PHLOEM DIFFERENTIATION

Like the primary xylem, primary phloem can be imprecisely designated as **protophloem** and **metaphloem,** using temporal criteria. Size cannot be used as a reliable basis to distinguish between protophloem and metaphloem sieve tubes. Developmentally, protophloem refers to the first-formed, very short lived phloem elements that mature before or during organ elongation at a time when tissues are subjected to stress. Protophloem cells are only rarely associated with companion cells or albuminous cells. Differentiation of the protophloem sieve tubes can occur prior to the time when the procambial strand attains its final size and form. The protophloem soon ceases to function and becomes stretched, crushed, and nondiscernible in older stems or becomes sclerified, usually forming long "bast fibers" that are positioned as bundle caps. The loss of protophloem is compensated for by the development of metaphloem, which is the later-formed primary phloem that matures following elongation. Both protophloem and metaphloem elements can possess thickened primary walls, with those of the protophloem characteristically thicker than those of the metaphloem. Primary phloem differentiation occurs in a uniformly centripetal direction in all organs and plant groups. In the stem, phloem differentiates along the longitudinal plane, like the procambium, in a largely continuous acropetal direction from the mature phloem. In grass leaves, the first-formed protophloem elements are isolated from the older stem vasculature. Subsequent differentiation of the phloem proceeds apically toward the leaf tip and basipetally toward the stem.

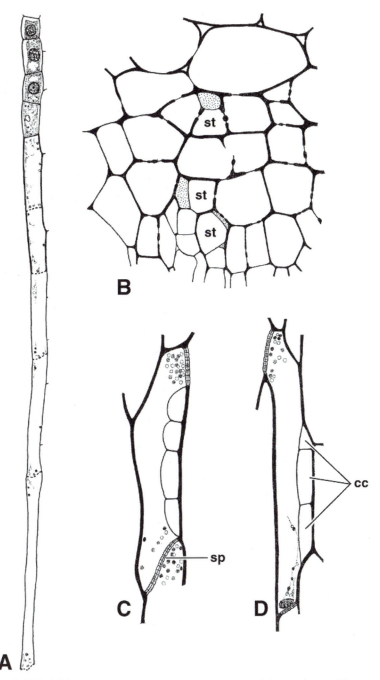

**FIGURE 2.50**   Root phloem of carrot, *Daucus carota*. (A) Series of protophloem sieve tube elements. Beginning at the top: three mother cells with nuclei; one cell with a disintegrating nucleus, two mature elements; and three elongated mature elements with thin walls. (B) Secondary phloem with sieve tubes (st), companion cells (stippled), and phloem parenchyma from the hypocotyl. (C, D) Secondary sieve tubes with companion cells (cc) in tangential view. Abbreviation: sp, sieve plate. A, B, ×605; C, D, ×472. Reprinted with permission from Esau (1940), *Hilgardia* **13,** 175–226. Copyright © Regents, University of California.

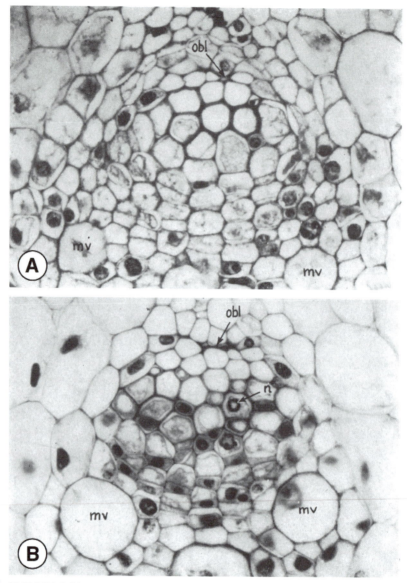

**FIGURE 2.51**  Primary phloem of *Zea mays*. (A, B) Transverse section of almost mature metaphloem from a bundle in an internode. The protophloem is obliterated (obl). The first metaphloem sieve tubes are smaller than the later. Abbreviations: mv, metaxylem vessel; n, nucleus. × 750. Reprinted with permission from Esau (1943), *Hilgardia* **15**, 327–368. Copyright © Regents, University of California.

## VASCULAR DIFFERENTIATION IN LEAVES

The development of leaf venation in dicotyledons is initiated within the young primordium as a single continuous and uninterrupted, medially positioned procambial strand that extends from the stem toward the primordium apex, with its growth keeping pace with the elongating primordium. This is followed

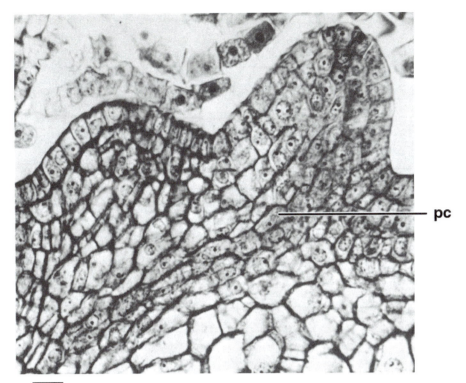

pc

**FIGURE 2.52** Longitudinal section of procambial strand (pc) extending into a young leaf primordium of *Eupatorium* (Asteraceae). × 1000. Courtesy of P. G. Mahlberg.

by secondary veins and high-order venation arising in sequence as soon as the lamina begins to form. The procambial strands forming the secondary veins of the leaf are continuous with the midvein and grow outward toward the leaf margins. The direction of secondary vein maturation can be either from leaf apex to base or from leaf base to apex or the earliest secondaries to differentiate can be positioned midway along the length of the young leaf, with maturation proceeding both toward the apex and toward the base. In dicotyledons the higher order veins originate in a basipetal direction within the expanding lamina. This results in developmentally more mature veins toward the leaf apex. Procambium forming the tertiary veins is reported to arise and extend simultaneously from one secondary to another. As increasingly higher vein orders differentiate, the lamina becomes subdivided into a series of areoles bordered by the minor veins. The freely ending veinlets within the system of areoles are the final level of foliar venation to become established.

Like dicotyledons, monocot leaf venation begins development with the formation of a single procambial strand that extends acropetally in the primordium. In monocots with more or less nearly parallel veined leaves (e.g., grasses), the major lateral bundles arise in succession from the medial vein outward. Foliar procambium arising near the base of the leaf primordium and in isolation from other vasculature distinguishes monocots from all other

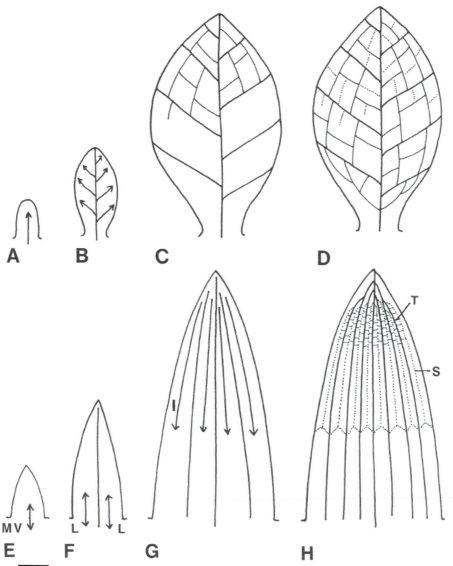

**FIGURE 2.53**  Leaf vascular pattern ontogeny in dicotyledons and monocotyledons. (A–D) *Arabidopsis*, a dicot; (E–H) Maize (*Zea*) a monocot. (A) Acropetal development of midvein provascular strand from stem vasculature (arrow). (B) Progressive formation of secondary vein provascular strands (arrows). (C) Simultaneous formation of tertiary vein network. Tertiary vein formation begins near the leaf apex and proceeds in a basipetal direction. (D) Formation of quaternary veins and freely ending vein-lets. The formation of minor order veins (dashed lines) also proceeds in a basipetal direction from the apex of the leaf toward the petiole. (E) Formation of midvein (MV) provascular strand in disk of inser-tion. The midvein extends acropetally into the leaf primordium and basipetally to connect to the stem vasculature (two-headed arrow). (F) Formation of large lateral vein provascular strands (L) in disk of insertion. Large lateral veins develop acropetally into the leaf primordium and later basipetally to con-nect to the stem vasculature. (G) Formation of intermediate longitudinal vein provascular strands (I) in distal portion of leaf. Only some of the intermediate longitudinal veins connect basipetally with the stem vasculature. (H) Formation of small longitudinal (S) and transverse (T) veins in leaf blade region. Note the basipetal pattern of transverse vein formation. Reprinted with permission from Nelson and Dengler (1997), *Plant Cell* **9**, 1121–1135. Copyright © the American Society of Plant Physiologists.

plants. The direction of differentiation of procambium in the major parallel bundles proceeds both acropetally into the growing leaf primordium and basipetally into the stem, until connection is established with a procambial strand of the stem. Parallel bundles of intermediate size differentiate in a basipetal direction after the period of leaf expansion. The last event of vascular differentiation in monocot leaves is the creation of transversely oriented commissural veins connecting the longitudinally extending parallel strands. This process occurs from the apical region of the leaf toward the base.

## SUMMARY

The postembryonic primary plant body arises from small, organized cell populations termed meristems that are located at the tips of stems and roots. Meristems are regions of active division, enlargement, and differentiation. The angiosperm vegetative shoot apical meristem has a stratified organization, with two or more outer tunica layers in which cell divisions are primarily all anticlinal, or oriented at right angles to the surface. Below the tunica layers is a region of corpus cells in which divisions occur in all planes. Active shoot apical meristems also exhibit cytohistological zonation, in which the cells in the central zone are relatively large and divide infrequently, as compared to the surrounding peripheral zone cells which divide more rapidly. A rib meristem underlies the central zone. In chimeric apices the different layers of the apex have a different genetic composition. Leaves are determinate organs initiated as groups of cells on the flanks of the shoot apex in precise and stable patterns. Leaf organogenesis commences with coordinated changes in the rate and pattern of cell division, and expansion in different regions of the developing primordium. Root apical meristems differ from shoot apical meristems in that they are subterminal and contribute cells to a covering root cap. No lateral organs are formed on the flanks of the root apex. The root apex also is characterized by a small aggregation of mitotically and metabolically inactive cells called the quiescent center. The quiescent center is located just behind the root cap. Cells lying within this region show a low rate of DNA synthesis and a correspondingly slow cell division cycle.

Three categories of primary meristematic tissues can be recognized in the undifferentiated region immediately behind the apical meristem. These undifferentiated tissues, termed protoderm, procambium, and ground meristem, mature into the primary tissue of the plant. Protoderm matures into the epidermis, procambium differentiates into the primary vascular tissue, and ground meristem becomes the ground tissue of the cortex and pith. The epidermis comprises the uniseriate surface layer of the primary body. It consists of a variety of cell types, including ordinary epidermal cells, guard cells, subsidiary cells, trichomes or emergences, and various idioblasts. The outer walls of epidermal cells are overlaid by a structurally complex waxy layer, the cuticle, that is largely impervious to liquids and gases. The cuticle is composed of cutin, a fatty substance that becomes oxidized and polymerized on the outer surface by a process known as cuticularizaton. Surface waxes are often deposited on the surface of leaves as epicuticular wax secretions. The epidermis of all aerial organs is typically provided with specialized openings called stomata. The pores are surrounded by a pair of specialized epidermal cells, the guard cells, and often by two or more distinct subsidiary cells. The patterns of stomatal development are very precise and often taxon specific.

The fundamental or ground tissue system is composed of cell types known as parenchyma, collenchyma, and sclerenchyma. Parenchyma and collenchyma are living, physiologically active cells with primary walls. Collenchyma is distinguished by having irregularly thickened walls of high tensile strength. Sclerenchyma includes a collection of cell types with thickened secondary walls that are usually lignified when mature and that generally lose a living protoplast at maturity. Two categories of sclerenchyma are arbitrarily recognized on the basis of their form, fibers and sclereids. A restricted number of flowering plants contain a distinctive and specialized cell known as a laticifer that contains a viscous fluid called latex. Depending upon their manner of origin and mature structure, laticifers are classified as articulated and nonarticulated types.

Specialized water- and food-conducting vascular tissues extend throughout the plant and make possible the long distance transport of water and inorganic and organic solutes. Xylem conducts large quantities of water and dissolved minerals from the roots to the crown. Phloem is the conduit for food solutes assimilated in photosynthesis. In addition to parenchyma and fibers, the two principal conductive cells of the xylem are tracheids and vessel elements, collectively termed tracheary elements. Tracheary cells are generally elongated cells that have lignified walls and are dead at maturity. Tracheids are imperforate cells, whereas vessel elements are perforate with the end walls forming open perforation plates. Phloem is composed of sieve elements, highly specialized living cells that are enucleate in angiosperms and have associated companion cells. Phloem also contains parenchyma and often fibers.

Differentiation of vascular tissues within the procambium occurs along both longitudinal and radial planes. Phloic elements are the first vascular cells to differentiate, followed by the xylem. Primary xylem and primary phloem are temporally and positionally divided into two categories of elements, protoxylem and metaxylem and protophloem and metaphloem. The cytodifferentiation of both xylem and phloem elements results in dramatic sequences of cellular changes that have profound functional significance.

## ADDITIONAL READING

1. Abeysekera, R. M., and McCully, M. E. (1993). The epidermal surface of the maize root tip. I. Development in normal roots. *New Phytol.* **125**, 413–429.
2. Aloni, R. (1987). Differentiation of vascular tissues. *Annu. Rev. Plant Physiol.* **38**, 179–204.
3. Aloni, R. (1992). The controls of vascular differentiation. *Int. J. Plant Sci.* **153**, S90–S92.
4. Arzee, T. (1953). Morphology and ontogeny of foliar sclereids in *Olea europaea* I & II. *Am. J. Bot.* **40**, 680–687; 745–752.
5. Bailey, I. W. (1936). The problem of differentiating and classifying tracheids, fiber tracheids, and libriform fibers. *Trop. Woods* **45**, 18–23.
6. Barthlott, W. (1990). Scanning electron microscopy of the epidermal surface in plants. *In* "Scanning Electron Microscopy in Taxonomy and Functional Morphology" (D. Claugher, Ed.), pp. 69–83. Systematics Association Special Volume No. 41. Clarendon Press, Oxford.

7. Barthlott, W., and Neinhuis, C. (1997). Purity of the sacred lotus, or escape from contamination in biological surfaces. *Planta* **202**, 1–8.

8. Barthlott, W., Neinhuis, C., Cutler, D., Ditsch, F., Meusel, I., Theisen, I., and Wilhelmi, H. (1998). Classification and terminology of plant epicuticular waxes. *Bot. J. Linnean Soc.* **126**, 237–260.

9. Beer, M., and Setterfield, G. (1958). Fine structure in thickened primary walls of collenchyma cells of celery petioles. *Am. J. Bot.* **45**, 571–580.

10. Behnke, H. D., and Sjolund, R. D. (Eds.) (1990). "Sieve Elements: Comparative Structure, Induction and Development." Springer-Verlag, New York.

11. Bierhorst, D. W., and Zamora, P. M. (1965). Primary xylem elements and element associations of angiosperms. *Am. J. Bot.* **52**, 657–710.

12. Bitonti, M. B., Chiappetta, A., Innoicenti, A. M., Liso, R., and Arrigoni, O. (1992). Quiescent center ontogenesis during early germination of *Allium cepa* L. *New Phytol.* **121**, 577–580.

13. Bloch, R. (1946). Differentiation and pattern in *Monstera deliciosa*. The idioblastic development of the trichosclereids in the air root. *Am. J. Bot.* **33**, 544–551.

14. Bokhari, M. H., and Burtt, B. L. (1970). Studies in the Gesneriaceae of the Old World XXXII: Foliar sclereids in *Cyrtandra*. *Notes Royal Bot. Gard. Edinburgh* **30**, 11–21.

15. Burat, R. (1989). "Ontogeny, Cell Differentiation and Structure of Vascular Plants." Springer-Verlag, New York.

16. Butterfield, B. G., and Meylan, B. A. (1972). Scalariform perforation plate development in *Laurelia novae-zelandiae* A Cunn.: A scanning electron microscope study. *Aust. J. Bot.* **20**, 253–259.

17. Carlquist, S. (1958). Structure and ontogeny of glandular trichomes of Madinae (Compositae). *Am. J. Bot.* **45**, 675–682.

18. Clark, S. E. (1997). Organ formation at the vegetative shoot meristem. *Plant Cell* **9**, 1067–1076.

19. Clowes, F.A.L. (1961). "Apical Meristems." Botanical Monographs. Blackwell Science Publ., Oxford.

20. Clowes, F. A. L. (1965). Meristems and the effect of radiation on cells. *Endeavour* **24**, 8–12.

21. Clowes, F. A. L. (1967). The quiescent center. *Phytomorphology* **17**, 132–140.

22. Clowes, F. A. L. (1994). Origin of the epidermis in root meristems. *New Phytol.* **127**, 335–347.

23. Cooper, D. C. (1932). The development of peltate hairs of *Shepherdia canadensis. Am. J. Bot.* **19**, 423–428.

24. Cosgrove, D. J. (1997). Relaxation in a high-stress environment: The molecular bases of extensible cell walls and cell enlargement. *Plant Cell* **9**, 1031–1041.

25. Crafts, A. S. (1932). *Cucurbita* phloem. *Plant Physiol.* **7**, 182–225.

26. Cronshaw, J. (1974). Phloem differentiation and development. *In* "Dynamic Aspects of Plant Ultrastructure" (A. E., Robards, Ed.), pp. 391–413. McGraw-Hill, London.

27. Croxdale, J. (1998). Stomatal patterning in monocotyledons: *Tradescantia* as a model system. *J. Exp. Bot.* **49**, 279–292.

28. Cunha, M. Da, Costa, C. G., Machado, R. D. and Miguens, F. C. (1998). Distribution of the laticifer system in *Chamaesyce thymifolia* (L.) Millsp. (Euphorbiaceae). *Acta Bot. Neerl.* **47**, 209–218.

29. Cutler, D. F., Alvin, K. L., and Price, C. E. (1982). "The Plant Cuticle." Academic Press, London.

30. Cutter, E. G. (1965). Recent experimental studies of the shoot apex and shoot morphogenesis. *Bot. Rev.* **31**, 7–113.

31. Dale, J. E. (1992). How do leaves grow? *BioScience* **42**, 323–332.

32. Dengler, N. G., Mackay, L. B., and Gregory, L. M. (1975). Cell enlargement and tissue differentiation during leaf expansion in beech, *Fagus grandifolia. Can. J. Bot.* **53**, 2846–2865.

33. Deshpande, B. P. (1975). Differentiation of the sieve plate of *Cucurbita*: A further view. *Ann. Bot.* **39**, 1015–1022.

34. Duchaigne, A. (1955). Les divers types de collenchymes chez le Dicotyledones; leur ontogenie et leur lignification. *Ann. Sci. Nat. Bot., Ser.* **11**, 455–479.

35. Eglinton, G., and Hamilton, R. J. (1967). Leaf epicuticular waxes. *Science* **156**, 1322–1335.

36. Esau, K. (1936). Ontogeny and structure of collenchyma and of vascular tissue in celery petioles. *Hilgardia* **10**, 431–476.

37. Esau, K. (1938). Ontogeny and structure of the phloem of tobacco. *Hilgardia* **11**, 343–424.

38. Esau, K. (1940). Developmental anatomy of the fleshy storage organ of *Daucus carota*. *Hilgardia* **13**, 175–226.

39. Esau, K. (1940). Structure of end walls in differentiating vessels. *Hilgardia* **13**, 229–244.

40. Esau, K. (1943). Ontogeny of the vascular bundle in *Zea mays*. *Hilgardia* **15**, 327–368.

41. Esau, K. (1943). Vascular differentiation in the vegetative shoot of *Linum*. III. The origin of the bast fibers. *Am. J. Bot.* **30**, 579–586.

42. Esau, K. (1948). Phloem structure in the grapevine, and its seasonal changes. *Hilgardia* **18**, 217–296.

43. Esau, K. (1961). "Plants, Viruses, and Insects." Harvard Univ. Press, Cambridge, Massachusetts.

44. Esau, K. (1965). Anatomy and cytology of *Vitis* phloem. *Hilgardia* **37**, 17–72.

45. Esau, K. (1965). "Vascular Differentiation in Plants." Holt, Rinehart and Winston, New York.

46. Esau, K. (1966). Explorations of the food conducting system in plants. *Am. Sci.* **54**, 141–157.

47. Esau, K., and Cheadle, V. I. (1959). Size of pores and their contents in sieve elements of dicotyledons. *Proc. Nat. Acad. Sci.* (USA) **45**, 156–162.

48. Esau, K., and Hewitt, W. B. (1940). Structure of end walls in differentiating vessels. *Hilgardia* **13**, 229–244.

49. Esau, K., Cheadle, V. I., and Gill, R. H. (1966). Cytology of differentiating tracheary elements. I. Organelles and membrane systems. *Am. J. Bot.* **53**, 756–764.

50. Fahn, A. (1979). "Secretory Tissues in Plants." Academic Press, London.

51. Fahn, A., and Leshmen, B. (1963). Wood fibers with living protoplasts. *New Phytol.* **62**, 91–98.

52. Feldman, L. J. (1984). Regulation of root development. *Annu. Rev. Plant Physiol.* **35**, 223–242.

53. Feldman, L. J. (1998). Not so quiet quiescent centers. *Trends Plant Sci.* **3**, 80–81.

54. Foard, D. E. (1959). Pattern and control of sclereid development in the leaf of *Camellia japonica*. *Plant Physiol.* **33**, (Suppl.) xli.

55. Fosket, D. E. (1994). "Plant Growth and Development. A Molecular Approach." Academic Press, San Diego, New York.

56. Foster, A. S. (1955). Structure and ontogeny of terminal sclereids in *Boronia serrulata*. *Am. J. Bot.* **42**, 551–560.

57. Fukuda, H. (1997). Tracheary element differentiation. *Plant Cell* **9**, 147–1156.

58. Gaudet, J. (1960). Ontogeny of foliar sclereids in *Nymphaea odorata*. *Am. J. Bot.* **47**, 525–532.

59. Gifford, E. M., and Corson, G. E (1971). The shoot apex in seed plants. *Bot. Rev.* **37**, 143–229.

60. Groover, A., and Jones, A. M. (1999). Tracheary element differentiation uses a novel mechanism coordinating programmed cell death and secondary wall synthesis. *Plant Physiol.* **119**, 375–384.

61. Hall, D. M. (1967). Wax microchannels in the epidermis of white clover. *Science* **158**, 505–506.

62. Hall, L. N., and Langdale, J. A. (1996). Molecular genetics of cellular differentiation in leaves. *New Phytol.* **132**, 533–553.

63. Harrar, E. S. (1946). Notes on starch grains in septate fibre-tracheids. *Trop. Woods* **85**, 1–9.

64. Hill, C.R., and Dilcher, D. L. (1990). Scanning electron microscopy of the internal ultrastructure of plant cuticle. *In* "Scanning Electron Microscopy in Taxonomy and Functional Morphology" (D. Claugher, Ed.), pp. 95–124. Systematics Association Special Volume No. 41. Clarendon Press, Oxford.

65. Hepler, P. K., and Fosket, D. E. (1971). The role of microtubules in vessel member differentiation in *Coleus*. *Protoplasma* **72**, 213–236.

66. Hepler, P. K., and Newcomb, E. H. (1963). The fine structure of young tracheary xylem elements arising by redifferentiation of parenchyma in wounded *Coleus* stem. *J. Exp. Bot.* **14**, 496–503.

67. Hepler, P. K., and Newcomb, E. H. (1964). Microtubules and fibrils in the cytoplasm of *Coleus* cells undergoing secondary wall deposition. *J. Cell Biol.* **20**, 529–533.

68. Hulbary, R. L. (1944). The influence of air spaces on the three-dimensional shapes of cells in *Elodea* stems, and a comparison with pith cells of *Ailanthus*. *Am. J. Bot.* **31**, 561–580.

69. Hunter, J. R. (1994). Reconsidering the functions of latex. *Trees* **9**, 1–5.

70. Jeffree, C. E., Dale, J. E., and Fry, S. C. (1986). The genesis of intercellular spaces in developing leaves of *Phaseolus vulgaris* L. *Protoplasma* **132**, 90–98.

71. Jeffrey, E. C. (1925). The origin of parenchyma in geological time. *Proc. Natl. Acad. Sci.* (USA) **11**, 106–110.

72. Juniper, B. E., and Jeffree, C. E. (1983). "Plant Surfaces." Arnold, London.

73. Kerstetter, R. A., and Hake, S. (1997). Shoot meristem formation in vegetative development. *Plant Cell* **9**, 1001–1010.

74. Kerstiens, G. (Ed.) (1996). "Plant Cuticles: An Integrated Functional Approach." BioScience Publ., Oxford.

75. Koslowski, T. T. (1971). "Growth and Development of Trees. Vol. I. Seed Germination, Ontogeny, and Shoot Growth." Academic Press, New York, London.

76. Larkin, J. C., Marks, M. D., Nadeau, J., and Sack, F. (1997). Epidermal cell fate and patterning in leaves. *Plant Cell* **9**, 1109–1120.

77. Ledin, R. B. (1954). The vegetative shoot apex of *Zea mays*. *Am. J. Bot.* **41**, 11–17.

78. Lemon, G. D., and Posluszny, U. (1998). A new approach to the study of apical meristem development using laser scanning confocal microscopy. *Can. J. Bot.* **76**, 899–904.

79. Lloyd, C. W. (Ed.). (1991). "The Cytoskeletal Basis of Plant Growth and Form." Academic Press, San Diego.

80. Lyndon, R. F. (1990). "Plant Development: The Cellular Basis." Unwin Hyman, London.

81. Mahlberg, P. G. (1959). Development of the non-articulated laticifer in proliferated embryos of *Euphorbia marginata* Pursh. *Phytomorphology* **9**, 156–162.

82. Mahlberg, P. G. (1963). Development of non-articulated laticifer in seedling axis of *Nerium oleander*. *Bot. Gaz.* **124**, 224–231.

83. Mahlberg, P. G. (1993). Laticifers: An historical perspective. *Bot. Rev.* **59**, 1–23.

84. Martin, J. R., and Juniper, B. E. (1970). "The Cuticles of Plants." Arnold, London.

85. McCann, M. C. (1997). Tracheary element formation: Building up to a dead end. *Trends Plant Sci.* **2**, 333–338.

86. Medford, J. I. (1992). Vegetative apical meristems. *Plant Cell* **4**, 1029–1039.

87. Neinhuis, C., and Barthlott, W. (1997). The tree leaf surface: Structure and function. *In* "Trees—Contributions to Modern Tree Physiology" (H. Rennenberg, W. Eschrich, and H. Ziegler, Eds.), pp. 3–18. Backhuys Publ., Leiden, The Netherlands.

88. Neinhuis, C., and Barthlott, W. (1997). Characterization and distribution of water-repellent, self-cleaning plant surfaces. *Ann. Bot.* **79**, 667–677.

89. Neinhuis, C., and Barthlott, W. (1998). Seasonal changes of leaf surface contamination in beech, oak, and ginkgo in relation to leaf micromorphology and wettability. *New Phytol.* **138**, 91–98.

90. Nelson, T., and Dengler, N. (1997). Leaf vascular pattern formation. *Plant Cell* **9**, 1121–1135.

91. Norris, R. F., and Bukovac, M. J. (1968). Structure of the pear leaf cuticle with special reference to cuticular penetration. *Am. J. Bot.* **55**, 975–983.

92. O'Brien, T. P. (1974). Primary vascular tissues. *In* "Dynamic Aspects of Plant Ultrastructure" (A. W. Robards, Ed.), pp. 414–440. McGraw-Hill, London.

93. O'Brien, T.P., and McCully, M. E. (1969). "Plant Structure and Development. A Pictorial and Physiological Approach." The Macmillan Co., London.

94. Pallarady, S. G., and Kozlowski, T. T. (1980). Cuticle development in the stomatal region of *Populus* clones. *New Phytol.* **85**, 363–368.

95. Pant, D. D. (1965). On the ontogeny of stomata and other homologous structures. *Plant Sci. Ser.* (Allahabad) **1**, 1–24.

96. Peterson, R. L., and Farquhar, M. L. (1996). Root hairs: Specialized tubular cells extending root surfaces. *Bot. Rev.* **62**, 1–40.

97. Poethig, R. S. (1997). Leaf morphogenesis in flowering plants. *Plant Cell* **9**, 1077–1087.

98. Popham, R. A. (1951). Principal types of vegetative shoot apex organization in vascular plants. *Ohio J. Sci.* **51**, 249–270.

99. Rao, T.A. (1991). "Compendium of Foliar Sclereids in Angiosperms: Morphology and Taxonomy." Wiley Eastern Limited, New Delhi.

100. Rasmussen, H. (1981). Terminology and classification of stomata and stomatal development—A critical survey. *Bot. J. Linnean Soc.* **83**, 199–212.

101. Reeve, R. M. (1946). Ontogeny of the sclereids in the integument of *Pisum. sativum* L. *Am. J. Bot.* 33, 806–816.58.

102. Romberger, J. A., Hejnowicz, Z., and Hill, J. (1993). "Plant Structure: Function and Development. A Treatise on Anatomy and Vegetative Development, with Special Reference to Woody Plants." Springer-Verlag, Berlin.

103. Rudall, P. J. (1987). Laticifers in Euphorbiaceae—A conspectus. *Bot. J. Linnean Soc.* **94**, 143–163.

104. Sachs, T. (1991). " Pattern Formation in Plant Tissues." Cambridge University Press, Cambridge.

105. Satina, S., and Blakeslee, A. F. (1941). Periclinal chimeras in *Datura stramonium* in relation to development of leaf and flower. *Am. J. Bot.* **28**, 862–871.

106. Satina, S., Blakeslee, A. F., and Avery, A. (1940). Demonstration of the three germ layers in the shoot apex of *Datura* by means of induced polyploidy in periclinal chimeras. *Am. J. Bot.* **27**, 895–905.

107. Sattler, R. (1982). "Axioms and Principles of Plant Construction." Martinus Nijhoff/Dr. W. Junk Publishers, The Hague.

108. Schiefelbein, J. W., Masucci, J. D., and Wang, H. (1997). Building a root: The control of patterning and morphogenesis during root development. *Plant Cell* **9**, 1089–1098.

109. Schmidt, A. (1924). Histologische Studien an phanerogamen Vegetationspunkten. *Bot. Arch.* **8**, 345–404.

110. Sjölund, R. D. (1997). The phloem sieve element: A river runs through it. *Plant Cell* **9**, 1137–1146.

111. Smith, L. G., and Hake, S. (1992). The initiation and determination of leaves. *Plant Cell* **4**, 1017–1027.

112. Steeves, T. A., and Sussex, I. M. (1989). "Patterns in Plant Development," 2nd ed. Cambridge University Press, Cambridge.

113. Steffensen, D. M. (1968). A reconstruction of cell development in the shoot apex of maize. *Am. J. Bot.* **55**, 354–369.

114. Stewart, R. N. (1978). Ontogeny of the primary body in chimeral forms of higher plants. *In* "The Clonal Basis of Development." (S. Subtelny and I. M. Sussex, Eds.) pp. 131–160. Academic Press, New York.

115. Stewart, R. N., and Dermen, H. (1970). Determination of number and mitotic activity of shoot apical initial cells by analysis of mericlinal chimeras. *Am. J. Bot.* **57**, 816–826.

116. Stewart, R. N., and Dermen, H. (1979). Ontogeny in monocotyledons as revealed by studies of the developmental anatomy of periclinal chloroplast chimeras. *Am. J. Bot.* **66**, 47–58.

117. Sussex, I. M. (1989). Developmental programming of the shoot meristem. *Cell* **56**, 225–229.

118. Torrey, J. G., and Clarkson, D. (Eds.) (1975). "The Development and Function of Roots." Academic Press, San Diego.

119. Walker, W. S. (1960). The effect of mechanical stimulation and etiolation on the collenchyma of *Datura stramonium. Am. J. Bot.* **47**, 717–724.

120. Wardlaw, C. W. (1965). The organization of the shoot apex. *Encyc. Plant Physiol.* **15**, 966–1076.

121. Wilkinson, H. P. (1979). The plant surface (mainly leaf). Part 1. Stomata. *In* "Anatomy of the Dicotyledons" (C. R. Metcalfe, and L. Chalk, Eds.), Vol. 1, 2nd ed., pp. 97–117. Clarendon Press, Oxford.

122. Williams, R. F. (1975). "The Shoot Apex and Leaf Growth." Cambridge University Press, Cambridge.

123. Wilson, K. J., and Mahlberg, P. G. (1980). Ultrastructure of developing and mature non-articulated laticifers in the milkweed *Asclepias syriaca* L. (Asclepiadaceae). *Am. J. Bot.* **67**, 1160–1170.

# 3

# TISSUE ORGANIZATION IN STEMS, LEAVES, AND ROOTS

In the preceding chapters, the study of plant anatomy was introduced by discussing cell structure, the very great variety of plant cells, the different cell aggregates that compose plant tissues, and the origin, development, modifications, and functions of cells and tissues. This chapter deals with the details of internal organization and the arrangement of cells and tissues within the three major organs of the primary plant body—the stem, the leaf, and the root. It is meant to provide a general picture of the manner of organization of different tissues as well as their distribution to maximize a particular physiological or mechanical function.

## STEMS

The stem of a typical plant functions in support of leaves, flowers, and fruit. The stem also conducts water and dissolved mineral salts upward and transports carbohydrates manufactured through the process of photosynthesis to various parts of the stem and root system. Stems vary considerably in structure, and these modifications, along with those of the leaves and roots, enable vascular plants to survive in a variety of habitats. The major anatomical differences among stems are largely determined by the manner of development and arrangement of the vascular and nonvascular tissues, and by the relative amounts of secondary growth. The features common to all stems are the external **nodes** and **internodes** and the lateral organs, the leaves, and associated axillary buds that arise exogenously and are positioned at the nodal regions. Stem anatomical variation is highly correlated with the development and organization of the primary vascular system, which in stems differentiates in relation to the developing leaf primordia. Stems thus differ from roots, in which no lateral organs are present.

## Vascular Bundles

In **herbaceous** or nonwoody shoots, the vascular tissues are found in close association and are generally organized into discrete columns or strands called **vascular bundles.** In the apical zone of maturation, the primary vascular tissues are differentiated from the **procambium** in such a way that the vascular bundles are oriented longitudinally along the plant axis, as well as extending into the foliar organs. In dicotyledons, commonly some procambium remains undifferentiated between the primary xylem and primary phloem and later functions as the vascular cambium. A vascular bundle that has the potential for secondary growth is referred to as an **open bundle.** In monocotyledons, all procambial cells within a bundle become primary vascular elements, so the bundle is termed **closed** to secondary growth.

The vascular system of the stem and leaf become interconnected at the nodal regions. The pattern by which the stem bundles are connected with the veins of the leaves and branches is varied and complicated. A vascular bundle in the stem that is recognizable as a continuation of a vascular bundle of the leaf base is called a **leaf trace.** At the level where the leaf trace is directed

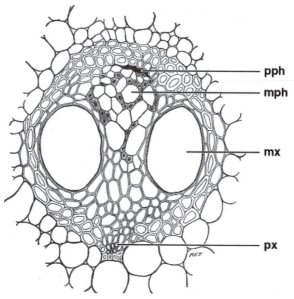

**FIGURE 3.1**    Vascular bundle of *Smilax herbacea* (Liliaceae). Abbreviations: mph, metaphloem mx, metaxylem tracheary element; pph, crushed protophloem; px, protoxylem. Reprinted with permission of the publisher from THE ANATOMY OF WOODY PLANTS by E. C. Jeffrey: The University of Chicago Press, Copyright © 1917.

toward the leaf, a parenchymatous region differentiates in the cauline vascular system. In classical descriptions this region of vascular interruption is called a **leaf gap.** Different plants in the primary state have different numbers of leaf traces per leaf and different numbers of leaf gaps per node. Each axial bundle and its associated leaf trace(s) is referred to as a **sympodium.** If the major sympodial bundles in the stem remain separate from one another at the level of leaf trace departure, or if the major stem bundles are only interconnected by minor bridge bundles, the primary vascular system is spoken of as being of the open type. If, however, a single leaf trace is connected to more than one sympodial bundle, or if the sympodial bundles are interconnected by large bridge bundles, the primary vascular system forms a reticulum or a closed pattern.

On the basis of the position of the xylem and phloem within individual bundles, the following distinct vascular bundle types are recognized. When the primary phloem and xylem are in the same radius of the stem, and the phloem is on the side toward the cortex and the xylem is confined to the side toward the pith, the bundle is called a **collateral vascular bundle.** When the phloem is positioned on both inner and outer sides of the xylem, as is found in some dicotyledons, the arrangement is called a **bicollateral vascular bundle.** Vascular tissues may completely surround one another and form a **concentric bundle.** Concentric bundles can be of two types. When the central phloem is surrounded by xylem, this is termed an **amphivasal vascular bundle.** This arrangement occurs in stems of sedges and some grasses as well as other monocotyledons. Bundles in which phloem surrounds the xylem are termed **amphicribral vascular bundles.**

Different types of collateral vascular bundles can be recognized in monocotyledons, primarily in relation to the arrangement of the late metaxylem conducting elements as viewed in transverse section. One type has the xylem elements meeting the phloem along a straight or slightly curved line. A second type has a conspicuous **V**-shaped arrangement of xylem with several larger cells in the arms of the **V.** Another common monocot bundle type is characterized by the presence of a single large metaxylem element on each side of the bundle. The xylem can be curved or straight and sometimes contains a large space or lacuna within the primary xylem, when protoxylem and metaxylem elements become torn and crushed during differentiation. The latter condition is common in grasses and can be seen in the often figured bundle of *Zea mays.* Many palms possess vascular bundles in which a single large metaxylem element is positioned in the center. Each bundle type can also have a characteristic arrangement of phloem, as well as a pattern of fiber development that gives bundles of certain groups a distinctive appearance. Because the vascular tissue may be immediately surrounded by several rows of fibers, the term **fibrovascular bundle** is frequently used. The arrangement of vascular tissue within a bundle can change at different levels of the stem. Some vascular bundles in the Araceae, Cyclanthaceae, and Pandanaceae are termed **compound bundles** because they are formed by the temporary union of separate bundles that retain their integrity during the association. These bundles appear in transverse section as three or more separate strands of xylem and phloem enclosed wholly or partly by a common bundle sheath.

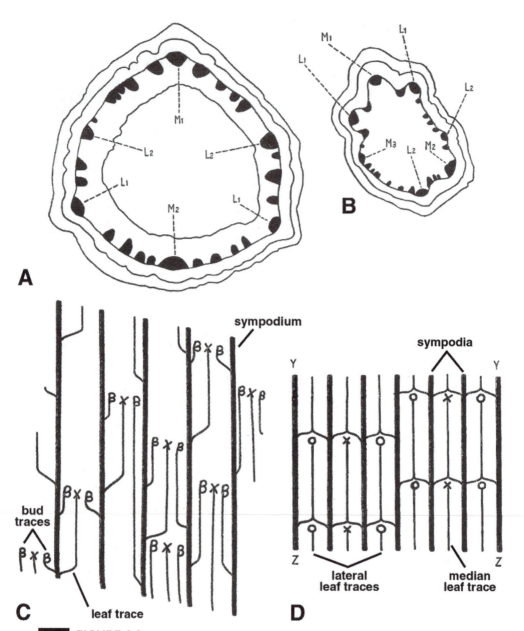

**FIGURE 3.2**   Diagrams of the primary vascular systems (shown in one plane) of Leguminosae. (A) Transverse section of stem of *Thermopsis montana*. $M_1$ and $L_1$, the median and lateral traces of the first leaf above the plane of section. $M_2$ and $L_2$, those of the second leaf above. (B) Transverse section of a young internode of *Gleditschia sinensis*. $M_3$, median trace of third leaf above. (C) Reconstructed vascular system of *Ulex europaeus* showing open type system. (D) Reconstructed vascular system of *Ornithopus sativus* showing closed type system. Abbreviations: β, lateral leaf trace; y–z, axial bundle. Reprinted with permission from Dormer (1945), *Ann. Bot. n.s.* **9**, 141–153.

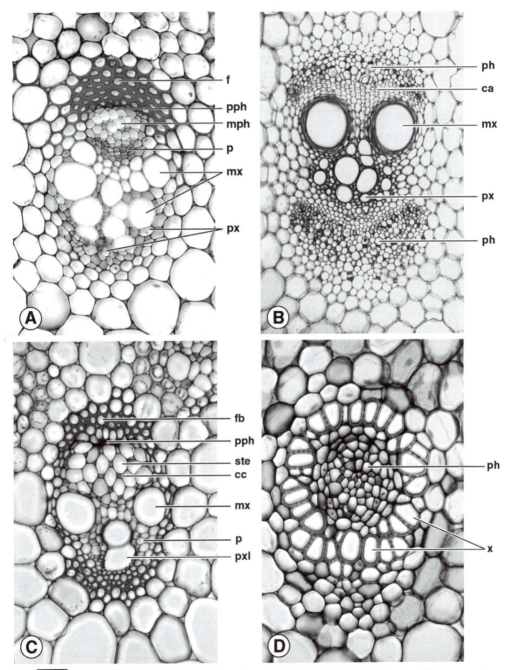

**FIGURE 3.3**  Transverse sections of vascular bundles. (A) Collateral bundle of *Ranunculus repens* (Ranunculaceae), an herbaceous dicotyledon lacking secondary growth. (B) Bicollateral bundle of *Cucurbita pepo* (Cucurbitaceae). (C) Collateral bundle of *Zea mays* (Gramineae). (D) Concentric amphivasal bundle from rhizome of Lily of the Valley, *Convallaria najalis* (Liliaceae). Abbreviations: ca, vascular cambium; cc, companion cell; fb, fiber; mph, metaphloem; mx, metaxylem; p, parenchyma from procambium; ph, phloem; pph, obliterated protophloem; px, obliterated protoxylem; pxl, protoxylem lacuna; ste, sieve tube element; x, xylem.

## Stem Structure

Among gymnosperms and dicotyledonous angiosperms, the primary vascular bundles are embedded in nonvascular ground tissue. The axial bundles are arranged in a virtual ring or circle near the periphery of the stem. In the early 20th century this structural pattern was termed a **eustele.** In some families cortical and medullary bundles are present on either side of the vascular cylinder. In the young stems of a few taxa, the procambium forms an almost continuous cylinder close to the apex, thereby resulting in an essentially continuous cylinder of primary vascular tissue.

The cells between the vascular bundles and the outer edge of the stem are primarily parenchyma; together they constitute the **cortex.** The cortical parenchyma is sometimes green and there can be numerous intercellular spaces. Various other cell types may comprise the cortical region, including sclerenchyma, collenchyma, and laticifers. In the stems of some plants, specialized photosynthetic parenchyma (chlorenchyma) or sclerenchyma can form a distinct subepidermal **hypodermis.** Among many ferns and other primitive vascular plants, the innermost cortical layer of the rhizome is differentiated into a special layer called an **endodermis.** As in roots, endodermal cell walls are distinguished by the presence of waxy Casparian thickenings (strips) composed of suberin. Within angiosperm stems, a true endodermis is only rarely present, occurring in a few (mostly herbaceous) families. Although aerial plant parts usually lack a recognizable endodermis, a layer with a recognizable Casparian strip can be induced to form in the rapidly elongating epicotyl internodes of pea seedlings (*Pisum sativum*) by transferring plants from the light to dark. The tropical family Bonnetiaceae provides an example of an endodermis in the stem of a woody group.

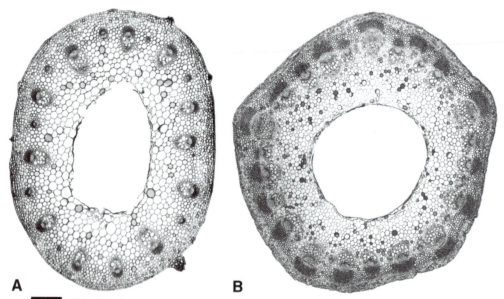

**A**                          **B**

**FIGURE 3.4**  Stem anatomy of herbaceous dicotyledons. (A) Transverse section of *Ranunculus* (Ranunculaceae), an herb without secondary growth. ×25. (B) Transverse section of *Delphinium* (Ranunculaceae). Note ring of collateral bundles. ×16. Courtesy of the Bailey-Wetmore Wood Collection, Harvard University.

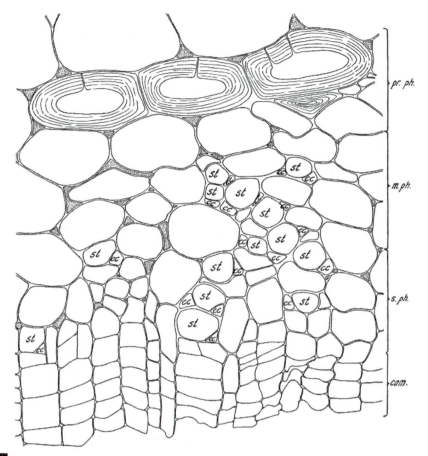

**FIGURE 3.5** Transverse section through external phloem of *Nicotiana tabacum* (Solanaceae) from an old stem. Intercellular spaces are stippled. Abbreviations: cam, vascular cambium; cc, companion cell; m. ph., metaphloem; pr. ph., protophloem; s. ph., secondary phloem; st, sieve tubes. ×492. Reprinted with permission from Esau (1938), *Hilgardia* **11**, 343–424. Copyright © Regents, University of California.

In some flowering plants, the innermost region of the cortex is composed of cells filled with conspicuous and relatively stable accumulations of starch. This region is referred to as an endodermoid layer or, more commonly, as the **starch sheath.** The starch sheath contains sedimenting amyloplasts, indicating that this cell layer functions as the gravity-sensing region in dicotyledonous stems. Genetic mutations that cause an absence of a starch sheath in hypocotyls result in the complete loss of any shoot gravitropic response. The central core of cells in the stem is commonly parenchymatous and is called the **pith.** In some plants, the pith remains a living tissue; in others, it breaks down, resulting in a cavity; in still others, a series of prominent hardened **diaphragms** alternate with large air cavities.

In contrast to dicotyledons, many monocotyledons possess vascular bundles that appear widely spaced and randomly scattered throughout the fundamental tissue of the solid stem. In the extensive literature on descriptive anatomy, this pattern is referred to as an **atactostele,** meaning literally, a stele without organization. The numerous vascular bundles tend to be crowded

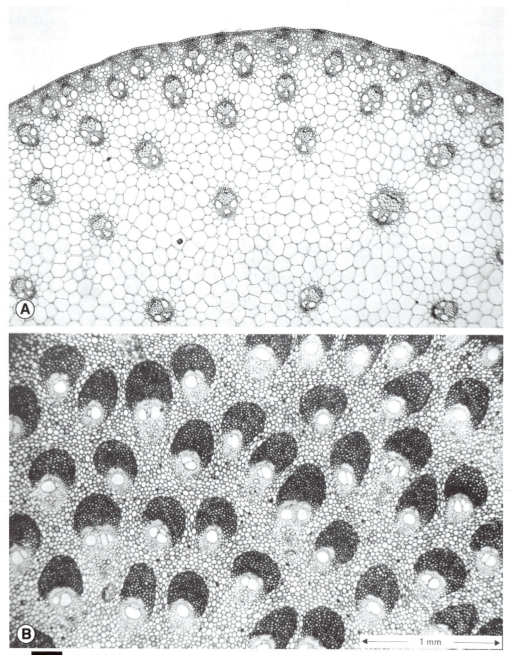

**FIGURE 3.6**   Stem anatomy of monocotyledons. (A) Transverse section of *Zea mays* (Gramineae) showing scattered distribution of vascular bundles. (B) Transverse section of palm *Geonoma umbraculifera* (Palmae) showing numerous vascular bundles extending through the ground tissue. Each bundle contains a prominent incompletely enveloping fibrous sheath. B, reprinted with permission from Tomlinson and Zimmermann (1967), *IAWA Bull.,* International Association of Wood Anatomists.

toward the stem periphery and actually form a highly organized network that provides an efficient system for long distance transport. Careful analysis has shown that the vascular construction of many monocot stems follows a basic groundplan. The monocotyledon vascular construction generally is referred to as the *Rhapis* principle because it was first observed in the small palm, *Rhapis excelsa*. In this construction, the many individual vascular bundles describe a uniform, shallow helix as they twist up the stem. The bundles are inclined toward the center of the stem, but at regular intervals they bend out sharply toward a leaf. As they bend, they divide, with one fork entering the leaf and providing a means of radial transport and the other continuing up the periphery of the stem as an axial bundle. As the neighboring bundles extend through the axis, they become interconnected by both major and minor bridge bundles. Individual bundles are either collateral or concentric in structure and are often surrounded by a sheath of fibers.

Although grasses and other monocots are often described as having vascular bundles embedded throughout the ground tissue of the stem, there is actually wide variation in monocot stem anatomy. In some grasses, for example, the principal vascular bundles are distributed in a single peripheral ring within the hollow stem and are bordered on the outside by thick-walled elements and cortex. Other grasses have the vascular bundles in two, three, or multiple, staggered concentric series, again bordered by or embedded in thick-walled cells. Vascular bundles can be of different sizes and outlines in various regions of the stem and among various subgroups of plants.

Secondary growth is known to occur in a limited number of monocotyledons, for example in the tree lilies with branched stems and a very unusual type of secondary thickening (see Chapter 4). In genera such as *Yucca, Cordyline, Dracaena, Aloë,* and others, a restricted lateral meristematic region called a thickening ring forms. This meristem lacks ray initials, and the majority of its derivatives are produced toward the inside as an anastomosing system of concentric vascular bundles and fundamental parenchyma. Some bundles of the primary system become confluent with the secondary system.

## LEAVES

In addition to stems, leaves are the other plant organs that, along with stems, constitute the **shoot** of the plant body. Leaves are the principal photosynthetic organ of vascular plants. As such, they exhibit an anatomy that facilitates this physiological process. Leaves also are an organ through which large amounts of water are lost both as vapor and as a liquid. Many of the wide array of tissue specializations of leaves are thus related to gas exchange, light absorption, prevention of water loss, and transport of water and manufactured carbohydrates. The leaves of seed plants are thought to be derived phylogenetically from modified branch systems that have undergone planation and that have subsequently assumed an expanded form by the extension of parenchymatous tissues between the vascular strands. The leaves of some plants have assumed unusual forms and functions, such as those specialized as storage organs, protective devices, climbing organs,

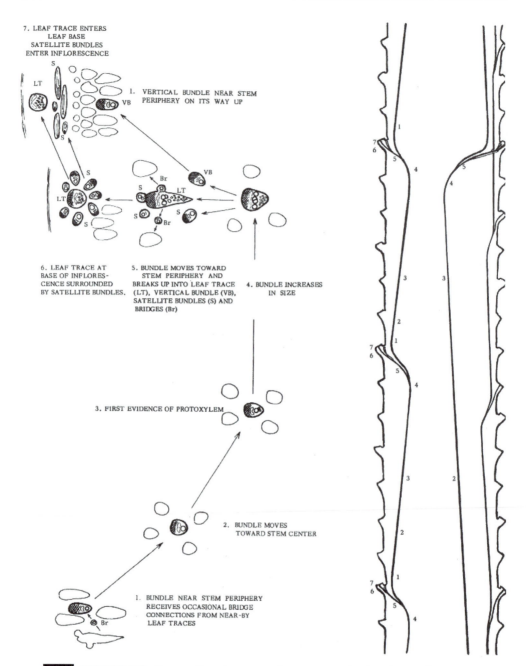

**FIGURE 3.7**   Pattern of long distance vascular continuity within the palm stem. Reprinted with permission from Tomlinson and Zimmermann (1967), *IAWA Bull.*, International Association of Wood Anatomists.

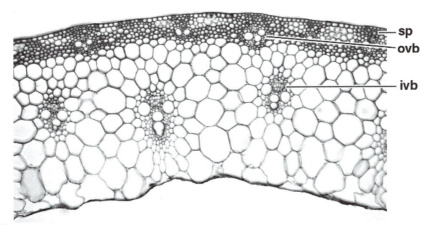

**FIGURE 3.8**  Partial transverse section of stem of wheat (*Triticum,* Gramineae). Abbreviations: ivb, inner vascular bundle; ovb, outer vascular bundle; sp, sclerified parenchyma.

vegetative reproductive structures, and especially animal traps. These latter types, known as insectivorous or carnivorous plants, include the pitcher plants, Venus fly trap, sundews, and bladderworts.

Leaves are initiated as lateral protuberances from rapidly dividing subsurface cells on the domelike stem apex. The **phyllotaxis** (the arrangement of leaves on the stem) varies in different plants and represents an important developmental and organizational expression of the plant. As previously noted, phyllotaxy has important consequences for the origin and differentiation of the primary vascular tissues. As a result of the elongation of very young foliar primordia, young leaves extend upward beyond the terminal apex. The early period of growth is followed by divisions in the intercalary regions to form the major leaf parts. The mature foliage leaf is composed of a **leaf base,** a stalklike **petiole,** and an expanded, flattened **blade** or **lamina.** Some taxa have small leaflike structures called **stipules** positioned between the petiole and stem. Monocotyledonous leaves are often linear; they ensheathe the stem at their point of attachment. A small **ligule** is positioned at the junction of sheath and stem.

The wide variation in leaf characters and their importance in systematic descriptions has resulted in the accretion of a large collection of descriptive terminology. Significant variation also is present in internal structure, although histologically the leaf is composed of the same tissue systems that are recognized in the stem, that is, dermal, fundamental, and vascular. The variability in plant structure that is commonly assumed to be correlated with habitat is particularly well emphasized in the structure of the epidermis, mesophyll, and vascular tissues of the leaf. In many plants, leaf morphology, anatomy, and orientation also can vary significantly between a young plant and an older, more mature individual. Leaves that occur on young plants are called **juvenile leaves,** whereas older foliage is referred to as **adult leaves.** These differences can relate not only to leaf size and shape but also to internal structure. In fact, it has been pointed out that in some instances, the leaves of a young plant may differ so much from those of the mature tree that they might appear to belong to different species.

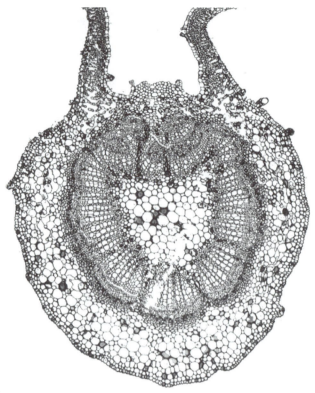

**FIGURE 3.9** Transverse of section of leaf midrib at base of lamina of *Euptelea* (Eupteleaceae). Courtesy of the Bailey-Wetmore Wood Collection, Harvard University.

The petiole of a typical foliage leaf possesses the vascular connection between leaf blade and stem. There is considerable anatomical diversity in petiolar vascularization among different angiosperms. By examining thin sections taken throughout the length of the petiole, it is possible to reconstruct the pattern of vascularization and understand the changes that occur from the level at which it connects to the stem until it becomes relatively stable in the leaf. The vascular system of the petiole can consist of numerous separate or fused vascular bundles, as well as scattered bundles and centrally positioned medullary bundles or cortical bundles. The outline of the vascular system can range from flattened to crescent shaped or circular. The location and abundance of sclerenchyma also vary.

## Lamina Structure

Considerable variation is present in the internal structure of the mature leaf blade. The outermost layers on both surfaces compose the upper (adaxial) and lower (abaxial) **epidermis.** The epidermal layers are continuous around the leaf and are usually constructed of flat, closely packed tabular cells. The outer wall of epidermal cells often are thicker than the inner wall, and the cells produce a waxy noncellular layer, the **cuticle,** over both upper and lower leaf surfaces. Gases are exchanged through pores (or **stomata**), in the epidermis. Each

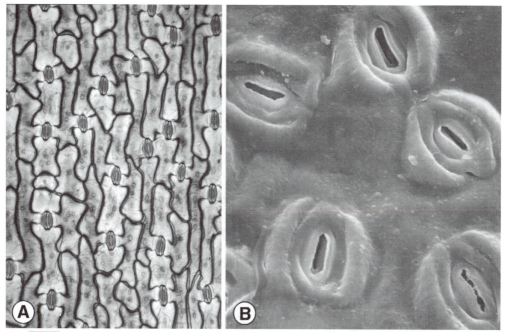

**FIGURE 3.10** Leaf epidermis structure. (A) Surface view of lower epidermis of *Tulipa* (Liliaceae) showing epidermal cells and stomata. (B) SEM view of lower epidermis of *Caraipa densifolia* (Clusiaceae) showing stomata surrounded by outer stomatal rims. × ca. 2000.

stoma consists of a pore and pair of guard cells, and the total stomatal complement is most commonly restricted to the lower surface of the lamina (**hypostomatous**). Stomata occasionally are located on both the upper and lower epidermis (**amphistomatous**), or they can be confined to the upper surface (**epistomatous**). Stomata are generally absent immediately over the leaf venation and in some cases fail to develop over certain underlying mesophyll cells. Guard cells have chloroplasts, whereas other epidermal cells typically lack them.

All the interior cells between the two outer layers compose the **mesophyll** region of the lamina, with the exception of the vascular bundles. The mesophyll is largely composed of thin-walled parenchyma cells containing abundant chloroplasts distributed over or within a labyrinth of free space. The mesophyll is usually differentiated into two regions. The columnar, evenly spaced cells toward the upper epidermis, elongated at right angles to that surface, make up the **palisade region.** This arrangement maximizes the efficiency rate of photosynthesis by placing cells at the optimum angle to the incoming sun rays. The palisade region can consist of one or multiple layers and is the primary photosynthetic zone of most leaves. The **spongy region** is the zone below the palisade that consists of irregularly shaped cells with much free surface area and that extends to the lower epidermis. Leaf blades in which palisade parenchyma is located on one side and spongy parenchyma lies on the other side are termed **bifacial.** Leaves having palisade parenchyma on both

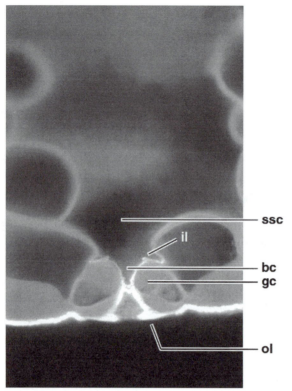

**FIGURE 3.11**   Transverse section of stoma of *Helleborus niger* (Ranunculaceae). Abbreviations: bc, back cavity; gc, guard cell; il, inner ledge; ol, outer ledge; ssc, substomatal cavity.

sides of the lamina are termed **isolateral (isobilateral)**. Isobilateral leaves are often amphistomatous and have thick mesophylls in response to receiving high levels of light. Some leaves, most notably those of the family Leguminosae, possess a layer of flat **paraveinal mesophyll** between the palisade and spongy regions. This anatomically distinct zone is a single or (less commonly) a double-layered region that is composed of cells with conspicuously extended arms and that extend horizontally between the veins of the mesophyll. The absence of mesophyll cells in the vicinity of individual stomata forms **substomatal cavities** that function in the exchange of gases between the mesophyll and the atmosphere. A subdermal **hypodermis** that is composed of a single layer or multiple layers of cells that are structurally distinct from other mesophyll cells also distinguishes some taxa.

## Leaf Venation

Leaves are vascularized by a venation system that extends through the petiole in an arrangement that is often taxon specific. Within the lamina the vasculature divides into one of several major venation patterns. Although the variation in leaf venation has been extensively cataloged, little is known about the functional significance of the various major venation types. Veins are primar-

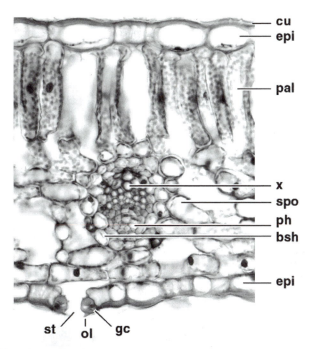

**FIGURE 3.12** Transverse section of leaf of *Helleborus niger* (Ranunculaceae). Abbreviations: bsh, vascular bundle sheath; cu, cuticle; epi, epidermis; gc, guard cell; ol, outer ledge; pal, palisade cell; ph, phloem; spo, spongy parenchyma; st, stoma; x, xylem.

ily cylindrical bundles of vascular tissue that occur mainly in the median plane of the mesophyll. Among dicotyledons, venation patterns consist of one or more branched **major veins** and a more or less dense network of smaller **minor veins** that frequently terminate in freely ending **veinlets** within the **areoles** (mesophyll islets bounded by veins of higher order). If freely ending veinlets are absent, the venation pattern is termed closed. Veinlets can be branched or unbranched and end in one or two tracheids of normal appearance, or the terminal elements can be greatly enlarged or thick-walled and sclerotic. Leaf veins are collateral in most plants, composed of adaxially positioned xylem and abaxially situated phloem. The xylem of moderately sized bundles con-

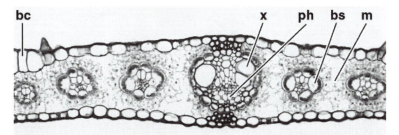

**FIGURE 3.13** Transverse section of leaf of *Zea mays*. Note conspicuous bundle sheath cells. Abbreviations: bc, bulliform cell; bs, bundle sheath; m, mesophyll; ph, phloem; x, xylem.

**FIGURE 3.14**   SEM view of lower leaf epidermis of *Caraipa punctulata* (Clusiaceae) showing papillate epidermal cells and epicuticular wax. × 600.

sists of a limited number of protoxylem and metaxylem cells. Phloem contains protophloem and metaphloem elements and in some grasses both thin-walled and thick-walled lignified sieve tubes are present, a feature believed to assist in diverting solutes from the xylem. In smaller veins protoxylem may be absent.

Vein orders are distinguished on the basis of vein size, with secondary veins branching from the primary vein, tertiary veins originating from secondaries, and so on. Secondary veins may terminate at the leaf margin or join with adjacent secondaries to form a series of closed marginal loops. In many leaves, the major veins are surrounded by extensive parenchymatous or supporting cells that project from the leaf surface as a ridge. An unusual system of mature disjunct foliar veins is present in several Hawaiian species of *Euphorbia*. In these plants, the minor leaf venation differentiates disjunctively, resulting in a pattern of isolated vein islets in which each short vein segment is surrounded by bundle sheath cells.

Monocot leaf venation is usually characterized by a series of longitudinal veins extending in a parallel and equidistant manner through the blade. The longitudinal veins can be of different size classes, and they characteristically converge and unite at the leaf apex. Notable exceptions to the generalized parallel pattern of monocot leaf venation occur. Reticulate veined leaves with freely

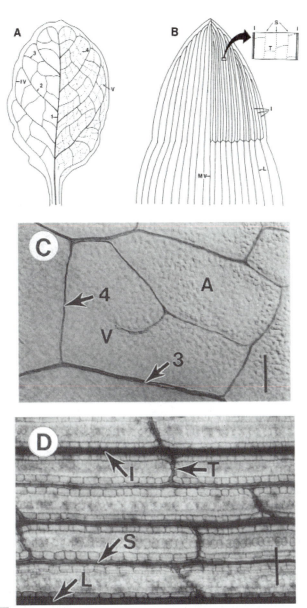

**FIGURE 3.15** Comparison of vascular pattern in mature leaves of *Arabidopsis* and *Zea mays*. (A) Diagram of vascular pattern in a rosette leaf of the dicotyledonous plant *Arabidopsis*. Note midvein (1), secondary veins (2), which are joined by an intramarginal vein (IV), tertiary (3), and quaternary (4) minor veins and freely ending veinlets (V). (B) Diagram of vascular pattern in seedling leaf of maize (*Zea*). Note midvein (MV), large (L), intermediate (I), and small (S) longitudinal veins and transverse veins (T). (C) Photograph of cleared *Arabidopsis* leaf. Note tertiary (3) and quaternary (4) minor veins surrounded an areole (A), and freely ending veinlets (V); bar = 100 μm. (D) Cleared maize leaf showing large (L), intermediate (I), and small (S) longitudinal veins and transverse (T) veins, bar = 100 μm. Reprinted with permission from Nelson and Dengler (1997), *Plant Cell* **9**, 1121–1135. Copyright © American Society of Plant Physiologists.

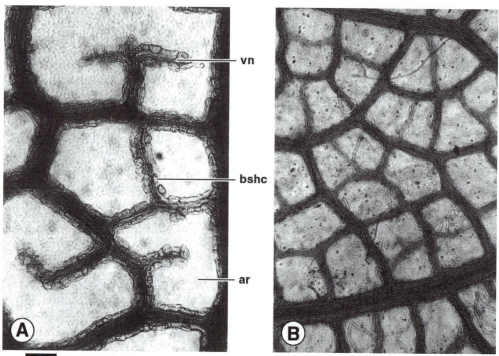

**FIGURE 3.16**   Minor leaf venation. (A) *Eucryphia milliganii* (Eucryphiaceae) showing areolation and veinlets. (B) *Dillenia papuana* (Dilleniaceae) showing closed vein pattern with regular areolation and a near absence of free vein endings. Abbreviations: ar, areole; bshc, vascular bundle sheath cell; vn, veinlet.

ending veinlets are found in members of such monocot families as Araceae, Dioscoreaceae, Smilacaceae, Alismataceae, Trilliaceae, and Orchidaceae, as well as other groups. Although the major veins in monocot leaves are interconnected by transversely oriented cross veins (**commissural veins**), the entire vascular network typically lacks the free vein endings that are characteristic of the leaves of dicotyledons.

Leaf veins are generally encircled by specialized **bundle sheath cells** that may or may not possess chloroplasts. These cells are physiologically important because they serve as flow channels between the veins and surrounding mesophyll. The bundle sheath may extend to the upper and lower epidermal layers and provide additional mechanical support for the blade. In large veins, sclerenchyma fibers can form all or part of the bundle sheath. In some leaves the major veins are surrounded by a double bundle sheath. In these cases the inner sheath, or **mestome**, is often composed of sclerenchymatous elements that have functionally important plasmodesmatal connections with the metaphloem sieve elements, especially in the small longitudinal vascular bundles. These cells also can possess suberized lamellae in their walls. The outer sheath is parenchymatous. It has been shown that in C$_4$ grasses, the mestome and parenchymatous sheaths have different developmental origins. The inner mestome sheath is procambial in origin, whereas the outer parenchymatous sheath originates in the ground meristem. It has been suggested that the suber-

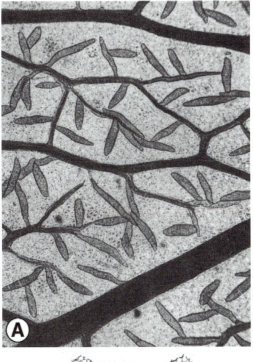

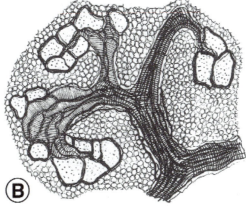

**FIGURE 3.17** Vein sheath structure and specialized vein endings. (A) *Xanthophyllum ellipticum* (Polygalaceae) showing enlarged tracheoid cells as terminal and subterminal tracheary elements. × 45. (B) *Pancheria hirsuta* (Cunoniaceae) showing vein endings with clusters of thick walled, pitted sclereids. × 55.

ized lamellae in mestome cell walls function in preventing water and solute leakage, forcing movement through plasmodesmata.

The development of an endodermis in leaves of angiosperms is uncommon. Among ferns a foliar endodermis is common, but it is less common in gymnosperms. A foliar endodermis, as identified by the presence of a Casparian strip, occurs mostly within a limited number of herbaceous flowering plants, especially legumes. The presence of an endodermal layer within the leaves of

woody angiosperms is rare. Etiolation has been reported to induce the formation of a Casparian strip in stems and petioles of some herbaceous taxa. Within
the few angiosperms that possess a foliar endodermis, the endodermal layer is
most frequently encountered surrounding the vascular tissue of the petiole, but
only rarely as a component of the vascular bundle sheathing in the lamina.

Among some monocotyledons, either fibers may completely surround a
vascular bundle and form a thick sheath that does not extend to the epider-
mises, or sclerenchymatous elements may form a conspicuous **girder** that
forms a connection between a vascular bundle and one or both of the epidermal layers. Incomplete girders may be present only as caps on the vascular
bundles without extending to the surface. A **strand** is a collection of fibers that
are free from the monocot vascular bundle sheath and often subepidermal.
Viewed in transverse section, leaf fiber patterns represent a major systematic
character in some monocotyledonous families. For example, considerable
emphasis has been placed upon the systematic importance of three main types
of sclerenchyma patterns associated with the vascular bundles in the leaves of
Velloziaceae. In this family the sclerenchyma typically takes the form of a Y-
shaped or inverted Y–shaped girder.

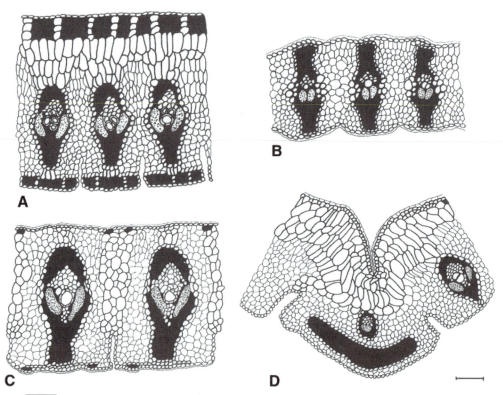

**FIGURE 3.18** Transverse sections of leaves of Velloziaceae showing a few of the many patterns of
distribution, types of sclerenchyma girders and strands (black areas), as well as the arrangement of mesophyll tissue. (A) *Vellozia swallenii.* (B) *Barbacenia fragrans.* (C) *Vellozia alata.* (D) *Vellozia resinosa.* Scale bar
= 100 μm. From Ayensu (1974), *Smithsonian Contributions to Botany* **15.**

## Gymnosperm Leaves

The leaves of *Pinus* are needle shaped and borne in bundles, or **fascicles,** of one to five. The number of needles in a fascicle aids in distinguishing species. Thus, there are one-needle, two-needle, three-needle, four-needle, and five-needle pines. In transverse section, the leaves of the pine show an epidermis with thick walls and cuticle. Stomata are sunken below the surface and often are overarched by cuticular ridges. Prominent subsidiary cells surround the guard cells. A subepidermal layer of hypodermal sclerenchyma is present. *Pinus* is unusual among conifers in that the mesophyll tissue is composed of lobed cells or cells with enfolded wall ridges that project into the cell interior. In transverse view, the mesophyll appears compact, but longitudinal sections reveal prominent intercellular spaces. The mesophyll tissue in pine is not divided into palisade and spongy zones, although such a condition can be seen in other Coniferales. Like some other, but not all, members of the Coniferales, resin ducts with specialized epithelial cells are distributed throughout the mesophyll of pine leaves. One or, more commonly, two vascular bundles are positioned centrally and are enclosed by an unusual tissue called **transfusion tissue.** The latter is composed of tracheids and living parenchyma cells. Transfusion tracheids are elongated or parenchymatous in shape and have thin secondary walls with bordered pits. The vascular and transfusion tissues are surrounded by a well-defined endodermis that has been reported to possess a Casparian strip.

*Ginkgo* belongs to a separate order from the conifers, the Ginkgoales, and is represented by only one species, *Ginkgo biloba.* Its leaves are flat and deciduous and have a dichotomously branched venation system. Epidermal cells have thin walls and are only slightly cutinized. There is no subepidermal sclerenchymatous layer. The stomata are somewhat depressed and are confined to the lower surface. The mesophyll on the adaxial side consists of one layer of palisadelike, lobed cells. The rest of the mesophyll can be referred to as spongy parenchyma with abundant air spaces. The mesophyll also is differen-

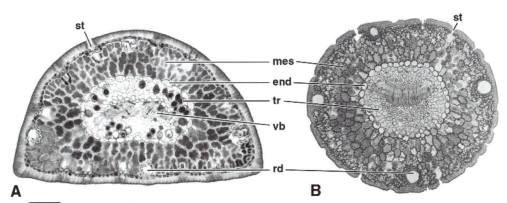

**FIGURE 3.19** Transverse section of *Pinus* leaves. (A) Scotch pine, *P. sylvestris.* (B) *P. monophylla.* Abbreviations: end, endodermis; mes, mesophyll; rd, resin duct; st, stomata; tr, transfusion tissue; vb, vascular bundle.

tiated into palisade and spongy cells in other gymnosperm genera, such as *Abies, Pseudostsuga, Dacrydium, Sequoia, Taxus, Torreya, Araucaria,* and *Podocarpus.* An indistinct, lignified endodermis surrounds each of the vascular bundles. Some transfusion tracheids are positioned along the sides of the vascular bundles. Resin ducts occur in the mesophyll, alternating with the vascular bundles.

   *Cycas* and *Zamia* belong to still a different order, the Cycadales. *Cycas* has evergreen, pinnately compound leaves with broad rigid pinnae. Both upper and lower epidermal layers are heavily cutinized in the two genera. The walls of epidermal cells are moderately thick. Subepidermal sclerenchyma is present and extends to the edges of the lamina. The guard cells and subsidiary cells are confined to the lower surface and are deep set. Palisade mesophyll occurs on the adaxial side. The lower mesophyll consists of short lobed cells, and abundant air spaces are present. The vascular tissue is associated with some transfusion tracheids. An endodermis is present but not clearly defined. In *Zamia* a sclerenchymatous bundle sheath extends above and below each vein to the epidermis.

## ROOTS

Roots serve several important functions in plants. They anchor the plant in the substratum and are the principal organ for water and mineral uptake from the surrounding environment. Roots also are important in the storage of food reserves. The roots produced in the soil by a plant are collectively called the **root system.** The root system begins its development from the embryonic radicle that grows out of the seed and forms the radially symmetrical primary root. Secondary lateral roots develop from the primary root, and each of these will in turn form new lateral roots of tertiary rank. The depth, degree of

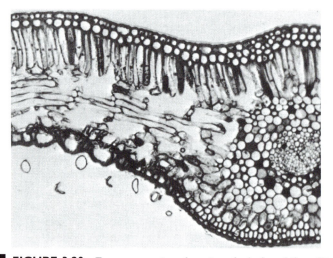

**FIGURE 3.20**  Transverse section of portion of a leaflet of *Cycas* (Cycadaceae) showing sunken guard cells with overarching subsidiary cells on the lower side and a hypodermis on the upper surface.

branching, and type of root spreading vary across species. Two morphological types of root systems are commonly distinguished: **fibrous** and **tap root systems.** Fibrous root systems are composed of large numbers of roots that are nearly equal in size. Root systems of this type are found in grasses. A tap root system is one in which the primary root, remains the largest root and a number of smaller roots are formed from it. All monocot roots are **adventitious** because they arise from a stem or leaf or older part of a root. The apex of the root is covered by a mass of loosely organized cells called the **root cap.** In addition to serving a protective function, the root cap is active in mucilage production and as a gravity-sensing structure. Just behind the root cap is the meristematic area in which new cells form and elongation follows.

Behind the root apex, discrete tissue systems differentiate in an acropetal and continuous manner from preexisting differentiated tissues beneath. These tissue regions are the uniseriate **epidermal layer,** a wide middle region of the **cortex,** and an inner core, the **stele.** In the roots of some plants, a narrow suberized **hypodermis** arises from periclinal divisions in layers beneath the protoderm. A hypodermis significantly reduces the rate of outward water movements in roots and thus prevents water loss. It is most characteristic of plants growing in xeric habitats or in roots growing close to the soil surface. In the apical region, files of protoderm and ground meristem cells become transformed into mature epidermal and highly vacuolated cortical cells. At maturity the root epidermis is a uniseriate covering composed of elongate, compact cells. Under the positive influence of the growth regulator ethylene, some individual epidermal cells, called **trichoblasts,** undergo controlled changes in cell shape and become extended as root hair cells resulting in a greatly increased absorptive area. Trichoblasts originate from an asymmetrical cell division of a protodermal cell and can be recognized from adjacent cells by their larger nuclei and nucleoli. They also contain plastids of altered structure and a distinct enzyme activity.

Immature root hair cells can be distinguished by their shorter length and delayed vacuolation as compared with ordinary epidermal initials. The root epidermis of most angiosperms, therefore, is composed of two cell types, root hair cells and nonhair cells. The root epidermis contains a two-dimensional system of patterning composed of these two cell types. Surveys of root surfaces have revealed that epidermal cells can be patterned in three basic configurations: (1) randomly positioned root hairs because any cell can potentially form a hair, (2) regularly spaced root hair cells that differentiate from a smaller product of an asymmetric cell division in a protodermal initial, and (3) root hair cells and nonhair cells that are located in alternating longitudinal files. In the experimental flowering plant *Arabidopsis thaliana,* longitudinal files of root hairs occur in which individual hair cell initials are uniformly positioned over the anticlinal walls of underlying cortical cells.

Cells of the cortex are thin-walled parenchyma that often contain starch grains. Water and dissolved solutes easily move between the cortical cells in the intercellular spaces. The innermost layer of the cortex is the **endodermis.** In their primary state, the radial walls of tangentially elongated endodermal cells are generally thought to become impregnated with ligninlike and suberin materials that are deposited in narrow bands called the **Casparian strip,** or

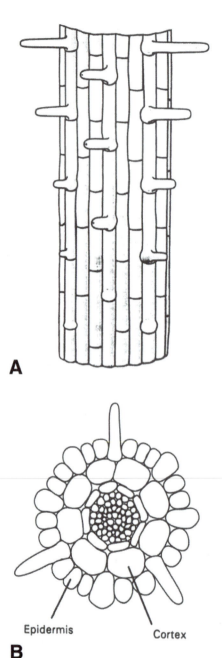

**A**

Epidermis          Cortex

**B**

**FIGURE 3.21**   Cell type determination in the epidermis of the *Arabidopsis* root. (A) Schematic drawing of an *Arabidopsis* root in surface view in the root hair developmental zone. Files of epidermal cells consist entirely of root hair cells, or hairless cells. (B) Schematic drawing of a transverse section from the mature portion of an *Arabidopsis* seedling root. Root hair cells are located in the clefts between adjacent cortical cells. Only three root hairs are shown in this figure because the length of epidermal cells generally prevents all eight root hair projections from being visible in a single transverse section. Reprinted with permission from Schiefelbein *et al.* (1997), *Plant Cell* **9**, 1089–1098. Copyright © American Society of Plant Physiologists.

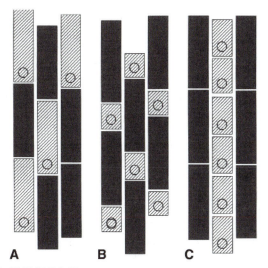

**FIGURE 3.22** Three types of root epidermal patterning in mature roots of vascular plants. Black cells are nonhair cells and hatched cells are root hair cells. The circle represents the position of the root hair base. (A) Type 1, differentiation in which any hair cell can form a root hair. (B) Type 2, differentiation in which root hair cells are the smaller product of an asymmetric cell division in the meristem. (C) Type 3, differentiation in which there are discrete files of hair cells and nonhair cells. Reprinted with permission from Dolan (1996), *Ann. Bot.* **77,** 547–553.

**Casparian band.** The Casparian strip girdles the entire endodermal cell, and at all points the plasma membrane of the endodermal protoplast is firmly attached to it. Because there are no intercellular spaces between the cells of the endodermis, and no movement can occur through the radial walls, substances entering or leaving the stele must pass through the protoplast of the endodermal cells. The endodermis thus functions to regulate the inward flow of ions and water into the central stele and to prevent outward leakage. No conclusive and satisfactory answers have been provided to the question of the precise

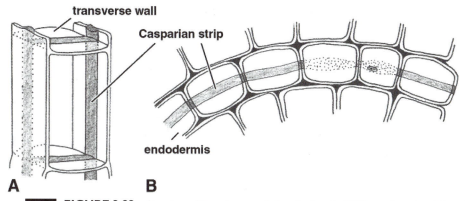

**FIGURE 3.23** Structure of the primary endodermis of roots. (A) Three-dimensional view of endodermal cell showing Casparian strip. (B) Transverse section of endodermis showing Casparian strips. Drawing courtesy of P. G. Mahlberg.

chemical nature of the Casparian strip naturally occurring in roots. From work during the last century, the Casparian strip has been thought of as a primary wall impregnated with a mixture of lignin, suberin, or cutin. Recent work has shown that the chemical nature of the strip is quite similar to that of a lignified cell wall.

The Casparian strip matures simultaneous with the maturation of the protoxylem elements in growing roots, and at maturity it is continuous from the main root to the lateral root and encloses the vascular tissue of the main and lateral roots. The Casparian strip undergoes considerable growth after its inception, and the enlargement is correlated with the development of the endodermal cell. Although the endodermis has long been considered a primary barrier to radial water movement, the developmental state of the endodermis is critical. In some plants, a small number of endodermal cells remain thin-walled and weakly suberized and are known as **passage cells.** These are typically located opposite the protoxylem poles. Traditionally, the **primary endodermis** refers to the developmental stage in which the Casparian strip is represented by a bandlike deposition within the anticlinal and periclinal endodermal walls. In a transverse section, therefore, the strip is detectable as dark and partially thickened points within the anticlinal walls. The **secondary endodermis,** by comparison, is distinguished by the deposition of a thin suberin lamella onto the inner surface of the anticlinal and periclinal walls. Developmentally, a thickened **tertiary endodermis** is present when additional layers composed of lignified wall materials are added onto the suberin lamella of the secondary endodermis. This final developmental stage can result in endodermal cells with extremely thick secondary cell walls. In many monocots, the endodermis has conspicuously thickened inner tangential and radial walls and a thin tangential wall. According to some authors, the tertiary stage of the endodermis is better interpreted as composed of two distinct developmental events. In the first of these, only cellulose is deposited into the inner surfaces of the radial and transverse walls, whereas in the final stage phenolic compounds are deposited as wall inclusions.

Some gymnosperms and a few species of angiosperms develop an additional distinct root cortical layer that contains cells with lignified wall thickenings that surround each cell in the way the Casparian strip surrounds each endodermal cell. This continuous or discontinuous layer is known as the **phi (φ), layer,** and the anticlinal wall thickenings are called **phi thickenings.** The phi layer can consist of a single cell layer or multiple cell layers and can develop in the outer region of the cortex or internally near the endodermis. In contrast to the endodermis, phi thickenings are composed of cellulose and lignins, and sometimes condensed tannins. Several functions have been suggested for the phi layer, including a mechanical supporting role to the primary root and a regulatory role in water and ion movement in the apoplast. The outermost layer of the stele is immediately internal to the endodermis and is represented by a layer of thin-walled cells called the **pericycle.** Pericyclic cells have the potential to resume active division and initiate the development of lateral roots. In older parts of the root, the pericycle cells often become sclerified.

The primary root body of dicotyledons contains a solid lobed core of xylem. Although discrete areas of vascular cells are initially produced from

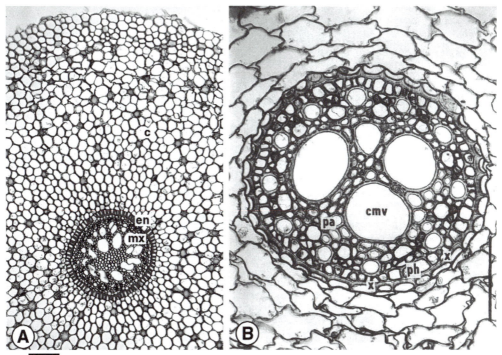

**FIGURE 3.24** Transverse sections of monocot roots. (A) Root of *Iris* (Iridaceae). Note thick-walled endodermal cells. (B) Root of barley (*Hordeum vulgare*, Gramineae). Abbreviations: c, cortex; cmv, central metaxylem vessel; en, endodermis; mx, metaxylem; pa, stelar parenchyma; ph, phloem; x, xylem. B, reprinted with permission from Luxová (1990), *Bot. Acta* **103**, 305–310. Copyright © Georg Thieme Verlag.

the procambium, further growth and differentiation of the stele in mature dicotyledonous roots result in a solid lobed core of primary xylem with alternately arranged primary phloem and associated parenchyma. Because the first-formed tracheary elements are located toward the outside next to the pericycle, and the later-formed xylem is formed internally, the root primary xylem differentiates in a centripetal direction, as does the primary phloem. Therefore, the primary xylem maturation is **exarch.** The number of protoxylem poles varies in angiosperm roots from two to many. The terms **diarch, triarch, tetrarch,** and **polyarch** are used to describe this variation in roots. Branch roots can differ from the primary root anatomically and possess just a single large metaxylem element in the center of the stele. The variation in the number of protoxylem poles is influenced by the level of growth regulatory hormones diffusing from the root apex. The primary root phloem is generally sparse in annual species and consists of slender strands of two to four sieve tube elements. Longitudinal vascular differentiation follows a strictly acropetal course, with the protophloem sieve tube elements maturing closer to the apex and continuing to divide longer than the first, fully differentiated primary xylem elements. In fast-growing roots, fully mature vascular cells are positioned more distally from the apex than slow-growing roots.

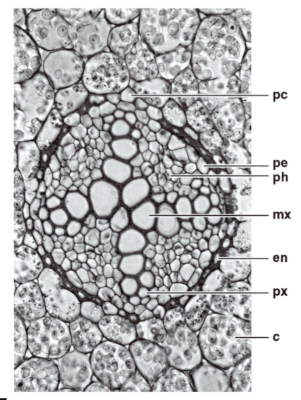

**FIGURE 3.25**    Transverse section of root of *Ranunculus repens* (Ranunculaceae). Abbreviations: c, cortex; en, endodermis; mx, metaxylem; pc, passage cell; pe, pericycle; ph, phloem; px, protoxylem.

Roots of monocotyledons typically have numerous protoxylem poles and are, therefore, **polyarch,** alternating with small phloem clusters. Smaller branch roots may deviate from this pattern. A sclerified pith also is present. In younger roots the endodermal cells have distinct Casparian strips in the form of bands, and the uniseriate pericycle is located between the endodermis and the outermost vascular elements. As monocot roots age, endodermal cells often develop thick secondary walls, and the pericycle often becomes multiseriate and thick-walled.

## Origin of Lateral Roots

An important distinction between stems and roots is found in the manner of origin of lateral organs. Leaves and axillary buds, as we have seen, arise on the surface of the shoot apex in an **exogenous** fashion. Lateral roots originate **endogenously,** deep within the parent root and at some distance removed from the root apex, in response to hormonal stimulation at the site of initiation. Because lateral roots can arise in a manner that makes water and nutrients more accessible to the plant, this aspect of plant growth has considerable economic importance because it can alter the vigor of crop and horticultural plants under changing growing conditions.

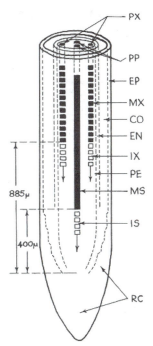

**FIGURE 3.26** Diagrammatic longitudinal view of a primary root tip of hops, *Humulus lupulus* (Moraceae), illustrating features of tissue differentiation. Abbreviations: CO, cortex; EN, endodermis; EP, epidermis; IS, immature sieve tube elements; IX, immature primary xylem; MS, mature sieve tube elements; PE, pericycle; PP, protophloem poles; PX, protoxylem poles; RC, rootcap. Reprinted with permission from Miller (1958), *Am. J. Bot.* **45**, 418–431.

Lateral roots form within the parent root pericycle. The first indication of branch root initiation is the presence of anticlinal divisions and increased cytoplasmic density in specific cells of this cell layer. In contrast to most cells, pericycle cells are arrested in the $G_2$ phase of the cell cycle prior to the completion of cell division during root initiation. Lateral roots generally arise and are arranged in linear rows opposite the protoxylem poles of the parent root, or opposite the phloem poles in some monocotyledons. The early anticlinal and subsequent periclinal divisions of the pericycle are highly programmed and ordered, and cells of the lateral root primordium assume their identities very early. During early ontogeny the primordium possesses a small, rudimentary apex that soon becomes organized into an apical meristem that resembles a mature root tip in organization. The young lateral root apical meristem usually does not become fully active until after the developing root has emerged from the parent root. Emergence of the lateral root follows its outward passage through the overlying endodermis and cortex, eventually rupturing the epidermis of the main root. The penetration and emergence of the primordium appears to occur primarily through cell expansion rather than apical cell division. Shortly after emergence of the lateral root, the immature vascular elements complete differentiation, and the lateral root becomes connected to the parental xylem and phloem. It is of interest that bare steles devoid of cortex are able to initiate and sustain new lateral roots from the pericycle, indicating that cortical tissue other than the endodermis is not

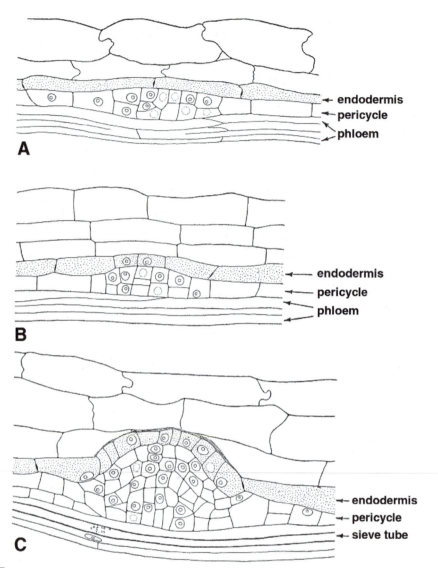

**FIGURE 3.27**   (A, B, C) Three successive stages of development of a lateral root in *Daucus carota* as seen in longitudinal section of a young tap root. × 421. Reprinted with permission from Esau (1940), *Hilgardia* **13**, 175–226. Copyright © Regents, University of California.

necessary for secondary and tertiary root development and that the endodermis is effective in preserving stelar functions. Because pericycle cells retain the potential to resume cell division, they also contribute to the development of the root vascular and cork cambia in those roots with secondary growth.

## Mycorrhizae

Most higher plants, both woody and herbaceous species, have fungi associated with their roots. The varieties of associations between fungi and roots are

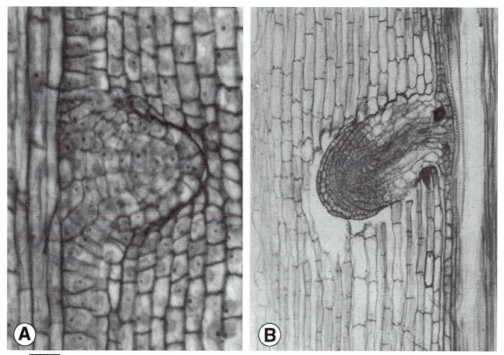

**FIGURE 3.28**   Lateral root development. (A) Longitudinal section through young branch root of *Pontederia cordata* (Pontederiaceae). (B) Longitudinal section through young lateral root of *Typha glauca* (Typhaceae). Courtesy of J. Seago.

called **mycorrhizae,** literally "fungus root," and represent classic examples of mutualistic symbiotic relationships in which both partners in the association derive some benefit. As a result of the extensive soil and fungal interface, the plant is benefited nutritionally by an increased surface receiving dissolved macro- and micronutrients such as phosphorus and nitrogen. Achlorophyllous plants may receive all their organic and inorganic requirements from the fungus. Mycorrhizae also are reported to increase plant resistance to diseases, drought, salinity, acid rain pollution, and other stresses. The fungus obtains most of its food supply from the plant in the form of carbohydrates and vitamins. In these associations the fungus takes over the absorptive function of root hairs; therefore, the formation of root hairs may be suppressed. The success of these relationships is evidenced by the fact that mycorrhizal symbioses are present in all but a few dicotyledons and monocotyledons, as well as in gymnosperms and lower vascular plants. In some communities more than 90% of the species are mycorrhizal. In addition to their contributions to the adult plant, mycorrhizal fungi also are essential for seed germination and seedling growth in some plants, for example, members of the Orchidaceae and many tree species. Mycorrhizae play an important role in improving the quality of nursery stock and agricultural crops and in regenerating forests. Many tree species cannot establish themselves unless a suitable fungal partner is present in the soil.

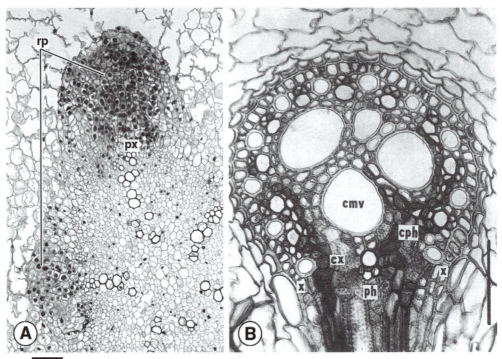

**FIGURE 3.29** Origin of lateral roots. (A) Transverse section of young root of *Vicia faba* (Leguminosae) showing two stages in the development of lateral roots. Note that the branch roots arise opposite the protoxylem poles. (B) Transverse section of root of barley (*Hordeum vulgare*, Gramineae) showing vascular connection with lateral root. Abbreviations: cmv, central metaxylem vessel; cph, connecting phloem; cx, connecting xylem; ph, phloem; px, protoxylem; rp, root primordium; x, xylem. B, reprinted with permission from Luxová (1990), *Bot. Acta* **103**, 305–310. Copyright © Georg Thieme Verlag.

A large number of species (representing all of the major groups of fungi) are capable of forming mycorrhizae in a host plant. Many mycorrhizal associations are nonspecific because a single species of fungus can colonize different plant species. The fungal composition of an association also can change over time. On descriptive and nutritional bases, mycorrhizae can be divided into three broad categories: **ectomycorrhizae, endomycorrhizae,** and **ectendomycorrhizae.** Ectomycorrhizae (ectotrophic mycorrhizae) are characterized by fungi forming a compact enclosing envelope, or hyphal mantle, around the rootlet. The fungal hyphae also penetrate intercellular regions of the root cortex by degrading the middle lamella of cortical cells at the growing hyphal tip. The cortical mycelial network is termed a **Hartig net.** Ectomycorrhizae are especially common on the roots of woody plants, but they can also occur in herbaceous species. Endomycorrhizae (endotrophic mycorrhizae) are distinguished by fungi living within root cortical cells. The fungus can also grow intercellularly but never forms a Hartig net. Various subtypes of endomycorrhizae are recognized; these are especially prominent in some members of the Ericales and Orchidaceae. Ectendomycorrhizae (ectendotrophic mycorrhizae) combine both intercellular and intracellular cortical hyphae but sometimes lack a mantle.

The most common type of endomycorrhizal symbiosis is referred to as an **arbuscular mycorrhiza,** or **vesicular arbuscular mycorrhiza.** This form of root

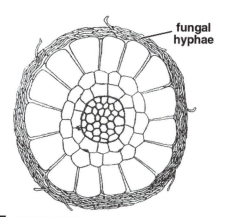

**FIGURE 3.30**   Diagrammatic view of root transection showing ectomycorrhiza.

and fungus association has been estimated to be present in approximately 80% of flowering plants. Arbuscular mycorrhizae originate when filamentous fungal hyphae penetrate between neighboring epidermal cells. Having entered the cortex, the fungal mycelium spreads axially in both directions within the cortex. The hyphae grow both intracellularly and intercellularly and eventually reach the inner region of the cortex. The fungus does not colonize the root vascular tissue or meristematic cells. Within the inner cortex, fungal hyphae enter cell walls and proliferate inside the cortical cells to form highly branched, terminal filaments termed **arbuscules.** The entry of the hyphal strand does not break the cortical cell plasma membrane that assumes the contour of the developing arbuscule. The movement of phosphate begins with uptake by external hyphae and subsequent transport to the internal portion of the mycelium. The large surface area of the membrane and fungal cell arbuscular interface is believed to function in the bidirectional movement of phosphate and carbon between plant and fungus. As the fungus enters the cortical cell, important cellular changes occur in these cells. These alterations include fragmentation of the vacuole, a major increase in the volume of the cytoplasm, and increases and changes in the number and structure of such cellular organelles as cytoskeleton components.

## Root and Bacteria Interactions

The roots of certain seed plants, chiefly legumes such as peas, beans, clover, and alfalfa, have small swellings or **nodules** on their roots within which certain species of nitrogen-fixing bacteria live. The bacteria secure food and a place to live from the host plant, and the plant in turn obtains nitrogen that has been fixed by the bacteria. Because nitrogen is the major limiting nutrient for most plant species, the availability of nitrogen stimulates plant growth. It is because of this symbiotic relationship that leguminous plants are grown to enrich the soil.

The bacteria living within the nodules take atmospheric nitrogen and combine it with other elements to form nitrogenous compounds that can be utilized by the host plant. The process of converting nitrogen directly from the

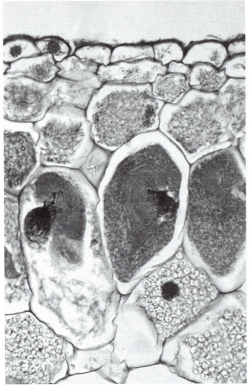

**FIGURE 3.31**    Transverse section of root of *Neottia nidus-avis* (Orchidaceae) showing endomycorrhizal fungi in the cortical cells.

air into compounds in which nitrogen is combined with other elements in the soil is termed **nitrogen fixation.** The principal nodule-forming nitrogen-fixing bacterium belongs to the genus *Rhizobium;* it can fix nitrogen only when in the nodules.

Nodule formation begins with the *Rhizobium* bacterium penetrating the rootlets of the host plant. This event is sometimes preceded by a deformation or curling of root hairs. Within the root the bacteria divide and ultimately reach and enter the cytoplasm of the cortical cells. Simultaneous with the release of bacteria into the cytoplasm, clusters of plant-derived membranes form that separate the bacteria from the cortical cell cytoplasm. These **peribacteroid membranes** act as selectively permeable barriers for the exchange of nutrients. In the next stage of development, the cortical cells are stimulated to begin abnormal growth and division to form the highly organized processes known as nodules within which the bacteria become established. In some cases these nodules can reach several millimeters in diameter.

Nodules are classified into two major groups based on their shape, meristematic activity within the nodule, and fixed nitrogen transport products. The first group contains plants with nodules that are elongate cylindrical, with indeterminate or persistent apical meristem activity. They transport fixed nitrogen as amides. Mature **indeterminate type nodules** have multiple bacte-

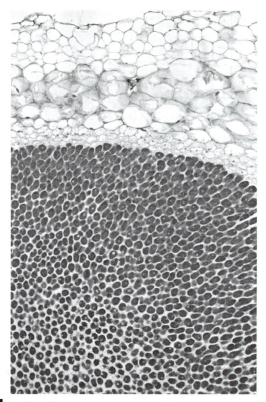

**FIGURE 3.32** Transverse section of root nodule of *Lupinus luteus* (Leguminosae) showing nodule bacteria.

ria enclosed within a single peribacteroid membrane. This category includes alfalfa, peas, and clover. The second group possesses spherical nodules with determinate or nonpersistent internal meristematic activity. They transport fixed nitrogen as ureides. In **determinate type nodules** individual bacteria are surrounded by a peribacteroid membrane. This group includes soybeans and the common pea. Meristem activity is initiated prior to bacterial entry.

In determinate nodulating plants, cell division in all planes begins in the outer root cortical cells and continues for several days during the early stages of nodule development. Subsequent cell enlargement produces the swollen mature nodule, within which no additional meristematic activity occurs. The meristem of indeterminate nodules, in contrast, arises within the inner cortex, and meristem activity persists for several months or years. Cell division in a single plane is followed by cell enlargement to form an elongate, cylindrical nodule. In both types, nodule primordia only form opposite the protoxylem poles.

## SUMMARY

This chapter describes the details of internal organization and arrangement of cells and tissues within the three major organs of the primary plant body—the stem, the leaf, and the root. All stems are characterized by repetitive sequences of external nodes and internodes. The leaves and associated axillary buds are positioned at the nodal regions. In herbaceous shoots the vascular tissues are organized into discrete strands called vascular bundles. On the basis of the position of the xylem and phloem within individual bundles, various types of vascular bundles can be recognized. A vascular bundle in the stem that is continuous with a vascular bundle of a leaf base is called a leaf trace. Leaf traces are associated with parenchymatous regions in the stem vascular system known as leaf gaps. Each axial stem bundle and its associated leaf trace is referred to as a sympodium. Among gymnosperms and dicotyledonous angiosperms, the vascular bundles in the stem are arranged roughly in a ring near the periphery of the stem. This structural pattern is a eustele. Many monocotyledons possess vascular bundles that appear widely spaced and randomly scattered throughout the fundamental tissue of the stem. This condition is referred to as an atactostele. The numerous vascular bundles of monocotyledons tend to be crowded toward the stem periphery and form a highly organized vascular network.

Leaves are initiated as lateral protuberances on the domelike stem apex and occur in a precise arrangement, or phyllotaxis, on the stem. The mature leaf is composed of a leaf base, a stalklike petiole, and an expanded, flattened blade or lamina. Some taxa have small, leaflike structures called stipules, positioned between the petiole and stem. The outermost layers on both surfaces of the blade constitute the upper and lower epidermis. With the exception of the vascular bundles, all the interior cells between the two outer layers make up the mesophyll region. The mesophyll itself is usually differentiated into two regions. The columnar cells toward the upper epidermis make up the palisade region. Below the palisade is a zone composed of irregularly shaped cells with much free surface area, called the spongy region. Some leaves have a subdermal hypodermis that is composed of one or more layers. Among dicotyledons, venation patterns consist of one or more branched major veins and a dense network of smaller minor veins that terminate in freely ending veinlets. Monocot leaf venation is usually characterized by a series of longitudinal veins extending in parallel through the blade and interconnected by transversely oriented cross veins. Leaf veins are generally encircled by specialized bundle sheath cells of various structure. The leaves of gymnosperms are structurally diverse. Many conifer leaves have resin ducts and an unusual tissue called transfusion tissue composed of tracheids and living parenchyma cells.

The roots of a plant are collectively called the root system. Roots are composed of three primary tissue regions—the outermost epidermal layer, a wide middle region called the cortex, and an inner core or stele. The innermost layer of the cortex is the endodermis, composed of cells with ligninlike wall material deposited in bands called the Casparian strip. The outermost layer of the stele is immediately internal to the endodermis and is represented by a layer of thin-walled cells called the pericycle. Pericyclic cells have the potential to resume active cell division and to initiate the endogenous development of lateral roots. The primary root of dicotyledons contains a solid lobed core of vascular tissue. The root primary xylem differentiates in a centripetal direction, as does the primary phloem. Roots of monocotyledons typically have numerous primary xylem regions and are referred to as polyarch.

Most higher plants have fungi associated with their roots. The varieties of associations between fungi and roots are called mycorrhizae and represent classic examples of mutualistic symbiotic relationships. The roots of certain seed plants, chiefly legumes, have small nodules on their roots in which certain species of nitrogen-fixing bacteria live. The bacteria secure food and a place to live from the host plant, and the plant in turn obtains nitrogen that has been fixed by the bacteria.

## ADDITIONAL READING

1. Ayensu, E. S. (1974). Leaf anatomy and systematics of New World Velloziaceae. *Smithsonian Contrib. Bot.* **15**, 1–125 p.
2. Biebl, R., and Germ, H. (1950). "Praktikum der Pflanzenanatomie." Springer-Verlag, Wien.
3. Boureau, E. (1954–1957). "Anatomie végétale." 3 vols. Presses Universitaire de France, Paris.
4. Chilvers, G. A., and Pryor, L. D. (1965). The structure of eucalypt mycorrhizas. *Aust. J. Bot.* **13**, 245–249.
5. Cohn, J. R., Day, B., and Stacey, G. (1998). Legume nodule organogenesis. *Trends Plant Sci.* **3**, 105–110.
6. Cutler, D. F. (1978). "Applied Plant Anatomy." Longman, London, New York.
7. Cutter, E. G. (1971). "Plant Anatomy: Experiment and Interpretation. Part 2. Organs." Addison-Wesley, Reading, Massachusetts.
8. Cutter, E. G., and Feldman, L. J. (1970). Trichoblasts in *Hydrocharis*. I. Origin, differentiation, dimensions and growth. *Am. J. Bot.* **57**, 190–201.
9. DeBary, A. (1884). "Comparative Anatomy of the Vegetative Organs of the Phanerograms and Ferns" (Bower and Scott, Engl. trans.). Clarendon Press, Oxford.
10. Dolan, L. (1996). Pattern in the root epidermis: An interplay of diffusible signals and cellular geometry. *Ann. Bot.* **77**, 547–553.
11. Dormer, K. J. (1945). An investigation of the taxonomic value of shoot structure in angiosperms with special reference to Leguminosae. *Ann. Bot. n. s.* **9**, 141–153.
12. Eames, A. J., and MacDaniels, L. H. (1947). "Introduction to Plant Anatomy," 2nd ed. McGraw-Hill, New York.
13. Engard, C. J. (1944). Morphological identity of the velamen and exodermis in orchids. *Bot. Gaz.* **105**, 457–462.
14. Esau, K. (1938). Ontogeny and structure of the phloem of tobacco. *Hilgardia* **11**, 343–424.
15. Esau, K. (1940). Developmental anatomy of the fleshy storage organ of *Daucus carota*. *Hilgardia* **13**, 175–226.
16. Esau, K. (1943). Ontogeny of the vascular bundle in *Zea mays*. *Hilgardia* **15**, 327–368.
17. Esau, K. (1977). "Anatomy of Seed Plants," 2nd ed. John Wiley & Sons, New York.
18. Fahn, A. (1982). "Plant Anatomy," 3rd ed. Pergamon Press, Oxford, New York.
19. Foster, A. S. (1947). "Practical Plant Anatomy," 2nd ed. Van Nostrand Co., New York.
20. French, J. C., and Tomlinson, P. B. (1986). Compound vascular bundles in Monocotyledonous stems: construction and significance. *Kew Bull.* **41**, 561–574.
21. Fukaki, H., Wysocka-Diller, J., Kato, T., Fujisawa, H., Benfey, P. N., and Tasaka, M. (1998). Genetic evidence that the endodermis is essential for shoot gravitropism in *Arabidopsis thaliana*. *Plant J.* **14**, 425–430.
22. Haas, D. L., Carothers, Z. B., and Robbins, R. R. (1976). Observations on the phi-thickenings and Casparian strips in *Pelargonium* roots. *Am. J. Bot.* **63**, 863–867.
23. Harley, J. L., and Smith, S. E. (1983). "Mycorrhizal Symbiosis." Academic Press, San Diego.
24. Harrison, M. J. (1997). The arbuscular mycorrhizal symbiosis: An underground association. *Trends Plant Sci.* **2**, 54–60.

25. Hayward, H. E. (1938). "The Structure of Economic Plants." Macmillan and Co., New York.
26. Herbst, D. (1971). Disjunct foliar veins in Hawaiian *Euphorbias. Science* 171, 1247–1248.
27. Hirsch, A. M. (1992). Developmental biology of legume nodulation. *New Phytol.* **122,** 211–237.
28. Jeffrey, E. C. (1917). "The Anatomy of Woody Plants." University of Chicago Press, Chicago.
29. Karahara, I. and Shibaoka, H. (1994). The Casparian strip in pea epicotyls: Effects of light on its development. *Planta* 192, 269–275.
30. Kaussmann, B. (1963). "Pflanzenanatomie." Gustav Fischer Verlag, Jena.
31. Korsmo, E. (1954). "Anatomy of weeds; Anatomical descriptions of 95 weed species with 2050 original drawings." Grondahl and Sons, Oslo.
32. Leavitt, R. G. (1904). Trichomes of the root in vascular cryptogams and angiosperms. *Proc. Boston Soc. Nat. Hist.* **31,** 273–313.
33. Lersten, N. R. (1997). Occurrence of endodermis with a Casparian strip in stem and leaf. *Bot. Rev.* **63,** 265–272.
34. Luxová, M. (1990). Effect of lateral root formation on the vascular pattern of barley roots. *Bot. Acta* **103,** 305–310.
35. MacKenzie, K. A. D. (1979). The development of the endodermis and phi layer of apple roots. *Protoplasma* **100,** 21–32.
36. Malamy, J. E., and Benfey, P. N. (1997). Down and out in *Arabidopsis:* The formation of lateral roots. *Trends Plant Sci.* **2,** 390–396.
37. Marks, G. C., and Kozlowski, T. T. (1973). "Ectomycorrhizaae." Academic Press, New York.
38. Mauseth, J. D. (1988). "Plant Anatomy." Benjamin/Cummings, Menlo Park, California.
39. Nelson, T., and Dengler, N. (1997). Leaf vascular pattern formation. *Plant Cell* **9,** 1121–1135.
40. Peterson, R. L., and Farquhar, M. L. (1996). Root hairs: Specialized tubular cells extending root surfaces. *Bot. Rev.* **62,** 1–40.
41. Peterson, R. L., and Peterson, C. A. (1986). Ontogeny and anatomy of lateral roots. *In* "New Root Formation in Plants and Cuttings" (M. B. Jackson, Ed.), pp. 2–30. Martinus Nijhoff, Dordrecht, the Netherlands.
42. Pizzolato, T. D., and Heimsch, C. (1975a). Ontogeny of the protophloem fibers and secondary xylem fibers within the stem of *Coleus.* I. A light microscope study. *Can. J. Bot.* **53,** 1658–1671.
43. Pizzolato, T. D., and Heimsch, C. (1975b). Ontogeny of the protophloem fibers and secondary xylem fibers within the stem of *Coleus.* II. An electron microscope study. *Can. J. Bot.* **53,** 1672–1697.
44. Prance, G. T. (1985). "Leaves: The Formation, Characteristics and Uses of Hundreds of Leaves Found in All Parts of the World." Crown Publishers, New York.
45. Pratikakis, E., Rhizopoulou, S., and Psaras, G. K. (1998). A phi layer in roots of *Ceratonia siliqua* L. *Bot. Acta* **111,** 93–98.
46. Roth, I. (1990). "Leaf Structure of a Venezuelan Cloud Forest," Encyclopedia of Plant Anatomy, 14. Borntraeger, Berlin.
47. Rudall, P. (1992). "Anatomy of Flowering Plants. An Introduction to Structure and Development," 2nd ed. Cambridge University Press, Cambridge.
48. Sachs, T. (1968). On the determination of the pattern of vascular tissue in peas. *Ann. Bot.* **32,** 781–790.
49. Schiefelbein, J. W., Masucci, J. D., and Wang, H. (1997). Building a root: The control of patterning and morphogenesis during root development. *Plant Cell* **9,** 1089–1098.
50. Schreiber, L., Breiner, H.-W., Riederer, M., Duggelin, M., and Guggenheim, R. (1994). The Casparian strip of *Clivia miniata* Reg. roots: Isolation, fine structure and chemical nature. *Bot. Acta* **107,** 353–361.
51. Spaeth, S. C., and Cortes, P. M. (1995). Root cortex death and subsequent initiation and growth of lateral roots from bare steles of chickpeas. *Can. J. Bot.* 73, 253–261.
52. Stover, E. L. (1951). "An Introduction to the Anatomy of Seed Plants." Heath and Co., Boston.
53. Tomlinson, P. B. (1984). Development of the stem conducting tissues in monocotyledons. *In* "Contemporary Problems in Plant Anatomy" (R. A. White and W. C. Dickison, Eds.), pp. 1–51. Academic Press, San Diego.
54. Tomlinson, P. B. (1990). "The Structural Biology of Palms." Clarendon Press, Oxford.

55. Tomlinson, P. B., and Zimmerman, M. H. (1967). The "wood" of monocotyledons. *IAWA Bull.* **2,** 4–24.

56. Waisel, Y., Eshel, A., and Kafkafi, U. (Eds). (1991). "Plant Roots. The Hidden Half." New Marcel Dekker, New York, Basel, Hong Kong.

57. Wilson, K. (1936). The population of root-hairs in relation to the development of the piliferous layer. *Ann. Bot.* **50,** 121–154.

58. Zeier, J., and Schreiber, L. (1998). Comparative investigation of primary and tertiary endodermal cell walls isolated from the roots of five monocotyledonous species: Chemical composition in relation to fine structure. *Planta* **206,** 349–361.

# 4
## ORIGIN AND STRUCTURE OF THE SECONDARY PLANT BODY

All gymnosperms and most dicotyledons, both large herbs and woody forms such as shrubs and trees, develop lateral meristems, which result in radial growth and are responsible for a continuous increase in the width of stems and roots. The growth that is derived from the lateral meristems is called **secondary growth.** Secondary growth is composed of **secondary tissues,** which form the **secondary plant body** as a result of a specific and rather complicated program of cell growth. The **vascular cambium** is a bifacial meristem, which adds to the girth of the stem and root axis in arboreal dicotyledons and gymnosperms by the production of **secondary xylem** internally and **secondary phloem** externally. Comparable cambia do not arise in monocotyledons. The vascular cambium varies greatly in its activities during different seasons of the year, in different plants, and on different parts of the same plant. In some tropical trees the cambium is more or less active the year around, although in tropical climates characterized by wet and dry periods cambial activity can be seasonal. In temperate regions the cambium is dormant for a period of time throughout the year. The resumption of cambial growth is associated with the renewed activity of buds and the formation of new leaves. Thus, new secondary growth is propagated downward in the stem, beginning at the buds. In temperate trees the dormant winter cambium is composed of thicker walled cells as compared to actively dividing cambial cells. As the cambium becomes reactivated, the cambial cells exhibit both a change in color and a slight swelling. A correlation between the reactivation of the vascular cambium and an increase in the hormone auxin has long been observed. It was long assumed that the cambium is stimulated by increasing levels of auxin production in the spring of the year.

Recent evidence suggests, however, that cambial cells resume division following changes in their sensitivity to a relatively high and constant auxin level throughout the year. Phloem cells are the first vascular elements to be produced by the reactivated cambium, followed by the xylem. In fact, phloem elements can be observed to overwinter in a partially differentiated condition. Although the vascular cambium (or simply cambium) is sometimes compared with the apical meristem, it differs in its fundamental organization and produces only tissues, not entire organs. The outer protective layers of the plant axis are called the **periderm** and are formed by another cambium known as the **cork cambium** or **phellogen**. The functioning of these two cambia represents an important component of woody plant growth and development. For this reason, the processes of growth and cell division in the cambia and derivative tissues will be considered in detail.

## VASCULAR CAMBIUM FORMATION

The vascular cambium develops in the region of the shoot and root apex, a short distance behind the apical meristem, although the precise ontogeny of the cambium varies with the type of organ and the plant group under consideration. The cambium usually becomes recognizable just prior to, or immediately following, axis elongation. The stem cambium usually develops during the first year of growth from undifferentiated procambial cells located

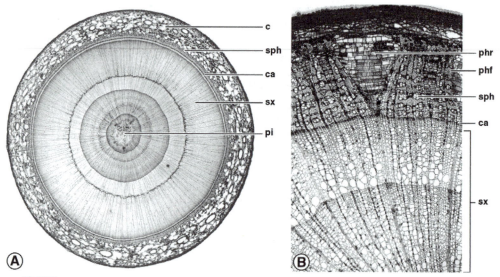

**FIGURE 4.1**   Secondary growth in stems. (A) Transverse section of stem of *Abies balsamea* (Pinaceae) with secondary tissues. (B) Transverse section of older stem of *Tilia americana* (Tiliaceae) showing vascular cambium and secondary tissues. Major tissue regions are indicated. Note conspicuous fibrous secondary phloem and dilated phloem rays resulting in wedge-shaped intrusions. Abbreviations: c, cortex; ca, vascular cambium; phf, phloem fiber; phr; phloem ray; pi, pith; sph, secondary phloem; sx, secondary xylem. A, courtesy of the Bailey-Wetmore Wood Collection, Harvard University.

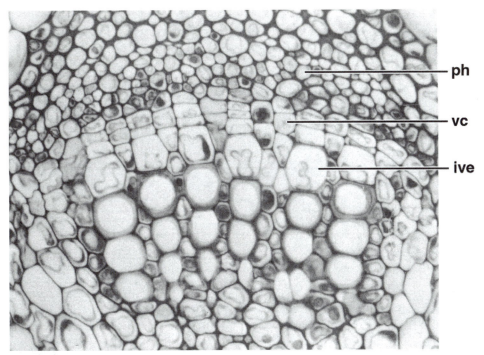

**FIGURE 4.2** Transverse section of vascular bundle from stem of *Medicago* (Leguminosae) showing vascular cambium and its early derivatives. × 550. Abbreviations: ive, immature vessel element; ph, phloem; vc, vascular cambium. Courtesy of P. G. Mahlberg.

between the primary xylem and phloem of individual vascular bundles. The cambium that forms within the vascular bundles is called the **fascicular cambium.** Somewhat later in the maturing portion of the elongating woody shoot, the **interfascicular cambium** is initiated. In stems with separate vascular bundles, this occurs from the renewed division of parenchyma cells located between the original vascular bundles. In both stems and roots, the vascular cambium differentiates in an acropetal direction (i.e., toward the apex) and is continuous with the cambium of branches. Both in stems with separate vascular bundles and in those with a continuous cylinder of primary vascular tissue, the activity of the fascicular and interfascicular cambia together produce an uninterrupted cylinder of secondary vascular tissue. Because the vascular cambium persists for the life of the plant, it continues to gain circumference after secondary growth begins. In some herbaceous dicotyledons the interfascicular cambium does not cause a woody cylinder to be formed by its activity. Instead, more bundles are formed in the middle of the medullary rays. Only in very ephemeral dicotyledonous herbs is the cambium totally nonfunctional or absent.

The formation of the interfascicular vascular cambium is an example of cellular dedifferentiation in which cells of one morphological type are transformed first into cells of another morphological type and then maintained in that state. Figure 4.6 illustrates different concepts that have been proposed to explain the induction of interfascicular cambium. **Induction** is the ability of one tissue to

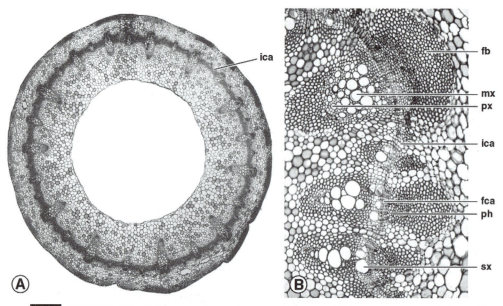

**FIGURE 4.3** Early stages of secondary growth in stems. (A) Transverse section of stem of *Sium* (Umbelliferae) showing herbaceous condition and development of interfascicular vascular cambium (ica). ×9. (B) Partial view of transverse section of stem of *Helianthus annuus* (Asteraceae) showing interfascicular cambium joined to fascicular cambium and early stage of secondary growth. Abbreviations: fb, fibers; fca, fascicular vascular cambium; mx, metaxylem; ph, phloem; px, protoxylem; sx, secondary xylem vessel. A, courtesy of the Bailey-Wetmore Wood Collection, Harvard University.

influence the development of another tissue. **Inductive effects** are the responses or changes produced within a group of cells or a tissue, when they are exposed to the influences of neighboring cells. At various times, induction of the interfascicular cambium has been attributed to the influence of the primary vascular bundles or the fascicular cambium, or to effects related to the axis surface. Evidence has accumulated showing that the position of future cambial cells and their radial polarity are determined during very early stages of development at the shoot apex, irrespective of whether they are histologically different from the surrounding parenchyma cells. This predisposition of cambial cells has been demonstrated in grafting experiments using stems of the castor bean (*Ricinus communis*). Sections of a young stem prior to the differentiation of interfascicular cambial cells can be removed and reinserted into the stem following a 180° reorientation of tissues. The inserted section will become grafted and eventually resume growth. Following the differentiation of interfascicular cambium, secondary xylem will be produced to the outside and secondary phloem to the inside, demonstrating that the cortical parenchyma to be transformed into cambial initials are developmentally predetermined. Under hormonal and other stimulation, these cells become transformed into the vascular cambium.

In some woody plants, the secondary xylem is not produced in a uniform manner around the circumference of the axis. The resulting asymmetric growth is particularly common in stems growing in exposed locations, in which the axis is subjected to strong winds. Under such conditions, the cam-

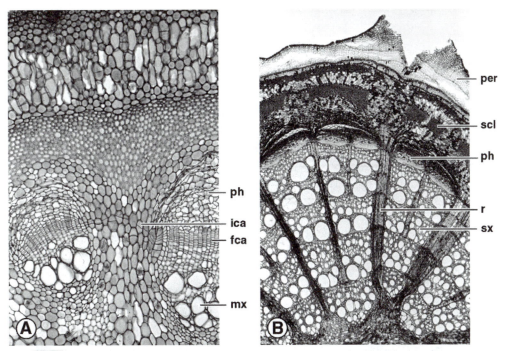

**FIGURE 4.4** Partial views of transverse sections of stem axes of *Aristolochia* (Aristolochiaceae). (A) Early stage in differentiation of interfascicular vascular cambium. The interfascicular cambium will subsequently join the fascicular cambia of both vascular bundles. (B) Late stage in secondary growth. Abbreviations: fca, fascicular vascular cambium; ica, interfascicular vascular cambium; mx, metaxylem; per, periderm; ph, secondary phloem; r, interfascicular secondary parenchymatous ray; scl, sclerenchyma; sx, secondary xylem.

bium forms more wood on the side facing away from the wind than it does on the side toward the prevailing wind. This manner of growth has the effect of strengthening otherwise weak areas of the stem and solving mechanical problems that result from uneven stress. Because the windward side of the stem is under tension and the leeward side is under compression, this pattern of cambial activity will equalize the stresses around the tree. More xylem also is produced on the underside of large branches in order to bear the weight of the branch. Some mature lowland tropical rainforest trees develop prominent platelike structures called buttresses at the base of the stem that help support the trunk by transferring mechanical stresses to the soil. In all these conditions, it is assumed that the cambium can detect and respond to the forces of stretching and compression by forming disproportionally more xylem elements in certain regions of the axis to increase strength and rigidity. This is called the constant stress hypothesis. It has been used to explain tree growth and morphology in varied habitats, although the mechanism by which such stresses stimulate wood formation is not well understood.

In older portions of the root, the vascular cambium originates between the primary xylem and the primary phloem and in the pericycle outside the protoxylem ridges. As secondary growth occurs, the star-shaped transectional outline of the first-formed cambium is altered into a circular outline.

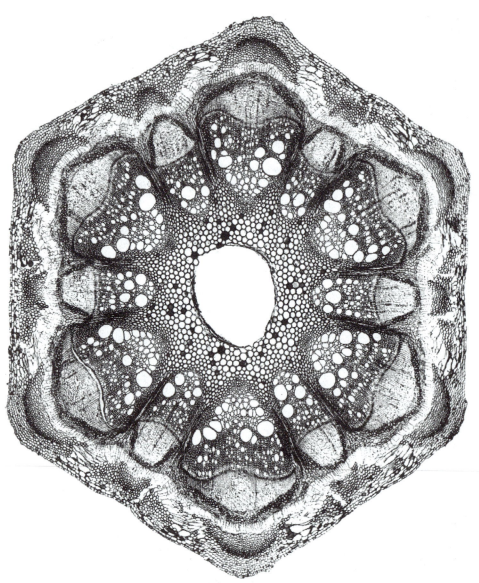

**FIGURE 4.5**    Transverse section of stem of *Clematis virginiana* (Ranunculaceae) illustrating the result of secondary growth. The original vascular bundles remain separated, although secondary vascular tissues have been formed. Note the combination of hexagonal, radial, and concentric symmetries in this transection. Courtesy of the Bailey-Wetmore Wood Collection, Harvard University.

## STRUCTURE AND FUNCTION OF THE VASCULAR CAMBIUM

At maturity the cambium normally forms a continuous cylinder around the stem and root and even extends into the leaf petioles of some plants. In contemporary seed plants, the cambium functions as a bifacial, uniseriate layer with new **derivatives** formed to either side of the **initials,** the xylem centripetal

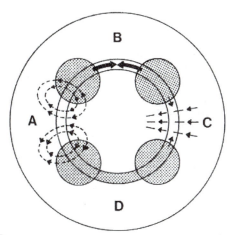

**FIGURE 4.6**   Schematic representation of hypotheses explaining the inductive influences on the initiation of the interfascicular cambium. (A) Inductive influence of hydrogen ion gradient originating from vascular bundles. (B) Inductive effect of the fascicular cambium. (C) Inductive effect of a gradient dependent originating on the stem surface. (D) Position and polarity of cambium determined at the procambial stage. Reprinted with permission from Siebers (1971); *Acta Bot. Neerl.* **20,** 211–220.

and phloem centrifugal. The divisions that form xylem and phloem mother cells involve periclinal (tangential) divisions and are of the additive type. The interesting developmental question, what causes cells on one side of the cambium to become xylem and those on the other side to become phloem, remains unanswered. The derivatives undergo great radial expansion and tip growth and are characteristically aligned in radial rows following periclinal divisions of the initials. The derivatives proceed to gradually differentiate as phloic and xylic elements following an ordered sequence of developmental events. As discussed previously, differentiation of vascular elements involves changes in cytology, wall structure, and wall chemistry. The differentiation of cambial derivatives also is accompanied by appreciable increase in cell size, resulting in radial and tangential expansion and cell elongation.

The existence of a single layer of initiating cambial cells has not always been recognized. Some workers have been proponents of a multiseriate concept, that is, that more than one layer of initiating cells is present. It is most convenient, however, to interpret the radial dimensions of the cambium as containing a functionally single layer of initials accompanied by actively dividing, undifferentiated xylem and phloem mother cells on either side. Because it often is not possible to pick out the single layer of initials with any certainty, the entire meristematic region is referred to as a **cambial zone** and is typically characterized by the appearance of one or several radial files of flattened cells. As a result, the cambial zone and its immediate xylem derivatives are composed of three regions: (1) a region of expanding derivatives prior to initiation of secondary wall thickenings, (2) a zone of active assimilation of secondary wall materials, and (3) the zone of mature tracheary elements. The width of the cambial zone can vary from 6 to 8 cells in slow growing trees, and from 12 to 40 cells in fast growing trees. Table 4.1 provides a summary of useful terminology relating to the cambium.

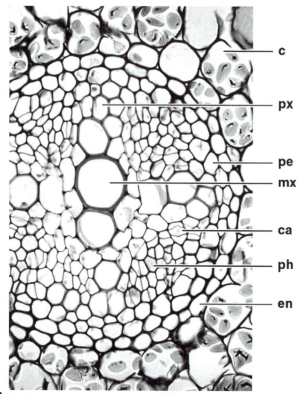

**FIGURE 4.7**   Transverse section of tetrach root of *Caltha palustris* (Ranunculaceae) showing early stage in formation of vascular cambium. Abbreviations: c, cortex; ca, vascular cambium; en, endodermis; mx, metaxylem; pe, pericycle; ph, phloem; px, protoxylem.

Most vascular cambia possess two categories of initials of distinct size and shape. These are the **fusiform initials** and **ray initials.** The ratio of ray and fusiform initials within the cambium varies from species to species and can be related to the growth rate of an individual. Fusiform initials are relatively large, considerably elongated, spindle-shaped elements with tapering, overlapping ends. They give rise to an axial system of radially aligned, vertically elongate tracheary elements, fibers, and sieve elements, as well as other cells of the axial system such as parenchyma. Fusiform initials are further distinguished by having a rich collection of organelles and by being highly vacuolated, but the form of the vacuome varies on a seasonal basis.

During actively growing months, the fusiform cells contain a large, elongated vacuole, and their walls have numerous primary pit fields, giving them a beaded appearance. In gymnosperms and primitive woody dicotyledons, the fusiform initials are one hundred to several hundred times as long as they are wide.  The fusiform initials also are arranged as adjacent, overlapping cells that vary considerably in length. In this structural condition the adjacent initials vary greatly in length and volume and form an **unstratified cambium.** In certain of the more highly specialized dicotyledons, in contrast, the fusiform initials are of nearly equal length and symmetrically grouped in parallel, hor-

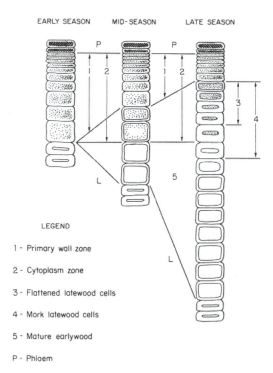

**FIGURE 4.8**   Diagrams in the transverse plane of cambial tissue zones in a conifer showing the cambial initials and derivatives in the radial files of tracheids at three times during the growing season. Reprinted with permission from *For. Sci.* **12**, 198–210. Published by The Society of American Foresters, 5400 Grosvenor Lane, Bethesda, MD 20814–2198. Not for further reproduction.

izontal rows. This latter type of highly **stratified cambium** is termed a **storied cambium.** The fusiform initials in storied cambia are much shorter, more or less hexangular, with parallel sides and abruptly tapering ends, and the elements of adjacent horizontal series do not overlap to any considerable degree. In old stems, the length of fusiform initials ranges from exceedingly long, 6000 μm in some gymnosperms such as *Sequoia,* down to rather short initials between 200 and 300 μm in some highly specialized dicotyledons (e.g., *Robinia*).

Ray initials divide periclinally to form aggregations of radially extended parenchymatous cells called **rays.** In most plants rays make up about 25% of the secondary plant body; the number of rays stabilize as the tree matures. Ray initials are smaller, more or less isodiametric cells that occur in groups corresponding in height and width to the radial (horizontal) system of the parenchymatous xylem and phloem rays to which they give rise. Within the actively dividing cambium, new ray initials are constantly being added, and fusiform initials, eliminated. Ray size increases with tree age in very young trees, until the xylem reaches its mature stage. The height and width of xylem rays also can change as a result of subdivision by the intrusion of fusiform initials, the reversion of ray initials to fusiform initials, and the merging of rays.

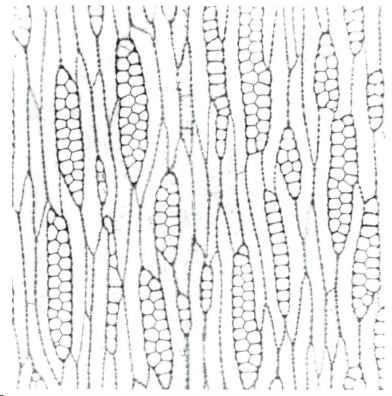

**FIGURE 4.9**  Tangential section of the vascular cambium of *Fraxinus americana* (Oleaceae) showing fusiform and ray initials. Note "beaded" appearance of walls of the fusiform initials as a result of crowded primary pit fields. Courtesy of the Bailey-Wetmore Wood Collection, Harvard University.

In some structurally specialized plants, cambial ray initials can be entirely lost, and the wood produced is **rayless**. It has been demonstrated that the existence of autonomous radial signal flows of developmental stimuli (e.g., the hormones auxin, ethylene, and gibberellin) participates in wood ray initiation and differentiation.

## Cell Division in the Cambium

The process of cell division in fusiform initials is among the most remarkable cytological events in the plant kingdom. Each fusiform cell contains a single ellipsoidal nucleus that is usually centrally located. Chromosomes are duplicated and separated during the mitotic phases of prophase, metaphase, and anaphase in the normal manner. At the end of anaphase, two daughter nuclei have moved apart from the equatorial plane but remain close together within the confines of the narrow fusiform cell. During telophase, however, the future cell wall, or **cell plate**, arises in the fibrous **phragmoplast** between the two new nuclei formed during nuclear division (**karyokinesis**). This marks the beginning of cytoplasmic division (**cytokinesis**). As cytokinesis proceeds, the two portions of the phragmoplast move apart and extend vertically toward opposite ends of the elongated cell. The phragmoplast plays an important role in the

██████  **TABLE 4.1   Terminology Describing the Vascular Cambium and Associated Tissues**

| TISSUE | CHARACTERISTIC EVENTS |
| --- | --- |
| Secondary phloem | Mature tissue |
| Differentiation phloem cells | Limited cell division, cell enlargement and secondary wall deposition, protoplasmic specialization |
| Cambium or cambial zone | |
|   Phloem mother cells | Usually periclinal cell division |
|   Cambial initials | Periclinal as well as anticlinal cell division |
|   Xylem mother cells | Usually periclinal cell division; rarely anticlinal |
| Differentiating xylem cells | Limited cell division, cell enlargement and secondary wall deposition, death of protoplasm, etc. |
| Secondary xylem | Mature tissue |

Reprinted with permission of Iqbal (1990), "The Vascular Cambium," Research Studies Press, Ltd., Taunton, Somerset, England.

formation of the cell plate, the initial partition dividing the mother cell in two. In elongated, periclinally dividing fusiform initials, cytokinesis is extended, and the cell plate becomes extraordinarily long, in some cases several millimeters in length. Cell division in the ray initials resembles division that occurs in other isodiametric cells such as parenchyma.

For the cambium to keep pace with the rapid increase in axis diameter, fusiform initials must occasionally divide in the radial, longitudinal plane. These anticlinal divisions provide for circumferential expansion of the cambium and are termed multiplicative divisions. In the nonstratified cambia of most woody dicotyledons, the orientation of the cell plate within dividing fusiform initials fluctuates between a nearly transverse position and varying degrees of obliquity. This means that the phragmoplast develops across the cell nearly transversely or only slightly obliquely. This is known as "pseudo-transverse" division. Following division, the daughter cells undergo apical elongation and slide by one another intrusively until a maximum length is attained, thereby producing an increase in the girth of the cambium. Sliding growth accounts for most of the enlargement of the cambial ring in unspecialized cambia. Fusiform initials occasionally divide anticlinally to form lateral derivatives off the sides of the initials. As fusiform initials phyletically decrease in length, the plane of cell division becomes more nearly "radiolongitudinal"; that is, the mitotic figure moves vertically toward opposite ends of the cell, progressively forming the new wall. In stratified cambia, an increase in circumference is not caused by the elongation and sliding growth of nearly transversely dividing cells; rather, an increase in cambial girth follows radio-longitudinal, anticlinal divisions of the fusiform initials, followed by the lateral expansion of the products of these divisions. Although the products of such divisions expand laterally, they do not undergo apical elongation to any appreciable extent. The transition from the nonstratified to the stratified arrangement of cambial cells represents a major trend in the structural specialization of flowering plants.

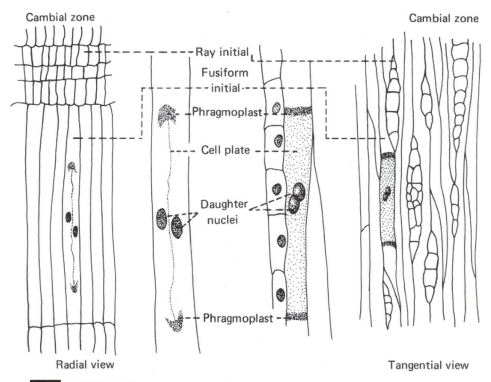

**FIGURE 4.10**    Diagram of tangential (periclinal) division of a fusiform initial in radial and tangential views. This type of division gives rise to radial rows of xylem and phloem derivatives. From Brown (1971). Used with permission of Springer-Verlag New York, Inc.

In both unspecialized and highly specialized cambia, the fusiform initials increase in length with the age of the stem, but they do not continue to increase in size throughout the entire life of the individual. The only exception to this rule is seen in the most highly specialized storied cambia, in which no increase in length occurs among the initials. Figure 4.14 demonstrates how the average length of fusiform initials has declined phyletically. In gymnosperms and primitively vesselless dicotyledons, moreover, the age-related fluctuations in cell size are considerably greater than in highly specialized dicotyledons, in which the initials undergo very little or no increase in length with age. Such **age on length curves,** therefore, become more depressed with increased specialization. Following a period of rapid increase in length, the size of cambial initials remains more or less constant during successive growth. In a stratified cambium, the length of fusiform initials is stabilized. The mean length of fusiform initials also can vary on a seasonal basis within one tree. The phyletic decrease in the length of fusiform initials from primitive dicotyledons to more specialized taxa closely parallels the formation of shorter tracheary elements because the length of tracheary elements is determined primarily by the length of fusiform initials.

Not only do the lengths of fusiform initials vary at different periods in stem development, but the secondary xylem tracheary elements produced by these initials also show developmental variation in length. This is documented by the

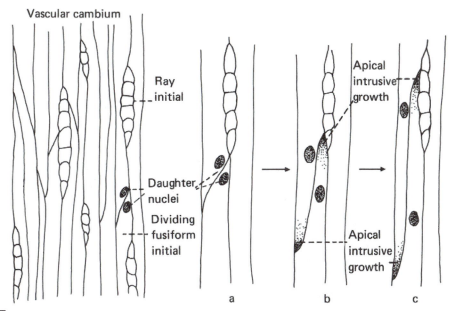

**FIGURE 4.11** Tangential view of vascular cambium showing (a) pseudotransverse division of fusiform initial followed by later stages of elongation by apical intrusive growth (b) and (c). This type of cell division gives rise to two daughter fusiform initials, thereby increasing the girth of the cambium. From Brown (1971). Used with permission of Springer-Verlag New York, Inc.

vesselless dicotyledon *Trochodendron aralioides* (Trochodendraceae), in which tracheid length increases over time up to an age when tracheid length remains more or less uniform (Figure 4.15). It also can be seen in this age on length curve that the first formed, helically thickened primary xylem elements (protoxylem) are longer on average than the later formed, pitted metaxylem cells. This pattern holds true for dicotyledons in general as determined statistically by measuring the lengths of tracheary cells in a wide range of angiosperms representing different levels of specialization (Table 4.2). Within any one organ, vessel element origin and specialization occur first in the late metaxylem and then progressively in the early metaxylem and protoxylem. As a result, vessels often are not present or are infrequent in the earlier formed part of the primary xylem. Tracheary elements of the first-formed secondary xylem are always shorter than the mean length of primary xylem tracheary cells.

## ANOMALOUS SECONDARY GROWTH

Some woody plants representing a wide range of unrelated families show variations from the normal pattern of cambial activity and are referred to as possessing "anomalous" secondary growth. These plants are generally vines (lianas) and are characterized by a variety of a structural patterns relating to cambial origin, mature structure, and function. A general survey of the xylem cylinders of 448 species of tropical lianas representing 35 families revealed that polystelic and multiple stems occur in nearly 80% of New World species

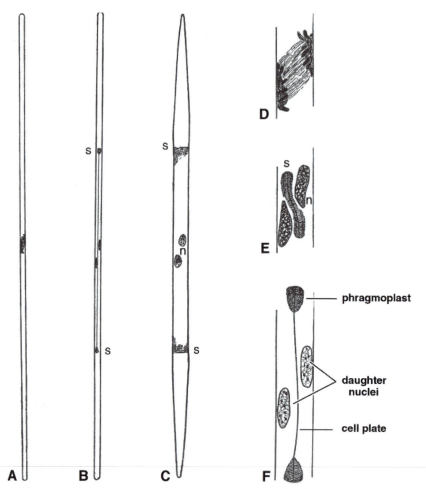

**FIGURE 4.12**   Cell division in fusiform initials of *Pinus strobus* (Pinaceae). (A) Fusiform initial in radial longitudinal view, showing oblique position of mitotic spindle. (B) Longitudinal view showing daughter nuclei and process of cell plate formation. (C) Cell B, viewed in tangential longitudinal extension. (D) Highly magnified view of mitotic spindle shown in A. (E) Central portion of fusiform initial showing the beginning of cell plate formation. (F) Later stage of cell plate formation. Abbreviations: n, daughter nuclei; s, aggregation of spindle fibers (microtubules). From Bailey (1919), *Proc. Natl. Acad. Sci.* (USA) **5**, 283–285.

and somewhat less than 50% of African taxa. Many basic questions remain with respect to the factors that induce and regulate anomalous cambial activity. The functional significance of different patterns of anomalous growth in lianas also remains unclear. Many functions have been suggested for these structural types: increase stem flexibility and strength, limit or predetermine cleavage patterns, facilitate cloning, confine disease to limited areas of the plant body, facilitate stem climbing, increase storage tissue, protect and maintain functional xylem and phloem under highly stressful conditions, facilitate rapid healing, and increase connections for adventitious roots.

The various types of anomalous secondary growth are difficult to classify into distinct groups because of their structural diversity and their integration

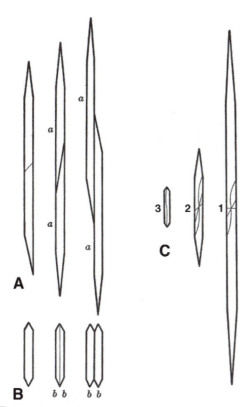

**FIGURE 4.13**   Types of cell division in fusiform cambium initials. (A) Fusiform initial from non-stratified conifer cambium dividing pseudotransversely and leading to the increase in girth of the cambium; a, a, products of this division which elongate and slide by one another. (B) Fusiform initial from stratified cambium; b, b, products of the radiolongitudinal division of this initial, which expand laterally but not longitudinally, leading to the increase in girth of the cambium. (C) Type of anticlinal divisions in fusiform initials; 1, conifer; 2, dicotyledon having nonstratified cambium; 3, dicotyledon having a stratified cambium. Reprinted with permission from Bailey (1923), *Am. J. Bot.* **10,** 499–509.

with normal patterns of cambial activity. One category of anomalous stem structure results from the abnormal functioning of the original cambium. An example is seen in the **cross vine** structure of some viny members of the Bignoniaceae as well as other families. Here, four arcs of the original cambium stop producing secondary xylem elements to the inside, while excessive secondary phloem is formed on the outside. Because xylem and phloem development is uneven, the cambium becomes **interrupted,** and the secondary xylem is broken up into discrete sectors. A cambial arc is positioned at the base of each furrow and at the crest of each ridge, resulting in a star-shaped woody cylinder. It has been suggested that the multiseriate rays that parallel each furrow are perhaps involved in the induction of this unusual growth pattern.

Another common type of anomalous secondary growth involves the formation of **successive cambia.** The gradual formation of successive cambia results in the sequential formation of either complete rings or bundles of vascular tissue arranged concentrically. In some plants a correlation exists between leaf formation and the development of successive cambia. The suc-

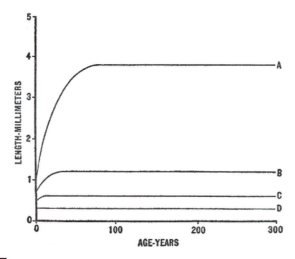

**FIGURE 4.14**    Age on length curves showing average lengths of cambial initials at successive stages in the enlargement of a stem. (A) Conifer or vesselless dicotyledon. (B) Less specialized type of dicotyledon. (C) Highly specialized type of dicotyledon. (D) Highly specialized type of dicotyledon having a stratified cambium. Reprinted with permission from Bailey (1923), *Am. J. Bot.* **10**, 499–509.

cessive vascular rings occur between the first-formed original vascular bundles and the more peripheral cork cambium. Whether the vascular cylinders are in continuous rings or in separate bundles depends upon the structure of the young stem. The original cambium is typically short lived, and each new cambium that is added usually functions normally, producing secondary xylem internally and secondary phloem externally. Careful observation has revealed that connections or anastomoses are sometimes present between the vascular cylinders. The thickness of the secondary xylem within each ring or bundle can vary from several centimeters to a few millimeters. The secondary phloem usually appears as strands rather than as a complete cylinder. Areas of conjunctive parenchyma are dispersed between adjacent secondary vascular cylinders in many taxa.

The establishment of secondary growth in plants with successive cambia occurs in at least two distinctive patterns. In the first, the initial cambium originates from procambium, and secondary growth is established prior to the formation of the second cambium. In the second developmental pattern, the initial cambium originates from the procambium but forms little or no secondary tissue within the primary bundle. Thus, the first complete secondary vascular cylinder develops from the next successive cambium and occurs very early in development. Over the years, studies to elucidate the place of origin of successive cambia have shown that dedifferentiation of cells may occur in different areas of the stem. Each new cambium may originate by tangential divisions within the primary phloem parenchyma, the outermost parenchymatous derivatives formed by the activity of the next youngest, inner cambium, or from the division of cortical cells.

The genus *Serjania* of the Sapindaceae is well known for containing species with a variety of anomalous growth patterns, some of which are diagnostic for certain species complexes. The various growth patterns have been

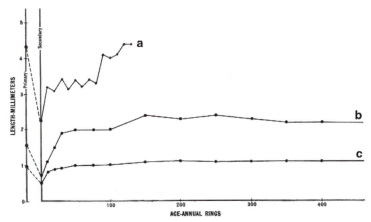

**FIGURE 4.15** Graphs illustrating variation in relative length of tracheary elements in passing from innermost to outermost secondary xylem of the stem. Lengths of primary xylem cells shown for comparison on left. (a) *Trochodendron aralioides* (Trochodendraceae), vesselless arborescent dicotyldeon; (b) *Liriodendron tulipifera* (Magnoliaceae), fiber tracheids; (c) *Liriodendron tulipifera*, vessel elements. Used with permission from Bailey and Tupper (1918). Copyright © the American Academy of Arts and Sciences.

referred to as simple, compound, divided, corded, and parted xylem masses. The **simple** (or **undivided**) **stem** contains a single xylem cylinder and in most respects resembles the typical dicotyledonous condition. Following the unequal production of xylem, however, the stem can become lobed in outline. In some species, vascular bundles originate within the cortex or bark regions. This condition is often called the **corded stem** type. The outer bundles are reported to be independent of the central cylinder. If the central cylinder of a mature stem becomes fractured into islands following the delayed production of parenchyma within the xylem, the condition is known as a **parted xylem**

**TABLE 4.2  Average Length of Tracheary Cells in Micrometers**

|  | PRIMARY XYLEM | | SECONDARY XYLEM | |
|---|---|---|---|---|
|  | **Helical** | **Pitted** | **Early** | **Late** |
| A[a] | 3500 (T) | 1910(T) | 4350 (T) |  |
| B[b] | 1950 (TV) | 1650 (TV) | 710 (V) | 1100 (V) |
| C[c] | 1020 (TV) | 830 (V) | 360 (V) | 730 (V) |
| D[d] | 540 (V) | 470 (V) | 260 (V) | 400 (V) |
| E[e] | 350 (V) | 240 (V) | 170 (V) | 170 (V) |

[a]Vesselless dicotyledon *Trochodendron*.
[b]Dicotyledons with transitional types of vessel members in the secondary xylem.
[c]Dicotyledons with relatively primitive vessel members in the secondary xylem.
[d]Dicotyledons with highly specialized vessel members in both the primary and secondary xylem.
[e]Dicotyledons with storied cambium (*Robinia*).
Abbreviations: T, tracheids; V, vessel members. Basis: 350 species of 89 families.
Reprinted with permission of Bailey (1944), *Am. J. Bot.* **31**, 421–428.

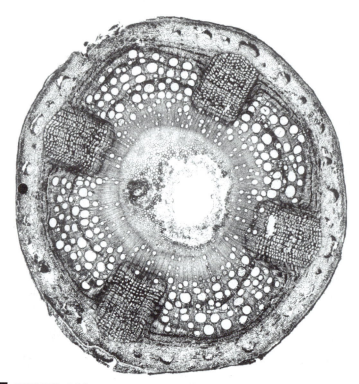

**FIGURE 4.16**   Transverse section of stem of *Macherium purpurascens* (Leguminosae) showing anomalous cross vine structure. Courtesy of the Bailey-Wetmore Wood Collection, Harvard University.

mass. A **multistelar** (or **compound**) **stem** is a very distinctive anomaly in which a single central cylinder (stele) becomes surrounded by 3 (most common), 5, or more (up to 10) independent peripheral vascular cylinders, each with its own cambium and secondary growth. Evidence from ontogeny suggests that the surface steles originate by the differentiation of interfascicular cambium within the primary plant body. In a few species of *Serjania*, a **divided stem** forms; the central cylinder becomes radially divided into four or five wedges by parenchymatous tissue.

### Secondary Xylem (Wood)

With the activity of the vascular cambium through time, a large amount of secondary xylem or **wood** is produced. In addition to xylem production, the vascular cambium also produces secondary phloem each year. The amount of phloem produced is usually much less than the amount of xylem produced. Ratios of the amount of xylem and phloem produced vary widely depending upon tree species, growth vigor, and environmental conditions. Ratios can range from a high of 15:1 to a low of 2:1, with a ratio of 4:1 being common. The increments of growth in the stems of temperate woody plants can be seen by differences in the cross-sectional areas of the tracheary and fibrous cells.

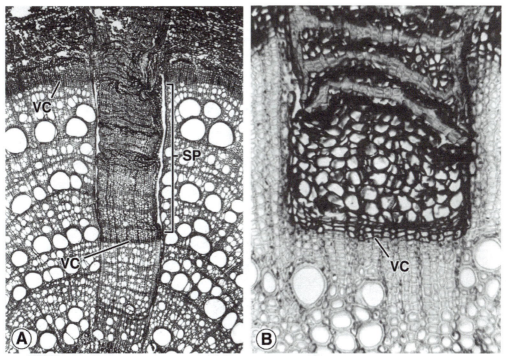

**FIGURE 4.17**  Partial view of transverse sections of stem of *Macfedyena unguis-cati* (Bignoniaceae) showing anomalous cambial activity. (A) The vascular cambium (VC) becomes discontinuous, and four arcs of the cambium produce little or no secondary xylem but extensive secondary phloem (SP) in a unidirectional manner. As secondary growth continues, the bidirectional and unidirectional arcs of the cambium become separated. (B) High-magnification view of arc of unidirectional vascular cambium. Courtesy of D. Dobbins.

Tracheary elements produced in the spring (**earlywood**) are usually noticeably wider than the narrower or radially flattened, thicker-walled elements produced in the summer (**latewood**). **Ring-porous** woods possess distinctly larger vessels that are concentrated in the earlywood, as compared to the latewood of the same growth ring. **Diffuse porous** woods have pores of more or less the same diameter throughout the growth ring.

Wood exhibits great diversity in structural and physical properties and finds its best expression in the conifers and arborescent dicotyledons, because they have prolonged cambial activity. The expressions **softwood** and **hardwood** refer, respectively, to wood produced by these two groups and not to the relative hardness of the xylem. The two kinds of wood show basic structural differences (Table 4.3), but degree of density and hardness do not necessarily characterize the two types; that is, there are soft and hard woods in both groups.

Wood consists of two positionally arranged tissue systems: the **axial** (or vertical) **system** of cells that arises from fusiform cambial initials, and the **ray** (or horizontal) **system** of cells that arises from the cambial ray initials. In dicotyledons, the axial system consists of different cell types (vessel elements, tracheids, fibers, axial parenchyma) oriented with their long axes parallel to

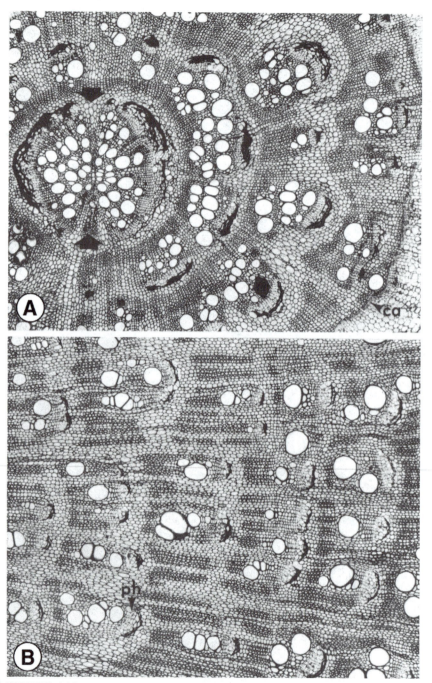

**FIGURE 4.18** Anomalous secondary growth in *Bougainvillea* (Nyctaginaceae). (A) Transection of root with one normal and three anomalous secondary growth increments. Arrows indicate orientation of diarch primary xylem plate. Darkly stained material, crushed phloem cells. (B) Transection of old stem showing several anomalous secondary growth increments. Both ×120. Reprinted with permission from Esau and Cheadle (1969), *Ann. Bot.* **33,** 807–819.

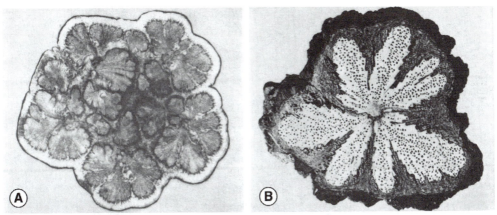

**FIGURE 4.19** Transverse sections of stems showing anomalous patterns of secondary growth. (A) Dispersed type of anomalous structure in *Passiflora multiflora* (Passifloraceae). ×2. (B) Interrupted type of anomalous structure in *Passiflora* sp. ×3.3. From Ayensu and Stern (1964), *Contributions from the United States National Herbarium* **34**, 45–73.

the long axes of the stem. The axial system of gymnosperm wood (excluding the genera *Ephedra*, *Gnetum*, and *Welwitschia*) consists almost exclusively of tracheids. The ray system is composed predominantly of living ray paren-chyma cells, oriented so that they form a system traversing the vertical system at right angles. Ray parenchyma translocates radially. Very rarely, rays possess radial files of cells that are connected through perforations in the tangential walls. Such cells are called **ray vessels.** Xylem ray cells commonly acquire a secondary wall and may remain alive for an extended period. Ray paren-chyma also may act as a storage tissue. A consequence of phylogenetic spe-cialization is the stratification of wood elements into distinctive **storied layers.**

Dicotyledonous woods are histologically described primarily on the basis of the type, size, abundance and distributional patterns of axial elements, coupled with the size, structure, and abundance of rays. Major features of value for wood identification and description are outlined next.

I. **TRACHEIDS** AND **FIBERS** (Imperforate tracheary elements)

Imperforate tracheary cells are described in terms of type, wall thickness, sculpture, and cell length.

    A. Vascular or vasicentric tracheids (present or absent).
    B. Ground tissue fibers (walls with simple pits, minutely bordered pits, or dis-tinctly bordered pits; elements may have **spiral thickenings,** i.e., helical ridges that are part of the inner face of the secondary wall).
    C. Septate fibers (present or absent).
    D. Fiber wall thickness (thin walled, thick walled, very thick walled).
    E. Fiber lengths (less than 900 μm, 900 to 1600 μm, greater than 1600 μm).

II. **VESSELS** (Perforate tracheary elements)

    A. Porosity (wood ring porous, semiring porous, diffuse porous).

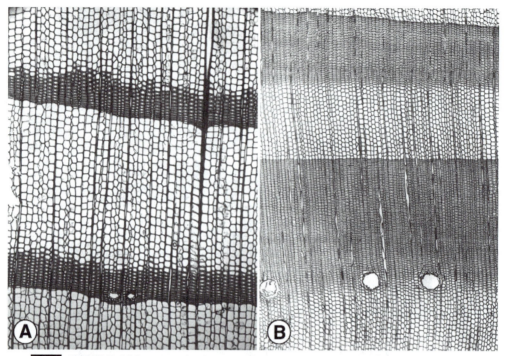

**FIGURE 4.20**   Wood anatomy of Pinaceae. (A) Transverse section of *Larix* showing conspicuous latewood. There is an abrupt transition from earlywood to the latewood resulting in a well-defined band of latewood. ×35. (B) Transverse section of *Pinus palustris* showing broad bands of conspicuous latewood. ×50. A, courtesy of E. A. Wheeler. B, courtesy of the Bailey-Wetmore Wood Collection, Harvard University.

B.  Pore (vessel) arrangement and groupings—vessels can be arranged in tangential bands, diagonal and/or radial patterns, or dendritic arrangements. With respect to groupings, vessels are **solitary** if completely surrounded by other elements as viewed in transverse section. **Radial multiples** are radial files of two or more adjacent vessels, whereas **clusters** are groups of three or more vessels having both radial and tangential contacts. Clusters and radial multiples can occur in combination. The most common vessel arrangement in wood is numerous radial multiples of two to four vessels, with a variable proportion of solitary vessels.

C.  Vessel element outline, wall thickness, size and frequency—vessel outline (as seen in transverse section) ranges from angular to circular with variation in wall thickness from thin to thick walled. The tangential diameter of the vessel lumina as measured in transverse section ranges from extremely small (up to 50 μm) to very large (over 200 μm). The length of vessel elements is recorded from element tip to tip. The frequency of vessels is a measure of the number of pores per square millimeter of wood.

D.  Perforation plates—principal types of perforation plates are scalariform and simple. In a scalariform plate the remnants of the plate between the openings are called **bars.** Significant variations include the number of bars composing the plate and whether the perforations are completely bor-

**TABLE 4.3  Comparison of Important Differences in the Wood of Gymnosperms[a] and Dicotyledons**

| GYMNOSPERMS | DICOTYLEDONS |
| --- | --- |
| True vessels mostly absent | True vessels mostly present |
| Tracheids present; form bulk of wood | Tracheids present or absent; nearly always subordinate |
| Ray tracheids present or absent | Ray tracheids absent |
| Wood fibers absent | Wood fibers present |
| Wood parenchyma generally absent or scarce | Wood parenchyma present, often very conspicuous |
| Ray parenchyma present, rays typically uniseriate | Ray parenchyma present, rays often multiseriate |

[a]Excluding the genera *Ephedra, Gnetum,* and *Welwitschia,* all of which possess vessels.

dered, bordered to the middle, bordered to the ends, or nonbordered. Perforation plate end walls range from highly oblique to transverse.

E.  Intervessel pits and vessel to ray pitting—intervessel pits are pits found between vessel members. The arrangement of these pits can be **scalariform** (pitting in which elongated or linear pits are arranged in a ladderlike series), **opposite** (multiseriate pitting in which the pits are in horizontal pairs or in short horizontal rows), or **alternate** (multiseriate pitting arranged in diagonal rows). The shape and diameter of the pits should be recorded.

## III.  WOOD RAYS

Xylem rays are classified on the basis of type and size. There are two principal types of rays, **heterocellular** and **homocellular.** A heterocellular ray is one in which the individual rays are composed of both procumbent cells and square or upright cells, whereas a homocellular ray is composed wholly of cells of the same morphological type (i.e., all procumbent or all square or upright). Rays are described as **uniseriate,** if they are one cell wide as seen in tangential section, and **multiseriate,** if they are more than one cell wide. Ray height is recorded either in terms of the number of cells or in micrometers. Multiseriate rays can have a multiseriate central body and uniseriate marginal wing extensions. Dicotyledonous woods can possess exclusively uniseriate rays, exclusively multiseriate rays, or, most commonly, a combination of uniseriates and multiseriates.

## IV.  AXIAL PARENCHYMA

Wood axial parenchyma cells are derived from fusiform cambial initials and can form an important but sometimes subtle wood character. Axial parenchyma distribution (as seen in transverse section) is classified using four major parameters.

A.  Axial parenchyma absent or extremely rare.

B.  **Apotracheal parenchyma** (axial parenchyma usually occurring independently of the pores or vessels).

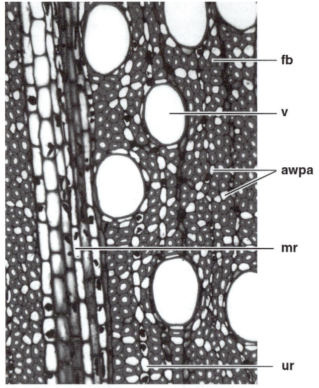

**FIGURE 4.21**   Transverse section of wood of *Dillenia pentagyna* (Dilleniaceae) showing cells of the axial and radial systems. Abbreviations: awpa, axial wood parenchyma (diffuse and diffuse in aggregates); fb, xylary fiber; mr, multiseriate ray; ur, uniseriate ray; v, vessel element. (See Color Plate.)

    1.  Axial parenchyma **diffuse** (single parenchyma strands or cells distributed irregularly among the fibers).

    2.  Axial parenchyma **diffuse in aggregates** (parenchyma cells that tend to be grouped in short tangential lines from ray to ray).

C.  **Paratracheal parenchyma** (axial parenchyma associated with the vessels or vascular tracheids).

    1.  Axial parenchyma **scanty** (incomplete sheaths or occasional parenchyma cells around the vessels).

    2.  Axial parenchyma **vasicentric** (parenchyma forming a complete sheath around a vessel, of variable width and circular or slightly oval in cross section).

    3.  Axial parenchyma **aliform** (parenchyma surrounding a vessel with winglike lateral extensions).

    4.  Axial parenchyma **confluent** (coalesced aliform parenchyma forming irregular tangential or diagonal bands from vessel to vessel).

D.  Banded parenchyma (axial parenchyma forming concentric lines or bands)—banded parenchyma is described as **broad banded** if it is more than three cells wide or **narrow banded** if three cells wide or less.

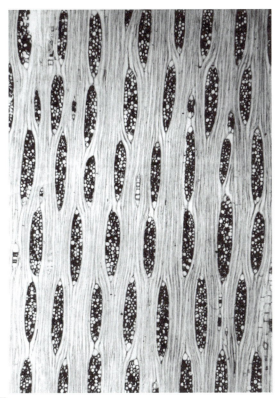

**FIGURE 4.22**  Storied xylem rays of *Entandrophragma cylindricum* (Meliaceae). × 35. Courtesy of E. A. Wheeler.

## Secondary Phloem

Like the secondary xylem, the secondary phloem arises from the vascular cambium and possesses an axially oriented cell system and a radially oriented ray system that originates from the cambial ray initials in the same manner as the xylem rays. As a result, phloem rays are comparable to xylem rays in general size and structure. The axial system is composed of secondary sieve elements, phloem parenchyma, and sometimes fibers. Overall, the secondary phloem is less sclerified than the xylem, although some plants develop conspicuous fibers and sclereids within the phloem tissue. The abundance and distributional patterns of secondary **phloem fibers** (seen in transverse section) are characteristic for some taxa. Fibers can be distributed singly, in small tangential rows, or in regular, wide tangential bands alternating with zones of nonsclerified elements. Phloem fibers have not been satisfactorily classified, in part because of their varied origin and their intergradation with sclereids. In some cases secondary phloem fibers are totally absent. The distribution of phloem parenchyma cells also shows considerable variation.

During secondary growth the older phloem is displaced outward by the activity of the vascular cambium and is subjected to tangential stresses resulting from the increase in stem diameter. As a result, the older phloem cells

**FIGURE 4.23**   Longitudinal section of wood of *Pseudotsuga* (Pinaceae) showing spiral wall thickenings in the secondary tracheids. Courtesy of the Bailey-Wetmore Wood Collection, Harvard University.

become nonfunctional and are deformed, displaced, or crushed. In most plants, only the most recent increment of secondary phloem is functional and suitable for the study of cell components regarding their size, shape, contents, and intercellular relationships. In some stems, wide rays become dilated in the nonconducting phloem so that the axial system appears as wedges, narrower toward the cambium and wider toward the stem periphery, with the fibers accentuating the pattern by their presence.

## STRUCTURE AND FUNCTION OF THE CORK CAMBIUM AND PERIDERM

In most woody perennial dicotyledons and gymnosperms, the original outer primary epidermal and cuticular covering is replaced in one year so that older stems and roots have an outer layer of secondary periderm. The periderm is usually initiated just after organ elongation is completed, although in some taxa it is much delayed. The periderm is not synonymous with the term **bark**. **Bark** is a nontechnical term that refers to all tissues exterior to the vascular cambium, therefore including secondary phloem. Bark refers to a number of tissue types of different origin, such as periderm and secondary phloem. In

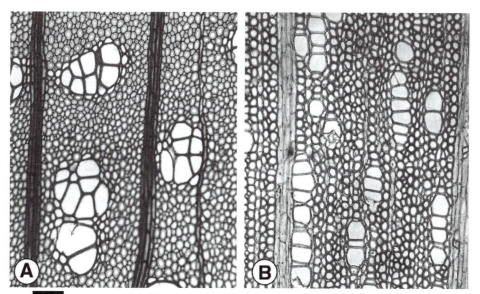

**FIGURE 4.24**   Vessel distribution as determined from wood transverse sections. (A) Pore clusters in the latewood of *Gymnocladus dioicus* (Leguminosae). × 180. (B) Radial pore multiples of four or more in *Ilex vomitoria* (Aquifoliaceae). × 180. Courtesy of E. A. Wheeler.

most woody plants, the vascular cambium is removed if the bark is stripped off. Bark formed during the early stages of the season's growth is often termed **early bark** or **soft bark** and consists mainly of sieve tubes and parenchymatous and suberized cells, but it does not include fibers or other strengthening cells. Toward the end of the growing season, **late bark** or **hard bark** is formed. It typically contains fibers and other sclerenchymatous cells, along with bark parenchyma and fewer and smaller sieve elements. Variation may be evident in the degree of development of periderm around the circumference of a single stem, which in extreme cases results in a "winged cork" condition. The mature periderm serves to protect the axis from attack by animals, fungi, and other pathogens. It also prevents the plant from drying out and insulates it from fire. The very thick (up to 1 ft) fibrous periderm of the redwood tree (*Sequoia*) is a particularly effective fire shield because it lacks combustible resins and thus does not easily burn.

Among different species, periderm morphology shows many differences in texture, color and thickness. Structurally the periderm consists of three components. The initiating layer of the periderm is the **phellogen,** or **cork cambium.** The cells produced to the outside of the phellogen are nonliving **cork cells,** technically called **phellem.** In some plants, the phellogen forms one or more layers of internal tissue composed of living, cortexlike parenchyma cells referred to as **phelloderm.**

The phellogen is a lateral meristem that forms at varying depths from the dedifferentiation of living, completely mature cells that resume division and become converted into a cambium. Depending upon the taxon, the phellogen may arise in the divisions of epidermal, cortical, or phloem parenchyma cells

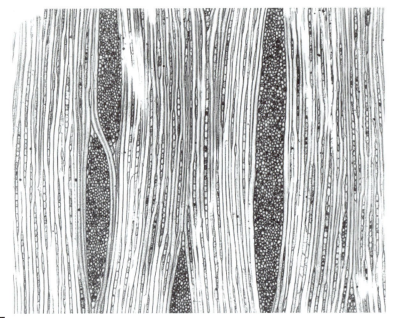

**FIGURE 4.25**   Tangential section of wood of *Ilex vomitoria* (Aquifoliaceae) showing rays of two distinct sizes. Courtesy of E. A. Wheeler.

of the stem. In the grapevine (*Vitis*), it originates as a result of cell divisions in the parenchyma of the metaphloem. In the vast majority of dicotyledons, the phellogen has its origin in subepidermal cortical layers, often in a discontinuous manner. This is referred to as a **superficial origin** of the cork cambium, as compared with a **deep-seated origin** within the inner cortex or secondary phloem. The root phellogen, in contrast, is always formed internally within the proliferating pericycle.

Unlike the vascular cambium, the phellogen contains initials of only one type. These are more or less isodiametric in shape, and in transverse section appear radially flattened. As noted earlier, periclinal divisions of the phellogen produce two types of derivative tissue: the protective phellem (or cork) toward the outside and the phelloderm (or green cork) toward the inside. As a general rule, more radial files of cork cells are formed centrifugally than phelloderm cells are formed centripetally. Collectively, the three layers constitute the periderm. With formation of the periderm, the primary epidermis becomes effectively isolated from internal tissues. The epidermis cracks and is sloughed off, although the dead and collapsed epidermis may cling to the periderm for some time.

Mature cork cells are usually arranged in compact, radial rows and are nonliving. They have an air space in the center that contributes to the buoyancy of cork in water. The air spaces also push compressed cork back to its original shape. Cell walls are thin or thick and without pitting and characteristically are full of a waxlike substance called **suberin.** The deposition of suberin typically occurs in layers alternating with other waxes, resulting in a

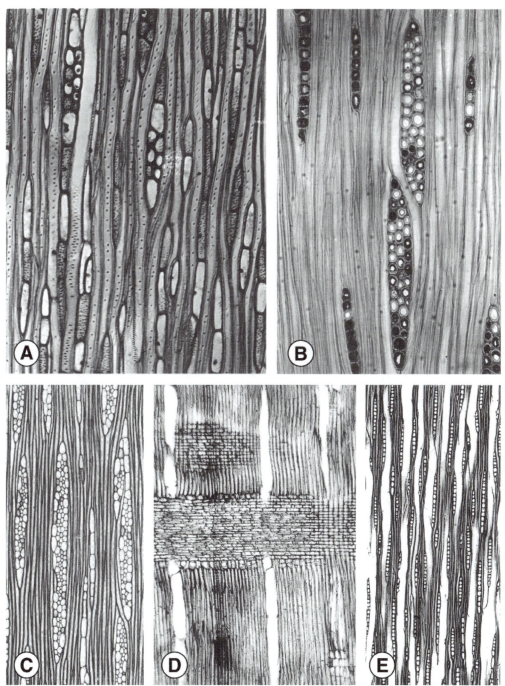

**FIGURE 4.26**  Wood ray structure as determined from tangential and radial wood sections. (A) Heterocellular rays of *Illicium floridanum* (Illiciaceae). (B) Homocellular rays of *Acer rubrum* (Aceraceae). Note all ray cells are procumbent. ×400. (C) Both multiseriate and uniseriate heterocellular rays of *Gastonia* (Araliaceae). (D) Heterocellular rays of *Pittosporum spathaceum* (Pittosporaceae) as viewed in radial section. Note one marginal row of upright and square cells at the top and bottom of ray and procumbent cells in ray center. × 55. (E) Exclusively uniseriate rays of *Eucryphia lucida* (Eucryphiaceae). A, B, C, courtesy of E. A. Wheeler.

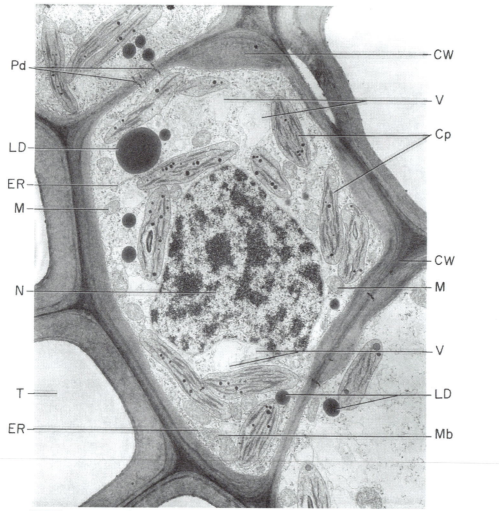

**FIGURE 4.27**    EM of living wood ray parenchyma cell of red pine (*Pinus resinosa*) showing the principal structure. Abbreviations: CW, cell wall; Cp, chloroplast; LD, lipid droplet; ER, endoplasmic reticulum; Mb, microbody; N, nucleus; V, vacuole; T, tracheid; Pd, plasmodesmata; M, mitochondrion. × 8000. Reprinted with permission from Kozlowski and Pallardy (1997), "Physiology of Woody Plants," 2nd ed., Academic Press.

lamellate structure of suberized walls when viewed in thin sections. In addition to suberin, all barks contain cellulose, hemicellulose, lignin, dark-staining polyphenols, and other extractives. Suberin and lignin are the principal components of cork, contributing about 40% and 22%, respectively, of its dry weight. The presence of large quantities of suberin in the walls of cork cells makes them poor conductors of heat and highly impervious to gases and liquids. These qualities make cork an effective barrier against the outside environment; thus cork is sometimes utilized in the manufacture of insulation board. The internal derivatives of the phellogen rarely comprise more than a few loosely organized rows that resemble cortical cells. In contrast to phellem,

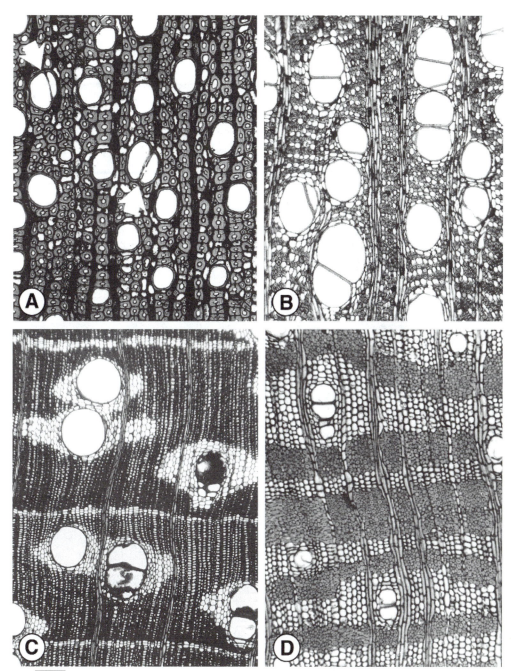

**FIGURE 4.28** Wood axial parenchyma distribution as determined from wood transverse sections. (A) Pores exclusively solitary and axial parenchyma apotracheal diffuse and paratracheal scanty in *Curtisia* (Cornaceae). Arrows indicate regions of vessel element overlap. ×180. (B) Axial parenchyma diffuse in aggregates and narrow banded in *Scytopetalum tieghemii* (Scytopetalaceae). × 55. (C) Aliform and confluent paratracheal parenchyma and zonate parenchyma associated with the growth ring boundaries in *Afzelia africanca* (Leguminosae), ×50. (D) Broad-banded wood parenchyma in *Andira inermis* (Leguminosae). ×35. A, C, D, courtesy of E. A. Wheeler.

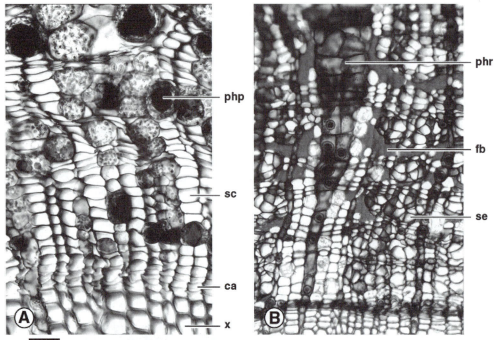

**FIGURE 4.29**   Secondary phloem. (A) Transverse section of secondary phloem of *Pinus sylvestris* (Pinaceae). Note radial arrangement of sieve cells. (B) Transverse section of secondary phloem of *Tilia americana* (Tiliaceae). Abbreviations: ca, vascular cambium; fb, fibers; php, phloem parenchyma; phr, dilated phloem ray; sc, sieve cell; se, sieve tube element; x, secondary xylem.

mature parenchymatous phelloderm cells retain their protoplasts with nuclei. The fact that these cells may have chloroplasts accounts for their designation as green cork. Phelloderm is not present in all woody plants.

To exchange gases and water, woody plants almost universally develop small structures known as **lenticels** within the periderm. Lenticels form on stems and even some fruits. Lenticels commonly arise beneath stomata or between stoma, usually just prior to or at the same time as the first periderm. As the lenticel develops, the epidermis is ruptured, and the mass of lenticel tissue is exposed. The inner boundary of the lenticel is a meristematic zone that is continuous with the phellogen. A fully developed lenticel is distinguished from the phellem by having a mass of loosely arranged, thin-walled, and unsuberized cells with abundant intercellular spaces. This tissue is called **complementary tissue.** At maturity the cells making up the complementary tissue lose their protoplasts and become colorless. Complementary tissue can occur as a somewhat uniform and compact tissue, or they can be rather loosely organized. In some plants (e.g., *Ulmus, Alnus, Betula,* and *Sorbus),* the complementary tissue is horizontally traversed by bands of densely arranged, suberized cells called **closing layers.** Beneath the lenticel, the cortical air spaces become continuous with those in the complementary region.

As axis diameter increases as a result of the production of secondary tissues, the periderm in most species is subjected to stresses that cause it to crack

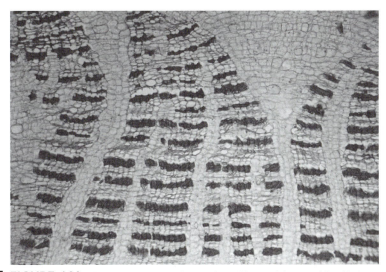

**FIGURE 4.30**  Transverse section of secondary phloem of *Annona glabra* (Annonaceae) showing V-shaped rays and stratified hard and soft bast. Courtesy of T. Terrazas-Salgado.

and split. Anticlinal divisions within the phellogen enable the cambium to increase in girth in concert with the increasing width of the vascular cylinder. As the axis grows, additional successive periderms can arise at any time interior to the one most recently produced. As this occurs, new segments of cork cambium develop in progressively older regions of the axis, often developing in overlapping layers. The isolated older tissues that are formed by the development of deeper and deeper cork cambia are called **rhytidome.** These tissues may include periderm and masses of cortical and phloem cells that have been successively cut off. Very little is known about the seasonal pattern of activity of the cork cambium, what stimulates the cambial cells to divide, and what controls the pattern of phellogen development. Bark surface features are

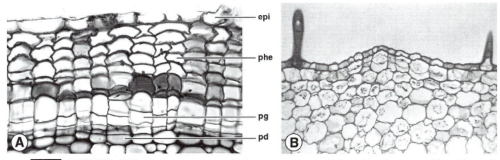

**FIGURE 4.31**  Periderm initiation. (A) Transverse section of stem of *Sambucus nigra* (Caprifoliaceae) showing an early stage in the development of periderm with phellogen and derivative tissues. (B) Early stage in subepidermal periderm development in *Pelargonium* (Geraniaceae). Note that the phellogen arises in a discontinuous manner. × 600. Abbreviations: epi, epidermis; pd, phelloderm; pg, phellogen; phe, phellem. B, courtesy of P. G. Mahlberg.

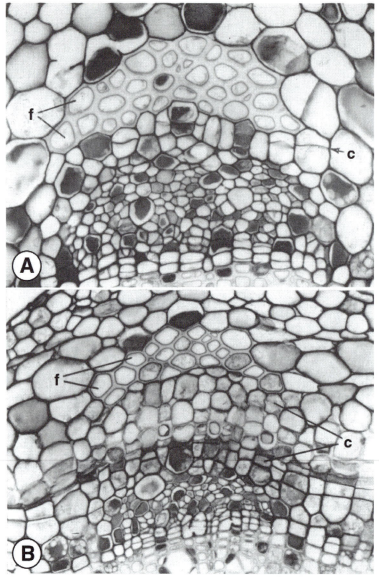

**FIGURE 4.32**   Transverse sections through a seedling stem of *Vitis vinifera* (Vitaceae) showing stages in the initiation of the deep-seated cork cambium. (A) Early stage of development showing initiation of cork cambium (c) in dividing phloem parenchyma cells. (B) Later stage of development of cork (c). Primary phloem fibers (f). Both ×290. From Esau (1948), *Hilgardia* **18**, 217–296. Used with permission of the Regents of the University of California.

determined by the developmental pattern of the cork cambium. In some trees, the bark is thin (0.5 in.), continuous, and smooth, and the original phellogen remains functional over an extended period. In these cases, the phellogen increases in girth by frequent anticlinal divisions. When trees develop successively more internal meristems, however, the outer bark becomes irregularly split and exfoliates in large sheets or in small, irregular segments. Trees in which phellogens are renewed in concentric, continuous layers shed their bark

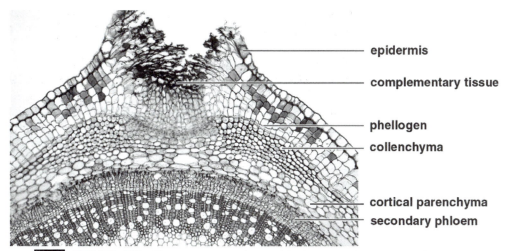

epidermis

complementary tissue

phellogen

collenchyma

cortical parenchyma
secondary phloem

**FIGURE 4.33**  Transverse section of stem of *Sambucus nigra* showing mature lenticel containing loosely arranged complementary cells.

in thin strips, papery layers or as cylinders. This type of bark pattern is called **ring bark.** Replacement phellogens that form in overlapping layers produce a **scale bark** surface pattern.

The most commercially valuable cork is derived from the outer bark of the cork oak (*Quercus suber*), an evergreen tree from the Mediterranean region. A uniform, high quality cork layer forms over a period of several years and is periodically stripped from the tree. If the harvesting is done carefully, the inner phellogen remains undamaged, and a layer of new growth cork is formed. Everyone is familiar with the most common use of cork, namely cork bottle stoppers. The anatomical characteristics of cork make it ideal for this purpose. Cork possesses lenticels, however, that extend through the tissue allowing for gas exchange between the outside and inner tissues. For this reason, bottle stoppers must be cut out at right angles to the surface of a piece of cork so that the lenticels extend horizontally rather than vertically, which would allow leakage of the contents. Barks also are used as a source of valuable fibers, resins, tannins, latex, medicines, poisons, flavors, and hallucinogenic substances. The remarkable properties of *Prunus serrula* bark has recently been described. The thin, tough, flexible, and semitransparent bark of this species has been compared to a strong natural polymer film. Unlike most barks that have cuboidal or slightly expanded cork cells among the tangential axis, the cork cells of *Prunus serrula* have considerable tangential enlargement many times the radial dimension. The bark also shows exceptionally high toughness values and high density.

## SECONDARY GROWTH IN MONOCOTYLEDONS

Most monocotyledons show a total lack of lateral cambial activity. This absence of any type of normal secondary thickening places severe restrictions on

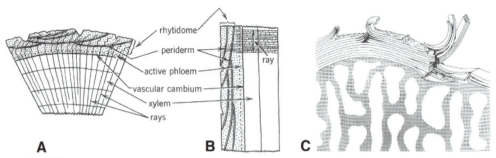

**FIGURE 4.34**   Rhytidome and its location with reference to vascular tissues. The rhytidome in this example is composed of periderm and nonliving secondary phloem. (A) Transverse section of part of stem. (B) Longitudinal section of part of stem. (C) Rhytidome development as observed in trunk from *Betula papyrifera* (Betulaceae). A, B reprinted with permission from Esau (1965), "Plant Anatomy," 2nd ed., John Wiley & Sons. C, courtesy of P. G. Mahlberg.

their size and growth habit. As a result, most monocots are small, herbaceous plants. There are some monocots, however, that form abruptly thickened stems in response to a special type of meristematic region located near the shoot apex. These are typically plants consisting of very short internodes and crowded leaves at the crown. Familiar examples of this condition are seen in the palms, rhizomatous irises, and bulbous lilies. There are also a few large, treelike monocots that have evolved an additional, unique type of secondary growth that is derived from a laterally positioned meristem. This latter category of taxa is sometimes referred to as the "woody" monocotyledons or "tree lilies."

It is a truism that one must bear the growth habits in mind when discussing large-bodied monocots. As a result, some authors have divided the arborescent monocots into three fairly natural groups: (1) the palms (Palmae), most of which have unbranched stems in the vegetative state and no lateral secondary thickening; (2) the pandans (Pandanaceae), with branched stems and no lateral secondary thickening; and (3) the somewhat miscellaneous assemblage of plants, represented by the treelike lilies, with branched stems and a very peculiar type of lateral secondary growth in thickness.

Many monocotyledons possess a meristematic region just below the shoot apex termed the **primary thickening meristem.** This narrow region originates

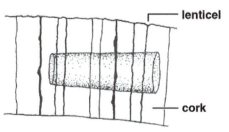

**FIGURE 4.35**   Diagram of cork bottle stopper cut from cork at right angles to the surface so that the lenticels extend horizontally across the stopper.

from ground tissue and is located just beneath the site of leaf attachment at the crown. It encircles the apical region and undergoes periclinal division to form radially aligned derivatives toward both the outside and inside. These derivatives form the bulk of the stem tissue. Parenchyma is produced toward the outside, whereas a combination of parenchyma and vascular bundles is formed internally. The vascular bundles formed in this way interconnect with the vasculature already supplying the leaves and stem. Adventitious roots also arise by the activity of the primary thickening meristem. The activity of the primary thickening meristem, followed by the considerable radial enlargement of its derivatives, is responsible for much of the increase in stem diameter just below the apical meristem. The further considerable increase in stem diameter among palms and other plants is primarily attributable to cell divisions and enlargement of dispersed ground parenchyma cells, a phenomenon known as diffuse secondary growth.

In a few large, woody members of the Liliflorae (*Yucca, Cordyline, Aloë, Dasylirion, Baeucarnea*) and some herbaceous taxa of Asparagales, a structurally and functionally unique type of cylindrical lateral meristem is present just external to the original primary vascular bundles. This meristem, or cambium, is termed the **secondary thickening meristem** (or thickening ring) and is a zone several cells in thickness within the cortex. This meristem may or may not be longitudinally continuous with the apical primary thickening meristem. The two meristems are sometimes regarded as representing different developmental phases of the same meristem. The initials are single layered and somewhat rectangular and tapering at the ends when seen in longitudinal view. No ray initials are present. The initials also are short lived and are continuously replaced with new dividing cells. The secondary thickening meristem gives rise to a little secondary parenchyma toward the outside, termed **secondary cortex,** and many secondarily derived vascular bundles embedded in parenchyma to the inside. The internal derivatives that differentiate into vascular bundles continue to divide longitudinally in various planes, and the maturing tracheary elements undergo extensive elongation. The secondarily produced vascular bundles differ from the primary bundles to which they make connection by being amphivasal in organization. The arborescent monocots possess the unusual anatomical feature of persistent and functional leaf traces that traverse the cambial region and secondary tissues. In some plants, the leaves thus remain attached and functional on the stem, for some distance below the level of initiation of the secondary thickening meristem. The only monocot that is known to develop secondary thickening in the root is *Dracaena*.

## PERIDERM FORMATION IN MONOCOTYLEDONS

The plant body of most monocotyledons is covered by the original primary epidermis; under some conditions it becomes lignified or suberized. In some monocots, hypodermal cells develop thick, lignified walls, adding additional external protection. Although few comprehensive comparative studies of periderm development in monocots have been completed, a number of arborescent monocotyledons (including some palms and Liliaceae) develop an

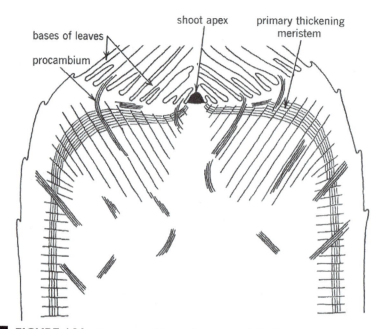

**FIGURE 4.36**  Upper part of shoot of a monocotyledon illustrating meristems concerned with its growth. Apical meristem produces axial tissue downward and leaf primordia laterally. Beneath the primordia, derivatives of apical meristem divide periclinally and form anticlinal rows (indicated by widely spaced parallel lines). Increase in axis thickness results. Periclinal divisions may be localized in a mantle-like tissue region, the primary thickening meristem. This meristem may be prolonged in the peripheral part of the axis and may be continuous with cambium-producing secondary tissues. The primary thickening meristem forms ground parenchyma and procambial strands. Reprinted with permission from Esau (1965), "Plant Anatomy," 2nd ed., John Wiley & Sons.

outer periderm that resembles that in dicotyledons and reportedly undergoes a similar development. The outer cortical layers of other monocots (e.g., members of the Bromeliaceae, Commelinaceae, and Zingiberaceae) form a protective region known as **storied cork.** This condition is characterized by bands of secondarily septated parenchyma cells in which the products of division do not enlarge. The divided cells are distributed among undivided cortical cells. All the cells in the outer cortical layers become suberized and collectively form the storied cork.

## SUMMARY

All gymnosperms and most dicotyledons develop lateral meristems, which result in radial growth and are responsible for an increase in the width of stems and roots. Growth that is derived from the lateral meristems is called secondary growth; it is composed of secondary tissues. Lateral meristems are termed cambia and function as bifacial zones composed of initials and derivatives. The vascular cambium adds to the girth of the axis by the production of secondary xylem internally and secondary phloem externally. The outer protective layers of the axis are called periderm and are formed by the cork cambium, or phellogen.

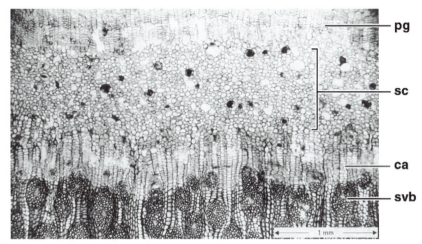

**FIGURE 4.37**  Transverse section of stem of the monocot *Cordyline indivisa*, the mountain cabbage tree of New Zealand, showing thickening lateral cambium and stages in development of secondary vascular bundles. A periderm originating from a cambium in the outer cortex is shown at the top. Abbreviations: ca, thickening cambium; pg, phellogen; sc, secondary cortex; svb, secondary vascular bundle. From Tomlinson and Zimmermann (1967), used with permission of the International Association of Wood Anatomists.

The vascular cambium develops a short distance behind the apical meristem. The cambium that forms within the vascular bundle is called fascicular cambium. The interfascicular cambium is initiated somewhat later from the renewed division of parenchyma cells located between the original vascular bundles. The formation of interfascicular cambium is an example of cellular dedifferentiation in which cells of one morphological type are transformed into another structural condition. Most vascular cambia possess two categories of initials of distinct size and shape. These are the fusiform initials and the ray initials. Fusiform initials are elongated, spindle-shaped elements that give rise to axial systems of tracheary elements, fibers, sieve elements, and parenchyma. Ray initials divide periclinally to form aggregations of radially extended parenchyma cells called rays.

A cambium in which the fusiform initials are overlapping and of variable length is termed unstratified. A specialized stratified or stored cambium occurs when the fusiform initials are of nearly equal length and grouped in parallel, horizontal rows. In order for the cambium to keep pace with the rapid increase in axis diameter, fusiform initials must occasionally divide in a radial, longitudinal plane. In nonstratified cambia the orientation of the cell plate within dividing fusiform initials fluctuates between nearly transverse and varying degrees of obliquity. Following division, the daughter cells undergo apical elongation and slide by one another intrusively until a maximum length is attained and the girth of the cambium is increased. In stratified cambia an increase in cambial girth follows nearly longitudinal, anticlinal divisions of fusiform initials, followed by the lateral expansion of the derivatives. The average length of fusiform initials has declined phyletically. Some woody plants deviate from the normal pattern of cambia activity and produce anomalous secondary growth.

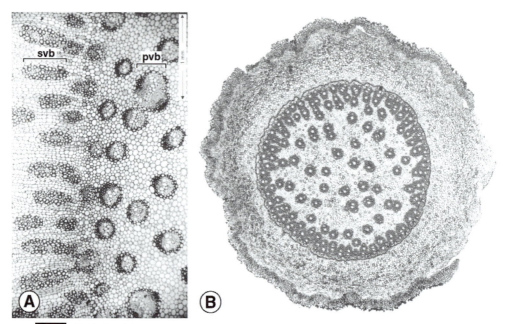

**FIGURE 4.38**   Secondary growth in Liliaceae. (A) Transverse section of stem of *Dracaena hawaiensis*. Note abrupt boundary between primary collateral bundles (pvb) at the right of section and secondary amphivasal bundles (svb) amidst radially seriated parenchyma at left of section. (B) Transverse section of root of *Dracaena hookeriana*. A, reprinted with permission from Tomlinson and Zimmermann (1967), *IAWA Bull.*, International Association of Wood Anatomists. B, courtesy of the Bailey-Wetmore Wood Collection, Harvard University.

The activity of the vascular cambium over time produces a large amount of secondary xylem, or wood. Wood consists of two positionally arranged tissue systems: the axial (or vertical) system and the ray (or horizontal) system. The axial system consists of vessel elements, tracheids, fibers, and axial parenchyma. The ray system is composed predominantly of ray parenchyma. The abundance, arrangement, and structure of the cells within these systems is of value for wood identification and description.

Structurally the periderm consists of three components: the initiating layer or phellogen, nonliving cork cells called phellem, and in some plants one or more internal layers composed of living parenchyma cells, called phelloderm. Mature cork cells are nonliving; their walls contain a waxlike substance called suberin. To exchange gases, woody plants develop small structures known as lenticels within the periderm. The isolated older tissues of periderm that are formed by the development of deeper and deeper cork cambia are called rhytidome.

Most monocotyledons show a lack of lateral cambial activity. Some monocots, however, form a functionally unique type of lateral meristem termed the secondary thickening meristem. The secondary thickening meristem gives rise to a little secondary parenchyma toward the outside and many secondarily derived vascular bundles to the inside.

## ADDITIONAL READING

1. Alfieri, F. J., and Evert, R. F. (1973). Structure and seasonal development of the secondary phloem in the Pinaceae. *Bot. Gaz.* **134**, 17–25.
2. Ayensu, E. S., and Stern, W. L. (1964). Systematic anatomy and ontogeny of the stem in Passifloraceae. *Contrib. U. S. Nat. Herb.* **34**, 45–73.
3. Bailey, I. W. (1919). Phenomena of cell division in the cambium of arborescent gymnosperms and their cytological significance. *Proc. Natl. Acad. Sci.* (USA) **5**, 283–285.
4. Bailey, I. W. (1920a). The formation of the cell plate in the cambium of the higher plants. *Proc. Natl. Acad. Sci.* (USA) **6**, 197–200.
5. Bailey, I. W. (1920b). The cambium and its derivative tissues, III. A reconnaissance of cytological phenomena in the cambium. *Am. J. Bot.* **7**, 417–434.
6. Bailey, I. W. (1923). The cambium and its derivative tissues. IV. The increase in girth of the cambium. *Am. J. Bot.* **10**, 499–509.
7. Bailey, I. W. (1944). The development of vessels in angiosperms and its significance in morphological research. *Am. J. Bot.* **31**, 421–428.
8. Bannan, M. W. (1955). The vascular cambium and radial growth of *Thuja occidentalis*. L. *Can. J. Bot.* **33**, 113–138.
9. Bannan, M. W. (1962). The vascular cambium and tree-ring development. *In* "Tree Growth" (T. T. Kozlowski, Ed.), pp. 3–21. Ronald, New York.
10. Bannan, M. W. (1967). Anticlinal divisions and cell length in conifer cambium. *For. Prod. J.* **17**, 63–69.
11. Brown, C. L. (1971). Secondary growth. *In* "Trees: Structure and Function" (M. H. Zimmermann and C. L. Brown, Eds.), pp. 67–123. Springer-Verlag, New York.
12. Butterfield, B. G., and Meylan, B. A. (1980). "Three-Dimensional Structure of Wood. An Ultrastructural Approach," 2nd ed. Chapman and Hall, London.
13. Caballé, G. (1993). Liana structure, function and selection: A comparative study of xylem cylinders of tropical rainforest species in Africa and America. *Bot. J. Linnean Soc.* **113**, 41–60.
14. Carlquist, S. H. (1988). "Comparative Wood Anatomy." Springer-Verlag, New York and Berlin.
15. Carlquist, S. (1991). Anatomy of vine and liana stems: a review and synthesis. *In* "The Biology of Vines" (F. E. Putz and H. A. Mooney, Eds.), pp. 53–71. Cambridge Univ. Press, Cambridge.
16. Catesson, A. M. (1994). Cambial ultrastructure and biochemistry: Changes in relation to vascular tissue differentiation and the seasonal cycle. *Int. J. Plant Sci.* **155**, 251–261.
17. Cheadle, V. I. (1937). Secondary growth by means of a thickening ring in certain monocotyledons. *Bot. Gaz.* **98**, 535–555.
18. Cooke, G. B. (1948). Cork and cork products. *Econ. Bot.* **2**, 393–402.
19. Core, H. A., Côté, W. A., and Day, A. C. (1979). "Wood Structure and Identification," 2nd. ed. Syracuse University Press, Syracuse.
20. Davis, J. D., and Evert, R. F. (1965). Phloem development in *Populus tremuloides*. *Am. J. Bot.* **52**, 627.
21. Dobbins, D. R. (1971). Studies on the anomalous cambial activity in *Doxantha unguis-cati* (Bignoniaceae). II. A case of differential production of secondary tissues. *Am. J. Bot.* **58**, 697–705.
22. Dobbins, D. R., and Fisher, J. B. (1986). Wound responses in girdled stems of lianas. *Bot. Gaz.* **147**, 278–289.
23. Esau, K. (1964). Structure and development of the bark in dicotyledons. *In* "The Formation of Wood in Forest Trees" (M. H. Zimmerman, Ed.), pp. 37–50. Academic Press, New York.
24. Esau, K., and Cheadle, V. I. (1969). Secondary growth in *Bougainvillea*. *Ann. Bot.* **33**, 807–819.
25. Ewers, F. W., Fisher, J. B., and Fichtner, K. (1991). Water flux and xylem structure in vines. *In* "The Biology of Vines" (F. E. Putz and H. A. Mooney, Eds.), pp. 127–160. Cambridge University Press, Cambridge.
26. Ghouse, A. K. M., and Hashmi, S. (1983). Periodicity of cambium and the formation of xylem and phloem in *Mimusops elengi* L., an evergreen member of tropical India. *Flora* **173**, 479–487.

27. Iqbal, M. (Ed.) (1990). "The Vascular Cambium." Research Studies Press Ltd., Taunton, Somerset, England.
28. Junikka, L. (1994). Survey of English macroscopic bark terminology. *IAWA J.* **15**, 3–45.
29. Kozlowski, T. T. (1964). Shoot growth in woody plants. *Bot. Rev.* **30**, 335–392.
30. Kozlowski, T. T. (ed.). (1971). "Growth and Development of Trees: Seed Germination, Ontogeny, and Shoot Growth," Vol. I. 443 pp. Academic Press, New York.
31. Kozlowski, T. T. (1971). "Growth and Development of Trees: Cambial Growth, Root Growth, and Reproductive Growth," Vol. II. 514 pp. Academic Press, New York.
32. Kozlowski, T. T., and Pallardy, S. G. (1997). "Physiology of Woody Plants," 2nd ed. Academic Press, San Diego.
33. Larson, P. R. (1994). "The Vascular Cambium: Development and Structure." Springer Series in Wood Science (T. E. Timell, Ed.). Springer-Verlag, New York and Berlin.
34. Lauchaud, S. (1989). Participation of auxin and abscisic acid in the regulation of seasonal variations in cambial activity and xylogenesis. *Trees* **3**, 125–137.
35. Lev-Yadun, S., and Aloni, R. (1995). Differentiation of the ray system in woody plants. *Bot. Rev.* **61**, 45–84.
36. Roth, I. (1981). "Structural Patterns of Tropical Barks." Encyclopedia of Plant Anatomy. Gebrüder Borntraeger, Berlin.
37. Rudall, P. (1991). Lateral meristems and stem thickening growth in Monocotyledons. *Bot. Rev.* **57**, 150–163.
38. Scott, F. M. (1950). Internal suberization of tissues. *Bot. Gaz.* **110**, 492–495.
39. Siebers, A. M. (1971). Initiation of radial polarity in the interfascicular cambium of *Ricinus communis* L. *Acta Bot. Neerl.* **20**, 211–220.
40. Srivastava, L. M. (1963). Secondary phloem in the Pinaceae. *Univ. Calif. Publ. Bot.* **36**, 1–142.
41. Stevenson, D. W. (1980). Radial growth in *Beaucarnea recurvata*. *Am. J. Bot.* **67**, 476–489.
42. Stevenson, D. W., and Fisher, J. B. (1980). The developmental relationship between primary and secondary thickening growth in *Cordyline* (Agavaceae). *Bot. Gaz.* **141**, 264–268.
43. Tomlinson, P. B. (1964). Stem structure in arborescent monocotyledons. *In* "The Formation of Wood in Forest Trees" (M. H. Zimmermann, Ed.), pp. 65–86. Academic Press, New York.
44. Tomlinson, P. B., and Zimmerman, M. H. (1967). The "wood" of monocotyledons. *Bull. IAWA* **2**, 4–24.
45. Tomlinson, P. B., and Zimmermann, M. H. (1969). Vascular anatomy of monocotyledons with secondary growth—An introduction. *J. Arn. Arb.* **50**, 159–179.
46. Waisel, Y., Liphschitz, N., and Arzee, T. (1967). Phellogen activity in *Robinia pseudoacacia* L. *New Phytol.* **66**, 331–335.
47. Waisel, Y., Noah, I., and Fahn, A. (1966). Cambial activity in *Eucalyptus camaldulensis* Dehn. II. The production of phloem and xylem elements. *New Phytol.* **65**, 319–324.
48. Wheeler, E. A., Baas, P., and Gasson, P. E. (Eds.) (1989). IAWA list of microscopic features for hardwood identification. *IAWA Bull. n.s.* **10**, 219–332.
49. Whitmore, F. W., and Zahner, R. (1966). Development of the xylem ring in stems of young red pine trees. *For. Sci.* **12**, 198–210.
50. Whitmore, T. C. (1962). Studies in systematic bark morphology. II. General features of bark construction in Dipterocarpaceae. *New Phytol.* **61**, 208–220.
51. Whitmore, T. C. (1963). Studies in systematic bark morphology. IV. The bark of beech, oak and sweet chestnut. *New Phytol.* **62**, 161–169.
52. Wilcox, H. (1962). Cambial growth characteristics. *In* "Tree Growth" (T. T. Kozlowski, Ed.), pp. 57–88. Ronald, New York.
53. Wilson, B. F. (1964). Structure and growth of woody roots of *Acer rubrum* L. *Harv. For. Pap.* **11**.
54. Wilson, B. F., Wodzicki, T., and Zahner, R. (1966). Differentiation of cambial derivatives: Proposed terminology. *For. Sci.* **12**, 438–440.
55. Wilson, K., and White, D. J. B. (1986). "The Anatomy of Wood: Its Diversity and Variability." Stobart and Son Ltd., London.
56. Worbes, M. (1989). Growth rings, increment and age of trees in inundation forests, savannas and a mountain forest in the Neotropics. *IAWA Bull. n.s.* **10**, 109–122.
57. Xu, X., Schneider, E., Chien, A. T., and Wudl, F. (1997). Nature's high-strength semitransparent film: The remarkable mechanical properties of *Prunus serrula* bark. *Chem. Matr.* **9**, 1906–1908.

# II

# EVOLUTIONARY, PHYSIOLOGICAL, AND ECOLOGICAL PLANT ANATOMY

# 5
# EVOLUTION AND SYSTEMATICS

**Systematics** is a broad field of inquiry that uses information from many disciplines to carry out its primary objectives of describing, identifying, naming, classifying, and determining relationships among plants. Systematics is based on the premise that, in the variation within evolutionary groups, discrete units that can be recognized, classified, circumscribed, and named occur, and that logical relationships exist among these units. The concept of **character** is fundamental to the discipline of taxonomy. Characters provide the basic information for classification and are the features used in identification. They are used in the determination of relationships, and complexes of characters associated with different plants provide the basis for the description and the naming of taxa. The characterization or description of plants and their component parts is the initial and primary process in systematics and clarifying patterns of evolutionary radiation. The assignment of attributes to plants, their parts, and their tissues is a prerequisite for the establishment of discontinuities in form and structure between organisms. Delimitation and definition of taxa are prerequisites for the development of identification and classification schemes. The fundamental unit of description is the **character state,** which is the particular expression of a character of a given object. Characters, in turn, are grouped together to form a type of **evidence.**

## SYSTEMATIC PLANT ANATOMY

Ideally, systematic and phylogenetic conclusions are reached by harmonizing different types of evidence. Anatomical data obtained from various cells and tissues are important evidence when elucidating the relationships among higher plants. Anatomy finds application when characterizing large natural complexes, such as the familiar structural differences between the vascular systems of monocotyledons and dicotyledons and the anomalous cambial activity in stems of the natural assemblage known as the Caryophyllidae (Centrospermae). Anatomy also has been used in some cases to separate species, although it has traditionally proven to be most useful at the genus level and higher.

During the first half of the 20th century, great strides were made in the study of systematic plant anatomy. The first comprehensive systematic summary of anatomical information was provided by Hans Solereder whose *Systematische Anatomie der Dicotyledonen* (2 vols.) was first published in 1899, with an English translation in 1908. In 1950 the monumental two-volume work *Anatomy of the Dicotyledons* by the Englishmen C. R. Metcalfe (1904–1991) and L. Chalk (1895–1979) appeared, again dealing with aspects of the vegetative anatomy of dicotyledonous families. Like other evidence, the use of anatomy for systematic purposes requires precision in the definition of character states, logic in the hierarchial ordering and arranging of characters and character states, and comprehensiveness of description. The better defined and restricted in occurrence a character is, the more useful it is for systematic delimitation. Thus, anatomy will not differ greatly among species of the same genus. In addition to an important role in phylogenetic analysis, anatomical data can be applied toward the independent resolution of such problems as helping to place systematically difficult taxa, evaluating the taxonomic homogeneity and naturalness of taxa, and revising families. The basic premise upon which most systematic anatomical investigation rests is that similarity of structure implies genetic relationship. However, as with the evaluation of other characters, caution is required when making anatomical comparisons because similarities in structural specialization do not necessarily imply close relationship but may be the result of **parallel** or **convergent evolution.** As a result, anatomical data often have proven most reliable in the refutation of claims of close relationship, rather than positive assertions of the relationship of taxa. It also is important to make the distinction between general (diagnostic) characters that enable one taxon to be separated or distinguished from another or that may imply phenetic relationship among taxa, and those characters that can be used phylogenetically. Therefore, anatomical information can be taxonomically useful without having obvious evolutionary or phylogenetic interpretation.

## Variation in Anatomical Characters

To use anatomical characters as a source of systematically useful information, it is necessary to understand the range and source of character variation within an individual plant, species, or group of related taxa. Variation can occur in both quantitative and qualitative aspects of structure. Some variation accompanies developmental change during the growth of an individual. Xylem cell size, for example, can vary considerably with the age of a tree and the location of a sample within the plant. This aspect of variation is seen in the increase in tracheary cell and fiber length and diameter along the radial direction from the pith to the stem periphery, that is, from the first formed to the more mature wood. A relatively stable tracheary element length is usually not reached for a period of several years, following the initiation of secondary growth and an initial period of rapid cell length increase. As a rule, vessel frequency decreases with cambial age. In woods with both simple and scalariform vessel element perforation plates, the incidence of scalariform plates can decrease with an increase in stem diameter. The age at which the xylem attains maturity varies with the taxon and in some trees is greatly delayed.

Tracheary cells and fibers also can show increases in overall lengths along the axial direction of the trunk. Cell length increases from the base of the stem upward through the tree, until a height is reached at which length begins to decrease. Cell lengths in a branch are consistently shorter than those found in the adjacent trunk. Ontogenetic developmental variation is further evidenced by extensive changes in wood ray size and morphology of their constituent cells between young and mature wood. Through successive stages of secondary growth, ray structure is often transformed in height and width from uniseriate or biseriate (or some combination of these two) in immature xylem to exclusively uniseriate or multiseriate in older wood. Not only do ontogenetic changes in ray width and height occur, but also phylogenetic specialization can become evident during transitions from the heterogeneous to homogeneous condition. Variation also can be attributed to seasonal changes in growth activity, as observed in ring porous trees in which the wide, earlywood vessel elements are often distinctly shorter in total length than the narrower, latewood cells. Variation in tracheid length also occurs across one growth increment in conifer wood. Significant differences in xylem structure also can be found between the stem and root of the same plant. Thus, anatomical descriptions should not be based on immature specimens.

Individuals or populations of plants within species with a wide geographic range often vary greatly in structure. As discussed in Chapter 8, this geographic variation may be due to complex environmental or habitat influences, such as the availability of water, light, altitudinal adaptation, and degree of crowding. Genetic factors also influence structure and frequently interact with other variables. In some cases, anatomical characters vary enormously without any evident ecological or developmental cause. In the genus *Cratoxylum* (Guttiferae), for example, the presence or absence of leaf

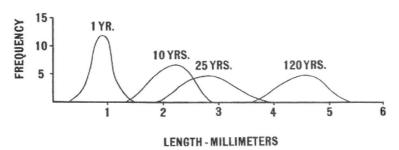

**FIGURE 5.1** Variations in length of tracheary elements with increasing age or diameter of the stem of *Pinus*. The basis of each frequency distribution measurement is 100 tracheids.

epidermal papillae and a hypodermis can vary within a single subspecies with as yet no obvious known pattern or explanation.

The implications of structural variation in anatomical characters are enormous when anatomy is used for systematic purposes. The most obvious conclusion that can be reached is that the assumption that a single sample of an organ or tissue provides reliable data upon which to base anatomical generalizations is invalid, particularly if the location of the sample within the plant is unknown. Also, because cell size varies with plant age, cells from immature tissues should never be used as a source of quantitative data. To describe the entire range of character variation accurately, it is necessary to investigate the range of variability within and across individuals in relation to ontogeny, the environment, and the location of tissues within the plant. Ideally, we examine multiple collections over the entire geographic range of a species. When measuring cell size, at least 50 measurements of any given character are recommended in order to obtain data representative of a sample.

In this chapter we consider several of the guiding principles of systematic and evolutionary anatomical study. Selected examples of the use of anatomical data in the solution of systematic problems at different levels in the taxonomic hierarchy are provided.

## Xylem Evolution

During the past half century or more, the value of wood anatomy and vessel element evolution in particular, has been firmly established in the study of phylogeny and classification of flowering plants. These early studies were sparked by interest in the evolution of fossil plants, but soon proved to be highly significant in elucidating higher level phylogeny of extant dicotyledons. In no other vegetative tissues are the trends of structural evolution as clearly defined as in the secondary xylem. Because these trends were recognized entirely without reference to existing taxonomic systems and thus without reference to the relative primitiveness or advancement of the plants in which they occur, wood anatomy in particular can be used as an independent test of systematic hypotheses. The vast accumulation of comparative data on wood structure has resulted in part from the economic importance of wood, the relative ease of specimen preparation, and the fact that the secondary xylem is often well preserved in dried specimens and fossil plant materials as a result of its rigid cell walls. Major wood characters of proven systematic and phylogenetic value include tracheid and fiber type, vessel element structure and distribution, ray histology, and axial parenchyma abundance and distributional patterns (see Chapter 4). Many of the major structural trends in wood character evolution have been regarded to be virtually irreversible. This conclusion has been challenged in recent years by assertions that many wood characters in different groups of flowering plants have undergone evolutionary reversals and evolved counter to the general trend of wood evolution.

One of the great achievements of evolutionary biology has been the elucidation of the origin and subsequent trends of specialization of the angiosperm vessel element. The study of vessel element evolution was initiated by the brilliant American plant anatomist Irving W. Bailey (1884–1967) and several collaborators early in the 20th century and has been continued and expanded by

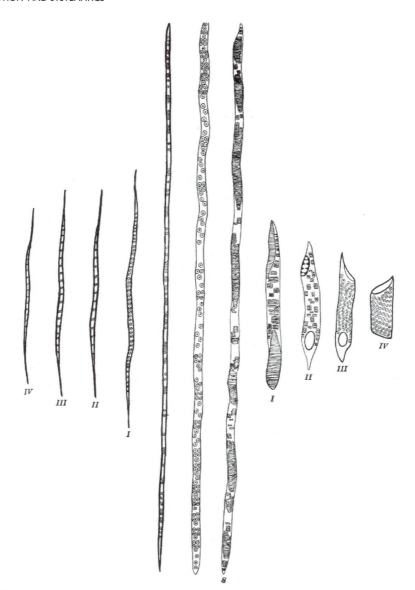

**FIGURE 5.2** Diagrammatic illustration of the major trends of specialization of tracheary elements and fibers in mature wood as reflected in size and structure. S, scalariformly pitted tracheid. I–IV on right shows trend in vessel element evolution associated with a decrease in cell length, reduction in inclination of end walls, change from scalariform to simple perforation plates, and from scalariform to alternate intervascular pitting. I–IV on left shows evolution of fibers associated with decrease in length, reduction in size of pit borders, and change in shape and size of pit apertures. Reprinted with permission from Bailey and Tupper (1918). Copyright © American Academy of Arts and Science.

many other individuals over the past 75 years. The evolutionary trends of the vessel element have been established through an extensive and intensive series of comparative anatomical studies of both fossils and living plants, statistical analyses, and ontogenetic studies. Therefore, the evolutionary trend from tracheids to vessel elements is one of the most reliable tools in the study of phylogeny. This is true because this trend is, to a large degree, both unidirectional and irreversible. It has been shown that angiosperm secondary xylem vessel elements have been phylogenetically derived from scalariform pitted tracheids by the progressive loss of pit membranes in the regions of tracheid overlap. Moreover, some primitive extant vessel element types have retained scattered remnants of pit membranes in the region of cell overlap. There is no evidence that pit membranes are ever regained once they are lost. Therefore, vessel elements that are most tracheidlike (i.e., elongate, narrow, angular in outline, with many scalariform perforations in the end walls) are believed to be the most primitive type. The relationship between element length and perforation plate condition is illustrated in Table 5.1 with data compiled from the long-term studies of monocotyledons conducted by Vernon I. Cheadle.

Other major trends of vessel element specialization are: (1) change in perforation plates from scalariform, many barred, and fully bordered to few barred and borders absent and then to more advanced simple (porous) perforation plates; (2) change in angle of end wall so that it becomes highly oblique and tapering with much overlap to transverse; (3) reduction in vessel element length; (4) change in appearance in transverse section from angular to circular; (5) increase in pore diameter; (6) change in intervascular pitting from scalariform to opposite and alternate multiseriate; and (7) change in vessel distribution from predominately or exclusively solitary to extensive aggregate groupings of vessels. The phylogenetic shortening of vessel elements is due primarily to a concomitant shortening of cambial fusiform initials. Vessels apparently first arose in stems of woody dicotyledons and only later evolved in the metaxylem and finally the protoxylem of the primary xylem. Like dicotyledons, all gradations between the most primitive vessel member type and the most advanced also can be found in extant monocotyledons. In monocots, vessels arose and became successively more specialized in the roots, stems, inflorescence axes, and leaves.

Character correlation has played a significant role in clarifying these trends. If one character can be demonstrated to be primitive (for example, scalariform perforation plates on the vessel element end wall), other charac-

**TABLE 5.1    Length of Vessel Elements (in millimeters) and Perforation Plate Type**

| END WALL INCLINATION | PERFORATION PLATE TYPE | | | |
| --- | --- | --- | --- | --- |
|  | Simple | 1–10 bars | 10–50 bars | Over 50 bars |
| Transverse | 0.73 | 1.22 | 1.97 | — |
| Slightly oblique | 0.96 | 1.31 | 2.14 | 1.50 |
| Oblique | 1.15 | 1.69 | 2.41 | 3.44 |
| Very oblique | — | — | 2.56 | 4.22 |

Reprinted with permission from Cheadle (1943a), *Am. J. Bot.* **30**, 11–17.

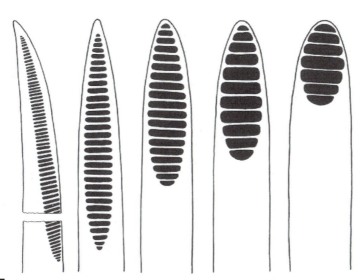

**FIGURE 5.3**  Stages in the evolution of scalariform perforation plates. From *Evolutionary Trends in Flowering Plants,* by A. Takhtajan © 1991 Columbia University Press. Reprinted with permission of the publisher.

ter states correlated with this feature are probably also primitive. An example of correlation is illustrated by intervessel pitting on the lateral or side walls (Table 5.2). The pattern of lateral wall pitting can be categorized into three main types of ascending specialization: **scalariform, opposite,** and **alternate.** Scalariform pitting is pitting in which elongated or linear pits are arranged in a ladderlike series on the side walls of tracheary elements. Multiseriate pitting in which the pits are in horizontal pairs or in short horizontal rows is called opposite. Alternate pitting is the most common type in dicotyledons and is characterized by pits in diagonal rows. A high degree of correlation exists between the presence of scalariform perforation plates and scalariform intervessel pitting and between simple perforation plates and opposite to alternate or alternate pitting. Similar correlations can be made between perforation plate type and other vessel element features, as well as other wood cell and tissue characters (Table 5.3). Contemporary studies of fossil woods are beginning to provide corroborative evidence relating to patterns of wood and vessel element evolution that were originally outlined from living plants. It has been found, for example, that Upper Cretaceous woods possessed fewer specialized characters, fewer short vessel elements in particular, compared to extant dicotyledonous trees.

The great value of wood anatomy as a guide to understanding flowering plant evolution lies in the fact that trends toward specialization are well defined, as well as that all stages in evolutionary specialization are observable in extant angiosperms. As we have seen, even intermediate stages of dissolution of the pit membrane in the region of tracheary element overlap are evident. Furthermore, a small number of genera of living woody dicotyledons are totally devoid of vessels and are widely considered to be primitively vesselless; that is, they never possessed vessels and still survive in that primitive

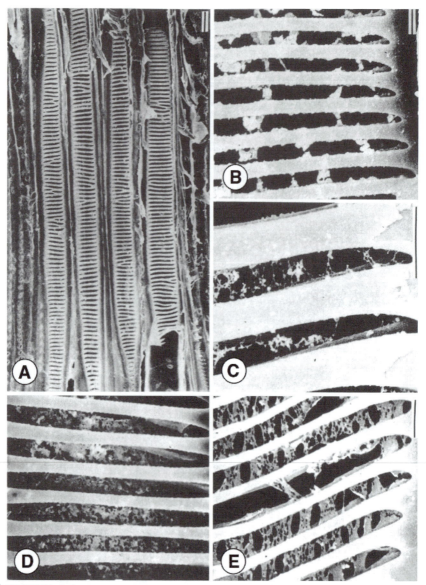

**FIGURE 5.4**  SEM photographs of scalariform perforation plates with membrane remnants. (A) Four perforation plates of *Aextoxicon punctatum* (Euphorbiaceae). Membrane remnants in perforation are inconspicuous at this magnification. Bar = 5 μm. (B) Perforations of *Aextoxicon punctatum* showing flakelike remnants. Bar = 1 μm. (C) Perforation of *Aextoxicon punctatum* showing strandlike remnants. Bar = 1 μm. (D) Perforation of *Illicium cubense* (Illiciaceae) showing nearly intact porous membranes in perforation. Bar in Fig. B. (E) Perforation in *Illicium cubense* showing porous bands and strands. Bar = 1 μm. Reprinted with permission from Carlquist (1992), *Am. J. Bot.* **79**, 660–672.

state today. The families and genera of extant vesselless dicotyledons are Amborellaceae *(Amborella)*, Chloranthaceae *(Sarcandra,* stem only), Tetracentraceae *(Tetracentron)*, Trochodendraceae *(Trochodendron)*, and Winteraceae *(Belliolum, Bubbia, Drimys, Exospermum, Pseudowintera,*

**TABLE 5.2** Correlations of Vessel Element Perforation Plates and Intervessel Lateral Wall Pitting Among Dicotyledons

| PERFORATIONS | INTERVESSEL PITTING | PERCENTAGE |
|---|---|---|
| Scalariform prevails | Scalariform and opposite | 86 |
| | Opposite and alternate or alternate | 14 |
| Intermediate, scalariform, | Scalariform and opposite | 80 |
| and simple | Opposite and alternate or alternate | 20 |
| Simple: vessel element end | Scalariform and opposite | 11 |
| walls tapering | Opposite and alternate or alternate | 89 |
| Simple: vessel element end | Scalariform and opposite | 6 |
| walls horizontal | Opposite and alternate or alternate | 94 |

Reprinted by permission from Stern (1978), *IAWA Bull* **2–3**, 33–39.

*Takhtajania, Tasmannia, and Zygogynum).* We cannot use the lack of vessels as the only evidence for concluding these are the most primitive angiosperms because evolution proceeds independently in the organs and tissues of plants.

Statistical correlations using vessel element features have revealed additional trends in phylogenetic specialization in the secondary xylem that are significant. These include the evolution of imperforate tracheary elements from tracheids to advanced libriform fibers through a progressive reduction in pit size and number, and eventual elimination of the tracheid pit border.

The classic work of David Kribs in the 1930s is an excellent example of the use of statistical data in anatomical research. Kribs utilized the established trends in vessel element specialization to clarify evolutionary trends in wood

**TABLE 5.3** Correlations of Vessel Element Perforation Plates and Ray Composition Among Dicotyledons

| TYPE OF VESSEL ELEMENT PERFORATION PLATE | RAY COMPOSITION | |
|---|---|---|
| | Heterogeneous (%) | Homogeneous (%) |
| Scalariform I (many barred) | 100.00 | — |
| Scalariform II (few barred) | 84.37 | 15.63 |
| Scalariform—porous | 86.56 | 13.44 |
| Porous—oblique | 89.63 | 10.37 |
| Porous—oblique and transverse | 44.5 | 55.42 |
| Porous transverse | 20.90 | 79.10 |

Reprinted by permission from Kribs (1935), *Bot. Gaz.* **96**, 547–557.

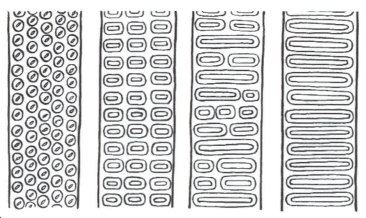

**FIGURE 5.5**    Stages in the evolution of lateral wall pitting in vessel elements from scalariform (right) through intermediate and opposite to the alternate pattern (left). From *Evolutionary Trends in Flowering Plants*, by A. Takhtajan © 1991 Columbia University Press. Reprinted with permission of the publisher.

ray structure. As can be seen in Tables 5.3 through 5.5, there is a strong correlation between long primitive vessel element types and the occurrence of rays in which the ray cells are oriented in both an erect and a procumbent (radially elongate) manner (**heterogeneous rays**). Conversely, advanced, porous vessel elements tend to be associated with rays in which the ray cells are all either upright or procumbent (**homogeneous rays**). These correlations enabled Kribs to recognize six structural categories or types of rays in dicotyledons of varying grades of evolutionary specialization. Each ray type is correlated with a particular level of vessel element specialization (Tables 5.4 and 5.5).

These statistical comparisons all point to a phylogenetic sequence of wood ray structure. Primitive woods possess the following features: (1) a combination of both uniseriate (one cell wide) and multiseriate (multicelled) rays are present, (2) both ray types are of marked vertical height and have a cellular composition consisting of both upright and procumbent cells, and (3) multiseriate rays have many rows of upright or square marginal cells. With evolutionary specialization this primitive ray composition has become modified so that the following changes can be observed: (1) reduction and elimination of multiseriate rays, (2) reduction and elimination of uniseriate rays, (3) reduction in height of the uniseriate marginal extensions on multiseriate rays, (4) formation of homogeneous rays composed of procumbent cells, and (5) elimination of all rays so that woods become rayless. The development of stratified rays formed by a storied cambium also represents an extreme advancement. Recent studies have resulted in the modification of older systems of ray classification and have emphasized the influence of ontogeny on ray structure.

The various types of axial parenchyma distribution patterns also can be treated statistically. In Table 5.6 we see the correlation between diffuse axial parenchyma that is independent of the vessels (**apotracheal parenchyma**) and primitive forms of vessel elements. Various patterns of banded parenchyma

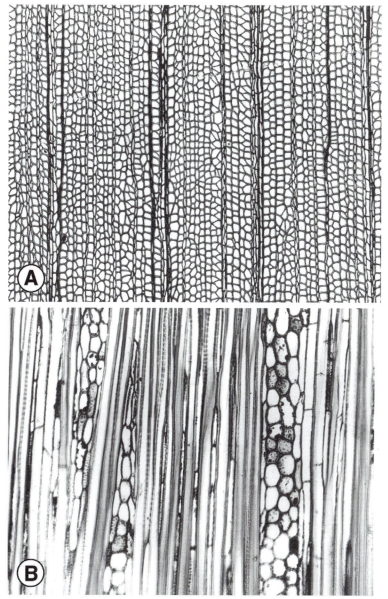

**FIGURE 5.6** Wood anatomy of the vesselless dicotyledon *Drimys* (Winteraceae). (A) Transverse section of *D. winteri*. (B) Tangential section of *D. axillaris* showing both uniseriate and multiseriate hetero-cellular rays. Both × 50. Courtesy of the Bailey-Wetmore Wood Collection, Harvard University.

and parenchyma that is associated with vessels (**paratracheal parenchyma**) are correlated with more advanced vessel element types.

Although it is reasonably clear that diffuse axial parenchyma represents the primitive state in dicotyledons from which a diversity of more advanced patterns were derived, the lines of structural specialization of wood parenchyma remain unclear.

**TABLE 5.4   Average Length of Vessel Elements for Each Ray Type**

| TYPE OF RAY | AVERAGE VESSEL ELEMENT LENGTH (mm) |
|---|---|
| Heterogeneous Type I | 0.81 |
| Heterogeneous Type III | 0.64 |
| Heterogeneous II | 0.58 |
| Homogeneous Type I | 0.52 |
| Homogeneous Type III | 0.38 |
| Homogeneous Type II | 0.35 |

Reprinted by permission from Kribs (1935), *Bot. Gaz.* **96**, 547–557.

## USE OF WOOD ANATOMY IN SYSTEMATICS

Flowering plant phylogeny and classification presents us with an outstanding example of the application of anatomical data to the solution of systematic problems. Which families occupy a primitive or basal position among living angiosperms? At the turn of the 20th century there were two opposing schools of thought regarding the origin of angiosperms and the most primitive floral characters. One school was led by the German botanist Adolph Engler (1844–1930), professor of botany at the University of Berlin and the most influential taxonomist of his time. The other school was fully developed by the American systematic botanist Charles E. Bessey (1845–1915). In the comprehensive Englerian system of classification, the woody plants with small,

**TABLE 5.5   Percentage of Ray Types in Each Vessel Type**

| | MULTISERIATE AND UNISERIATE RAY TYPES | | | | UNISERIATE RAY TYPES | |
|---|---|---|---|---|---|---|
| | Heterogeneous | | Homogeneous | | Heterogeneous | Homogeneous |
| TYPE OF VESSEL ELEMENT PERFORATION PLATE | Type I | Type II | Type I | Type II | Type III | Type III |
| Scalariform I | 79.36 | 15.87 | — | — | 4.77 | — |
| Scalariform II | 53.12 | 21.87 | 12.50 | 3.13 | 9.38 | — |
| Scalariform—porous | 47.77 | 35.81 | 5.98 | 4.48 | 2.98 | 2.98 |
| Porous—oblique | 44.09 | 43.18 | 5.45 | 3.56 | 2.36 | 1.36 |
| Porous—oblique and transverse porous | 9.45 | 33.11 | 34.45 | 9.49 | 2.02 | 11.48 |
| Porous—transverse | — | 19.09 | 27.28 | 44.10 | 1.81 | 7.72 |

Reprinted by permission from Kribs (1935), *Bot. Gaz.* **96**, 547–557.

**TABLE 5.6** **Percentage of Parenchyma Types in Each Vessel**

| TYPE OF VESSEL ELEMENT PERFORATION PLATE | DIFFUSE | DIFFUSE AGGREGATE | SCANTY | NARROW BANDED | WIDE BANDED | VASICENTRIC |
|---|---|---|---|---|---|---|
| Scalariform I | 69.84 | 19.04 | | | | |
| Scalariform II | 59.37 | 15.62 | | | | |
| Scalariform—porous | 13.23 | 41.18 | 20.59 | 8.82 | | |
| Porous oblique | 11.88 | 32.65 | 15.78 | 8.41 | | 11.96 |
| Porous oblique and transverse | 4.16 | 14.06 | 7.81 | 17.18 | 10.42 | 31.28 |
| Porous transverse | — | 2.69 | — | 6.66 | 9.33 | 70.22 |

Reprinted by permission from Kribs (1937), *Bull. Torrey Bot. Club* **64**, 177–186.

apetalous, unisexual, anemophilous or wind-pollinated flowers borne in aments or catkins (e.g., willows, walnuts, birches, oaks, etc., often called Amentiferae) were considered the most primitive dicotyledons. These taxa are the first apetalous families in Engler's classification. According to this view, the ancestors of the angiosperms were coniferoid or gnetoid gymnosperms with unisexual strobili. On the other hand, Bessey considered the early flower to be relatively large, solitary, bisexual and insect pollinated, with an apocarpous gynoecium and a perianth of many free parts. Flowers of this general construction are represented today by some members of the magnoliaceous alliance (magnolias, annonas, buttercups), and Bessey grouped them under a category called the Ranales. The ancestors of the angiosperms were believed to be cycadophytes with bisexual strobili.

The comparative anatomist William L. Stern, used wood anatomical evidence to test these two hypotheses. The main features of the secondary xylem of amentiferous and ranalean families are summarized and compared in Table 5.7. Whereas the amentiferous families show a moderately advanced level of xylem evolution, many of the features of the secondary xylem are regarded as primitive in the extant ranalean families. Furthermore, the latter complex includes woody taxa that are entirely vesselless, suggesting that the immediate ancestors of the flowering plants also lacked vessels. Xylem anatomy, therefore, has been found to be correlated generally with other lines of evidence in supporting the view that a component of the Besseyan Ranales (the Magnolidae or Annoniflorae of contemporary phylogenists) should be considered primitive in the classification system. According to the basic tenets of secondary xylem evolution it would be most unlikely (or impossible) that a complex with more advanced xylem features (i.e., Amentiferae) gave rise to one with the least advanced representatives (i.e., Magnoliales). Despite the importance of xylem anatomy in building a phylogenetic system of classification, caution is required when basing systematic

**TABLE 5.7  Comparison of Selected Anatomical Characters of the Xylem in Amentiferous and Ranalean Families**

| ENGLERIAN AMENTIFEROUS FAMILIES | BESSEYAN RANALEAN FAMILIES |
| --- | --- |
| 1. Perforations tending toward simple or simple and scalariform | Perforations predominantly scalariform |
| 2. Vessel elements relatively short | Vessel elements relatively long |
| 3. Lateral wall pitting opposite to alternate | Lateral wall pitting commonly scalariform and opposite |
| 4. Vessels broad and rounded in trans-section | Vessels narrow and angular in trans-section |
| 5. Axial xylem parenchyma both apotracheal and paratracheal | Axial xylem parenchyma usually apotracheal |
| 6. Vascular rays tending toward homogeneous | Vascular rays basically heterogeneous |
| 7. Imperforate tracheary elements (fibers) often with minutely bordered or unbordered pits | Imperforate tracheary elements (fibers) often with conspicuously bordered pits |

Reprinted by permission from Stern (1978), *IAWA Bull.* **2–3**, 33–39.

and phylogenetic conclusions upon similarities and differences in xylem structure. Because both quantitative and qualitative features of the wood are strongly influenced by various climatic variables, the ecology and habitats of plants also must be considered.

As with other anatomical features, wood characters of restricted distribution assume increased systematic value and are especially useful in the circumscription of taxonomic groups. For example, the constant combined occurrence of **vestured pits** and **included phloem** (both uncommon derived features) can be used to characterize the core families of the woody order Myrtales, comprising the families Combretaceae, Lythraceae, Melastomataceae (including Crypteroniaceae), Myrtaceae, Oliniaceae, Onagraceae, Penaeaceae, Punicaceae, Psiloxylaceae, and Sonneratiaceae. Included phloem is secondary phloem located in the secondary xylem (also known as interxylary phloem). Broadly based surveys of the order Myrtales have shown that the shared possession of these two characters is strong evidence of relationship. The lack of internal phloem and vestured pits, coupled with other vegetative and reproductive features, has provided an equally strong argument for the removal of such families as the Lecythidaceae and Rhizophoraceae from the core Myrtales. Although each of these features occurs individually in a number of other widely unrelated taxa, their occurrence together is a good indicator that the taxa possessing them have mutual affinities.

Careful investigation of the structural variation among vestured pits at the scanning electron microscope level provides additional evidence for the recognition of two major types of vesturing that can be used to distinguish subfamilies of the Combretaceae. Vestured pitting is generally characteristic of entire families, or groups within a family. The number, size, and distribution of vestures vary considerably, and these variations may be of diagnostic value. However, caution is again required. Although most Leguminosae possess

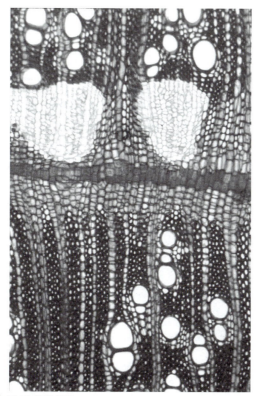

**FIGURE 5.7** Transverse section of wood of *Avicennia germinans* (Avicenniaceae) showing isolated secondary phloem strands of included phloem. Courtesy of Teresa Terrazas.

wood elements with vestured pits, a few lack this feature. Therefore, even though vestured pitting is characteristic of the Leguminosae, it is not a completely reliable diagnostic feature of the family.

A special type of apparently empty upright or square xylem ray cell occurring in intermediate horizontal series and usually interspersed among the procumbent cells is called a **tile cell.** These cells were originally described by the term *ziegelsteinformig* in reference to their resemblance to tile work when viewed in radial section. Tile cells are a distinctive wood anatomical feature and are restricted to members of the order Malvales. The identification of tile cells in either extant or fossil woods allows such specimens to be reliably grouped with families of the Malvalean complex. **Internal phloem** is primary phloem located internally to the primary xylem. Members of the order Gentianales are nearly always characterized by internal phloem, either as a continuous ring or as separate strands at the margin of the pith.

## Nodal Anatomy

Leaves, as organs of the plant body, arise successively from the stem apical bud, which forms a series of nodes and internodes along the primary axis.

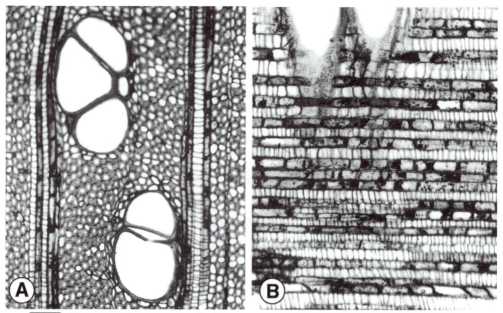

**FIGURE 5.8**  Tile cells in wood rays of *Guazama* (Sterculiaceae). (A) Transverse section. The cells can be distinquished from procumbent ray cells in cross section. The procumbent ray cells have long axes in the radial direction. (B) Radial view of ray with tile cells that lack contents and look like small upright cells. Both × 180. Courtesy of E. A. Wheeler.

Several patterns of leaf arrangement (phyllotaxy) occur in different plants or occasionally in the same plant. Some common phyllotactic patterns include whorled, in which three or more leaves are inserted at a node; opposite, in which a pair of leaves is positioned at a node; decussate, in which pairs of leaves are arranged at right angles to the preceding pair; distichous, in which leaves form two parallel ranks along the stem; and alternate, in which the leaves are arranged spirally on the stem.

Development of vascular tissues is intimately associated with leaf formation and in the higher plants occurs as strands or bundles that connect the leaf and stem. **Nodal anatomy,** accordingly, describes the pattern of vascular continuity between these two organs. The vascular connection between leaf and stem is maintained by vascular bundles or **leaf traces,** which are associated with parenchymatous interruptions in the stem vascular cylinder called **leaf gaps.** A leaf trace may be defined as a vascular bundle in the stem, extending between its connection with a leaf and with another vascular unit in the stem. Early in the 20th century it was observed that the number of leaf traces and associated gaps at a node was almost always constant within an individual species, although often variable in different higher level plant groups. As a result, the number was considered diagnostic when studying the systematic and phylogenetic relationships between different taxa. Depending upon whether one, three, or five or more leaf gaps are left in the stele by the departure of leaf traces, nodes are described as **unilacunar, trilacunar,** or **multilacunar.** The nodal pattern is often expressed in terms of the number of traces and gaps. For example, a unilacunar node with a single trace would be described as 1:1; a unilacunar node with two diverging traces would be 2:1; a trilacunar, three-trace node would be 3:3, and so on. The first figure represents the number of traces, whereas the second figure reveals the number of gaps in the vascular system of the stem. Trilacunar, three-trace nodes are distinguished by having a median gap and an associated median trace that is flanked on either side by a lateral gap and an associated lateral trace. The two lateral traces extend obliquely outward around the cortex before entering the leaf base. It must be emphasized that in studies of nodal anatomy, descriptive words of motion indicating origins and pathways of vascular bundles are frequently used as a matter of convenience, even though leaf traces and gaps result from different patterns of differentiation of procambial strands.

Most families tend to have a uniform nodal anatomy. Variability in nodal structure is important and may be of systematic value in intrafamilial classification. Trilacunar nodes occur in the majority of dicotyledons. Studies show that the unilacunar, two-trace type is present in ginkgophytes, various gymnosperms, and some dicotyledons. Multilacunar nodes are relatively uncommon in dicotyledonous families but are found in members of some primitive as well as some advanced orders. The unilacunar node is scattered among various groups but can distinguish an entire family, such as Theaceae. As leaves are reduced in size as a result of more xeric habitats, nodal anatomy is sometimes concomitantly reduced from trilacunar to unilacunar. At the base of the petiole, the leaf traces often subdivide or fuse to form the petiole vascular supply. The variability of vascular patterns in the petiole is shown in Figure 5.11.

## Foliar Epidermis

Leaves are anatomically highly variable organs, and the variation is sometimes specific for genera, species, or even families. Numerous anatomical characters within the leaf have proven to be of systematic value in different lineages. The leaf surface is a particularly useful source of systematically valuable characters. The waxy outer covering, or cuticle, may possess depositions in the form of papillae, striae, or rods of diagnostic distribution and orientation. Epidermal cell size, shape, contents, and wall thickness, including the occurrence of papillae, are potentially important systematic features.

Diversity of stomatal types offers one of the most important and readily observable epidermal characters. Stomatal type, for example, is a key character in the circumscription of the major groups of monocotyledons. As discussed in Chapter 2, stomata can be categorized on the basis of the developmental origins of subsidiary cells relative to the guard cells. Stomata also can be classified according to their mature appearance. These latter schemes are based on the presence or absence of subsidiary cells and the number and shape of subsidiary cells in relationship to the guard cells. Classifications of mature stomatal types have long been utilized. The following represent a few of the readily recognizable major types of angiosperm stomata that are based on mature appearance:

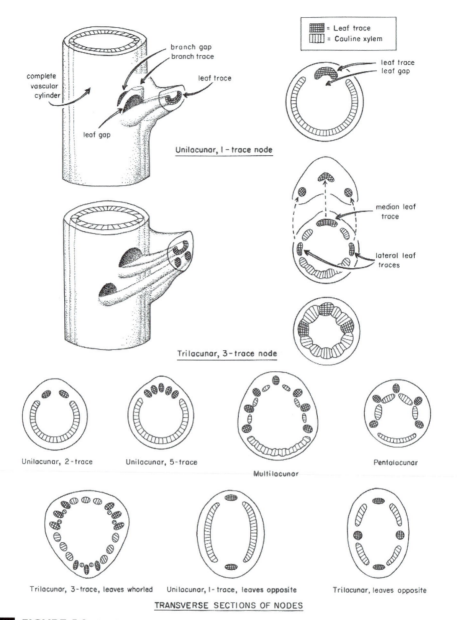

**FIGURE 5.9**  Nodal anatomy and nodal types. From *Vascular Plant Systematics* by Radford et al., copyright © 1974 by Albert E. Radford, William C. Dickison, James R. Massey and C. Ritchie Bell. Reprinted by permission of Addison-Wesley Educational Publishers.

**Anomocytic** (irregular-celled or ranunculaceous) stoma surrounded by a limited number of cells that are indistinguishable in size, shape, or form from those of the remainder of the epidermis.

**Anisocytic** (unequal-celled, cruciferous type) stoma surrounded by three subsidiary cells of which one is distinctly smaller than the other two.

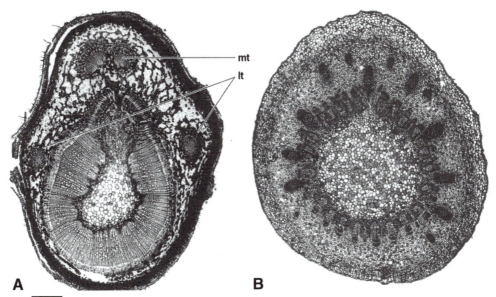

**FIGURE 5.10** Nodal anatomy. (A) Transverse section of node of *Corylus* (Betulaceae) showing a pair of lateral leaf traces (lt) and subdivided median leaf trace (mt). (B) Transverse section of node of *Magnolia acuminata* (Magnoliaceae) showing multilacunar, multitrace condition. Courtesy of the Bailey-Wetmore Wood Collection, Harvard University.

**Paracyctic** (parallel-celled, rubiaceous type), stoma accompanied on either side by one or more subsidiary cells parallel to the long axis of the pore and guard cells.

**Diacytic** (cross-celled, caryophyllaceous) stoma enclosed by a pair of subsidiary cells whose common wall is at right angles to the guard cells.

**Tetracytic** type, four subsidiary cells are present, two lateral and two terminal.

**Actinocytic** type, stoma are surrounded by a circle of radially elongated subsidiary cells which form a ring around each stoma.

**Cyclocytic** type, stoma surrounded by four or more subsidiary cells which form a ring around each stoma.

**Hexacytic,** stoma accompanied by six subsidiary cells consisting of two lateral pairs parallel to the long axis of the pore and two polar (terminal) cells; the second lateral pair are as long as the stomatal complex.

Most taxonomic groups can be characterized on the basis of one stomatal type, although some taxa show a combination of two or more patterns.

Trends of evolution among stomatal types have long been unclear. At one time both stomata with subsidiary cells and without subsidiary cells were regarded as primitive in angiosperms. Emphasizing development led to the hypothesis that a pair of guard cells without subsidiary cells is the simplest stomatal structure and probably represents the primitive condition in vascular plants. The formation of subsidiary cells represents a higher level of development and evolutionary advancement.

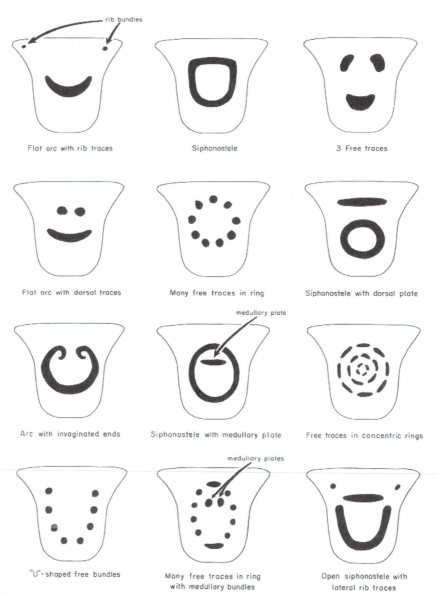

**FIGURE 5.11**   Common petiole vascular patterns as seen in transverse sections at the base of the lamina. Blackened areas represent vascular tissue. From *Vascular Plant Systematics* by Radford et al., copyright © 1974 by Albert E. Radford, William C. Dickison, James R. Massey, and C. Ritchie Bell. Reprinted by permission of Addison-Wesley Educational Publishers.

In the vegetative organs of the grass family, the most important diagnostic characters are found as microsopic features of the leaf. The epidermis of grass leaves possesses constituent cells that have long been used in identifying and classifying the Gramineae. Even fossil grass leaf fragments can be successfully identified using epidermal cell characters. The foliar surface is composed of cells of two distinct sizes, so-called **long cells** and **short cells.** The

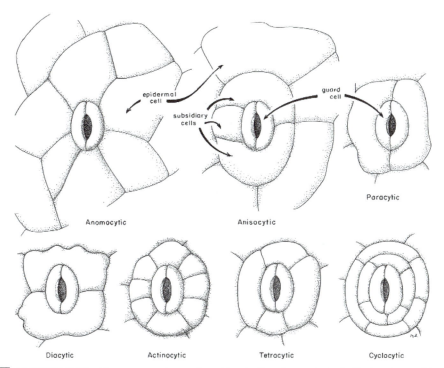

**FIGURE 5.12** Mature stomatal types. From *Vascular Plant Systematics* by Radford et al., copyright © 1974 by Albert E. Radford, William C. Dickison, James R. Massey, and C. Ritchie Bell. Reprinted by permission of Addison-Wesley Educational Publishers.

long cells are elongated parallel with the long axis of the leaf, whereas the short cells are more or less isodiametric in shape. The frequency of distributional pattern of these two cell types (singly or in pairs) is informative. Short cells are further classified either as **silica cells,** those containing silica grains, or as **cork cells,** those with suberized walls. The silica bodies within the silica cells assume characteristic forms that are also of diagnostic and systematic value. The grass stomatal apparatus has a unique appearance. The elongated pair of guard cells are enlarged at either end and constricted in the middle. The bulbous ends of the guard cells are thin walled, whereas the middle part of the walls are unevenly thickened. The two guard cells are bordered on either side by a pair of large hemispherical subsidiary cells. This stomatal structure is apparently restricted to grasses. In addition to surface cell type and distribution, dermal appendages are very common in grasses and are conveniently divided into **macro-hairs, micro-hairs, prickle-hairs,** and **papillae.** Leaf anatomy provides evidence for the existence of well-defined groups of grasses.

## Trichomes

Careful study of trichome morphology and distribution can yield important clues regarding specific, generic, tribal, and subfamilial relationships, although

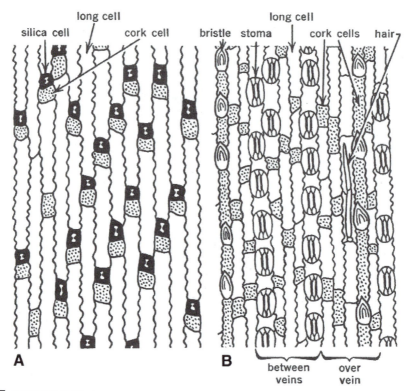

**FIGURE 5.13**    Epidermis of sugarcane, *Saccharum* (Gramineae), in surface view. (A) Stem epidermis showing alternation of long cells with pairs of short cells: cork cells and silica cells. × 500. (B) Lower epidermis from leaf blade showing distribution of stomata in relation to various kinds of epidermal cells. × 320. Reprinted with permission from Esau (1965), "Plant Anatomy," 2nd ed., John Wiley & Sons.

the systematic value of certain hair types is lessened by their independent evolution in different phyletic lines. Most plants have trichomes (hairs) on some part of their stems, leaves, or flowers. A plant without trichomes is said to be **glabrous.** If trichomes are present, the plant is **pubescent.** When describing trichomes, a distinction must be made between the nature of the surface vestiture, or indument, and the structure (and sometimes color) of the individual hair. Although trichomes were used in classical systematics, a classification that satisfactorily accounts for their great structural diversity has yet to be proposed. Current terminology of mature hair forms is far from exact. Generally speaking, trichomes are subdivided into **unicellular** and **multicellular** forms, whether they consist of a single row of cells (uniseriate) or of several to many rows of cells (multiseriate), and whether the hairs are branched or unbranched. The occurrence of **glandular hairs,** with a knoblike secretory apical swelling, is particularly important taxonomically.

Entire families or genera may be delimited by a single hair type, but families also can possess numerous trichome morphologies. The commonly cultivated tomato and potato plants form unicellular, multicellular, and glandular hairs on the same leaf. Multicellular branched hairs are characteristic of *Verbascum thapsus* and *Platanus*. Members of the Urticaceae form **stinging hairs. Scale hairs** are typical of the Bromeliaceae, whereas certain Malvaceae have stellate

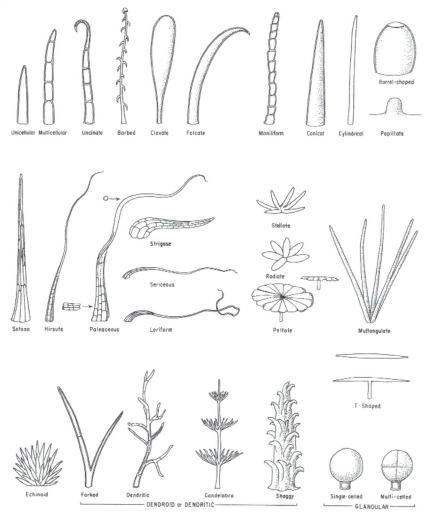

**FIGURE 5.14** Trichome structures. From *Vascular Plant Systematics* by Radford et al., copyright © 1974 by Albert E. Radford, William C. Dickison, James R. Massey, and C. Ritchie Bell. Reprinted by permission of Addison-Wesley Educational Publishers.

and tufted hairs. The tropical family Malpighiaceae is especially well characterized, although not uniquely so, by so-called **T-shaped** unicellular hairs ("Malpighian" type) composed of one or two roughly horizontal arms, attached to the plant by a more or less vertical stalk. In a number of instances, hairs are species specific and are of value in the analysis of hybrids. The use of pubescence characters to support generic relationships or differences established on the basis of other characters has been clearly established in a number of diverse families, including Asteraceae, Icacinaceae, Goodeniaceae, Ericaceae, and Gramineae.

## Mineral Inclusions

Crystalline substances of varied form and chemical composition are found in the cells, cell walls, and intercellular spaces of plant tissues. An unusual crystal cell that is of special taxonomic significance is the **cristarque cell.** At maturity these cells are dead and have a lignified U-shaped wall thickening on their inner walls, with an unthickened outwardly directed wall. In the lumen or space of each cell is a solitary and usually spherical crystal. Cristarque cells are highly diagnostic because they occur in only a few families, namely some members of the Melastomataceae, Scytopetalaceae, and especially the Ochnaceae, as well as a few other taxa. Among the major crystal types, druses and prismatic crystals are the most frequently occurring forms in dicotyledons, often being found together in the same plant. Raphides are much more common in monocotyledons than in dicotyledons, but they are characteristic of certain dicotyledonous families (i.e., Aizoaceae, Onagraceae, Nyctaginaceae, Rubiaceae, Dilleniaceae, and Vitaceae). Styloids are frequently encountered in members of the monocotyledonous families Liliaceae, and Iridaceae. Crystal sand also is much less common among dicotyledons than either druses and prismatic forms. Because not only the frequency but also the morphology of crystals can change during the life history of a plant, special care is required when utilizing crystal evidence to reach systematic conclusions.

The study of plant crystals can be applied to the recognition of altered plant products. Jute fibers derived from *Corchorus capsularis* and *C. olitorius* (Tiliaceae) can be separated from inferior malvaceous substitute fibers such as Kenaf or Bilimbi jute *(Hibiscus cannabinus)* and Roselle *(Hibiscus sabdariffa)* by the presence of chains of solitary crystals. After reducing the fiber to ash, the more valuable commercial fibers are associated with solitary prismatic crystals, whereas druses are present in the preparation of the substitute fibers.

Members of the monocotyledonous orders Juncales and Restionales provide one of the best known examples of the systematic usefulness of silica. Members of these complexes are often similar in general habit and have a superficial resemblance to Cyperaceae (sedges) and Gramineae (grasses). Significantly, silica bodies are always absent in the rush family Juncaceae and related Centrolepidaceae, a family of semiaquatic herbs from the Southern Hemisphere. Notably, silica bodies are present in some of the cells of the small, putatively related family Thurniaceae. A further distinguishing feature of the Thurniaceae is the occurrence of vascular bundles in vertical pairs in the leaf, the lower and smaller bundle of the pair inverted so that the phloem is

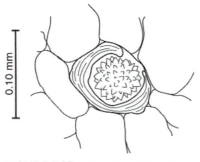

**FIGURE 5.15**    Crystal-containing cristarque cell from leaf of *Ochna multiflora* (Ochnaceae). Drawing by the author.

positioned adaxially to face the phloem of the upper bundle. This distinctive arrangement is unknown in the leaves of any other plants. Spheroidal, somewhat druselike silica bodies also are common in the stems of another rushlike family of the Southern Hemisphere, the Restionaceae. Silica deposits also are of almost universal occurrence throughout the Cyperaceae and Gramineae. Among sedges the distinctive silica bodies virtually always take the form of cones, or they are conical with flattened bases. Silica body shape, on the other hand, is quite variable in grasses. Many terms have been employed to refer to this variation (for example, cubical, round, elliptical, saddle shaped, dumbbell shaped and crescent shaped). Silica bodies in palms are of two main types, hat shaped or conical with a flattened or irregular base, and spherical, usually rather irregular and often somewhat ellipsoidal. One type occurs throughout all the organs of an individual species, so the shape of the silica body is a useful diagnostic feature.

## Ultrastructure

Although systematic anatomical evidence has historically been obtained at the light microscope level of observation, the subcellular level of analysis has assumed increasing importance. The widespread use of the electron microscope has shown the cell cytoplasm to be a highly organized region composed of many structurally complex and interrelated membranous organelles. Broadly based comparative ultrastructural investigations have successfully contributed to the field of plant systematics.

The significance of electron microscopy for identifying characters for angiosperm classification is illustrated by the cells of the phloem tissue. It has been shown that sieve tube elements can be separated on the basis of the contents of their plastids. Sieve tube elements contain plastids that have been designated as either S-type or P-type. S-type plastids accumulate starch only, whereas plastids of P-type contain protein alone or protein and starch. Numerous P-subtypes also can be recognized and are helpful in the circumscription of higher taxa. Uniformity of plastid type can be found as high as the rank of class. Monocotyledons are thus far known to possess only P-type plastids. A more specific case is seen in the Caryophyllales (Centrospermae), in which sieve element plastids provide a reliable character to delimit the order. Members of the complex have P-type plastids distinguished by peripheral ring-shaped bundles of proteinaceous filaments, often encircling an additional central core of proteinaceous material. Subtypes are recognized on the basis of the central protein core, which can be either polygonal or globoid in outline, or absent.

## PHYLOGENETIC SYSTEMATICS

The objectives of phylogenetic systematics are to reconstruct the evolutionary history of a group of organisms, to study character evolution in reference to the reconstruction, and to develop a classification that reflects these relationships. If a classification is veridical and reflects the evolutionary history of a group, it should be able to accommodate new information and be predictive. **Cladistic analysis** is the modern approach to phylogenetic reconstruction because it provides a rigorous, logical, and repeatable method for developing hypotheses and analyzing relationships among taxa. The group under study must be **monophyletic.** A monophyletic group is one that is hypothesized to have a single common ancestor and includes all the taxa derived from that ancestor. An hypothesis of relationship for two or more taxa is based on their shared possession of a derived character state. Shared, derived character states are called **synapomorphies.**

Character states change during evolution, and the transformation of states results in variation in character states and character state polarity. When a character state resembles the condition present in the ancestor of the group being studied, it is called the primitive or **plesiomorphic** state. A transformed character state that is possessed only by the present-day group under study and is not present in its ancestor is called the derived or **apomorphic** state. Character states thus change during evolution from plesiomorphic to apomorphic states. Like other characters, anatomical character states are potentially reversible to previous expressions. The extent of reversibililty or loss of anatomical features is a matter of considerable discussion. For example, the question of whether vessellessness and scalariform perforation plates are always primitive features in flowering plates (or can represent derived conditions) is of particular importance in cladistic phylogenetic studies. Key morphological and anatomical innovations, or synapomorphies, in land plant evolution are diagrammed in Figure 5.16. These innovations are characters that have become structurally and functionally important characteristics firmly established as defining features of the clade where they originated.

Assignment of polarity involves determining the primitive and derived status for the character states under study. In the absence of an adequate fossil record, one method used to form hypotheses about character state polarities is the use of **outgroups,** or monophyletic taxa similar to the ancestor but outside the group being studied. Character states are polarized when one state of a particular character that is found in the study group is also present in the outgroup. When this occurs, it can be hypothesized that the common character state is plesiomorphic, whereas the character state that is present only in the study group is the apomorphic state. In addition to the accurate interpretation of character state polarity, the characters being compared must be **homologous,** that is, share a common ancestry or origin. The assessment of homology among anatomical features is often difficult and may require careful developmental study. For example, superficially similar mature stomatal types can arise by different ontogenetic routes and thus may not be homologous. Scalariform intervessel pitting is another character state that can have different origins, and it would be a mistake to regard all scalariform pits as comparable. In most species scalariform pitting represents an ancestral condition; in few taxa it is probably a derived condition.

Another critical aspect of character analysis is the accurate coding or scoring of variable character states. When character states exist as discrete and well-defined units, the assignment of character codes is often readily accomplished. Characters that show a continuum of states are more problematic and

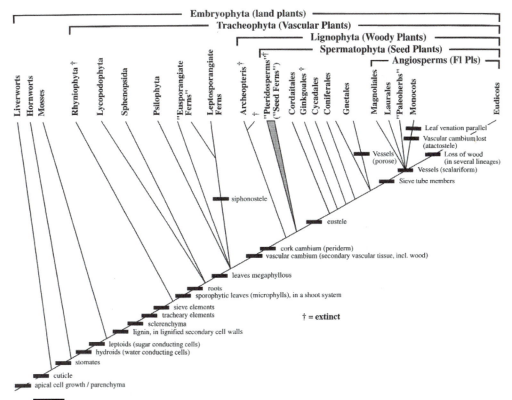

**FIGURE 5.16**    Cladogram showing hypothetical land plant phylogeny and major structural synapomorphies. Courtesy of Michael G. Simpson.

may not be useful for the creation of a data matrix. Quantitative wood anatomical characters represent a potentially valuable data set but may not be divisible into discrete states that can be accurately coded. Many anatomical characters are affected by the environment or plant habit and should be used with caution as characteristics to evaluate phylogenetic relationships. Accurate scoring of anatomical character states requires precise character state description and relies on a complete knowledge of the total range of structural variation as determined from all members of a taxon. Branching tree diagrams that specify hierarchial relationships among taxa based on shared, derived character states are called **cladograms.** Two closely related lineages within a cladogram that share the same common ancestor from which no other lineage has originated form a **sister group.**

An example of how anatomical characters can be used in the solution of systematic problems is provided by the small dicotyledonous family Alseuosmiaceae. In a recent systematic redefinition, the family was envisioned to consist of three well-defined genera scattered throughout the southwest Pacific area. The component taxa previously were assumed to have diverse affinities. Wood anatomy provided additional evidence to test the validity of this new realignment. Observations have shown that all three genera possess narrow vessel elements with primitively scalariform perforation plates. However, such plesiomorphic character states are not convincing evidence in establishing phylogenetic relationships. More meaningful in this case is the shared possession by all taxa of the clearly apomorphic wood character state of living septate fibers that store starch at maturity. The scarcity or absence of axial wood parenchyma and pores that are distributed as a combination of both solitary and grouped vessels are further derived features. Septate fibers tend to be common throughout natural groups and, when coupled with the retention of a living protoplast, form an extremely useful index of affinity. Furthermore, the family shows a strong phyletic shift toward raylessness. All

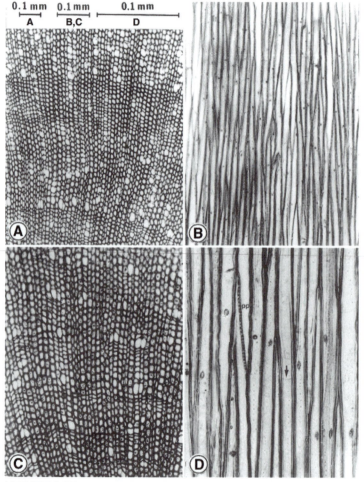

**FIGURE 5.17**   Wood anatomy of Alseuosmiaceae. (A) *Alseuosmia macrophylla,* transverse section. (B) *A. macrophylla,* tangential section showing an absence of rays. (C) *A. pusilla,* transverse section illustrating narrow, angular pores and weakly defined growth rings. Note starch grains in fibers. (D) *A. pusilla,* tangential section showing an absence of rays, the presence of nucleated, septate fibers (arrow) and oblique scalariform perforation plate (pp). Reprinted with permission from Dickison (1986), *Syst. Bot.* **11,** 214–221.

members of the family also have the unusual anatomical feature of a differentiated endodermis in the stem that is composed of cells with a Casparian band. The same primary form of endodermis also is continuous around each vascular bundle in the petiole. Vegetative anatomy, therefore, provides a valuable extra measure of confidence that can be used to validate the conclusion that the Alseuosmiaceae form a coherent group, defined by a combination of anatomical synapomorphies.

# SUMMARY

Systematics is a broad field of inquiry that uses characteristics and data from many disciplines to carry out its primary objectives of describing, naming, classifying, identifying, and determining relationships among plants. Anatomy broadens the base of systematic investigations by providing another set of characters that indicate relationships in harmony with external features. As with morphological characters, some anatomical features are useful in determining phylogenetic trends, whereas others delimit families, genera, or species. To use anatomical characters as a source of systematically useful information, it is necessary to understand the range and source of character variation within an individual plant or species and among a group of related taxa. Anatomical characters are neither more nor less reliable than characters from other parts of the plant, yet they play a central role in formulating systematic hypotheses and in constructing a phylogenetic system of angiosperm classification. Evolutionary modifications of anatomical characters are not necessarily closely synchronized. Primitive wood structure, for example, can be associated with specialized floral and other features.

The secondary xylem has proved to be highly significant in elucidating higher level plant phylogenies. In no other vegetative tissues are the trends of structural evolution as clearly defined. One of the great achievements of plant anatomy has been the elucidation of the origin and subsequent trends of specialization of the angiosperm vessel element. It was shown that angiosperm vessel elements have been phylogenetically derived from scalariform-pitted tracheids by the progressive loss of pit membranes in the regions of tracheid overlap. Long vessel elements, which are most tracheidlike, are believed to be the most primitive. Short vessel elements with simple perforation plates represent an advanced type. A small number of living woody dicotyledons are totally devoid of vessels and are widely considered to be primitively vesselless. Correlated with primitive vessel elements are occurrences of primitive heterogeneous wood ray types and diffuse axial parenchyma. Advanced vessel elements tend to be correlated with homogeneous wood rays and paratracheal axial parenchyma. Comparisons among extant dicotyledons show that families of the Magnoliales possess many primitive wood features.

Nodal anatomy describes the pattern of vascular continuity between the leaf and stem. Depending upon whether one, three, or five or more leaf gaps are present in the stem stele, nodes are described as unilacunar, trilacunar, and multilacunar. Variation in nodal anatomy is of systematic value. Other systematically useful evidence is obtained from foliar epidermal features such as stomatal type, trichome morphology, and the type and distribution of mineral inclusions. The significance of the electron microscope in providing important characters for angiosperm classification is illustrated by the fact that sieve tube elements can be distinguished on the basis of the contents of their plastids.

The methodology of cladistic analysis has become the most widespread approach to modern phylogenetic reconstruction. An hypothesis of relationship between two or more taxa is based on their shared possession of a derived character state. Anatomy provides important derived character states for the recognition of closely related taxa. Cladistic analysis involves the assignment of polarity or the determination of primitive and derived states of the characters, requires precise character state description, and relies on a complete knowledge of the total range of structural variation of a character. As is the case with other characters, the assessment of homology among anatomical features is often difficult and requires careful study.

## ADDITIONAL READING

1. Ayensu, E. S. (1972). "Anatomy of the Monocotyledons. VI. Dioscoreales." The Clarendon Press, Oxford.
2. Baas, P. (1982). Systematic, phylogenetic, and ecological wood anatomy—History and perspectives. *In* "New Perspectives in Wood Anatomy" (P. Baas, Ed.), pp. 23–58. Martinus Nijhoff/Dr. W. Junk. Publ., The Hague.
3. Bailey, I. W. (1944). The development of vessels in angiosperms and its significance in morphological research. *Am. J. Bot.* **31**, 421–428.
4. Bailey, I. W. (1953). Evolution of the tracheary tissue of land plants. *Am. J. Bot.* **40**, 4–8.
5. Bailey, I. W. (1957). The potentialities and limitations of wood anatomy in the phylogeny and classification of angiosperms. *J. Arn. Arb.* **38**, 243–254.
6. Bailey, I. W., and Tupper, W. W. (1918). Size variations in tracheary cells. I. A comparison between the secondary xylems of vascular cryptogams, gymnosperms, and angiosperms. *Proc. Am. Acad. Arts Sci.* **54**, 149–204.
7. Carlquist, S. (1961). "Comparative Plant Anatomy." Holt, Rinehart and Winston, New York.
8. Carlquist, S. (1988). "Comparative Wood Anatomy. Systematic, Ecological, and Evolutionary Aspects of Dicotyledon Wood". Springer-Verlag, Berlin.
9. Carlquist, S. (1992). Pit membrane remnants in perforation plates of primitive dicotyledons and their significance. *Am. J. Bot.* **79**, 660–672.
10. Cheadle, V. I. (1942). The occurrence and types of vessels in the various organs of the plant in the Monocotyledoneae. *Am. J. Bot.* **29**, 441–450.
11. Cheadle, V. I. (1943a). The origin and certain trends of specialization of the vessel in the Monocotyledoneae. *Am. J. Bot.* **30**, 11–17.
12. Cheadle, V. I. (1943b). Vessel specialization in the late metaxylem of the various organs in the Monocotyledoneae. *Am. J. Bot.* **30**, 484–490.
13. Cheadle, V. I. (1944). Specialization of vessels within the xylem of each organ in the Monocotyledoneae. *Am. J. Bot.* **31**, 81–92.
14. Cheadle, V. I. (1953). Independent origin of vessels in the monocotyledons and dicotyledons. *Phytomorphology* **3**, 23–44.
15. Core, H. A., Côté, W. A., and Day, A. C. (1979). "Wood Structure and Identification," 2nd ed. Syracuse University Press, Syracuse.
16. Cutler, D. F. (1969). "Anatomy of the Monocotyledons. IV. Juncales." The Clarendon Press, Oxford.
17. Desch, H. E. (revised by J. M. Dinwoodie) (1981). "Timber: Its Structure, Properties, and Utilization," 6th ed. Timber Press, Forest Grove, Oregon.
18. Dickison, W. C. (1975). The bases of angiosperm phylogeny: Vegetative anatomy. *Ann. Missouri Bot. Gard.* **62**, 590–620.

19. Dickison, W. C. (1986). Wood anatomy and affinities of the Alseuosmiaceae. *Syst. Bot.* **11**, 214–221.

20. Dilcher, D. L., (1974). Approaches to the identification of angiosperm leaf remains. *Bot. Rev.* **40**, 1–157.

21. Edlin, H. L. (1977). "What Wood is That? A Manual of Wood Identification." Stobart and Son Ltd., London.

22. Esau, K. (1965). "Plant Anatomy," 2nd ed. John Wiley & Sons, New York, London.

23. Flynn, J. H., Jr. (Ed.) (1994). "A Guide to Useful Woods of the World." King Philip Publ. Co., Portland, Maine.

24. Grosser, D. (1977). "Die Holzer Mitteleuropas." Springer-Verlag, Berlin, Heidelberg, New York.

25. Hall, D. M., Matus, A. I., Lamberton, J. A., and Barber, H. N. (1965). Intraspecific variation in wax on leaf surfaces. *Aust. J. Biol. Sci.* **18**, 323–332.

26. Hoadley, R. B. (1980). "Understanding Wood, A Craftsmen's Guide to Wood Technology." The Taunton Press, Newton, Connecticut.

27. Hoadley, R. B. (1990). "Identifying Wood, Accurate Results with Simple Tools." The Taunton Press, Newtown, Connecticut.

28. IAWA Committee. (1989). IAWA list of microscopic features for hardwood identification. *IAWA Bull. n. s.* **10** (3), 219–332.

29. Ilic, J. (1987). "The CSIRO Family Key for Hardwood Identification," Technical Paper No. 8. CSIRO Division of Chemical & Wood Technology, Clayton, Victoria, Australia.

30. Ilic, J. (1990). "The CSIRO macro key for hardwood identification." CSIRO Division of Forestry & Forest Products, Highett, Victoria, Australia.

31. Ilic, J. (1991). "CSIRO Atlas of Hardwoods." Crawford House Press, Bathurst, Australia.

32. Jane, F. W. (1970). "The Structure of Wood," 2nd. ed. Adam and Charles Black, London.

33. Kribs, D. A. (1935). Salient lines of structural specialization in the wood rays of dicotyledons. *Bot. Gaz.* **96**, 547–557.

34. Kribs, D. A. (1937). Salient lines of structural specialization in the wood parenchyma of dicotyledons. *Bull. Torrey Bot. Club* **64**, 177–186.

35. Kribs, D. A. (1968). "Commercial Foreign Woods on the American Market." Dover Publishers, New York.

36. Kubler, H. (1980). "Wood as Building and Hobby Material." John Wiley & Sons, New York.

37. Kukachka, F. (1960). Identification of coniferous woods. *Tappi* **43**, 887–896.

38. Metcalfe, C. R. (1960). "Anatomy of the Monocotyledons. I. Gramineae." The Clarendon Press, Oxford.

39. Metcalfe, C. R. (1971). "Anatomy of the Monocotyledons. V. Cyperaceae." The Clarendon Press, Oxford.

40. Metcalfe, C. R., and Chalk, L. (1950). "Anatomy of the Dicotyledons," 2 vols. The Clarendon Press, Oxford.

41. Metcalfe, C. R., and Chalk, L. (1979). "Anatomy of the Dicotyledons." Systematic Anatomy of Leaf and Stem, with a Brief History of the Subject," Vol. I, 2nd ed. 294 pp. The Clarendon Press, Oxford.

42. Metcalfe, C. R., and Chalk, L. (1983). Anatomy of the Dicotyledons. Wood Structure and Conclusion of the General Introduction," Vol. II, 2nd ed. 309 pp. The Clarendon Press, Oxford.

43. Miles, A. (1978). "Photomicrographs of World Woods." Princess Risborough Laboratory, Her Majesty's Stationery Office, London.

44. Miller, R. B. (1980). Wood identification via computer. *IAWA Bull. n. s.* **1**, 154–160.

45. Palmer, P. G., and Gerbeth-Jones, S. (1988). "A Scanning Electron Microscope Survey of the Epidermis of East African Grasses." Smithsonian Institution Press, Washington, D. C.

46. Panshin, A. J., and deZeeuw, C. (1980). "Textbook of Wood Technology," 4th ed. McGraw-Hill, New York.

47. Prat, H. (1932). L'epiderme des graminees, etude anatomique et systematique. *Ann. Sci. Nat. Bot., Ser. 10,* **14**, 119–324.

48. Saiki, H. (1982). "The Structure of Domestic and Imported Woods in Japan. An Atlas of Scanning Electron Micrographs." Japan Forest Tech. Assoc., Tokyo.

49. Schiffer, N., and Schiffer, H. (1977). "Woods We Live With. A Guide to the Identification of Wood in the Home." Schiffer Ltd., Exton, Pennsylvania.

50. Schweingruber, F. H. (1978). "Microscopic Wood Anatomy. Structural Variability of Stems and Twigs in Recent and Subfossil Woods from Central Europe." Swiss Federal Institute of Forestry Res.

51. Schweingruber, F. H. (1990). "Anatomy of European Woods. An Atlas for the Identification of European Trees, Shrubs and Dwarf Shrubs." P. Haupt, Bern and Stuttgart.

52. Sharp, J. B. (1991). "Wood Identification. A Manual for the Non-Professional." Univ. Tennessee, Agri. Ext. Ser., Knoxville, Tennessee.

53. Solereder, H. (1908). "Systematic Anatomy of the Dicotyledons" (L. A. Boodle and F. E. Fritsch, Transl.), 2 vols. The Clarendon Press, Oxford.

54. Stern, W. L. (1978). A retrospective view of comparative anatomy, phylogeny, and plant taxonomy. *IAWA Bull. n. s.* **2–3**, 33–39.

55. Takhtajan, A. (1991). "Evolutionary Trends in Flowering Plants." Columbia University Press, New York.

56. Tomlinson, P. B. (1961). "Anatomy of the Monocotyledons. II. Palmae." The Clarendon Press, Oxford.

57. Tomlinson, P. B. (1969). "Anatomy of the Monocotyledons. III. Commelinales-Zingiberales." The Clarendon Press, Oxford.

58. Tomlinson, P. B. (1982). "Anatomy of the Monocotyledons. VII. Helobiae (Alismatidae)." The Clarendon Press, Oxford.

59. Wheeler, E. A., and Baas, P. (1991). A survey of the fossil record for dicotyledonous wood and its significance for evolutionary and ecological wood anatomy. *IAWA Bull. n. s.* **12**, 275–332.

60. Wheeler, E. A., and Baas, P. (1998). Wood identification—A review. *IAWA Bull. n. s.* **19**, 241–264.

61. White, M. S. (1980). "Wood Identification Handbook: Commercial Woods of the United States." Charles Scribner's Sons, New York.

# 6
# MACROMORPHOLOGY

Broadly defined, **morphology** deals with the study of plant form and structure. It is a biological discipline that attempts to interpret the evolution of shape and structure, the development of form in the individual and its component organs, and the causal factors underlying these phenomena. Morphologists are not content merely to examine the adult form of a particular plant or organ, but rather they seek to trace the evolution of form from that of a more generalized or primitive ancestor. They want to understand how variously modified structures arose evolutionarily and, where possible, to outline the successive developmental transformations that produced the complexity of the present form. In a narrow sense, **anatomy,** as reviewed in previous chapters, deals with internal structure at the cellular and tissue levels of organization. By strict definition, anatomy (histology) and morphology remain separate activities, although in reality the distinction between the two subjects is not easily made. As an example of the often close relationship between morphology and anatomy, one can point to the strong general correlation among shoot symmetry, leaf development, and vascular differentiation within dimorphic shoots. Within some families, the vascular architecture in habitually anisophyllous (heterophyllous) species differs dramatically from the sometimes closely related isophyllous taxa. Among anisophyllous plants with unequal leaves, the consistent differences between larger dorsal and smaller ventral leaves are associated with differences in the pattern of vascular supply and differentiation.

The integration of these two levels of organization has unfortunately often been neglected or disregarded completely. Theoretical considerations of how the cellular makeup of an organism relates to the developmental origin of the adult form have been topics of renewed discussion. The contemporary American and German plant morphologists Donald R. Kaplan and Wolfgang Hagemann have argued persuasively that histology and morphology are ontogenetically independent of one another and represent separate levels of organization that are under different morphogenetic control. If this is true, extant plant form cannot, or should not, be described or interpreted solely on the basis of internal cellular organization. In their view, an understanding of one level does not necessarily lead to a complete understanding of other levels. With respect to higher plants, therefore, it is the whole organism that is important in understanding form, not the individual cells.

Historically, the broad objectives of many morphological investigations have been to clarify the homologies in structure among diverse taxa. **Homologous structures** or organs are those based upon structural or developmental similarity that result from common ancestry. The recognition and comparison of homologous structures is basic to any phylogenetic or systematic study, and the identification of homologous structures provides a foundation for developing a phylogenetic system of classification. To better interpret the origin and nature (i.e., morphology) of mature structures, different sources of evidence can be employed, such as comparative, developmental, morphogenetic, paleobotanical, and anatomical. In this chapter we will see how the careful examination of anatomical detail, particularly the number, arrangement, and course of vascular bundles, sometimes provides insight into the possible origin, development, modification, and function of the organs of higher plants.

## FLORAL MORPHOLOGY AND ANATOMY

Flowers are the reproductive structures of angiosperms and represent one of the primary distinguishing features of the group. Morphologically, a "complete" flower is one with an aggregation of four sets of concentrically arranged floral units: sepals, petals, stamens, and carpels. The **sepals** are the outermost parts of the flower and may encase the developing bud. They are typically green and nonshowy, but if they are colored like the petals, they are described as being petaloid. Collectively, the sepals are termed the **calyx.** The **petals** form a whorl just inside the sepals. Petals are commonly conspicuously colored and may be shed soon after the flower opens. Their color and sometimes fragrance serve to attract specific insects or birds that act as pollinators. The entire collection of petals is called the **corolla.** The sepals and petals together constitute the **perianth** of the flower. If the perianth parts are not differentiated into sepals and petals, the term **tepal** is used to refer to the individual perianth parts. Flowers that lack one or both of the above perianth components are said to be "incomplete."

The **stamens** are the male reproductive parts of the flower and are found inside the perianth. Two parts make up the typical stamen. The **anther** is a saclike region composed of microsporangia that produce pollen. The slender **filament** is the supporting stalk. The stamens collectively are called the

**androecium. Carpels** are the female reproductive units and are centrally located within the flower. All the carpels in a flower compose the **gynoecium.** The gynoecium may consist of a single carpel or many carpels. Each carpel generally consists of three regions. The **stigma** receives the pollen, the **style** is the stalk or extended region that elevates the stigma, and the **ovary** is an enlarged basal portion that contains **ovules** or immature seeds.

Flowers show great morphological diversity and thus a vast number of descriptive terms are used to refer to various conditions. Differences in floral structure are evident in the symmetry, position and numbers of parts, and in the degree of fusion of floral members. Fusion can occur between like

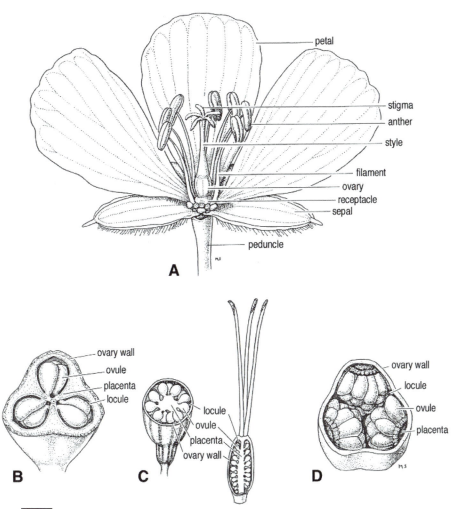

**FIGURE 6.1** Floral morphology. (A) A typical dicotyledonous flower. *Geranium maculatum* (Geraniaceae). One sepal, two petals, and three stamens have been removed. (B) Transverse section of compound ovary showing partitioned ovary with placentation on central wall. (C) Transverse and longitudinal sections depicting free central placentation (partitions lost). (D) Transverse section showing placentation on outside wall (partitions lost).

members (connation) or unlike members (adnation). Ovary position is particularly important in describing flowers. If the ovary is inserted above the level where the other floral parts are attached, it is termed a superior ovary, and the flower is **hypogynous.** If it is inserted below, it is an inferior ovary and the flower is **epigynous.** Not all flowers possess all four categories of floral units. Some presumably specialized and reduced flowers may lack sepals, petals, stamens, carpels, or any combination thereof.

The flower has long been the principal source of systematically useful information. One of the most fundamental and long-standing problems of floral biology is the question of the nature of the flower itself, along with the phylogenesis of its component parts. Interpreting the gynoecium has been especially difficult. Some basis for understanding the evolution and modification of floral form is provided by comparative and developmental anatomy. Despite opinions to the contrary, careful analysis of the position and course of floral vascular bundles is valuable (and sometimes the only) evidence for the recognition of homologies. Such analysis can aid in delimiting the former boundaries, relative positions, numbers, and categories of organs or their parts, which may be obscured by reduction, cohesion, and adnation. For example, most members of the mint family (Labiatae) have zygomorphic corollas with four, or more commonly five, distal lobes. In some taxa, however, the top lobe, which superficially appears to represent a single petal, is supplied by a pair of vascular bundles (traces) instead of a single vascular strand that is typical of most petals. This evidence suggests that the uppermost corolla segment may represent the fusion of two petals. Furthermore, the lower corolla lip is bifid and appears to represent two petals; however, it is vascularized by only a single trace and, as a result, most likely represents a single petal that has become secondarily lobed. The use and interpretation of mature floral vascular patterns, albeit often in rather controversial ways, has been a major cornerstone in attempts to explain floral evolution. During a discussion of the use of floral anatomy in the solution of problems of morphology and phylogeny, one floral anatomist once remarked, "Investigators who base phylogeny on floral vasculature are known to be imaginative and critical—imaginative when writing their own contributions, critical when evaluating someone else's." Recent studies of floral vascular systems have called attention to relationships between venation and function as well as to the need to understand the morphogenetic controls that result in the production of particular vascular patterns.

## HOMOLOGY OF FLORAL APPENDAGES

The oldest and most widely accepted interpretation of the flower is that it represents a determinate vegetative branch or shoot, with a condensed axis bearing a series of spirally arranged and progressively modified organs that are **phyllomic,** or leaflike, in nature. The appendages are either sterile (sepals and petals) or fertile (stamens and carpels). This classical concept of the flower therefore interprets the floral parts to be variously modified leaves. Like leaves, floral members are arranged in regular phyllotactic patterns, although the pattern may change at the junction of the vegetative and reproductive axes. Although there is some disagreement about the precise mode of organ inception, development, and mature structure, floral parts and leaves are viewed as homologous structures. This generalization has usually been broadly applied to the reproductive structures of all angiosperms. Comparative investigations of the floral vascular system have been widely used to defend this view. For example, the preeminent American floral anatomist Arthur J. Eames (1881–1967) commented that "flowers, in their vascular skeleton, differ in no essential way from leafy stems." Some variation has been reported in the patterns of differentiation of procambium and xylem in the reproductive shoot axis, although it is not dissimilar to the pattern in vegetative shoot apices. These observations have led to the view that the vascular construction of the floral axis should be described in the same manner as the vegetative stem by using the concepts of a receptacular stele, with gaps and vascular trace connections to the various floral members arising in succession. The compressed nature of the floral axis, when coupled with morphological complexities brought about by varying degrees of union of parts and epigyny, often complicates the interpretation of the dissected floral vascular pattern. Following the convenient tradition of descriptive nodal anatomy, the traces supplying the various floral members are usually described using words of motion to indicate the origins and pathways of vascular bundles, rather than using developmental terms.

If this concept of the flower is correct, then the mature vascularization of the floral axis should be comparable to a leafy stem. The predominance of three traces supplying the sepals is the common pattern in foliage leaves and suggests that sepals are the equivalent of subtending sterile foliar bracts. Petals, on the other hand, are most frequently supplied by a single trace, or in some cases three or more traces, that subsequently form a branched network in the petal. In some taxa (e.g., the primitive Winteraceae), petals may have three traces, or the traces may be reduced to one or amplified to more than three. Petals have apparently arisen in two ways. In some flowers, the petals have originated from the modification of tepals, in which case sepals and petals are morphologically equivalent. In other plants, petals appear to have resulted from the sterilization and subsequent elaboration of stamens. The frequently cited transitional series in the Nymphaeaceae is representative of this sequence. That floral parts may be derived from leaves is also supported by ontogenetic resemblances and by the fact that perianth parts are basically leaflike in form with the same general anatomical features as leaves. Epidermal cells are covered with a cuticle; they overlie a central mesophyll of undifferentiated ground parenchyma. Stomata are occasionally present on the abaxial surfaces. Unlike leaves, however, perianth parts generally lack mechanical supporting tissue.

The vast majority of angiosperms possess stamens vascularized by a single vascular bundle. A stamen type interpreted as a broad, three-veined microsporophyll possessing two pairs of linear, nonmarginal sporangia deeply embedded in its surface is found in the primitive woody families Degeneriaceae and Himantandraceae, as well as in several tropical members of the Magnoliaceae. To some, this seemingly leaflike stamen form represents a relatively unmodified condition and suggests comparison with an ancestral

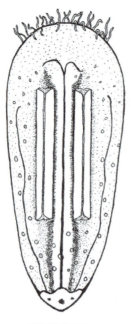

DEGENERIA

**FIGURE 6.2**  Diagrammatic representation of the stamen of *Degeneria vitiensis* (Degeneriaceae). Note three-veined condition. Courtesy of J. E. Canright.

sporophyll, further supporting the view that stamens are modified phyllomic structures.

The carpel has been the subject of more phylogenetic speculation than any other floral organ. The major opposing theories on the morphological nature of the carpel fall into three categories: (1) appendicular (foliar) theories based on the idea that the carpel is fundamentally a laminar organ, (2) axial theories, which are based on the concept that ovules were originally, and in some instances still are, axis borne (cauline), and (3) *sui generis* theories, which sidestep the question by assuming that floral parts do not find homologies in either leaves or stems but are entirely new entities. Like other floral members, the carpel often shows compelling anatomical evidence of its foliar origin. For many plants, the remaining unresolved issue concerns exactly the type of leaf and associated developmental sequence to which the carpel is most comparable. Some authors have suggested that the basic carpel type has a peltate or ascidiform construction, either basally or throughout, that is comparable to a peltate leaf type. Others have described the primitive carpel as a conduplicately folded foliar structure with open ventral margins. Still others argue that some developmental combination of these two morphologies can be observed in the angiosperm carpel.

Like leaves, bracts, and sepals, the carpel is most often vascularized by three major vascular bundles: a midvein (or dorsal carpellary bundle) and two marginal veins (or ventral carpellary bundles). Significantly, the carpellary bundles can sometimes be used to distinguish carpel boundaries of compound

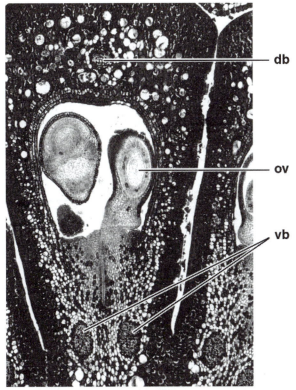

**FIGURE 6.3** Transverse section of carpel of *Dillenia philippinensis* (Dilleniaceae), showing ovules (ov) and three major carpellary vascular bundles, a pair of ventral bundles (vb) and a single median or dorsal bundle (db).

gynoecia, if other indicators such as styles and placentas are ambiguous. This is possible because the vascular bundles are initiated at an early stage of gynoecial development, and they can continue to differentiate as bundles while other primordial features are obscured by subsequent developmental changes, such as the loss of a member or the fusion of adjacent units.

Not all carpels are vascularized by three major bundles. A number of families have multitrace carpels, and some taxa have carpels with fewer than three veins. It has been suggested, for example, that the basic pattern in the Rosaceae consists of five vascular bundles. Comparative studies have shown that the primitive vascular pattern in the Rosaceous carpel is most likely a five-trace condition, composed of a dorsal bundle, a pair of ovular bundles to supply the ovules, and a pair of so-called wing bundles. With evolutionary advancement, the ovular and wing bundles have fused to form two ventral bundles, giving rise to the three-bundle condition. Similar five-bundle carpels or presumed derivatives of the five-bundle condition are known in other families of putative rosalean affinities.

Clearly, however, some flowers present a challenge to classical theory. For example, ovules that are seemingly cauline and borne on the floral axis have

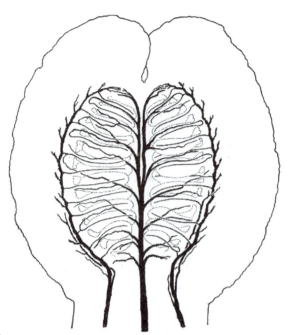

**FIGURE 6.4**    Unfolded carpel lamina of *Drimys* (Winteraceae) showing pattern of vascularization by three major veins. Reprinted with permission from the Bailey-Wetmore Wood Collection, Harvard University.

been described in a number of widely unrelated flowering plants and have influenced interpretations of the gynoecium as either phyllomic or cauline in nature. In *Illicium floridanum*, the ovules have been described as originating directly from the floral axis. These carpels have been interpreted as compound structures composed of an ovule with an enclosing sterile leaf.

## CONSERVATIVE AND VESTIGIAL VASCULAR BUNDLES

Certain fundamental and even controversial assumptions regarding the nature of floral vascular systems underlie much floral anatomical study and have been very influential in explanations of floral structure. The first of these beliefs is that the floral vascular system is nearly always more phyletically conservative than the organs it supplies. It follows, therefore, that changes in the vascular supply will tend to lag behind outer morphological alterations in floral form. This implies that changes in form, often not discernible by gross morphological observation, may be revealed by the persistence of certain internal characters of the vascular anatomy. It is possible to point to flowers, such as those of certain Theaceae, in which the fasciculate or fused condition of the androecium is no longer discernible in the external morphology of the mature flower but is only detectable in the presence of stamen vascular traces that remain united in the floral axis. Also of importance in this regard are lacunae number and position in the torus, bundle orientation, and the presence of "vestigial" traces. A **vesti-**

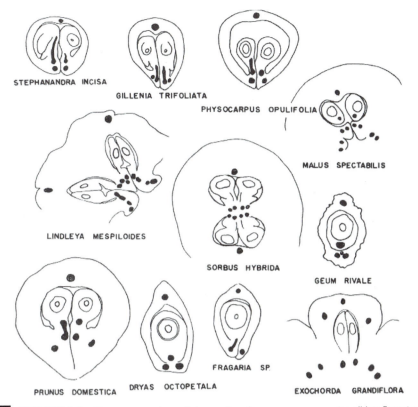

**FIGURE 6.5**   Transverse sections of the ovaries of some rosacean species, all but *Fragaria* showing the five-trace pattern as a constant feature of carpellary vascularization. Reprinted with permission from Sterling (1953), *Bull. Torrey Bot. Club* **80**, 457–477.

**gial trace** is one that is currently reduced or rudimentary and of apparent marginal or no use to the flower but that is believed to have functioned more fully in ancestors by vascularizing a now missing or highly modified part of the flower. The presence of such vestigial bundles, it is reasoned, can provide valuable information about the morphology of ancestral forms and thus provide evidence of relationships to other groups. The second premise is that the pattern of the vascular system may mark former boundaries or arrangements of organs or their parts, which are now lost or morphologically obscured. For example, a vascular bundle may document the position of a lost appendage that is not otherwise obvious. Finally, there is a widespread belief that a condition of united floral vascular bundles is almost always more advanced than a condition of unfused bundles. During floral specialization, vascular bundle fusions can potentially occur between adjacent members of the same whorl or between vascular strands of adjacent unlike floral members. In these cases, vascular bundles may be united completely or only partially along their length. These widely held generalizations have found strong opposition, although the examination of floral vascular anatomy sometimes provides important evidence for morphological change that is not detectable by other means.

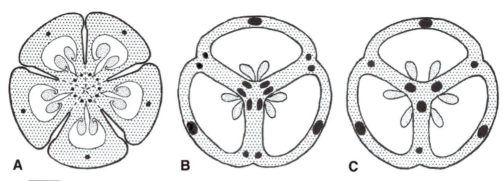

**FIGURE 6.6**   Diagrammatic transections of carpels and gynoecia showing vascular patterns (blackened areas) and stages in the union of carpels. (A) Gynoecium showing concrescence of ventral carpel surfaces. Each carpel contains a dorsal and two distinct ventral carpellary bundles. (B) Stage in the development of syncarpy and union of ventral and lateral bundles of adjacent carpels. (C) Complete syncarpy showing union of ventral and lateral bundles of adjacent carpels.

## NECTARIES OF *SALIX*

The small, largely anemophilous flowers of the Englerian Amentiferae, with no or only a poorly developed perianth, pose many morphological questions that have been solved by using floral vascular systems. The homology of the nectaries in the Salicaceae is a good example. **Nectaries** are highly specialized floral parts that produce nectar, a sugary liquid that is collected by insects and other small animals. The willows are dioecious woody plants with the flowers of both sexes borne in catkins. The unisexual flowers lack perianth members and possess a cup-shaped and often glandular subtending disk near the base of the ovary or stamens. Among different species, the disk organs range in form from flat, petaloid nonnectariferous structures, as in a few tropical species, to inconspicuous and fleshy nectariferous glands. In some species of *Salix*, these organs are deeply bilobed or trilobed at their distal ends. They also are characterized by the absence of vascular tissue, the presence of from one to three traces, or isolated vestiges of bundles.

The nectaries of *Salix* have been regarded variously as: (1) glands that arose as new emergences in response to a shift toward entomophilous pollination, (2) modified and reduced stamens, (3) modified bracts or bracteoles, or (4) reduced perianth parts. Examination of the internal anatomy of the flower facilitates the rejection of some of these possibilities. The nectaries of *Salix* and *Populus* most likely represent reduced remnants of a primitive ancestral perianth that was formerly lobed and essentially fused. This interpretation has received common acceptance and is supported by nectary position and, in some species, by the retention of a presumably ancestral sepal or petallike three-trace vascular supply that arises at a level in the flower where one would expect perianth traces to diverge from the receptacular stele. The occurrence of vestigial traces in the nectariferous tissue of some species has been traditionally viewed as representing a transitional stage in perianth reduction and transformation, with the vascular supply phyletically lagging behind external morphological changes that ultimately culminate in complete loss of venation. In other families, the nectariferous disc has been interpreted as carpellodal or staminodal in origin.

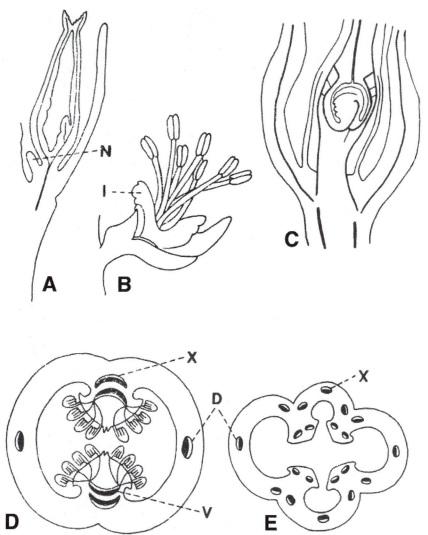

**FIGURE 6.7**   Floral vascularization. (A) Median longitudinal section of a female flower of *Salix alba* (Salicaceae) showing a nectary without a vascular supply. (B) Male flower of *Salix safsaf* showing a structure intermediate between a corolla and the nectaries found in many other species. (C) Longitudinal section of flower of *Myristica fragrans* (Myristicaceae) showing course of ovule vascular supply. (D) Diagrammatic view of transverse section of a typical bicarpellate ovary of *Crataeva religiosa* (Capparidaceae) at about its middle where the locules are confluent. (E) Diagram of a tetracarpellate ovary of *Crataeva religiosa* about in its middle. Abbreviations: D, dorsal carpellary bundle; I, intermediate structure between corolla and nectary; N, nectary; X, lateral strand or dorsal carpellary bundle; V, ventral carpellary bundle. Courtesy of M. Moseley.

## PSEUDOMONOMEROUS GYNOECIA

The number of parts within a whorl of floral members is not always obvious. In some apparently simple gynoecia, the seemingly single carpel may actually be compound in nature and a reduction product of two or more carpels. If the ovary appears to be unicarpellate, but more than one carpel actually is present, we call it a **pseudomonomerous gynoecium.** Some external evidence of carpel number may be present within the gynoecium (e.g., extra styles or stigmas or the presence of abortive carpels connate to the fertile one). In some instances, however, reduction may have progressed to the point of complete external loss of the component carpels. A reduction in ovular number usually accompanies the decrease in carpel number so that the ovary has a solitary ovule. Often the number, origin, and position of independent carpellary vascular bundles within the compound structure are relevant in trying to trace the derivation of these gynoecia.

The pseudomonomerous condition is found in a number of angiosperms, both monocotyledons and dicotyledons. The genus *Phryma* (Phrymaceae), an herbaceous perennial of somewhat uncertain relationship that inhabits rich mesic woods of the eastern United States, Japan, and India, can be used to illustrate this morphological condition. Flowers of *Phryma* are zygomorphic, with a sympetalous five-part perianth and didynamous stamens. The persistent calyx is bilabiate. The pistil is unilocular, containing one ovule in a subbasal position. Two stigmatic surfaces terminate the style. Descriptions of transverse sections through the ovarian region of *Phryma* recognize the presence of two prominent dorsal bundles and four lateral (ventral) carpellary bundles. The dorsal bundles extend the length of the pistil. Because carpels are typically supplied by a single major dorsal bundle, the occurrence of two strong dorsal bundles in the ovary of *Phryma* is positive evidence for the fundamentally bicarpellate nature of the gynoecium. On the basis of the position of these bundles, one can deduce that a suppression of the abaxial carpel has occurred. The ovular vascular supply, furthermore, consists of a pair of bundles that unite prior to supplying the solitary ovule. This pattern further indicates that the ovary of *Phryma* derives from an ancestral biovulate condition.

**FIGURE 6.8**   Diagram of the pseudomonomerous gynoecium of *Phryma leptostachya* (Phrymaceae). From Lipscomb (1968), Thesis, University of North Carolina.

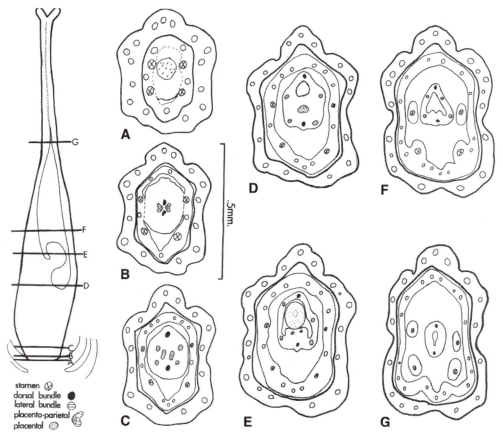

**FIGURE 6.9** Gynoecial vascularization of *Phryma leptostachya* (Phrymaceae). From Lipscomb (1968), Thesis, University of North Carolina.

A pseudomonomerous gynoecium is especially common among members of the order Urticales, and another excellent example of an apparent pseudomonomerous condition occurs in gynoecia of some Urticaceae. Reports of vestigial vascular bundles in the gynoecia of *Laportea canadensis* and *Urtica gracilis* have led to the suggestion that the unicarpellate ovary, often listed as a family character, may have been derived through the loss of a second carpel. There also is evidence from vascular anatomy for the phyletic shift in ovule position from either the side or apex of the locule to its base. This opinion receives support from the ovular vascular supply in *Boehmeria cylindrica* that initially ascends the ovary wall for a short distance prior to recurving downward to enter the ovule at the base of the ovary. A similar situation is present in *Myristica fragrans* (Myristicaceae), in which the vascular supply to the subbasal ovule forms at a high level in the ovarian wall from both dorsal and ventral carpellary bundles. This again seems to indicate that the ovule has gradually descended to its present position phyletically, and that the venation pattern is a reflection of a more ancient vascular supply.

## INVERTED VASCULAR BUNDLES

By way of further example, the morphological interpretation of the complex gynoecia within the Capparaceae and Cruciferae line of evolution has provoked intense debate and remains an unsettled issue. At maturity the crucifer gynoecium normally appears as a bicarpellate structure with parietal placentation. Examination of the vascular system of *Crataeva religiosa*, a member of the Capparaceae, provides a valuable indicator of the ancestral condition in the complex. An important feature of the floral venation is the presence of "inverted vascular bundles," bundles with phloem to the inside and xylem to the outside.

The gynoecium of *Crataeva religiosa* is composed of what appears to be two carpels. As viewed transversely, the pair of carpels are positioned lateral to the median posterior axis with their locules confluent distally. Placentation is axile in the basal region of the ovary but parietal at midlevels. The dorsal carpellary bundles are normally positioned, and the xylem and phloem have the usual orientation. The principal ventral carpellary bundles of adjacent carpels, however, are inverted and appear to be united, forming an arc on either side of the locule. These bundles originated in a ring of inverted vascular tissue at the flower base that is not continuous with the rest of the vascular system of the flower. Some degree of inversion is the common condition of ventral carpellary bundles and is characteristic of gynoecial vascular patterns associated with axile placentation. The ovular traces arise from these bundles. Immediately to the outside of each ventral bundle is another arclike bundle or group of bundles, both with normally oriented vascular tissue. Whatever their current function might be, the idea has been advanced that these bundles are vestigial dorsal carpellary wall bundles, derived from the fusion of an additional pair of carpels that have been lost during reduction. This deduction receives support from the occasional presence in some flowers of tri- or tetracarpellate gynoecia, with independent and normally oriented vascular strands in the dorsal or lateral walls of the extra fertile carpels. On the basis of these observations, it seems plausible that the ancestors of these families had gynoecia with axile placentation and that the unique capparalean bundle arrangement resulted from a change from axile to parietal placentation. Inversion of the vascular bundles in the parietal placentae of *Crataeva* can thus be seen as the retention of an ancestral feature; this has led to the morphologically important conclusion that parietal placentae have been derived from axile placentae.

## THE INFERIOR OVARY

The question of the origin and organization of the inferior ovary of flowers is a very important one from a phylogenetic standpoint. It is apparent that the inferior ovary (a floral form of seemingly similar morphological construction) has actually evolved many times among different groups of angiosperms and in distinctly different ways. Historically, two major theories have developed to explain how inferior ovaries arose. Proponents of both theories have paid close attention to the organization and course of the vascular bundles supplying the ovules. The **appendicular theory** holds that an extensive fusion (both connation and adnation) of the outer lower portions of the surrounding floral whorls to one another and to the ovary wall has occurred, resulting in the epigynous floral condition. The appendicular mode of origin has been the more prevalent pathway, as evidenced by the following anatomical observations. Flowers with this morphological type possess an ovular vascular supply that originates in acropetal fashion moving up the ovary wall. Ovules, accordingly, are supplied from traces branching from an ascending bundle. Figure 6.10 represents a hypothetical flower with an inferior ovary of appendicular origin with parietal placentation. The dorsal sides of the carpels are indicated by broken lines in this figure. The diagram illustrates two significant aspects that generally characterize this gynoecial type: (1) the dorsal sides of the carpels are surrounded by and united to a floral tube; and (2) the carpel bases, where both dorsal and ventral carpellary bundles diverge, are positioned below the ovarian cavity.

The **receptacular (axial) theory**, in contrast, assumes that carpels have phyletically "sunk" into the tissue at the end of the cauline axis, followed by the fusion of the receptacular tissue to the dorsal carpel wall. The ovary is thus partially enclosed in receptacular tissue. Inferior ovaries of receptacular origin are found in the minority of all angiosperm families, although they are characteristic of the Cactaceae and reportedly also of the Santalaceae. A

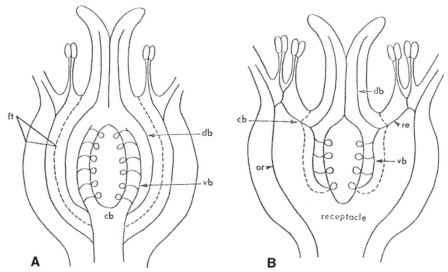

**FIGURE 6.10** Vascularization of inferior ovaries. (A) Diagram of a hypothetical flower with an inferior ovary and parietal placentation. The floral tube is adnate to the ovary. Note the ascending course of the floral vascular bundles. (B) Diagram of a cactus flower. The ovary is sunken into the receptacle, and there is no adnation of floral tube to ovary. Note the descending floral vascular bundles. Abbreviations: ar, ascending receptacular bundle; cb, carpel base; db, dorsal carpel bundle; ft, floral tube; re, recurrent receptacular bundle; vb, ventral carpel bundle. Dashed lines represent the theoretical limits of the gynoecium.

major argument in favor of the receptacular origin of the inferior ovary is derived from the prolongation of the major vascular bundles to a level just below the insertion point of sepals, petals, and stamens, and the subsequent descending course of the ovular vascular supply toward the morphological bases of the carpels. Such "recurrent" bundles typically have an inverted orientation of xylem and phloem and represent the principal evidence that the inferior ovary evolved by invagination of the floral axis. The presence of residual vascular tissue at the base of the ovary provides further evidence for this phyletic history.

The gynoecium of a typical nonleafy cactus is composed of multiple carpels enclosed in a floral cup, with a unilocular ovary and parietal placentation. Although the inferior ovary appears to conform to the receptacular type of origin, it is different from most other families. It seems entirely correct to suppose that the cactus ovary is embedded at the terminus of a vegetative axis, making the tissue surrounding the ovary wall receptacular in nature. The origin of this condition, however, is unusual. In *Pereskia pititache*, one of the large and leafy tree cacti of southern Mexico, a more primitive multicarpellate gynoecium that is phylogenetically significant is present. The gynoecium is superior, more or less multilocular with axile placentation. The carpels are fused laterally and are adnate to a central conical receptacle. This species is of special interest because it forms the basis for a novel hypothesis that explains the origin of the more typical cactus gynoecium. Very briefly stated, it has been assumed that the original ovular bearing axile ventral carpel sutures were deformed downward as a result of differential growth, until they came to occupy a position on the wall of the ovarian cavity. The course of the vascular bundles supports this hypothesis. A series of bundles move upward through the floral wall tissue to the approximate level of attachment of the perianth and androecium. Near the level of insertion of stamens and perianth parts, recurrent bundles form and descend toward the gynoecium. At a level still above the ovarian cavity, dorsal carpellary bundles and ventral carpellary bundles arise from the recurrent veins that are positioned near the outer edges of the flattened ovary. The dorsal and ventral carpel bundles subsequently move in opposite directions. The dorsal bundles extend upward through the styles, whereas the ventral bundles descend over the course of evolution, and become the source of the ovular traces. This explanation implies that the site of divergence of dorsal and ventral bundles represents the morphological bases of the carpels. Thus, the dorsal sides of the carpels are covered by the ovary roof, style, and stigma and are not invested by extracarpellary tissue as most inferior ovaries are. The bottom of the ovarian cavity, therefore, is receptacular tissue and not of carpellary origin. If this is correct, then the placenta will be parietal and will be confined to a single carpel, not to the united margins of two adjacent carpels.

## LEAVES AND STIPULES

**Stipules** are leaf-associated basal appendages accompanying one or two leaves. They have served as an important vegetative character for the delimitation and subdivision of various dicotyledons. Among dicotyledons, both the evolutionary origin and the morphological interpretation of the structures called stipules are debatable. Stipules generally have either been regarded as a part of the leaf or as independent structures of various origin. The classical leaf concept accepts stipules to be only the basal outgrowths of the foliage leaf. In this view, stipules always originate with the lower leaf zone as opposed to the upper zone that contributes the foliar blade and petiole. Stipules are normally vascularized by bundles that branch from the corresponding foliar traces (lateral leaf traces) that depart from the vascular cylinder of the stem, although in some cases they are not vascularized at all. Because stipules usually receive their vascular supply directly from a bundle that supplies a leaf, vascular anatomy can be used to defend the classical view that stipules are leaf subunits.

A few dicotyledons possess interfoliar (interpetiolar) stipules that occupy a position between the insertion areas of opposite or whorled leaves. These structures have classically been regarded as dual structures that have resulted from the union of two adjacent lateral stipules. They are significant, however, because in some groups they provide evidence in support of an alternate view of the leaf and stipule association. This opposing view is called the modified leaf-stipule concept. According to the modified hypothesis, the leaves and stipules of the same node are interpreted to be partially homologous with each other. That is, foliage leaves and stipules are the result of divergent developmental pathways of positionally equivalent primordia within the same node or whorl. In very general terms, one can consider the node or shoot apical ring as a locus of meristematic activity capable of producing one or more than one appendage category. Support for this idea comes from the stipular independence from the foliage leaves of the same whorl that occurs in some taxa. Not only can stipules be detached or free from the adjacent leaves, but they also can have a simultaneous and independent inception. Furthermore, in contrast to the most common condition in dicotyledons in which stipular traces arise as branches from lateral leaf traces only, the pattern of nodal vascularization of a few species reveals the presence of pure stipular traces that depart directly from the cauline stele and that exclusively vascularize the stipules. This indicates a certain degree of vascular independence of the stipules.

According to the older classical interpretation, among a whorl of leaflike members in the rubiaceous genus *Galium*, there is only a single pair (rarely three) of true opposite leaves, whereas the other whorl members are leaflike interfoliar stipules. Developmental evidence, however, has shown that in some species all members of a whorl are probably best considered to be true leaves, with the stipules reduced to mucilage-secreting "colleters." Furthermore, in species of *Galium* (Rubiaceae), *Hippuris* (Hippuriaceae), *Hydrothrix* (Pontederiaceae), and *Acacia* (Leguminosae), one or more leaves of a whorl can receive a trace that is connected directly to the stele, whereas other members of the whorl are vascularized by branches that arise from a vascular ring that encircles the nodal region and is not continuous with the cauline vascular system. These species all possess whorled phyllotaxis, and the number of traces extending from the stele ("primary traces") to vascularize each whorl is nearly always fewer than the number of phyllodes found in the whorl. Primary traces branch dentritically to form a nodal vascular ring from which each whorl member normally receives one vascular branch ("secondary trace"). An important implication of this observation is that leaflike members of a whorl at the same node can be regarded as phyllomes, whether or not they possess a direct vascular connection with the cauline stele. This structural condition leads to the important conclusion that the morphological nature of whorl members is not related to the manner of their vascularization but rather to developmental differences among whorl members during ontogeny.

## TWO-BUNDLED STRUCTURES IN MONOCOTYLEDONS

Among many but not all monocotyledons, there occur structures of uncertain homology and origin that tend to be vascularized by two principal vascular bundles of nearly equal dimensions. This pattern of vascularization has been widely utilized in attempts to reconstruct ancestral history.

One characteristic feature of lateral branch systems of monocotyledons is the presence of small phyllomic structures, called **prophylls,** that are borne near the base. In monocotyledons there typically is a single adaxial prophyll, although it may be obscured by the subtending foliage leaf. The monocot prophyll is frequently two keeled, bifid, and vascularized by a pair of unequal major bundles, one positioned on each keel. In the prophylls of some taxa, the two bundles extend in parallel along the length of the appendage, with each one terminating in an apical tooth. That prophylls are not the equivalent of stipules of the subtending leaf is indicated by the fact that the prophyll vascular supply arises from the lateral axis, rather than from a leaf trace or directly from the cauline stele. There are at least two views of the derivation of the monocotyledonous prophyll. Some authors have viewed the prophyll as a fundamentally solitary or entire structure and point out that the appendage is sometimes modified or reduced and not two keeled. Some prophylls are multitrace structures, whereas others are avascular. Others argue that the prophyll arose by the confluence of two phyllomic appendages (bracts). That the latter interpretation of the morphology of the prophyll may be correct is indi-

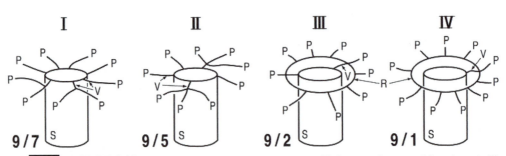

**FIGURE 6.11**    Comparison of whorls with more leaves (P) than vascular traces (V) to the stele (S) showing numbers for P/V. R = girdling vascular ring. (I) *Hippuris vulgaris;* (II) *Platytheca galioides;* (III) *Galium mollugo;* (IV) *Hydrothris gardneri.* Courtesy of R. Rutishauser.

cated by the tendency of prophylls to be bifid in form and by the presence of two major independent vascular bundles that differentiate from two procambial strands at a very early stage of prophyll development. Each bundle is believed to represent the median bundle of an ancestral foliage appendage. When prophylls have multiple bundles, one major bundle lies nearly or exactly opposite a cluster of opposing bundles.

Another convincing example of a two-bundled structure that has apparently evolved from the concrescence of independent appendages is the **palea** of the grass flower. The grass flower, or floret, is devoid of perianth members. A collection of florets constitute a **spiklet** that is subtended by two dry, empty bracts called **glumes.** Each floret has at its base two additional, thin-textured bracts that envelope the floret. The lower is the **lemma,** and the upper is the **palea.** The palea is often provided with two longitudinal ridges or keels and is generally vascularized by two major bundles that do not converge at the apex. In some presumably primitive grasses, the palea is bifid and arises from nearly separate primordia. These lines of evidence advance the view that the palea is the fusion product of two separate sepals of an ancestral flower.

## SUMMARY

Morphology deals with the study of plant form and structure. Morphologists attempt to interpret the evolution of shape and structure, the development of form in the individual and its component organs, and the causal factors underlying these phenomena. One objective of morphological investigation has been to clarify the homologies in diverse structures among taxa. Homologous structures or organs are those based upon structural or developmental similarity that result from common ancestry. Internal structure may show the nature of organs that are obscure or that cannot be determined externally. One major source of evidence to interpret the origin, development, modification, and function of structures is the number, arrangement, and course of vascular bundles supplying these structures. When a shoot ceases vegetative growth, the apical meristem undergoes changes in growth to produce the reproductive organ. These changes are reflected in the production of floral parts instead of leaves. Among angiosperms, flowers show a vast variation in structure, although commonly the flower has the usual complement of parts: sepals, petals, stamens, and carpels (gynoecium) in spiral or whorled arrangement. Various unions of floral parts can occur. Anatomically sepals and petals are essentially leaflike in structure with a mesophyll and epidermis. The stamens and carpels are also composed primarily of parenchymatous tissue with an epidermis on the outside as well as lining the locules of the ovary. The vascular bundles of the various floral parts are recognizable below the floral axis (receptacle) as traces extending to these various parts, (i.e., sepal traces, petal traces, stamen traces and carpellary and ovular traces). Serial transverse sections permit recognition of these traces and their divergence into the floral parts. Like leaves, sepals most commonly are supplied by three traces that form a branched vascular system. Petals and stamens most frequently are supplied by a single trace, and carpels are generally three-trace structures: a midvein, or dorsal carpellary bundle, and two marginal veins, or ventral carpellary bundles. Some flowers possess vestigial traces, which are interpreted to be currently reduced or rudimentary vascular supplies that have apparently marginal or no use to the flower, but which are believed to have functioned more fully in ancestors by vascularizing a now missing or highly modified part of the flower. The pattern of the floral vascular system may mark the former boundaries, arrangements of organs, or their parts, which may now be lost or morphologically obscured. A condition of united floral bundles is almost always considered to be more advanced than a condition of unfused bundles. The organization and course of the vascular bundles supplying the ovules have proven especially important in interpreting the question of the origin and organization of the inferior ovary.

Among dicotyledons, the evolutionary origin and morphological interpretation of stipules are debatable. Stipules generally have either been regarded as a part of the leaf or as independent structures of various origin. Stipules are normally vascularized by bundles that branch from the corresponding foliar traces that depart from the vascular cylinder of the stem, thus supporting the view of stipules as leaf subunits. A few dicotyledons have stipules that are detached, or free, from leaves, and that are supplied by traces that depart directly from the stem stele. This indicates a degree of independence of the stipules.

## ADDITIONAL READING

1. Bailey, I. W., and Swamy, B. G. L. (1951). The conduplicate carpel of dicotyledons and its initial trends of specialization. *Am. J. Bot.* **38**, 373–379.
2. Brown, C. L. (1980). Growth and form. *In* "Trees, Structure and Function" (M. H. Zimmermann and C. L. Brown, Eds.), pp. 125–167. Springer-Verlag, New York, Heidelberg, Berlin.
3. Burger, W. (1996). Are stamens and carpels homologous? *In* "The Anther: Form, Function, and Phylogeny" (W. G. D'Arcy and R. C. Keating, Eds.), pp. 111–117. Press Syndicate, Cambridge University Press, Cambridge.
4. Canright, J. E. (1952). The comparative morphology and relationships of the Magnoliaceae. I. Trends of specialization in the stamens. *Am. J. Bot.* **39**, 484–497.
5. Carlquist, S. (1970). Toward acceptable evolutionary interpretations of floral anatomy. *Phytomorphology* **19**, 332–362.
6. Eames, A. J. (1953). Floral anatomy as an aid in generic limitation. *Chronica Bot.* **14**, 126–132.
7. Eames, A. J. (1961). "Morphology of the Angiosperms." McGraw-Hill Book Co., New York.
8. Endress, P. K. (1994). "Diversity and Evolutionary Biology of Tropical Flowers." Cambridge University Press, Cambridge.
9. Eyde, R. H. (1975). The bases of angiosperm phylogeny: Floral anatomy. *Ann. Missouri Bot. Gard.* **62**, 521–537.
10. Eyde, R. H. (1982). Flower. *In* "McGraw-Hill Encyclopedia of Science and Technology," pp. 478–485. McGraw-Hill Book Co., New York.
11. Gifford, E. M., and Foster, A. S. (1989). "Morphology and Evolution of Vascular Plants," 3rd ed. W. H. Freeman and Co., New York.
12. Hjelmqvist, H. (1948). Studies on the floral morphology and phylogeny of the Amentiferae. *Bot. Notiser Suppl.* **2**, 1–171.
13. Kaplan, D. R., and Hagemann, W. (1991). The relationship of cell and organism in vascular plants. *BioScience* **41**, 693–703.
14. Lipscomb, H.A. (1968). "An Anatomical and Morphological Study of *Phryma leptostachya* L. with Possible Systematic Implications." Thesis, University of North Carolina, Chapel Hill, North Carolina.

15. Moseley, M. F., Jr. (1967). The value of the vascular system in the study of the flower. *Phytomorphology* **17,** 159–164.

16. Moseley, M. F., Schneider, E. L., and Williamson, P. S. (1993). Phylogenetic interpretations from selected floral vasculature characters in the Nymphaeaceae sensu lato. *Aquatic Bot.* **44,** 325–342.

17. Puri, V. (1951). The role of the floral anatomy in the solution of morphological problems. *Bot. Rev.* **17,** 471–553.

18. Puri, V. (1952a). Floral anatomy and the inferior ovary. *Phytomorphology* **2,** 122–129.

19. Puri, V. (1952b). Placentation in angiosperms. *Bot. Rev.* **18,** 603–651.

20. Rao, V. S. (1951). The vascular anatomy of flowers. A bibliography. *J. Univ. Bombay* **29,** 38–63.

21. Rury, P. M., and Dickison, W. C. (1984). Structural correlations among wood, leaves and plant habit. *In* "Contemporary Problems in Plant Anatomy" (R. A. White and W. C. Dickison, Eds. ), pp. 495–540. Academic Press, San Diego.

22. Sterling, C. (1953). Developmental anatomy of the fruit of *Prunus domestica* L. *Bull. Torrey Bot. Club* **80,** 457–477.

23. Sterling, C. (1969). Comparative morphology of the carpel in the Rosaceae. X. Evaluation and summary. *Österr. Bot. Z.* **116,** 46–54.

24. Weberling, F. (1989). "Morphology of Flowers and Inflorescences" (R. J. Pankhurst, Transl.). Cambridge University Press, Cambridge.

# 7
## STRUCTURE AND FUNCTION

In maintaining themselves as living systems and in performing their characteristic activities, plants carry out numerous complex processes that together constitute their **physiology.** Included among these basic processes are photosynthesis, respiration, metabolism, nutrition, and translocation and mobilization of water and solutes. Through increasingly more sophisticated experimentation, we now understand many aspects of organismal plant physiology. Nevertheless, a number of problems relating to the functional behavior of plants remain unanswered. It also is clear that answers to many of these questions cannot be gained in the absence of careful attention to many structural variables. Investigations of structure–function relationships have been carried out at various levels, from whole organisms to tissues, to individual cells, and finally to subcellular organelles and genes. The fact that there exists an intimate interrelationship between structure and function is certainly not a new or novel concept. While laying an important foundation for physiological anatomical study, the German botanist Gottlieb Friedrich Johann Haberlandt (1854–1945) remarked in his monumental work entitled *Physiologische Pflanzenanatomie* that investigations of the various cells and tissues of the plant body clearly demonstrate that physiological activity reflects the general structure of the plant and its individual anatomical features, just as the specific action of every machine is the result of its particular mode of construction. Despite this general understanding, there are many structural conditions for which it has been difficult to assign a physiological function, or the mechanism of function has never been satisfactorily elucidated. It also is clear that the normal physiology of a plant involves a consideration not only of its structure, composition, and functioning but also of external environmental influences.

## VASCULAR TISSUES

The vegetative body of self-supporting land plants is adapted to terrestrial life in various ways. As already discussed, one feature that separates the land plants (vascular plants) from the lower bryophytes (and algae) is the possession of a specialized supporting and water-conducting tissue called **xylem** and a principal food-conducting tissue called **phloem.** Before the evolutionary development of these specialized conducting tissues, called **vascular tissues,** large plants existed only in aquatic environments where support and water conduction were not necessary. Xylem and phloem occur in close spatial relationship and form a continuous integrated long-distance transport system throughout the plant body. Anatomically, each tissue is a reflection of the specific tasks it carries out. Not only do xylem and phloem enable rapid axial transport over great distances, but substantial lateral exchange of solutes can take place within the xylem and phloem and neighboring tissues along their length. A substantial amount of research has been historically devoted to clarifying the relation between structure and transport phenomena in these tissues, in part because of their economic importance. Although tremendous progress has been made in interpreting the interaction between anatomy and physiology in conducting tissues, significant gaps remain. Over the years, investigators have reached particularly controversial conclusions relating to the development, mature structure, and functioning of the phloem sieve element. To completely clarify growth processes in plants, important facets of the vascular tissues and the functional consequences of their cellular specialization must be understood. For many years the phloem was subjected to extensive study by the distinguished Russian-born American plant anatomist Katherine Esau, who provided unparalleled insight into the structure and function of this particular tissue.

## PHLOEM STRUCTURE AND TRANSLOCATION

Experiments have conclusively demonstrated that the rapid movement of organic material takes place through the phloem. The principal functional units of the phloem tissue of flowering plants include the major conducting conduit called the **sieve tube element** (sieve tube member), along with its associated **companion cells,** and a variety of intergrading **phloem parenchyma cells.** Wall characteristics and unusual cytological features characterize the phloem elements and are important for understanding long-distance food translocation (transport) in this most complex of all plant tissues. As described in Chapter 2, the end walls of sieve tube elements have not disappeared during evolution as they have in vessel elements, but they bear one or more highly differentiated and conspicuous sieve areas.

As a result of an unusual developmental pathway, the sieve tube element is enucleate at maturity; however, it retains a variable number of structurally disorganized plastids, mitochondria, and modified smooth endoplasmic reticulum, all distributed in a parietal fashion that results in a cell with a thin layer of cytoplasm. Sieve plates begin development at roughly the same time as the

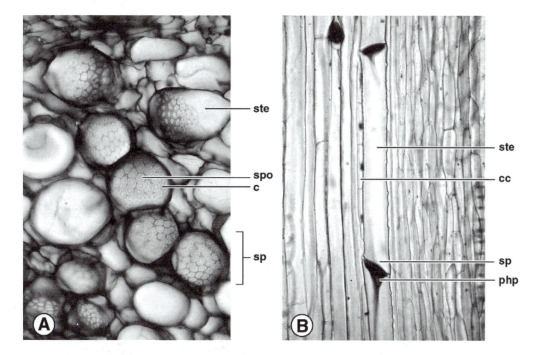

**FIGURE 7.1** Light micrographs of phloem anatomy of Cucurbitaceae. (A) Transverse section of phloem of *Bryonia* showing sieve tube elements with partial or complete simple sieve plates in surface view. (B) Longitudinal section of stem of *Cucurbita pepo* showing sieve tube element and associated companion cells. Note accumulations of phloem protein on the sieve plates. Abbreviations: c, callose; cc, companion cell; php, phloem protein; sp, sieve plate; spo, sieve pore; ste, sieve tube element.

nucleus begins to degenerate and cell autonomy diminishes, resulting in wall perforations at the pore sites, and finally in the formation of enlarged, unoccluded sieve plate pores. The great continuity of protoplasts brought about through the development of sieve pores and modification of the protoplast presumably are features that facilitate transport of photosynthetic products from cell to cell. Because companion cells and phloem parenchyma are nucleate at maturity, it is assumed that they also are actively involved in the translocation process in ways not yet fully understood. It is likely, however, that these cells serve in one capacity as energy sources for the active movement of assimilates. It also is important to keep in mind that the functional life of a sieve tube element is often short, and that dicotyledons must regularly form new phloem cells to ensure continuous transport.

Composed primarily of sugars and other organic constituents, ions, and hormones, solutes generally move from regions where they are synthesized to regions where they are utilized. That is to say, sugars are transported from the leaves to nonphotosynthetic and metabolically active tissues such as meristems, roots, or developing flowers and fruits. For the most part, sugar movement within the plant occurs in a downward direction, although upward transport away from sites of storage also takes place. From a functional standpoint, the movement of sugars in the plant body involves the

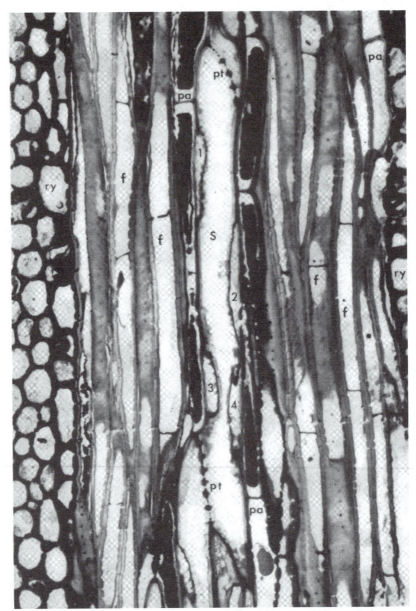

**FIGURE 7.2**   Tangential longitudinal section of secondary phloem of *Vitis vinifera* (Vitaceae). Sieve element(s) with four companion cells (1–4) and compound sieve plate (pt) covered with callose. × 440. Abbreviations: f, septate fiber; pa, parenchyma cell; pt, sieve plate; ry, ray cell; s, sieve element. From Esau (1965), *Hilgardia* **37,** 17. Copyright ©, Regents, University of California.

loading of assimilates into sieve elements at the sites of production, a subsequent long-distance transport of sugars between the green source organs and the sites of sugar consumption or storage, and, lastly, the removal or unloading of the translocates in the cells of the sink organs. Some authors use the terms **collection phloem, transport phloem,** and **release phloem** for

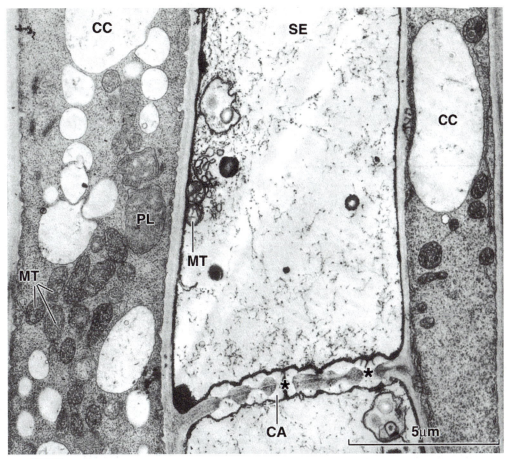

**FIGURE 7.3** Longitudinal section of a sieve tube and associated companion cells of *Platanus occidentalis* (Platanaceae). The phloem protein is evenly dispersed, and the sieve pores (stars) are blocked with callose. Abbreviations: CA, callose; CC, companion cell; MT, mitochondrion; PL, plastid; SE, sieve tube element.

phloem adapted for these three specialized functions. A structure and function relationship is evident within the collection, transport, and release phloem, where a decreasing correlation of the volume ratio between companion cells and sieve tube elements exists. Specifically, companion cells have wider dimensions than sieve tube elements in the collection phloem, comparatively smaller sizes in the transport phloem, and still smaller sizes in the release phloem. This presumably reflects decreasing energy requirements for photosynthate retention along the pathway.

Clarifying the exact mechanism of solute translocation has proven to be one of the most difficult problems in all of plant physiology. Although the concept of translocation velocities has been used inconsistently by different authors, it is known with certainty that the rate of movement of phloem solute is sufficiently rapid (typically 50–100 cm/hr in angiosperms) that it cannot be explained solely by diffusion. It is also well established that the photosynthate

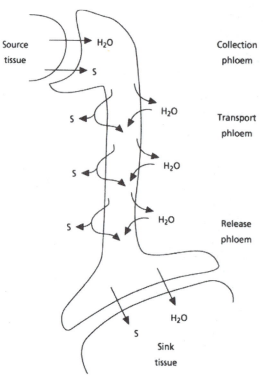

**FIGURE 7.4**   Current view on phloem translocation in higher plants. Photoassimilates are loaded into the collection phloem in source tissues, producing an osmotic potential inside the sieve element and companion cell complexes sufficient for local attraction of water from the apoplast. The generated turgor gradient gives rise to a source to sink pressure gradient, creating a mass flow through the sieve tube system. In the transport phloem, solutes escaping from the sieve tubes are partly retrieved by carrier systems located in the sieve element and companion cell complexes. The permanent and intense release and retrieval processes are accompanied by the continuous exchange of water. The retrieval serves to maintain the turgor gradient toward the sinks. In the release phloem in the sinks, the turgor potential is lower than in the source, not in the least by intense solute release from the sieve elements sustained by steady accumulation by the sink tissues. Reprinted with permission from Van Bel *et al.* (1993), "Progress in Botany," *Vol. 54,* 134–150. Copyright © Springer-Verlag.

moves under substantial positive pressure. If a sieve tube element is cut or punctured, the exudate flows out forcefully indicating high turgescence. Various forms of evidence suggest that a positive gradient of solute concentration exists in the sieve tubes from the leaves to the sites of utilization, although more precise evidence for a turgor gradient in the sieve tubes in the direction of movement is needed.

The mechanism of phloem translocation is controversial, but the explanation that fits the facts of plant anatomy and physiology best is the "pressure flow" model, first proposed by the German botanist E. Münch. The plant maintains a descending solute concentration gradient and a turgor pressure gradient within the phloem from the leaves (source) to the roots (sink); consequently, there is a mass flow of solution through the sieve tubes according to this model. Conclusive evidence for such a gradient is lacking, however, and more data are required on the resistance to flow in the sieve tubes. Nevertheless, this is the most satisfactory explanation of solute movement. At maturity, the highly specialized sieve tube elements of flowering plants show a structure and function relationship that is well suited for the high-velocity movement of sugars. With their large, unoccluded sieve plate pores, a cell lumina lined by a thin parietal layer of cytoplasm, and a differentially permeable plasmalemma, sieve tube elements are ideally structured for the long-distance transport of assimilates without major obstruction. The presence of a differentially permeable membrane around the side walls of the mature sieve tube element is essential in order to prevent osmotically active sugars from moving out of the system. The membrane results in the high turgescence that is evident in intact sieve tubes. Permeability also is evident between the sieve tube elements of a sieve tube. It is not known whether bidirectional phloem transport occurs through a single sieve tube. Bidirectional transport of assimilates would invalidate the pressure flow model, but to date there has been no conclusive demonstration that bidirectional phloem transport occurs.

## MINOR LEAF VEINS AND PHLOEM LOADING AND UNLOADING

Leaves are the primary sites of photosynthesis in higher plants. The flattened leaf blade of angiosperms is highly specialized for the surface-dependent processes of gas exchange and conduction of water and assimilates. This adaptation is most often reflected in its system of major conducting vascular bundles and an extensive hierarchy and reticulation of buried minor veins that extend through the photosynthesing mesophyll tissue. Minor veins (composed of both sieve elements and tracheary cells) typically possess an encircling sheath of parenchymatous cells that also encloses the terminal vascular elements. Minor veins serve an important function by collecting photosynthetic assimilates from nearby mesophyll cells where they are synthesized and depositing them into the transport sieve tubes. The process by which the major products of photosynthesis are selectively and actively delivered to the sieve elements in the source region is known as **phloem loading**. Sugars are subsequently transported to the major veins and away from the leaf where they are delivered to the sink tissues and then released in a process called **phloem unloading**. The principal function of protophloem cells, for example, is to provide a one-way path for unloading sugars in target regions of undifferentiated tissues such as apical meristems and immature leaves. The role of the metaphloem elements of minor veins is to collect sugars from the mature leaf mesophyll tissue and provide a transport conduit to the sites of release. Secondary sieve elements are primarily involved in long-distance transport but, in certain instances, also participate in the unloading process. The mechanism of phloem loading and unloading, and therefore the structure of minor leaf veins, has been the focus of intensive research in recent years. From a commercial standpoint, the efficiency of these processes can be important in determining the yield of crop plants and the economic viability of new crop varieties.

In one model of phloem loading that fits the available data and has gained wide acceptance, photosynthetic assimilates move across the mesophyll and

into the bundle sheath cells and sieve elements by traveling through intervening protoplasts and interconnecting plasmodesmata (the metabolic space, or
symplast). In some species, assimilates travel against a concentration gradient.
Sugars entering the companion cell move into the sieve tube element through
the plasmodesmata pore connections between the two cell types. The accumulation of high sugar concentrations in the sieve elements of the ultimate
leaf veinlets creates the concentration gradient of assimilates between two
regions along the length of the same sieve tube. Photosynthate also can pass
into the cell walls and intercellular space (the apoplast) and then be actively
transported across the cell membrane of the sieve tube element and companion cell complex.

The structure of the phloem of minor veins clearly plays an important role
in the process of loading photosynthetic products in the leaves. The minor leaf
bundles of dicotyledons possess specific functional adaptations, such as small
sieve tube elements and correspondingly large phloem parenchyma cells
(including companion cells). The parenchyma cells have a dense assemblage

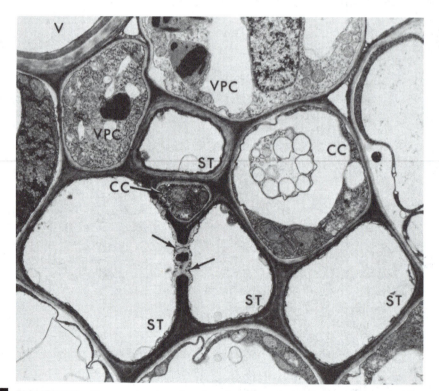

**FIGURE 7.5**   Transverse section of portion of small vascular bundle from *Zea mays* leaf. This bundle contains one thick-walled sieve tube (separated from a vessel by vascular parenchyma cells) and
three thin-walled sieve tubes. Unlabeled arrows point to callose-plugged sieve plate pores in lateral wall
between two thin-walled sieve tubes. Abbreviations: CC, companion cells; ST, sieve tube; VPC, vascular
parenchyma cell. ×9500. Reproduced with permission from Evert (1984), "Contemporary Problems in
Plant Anatomy," pp. 145–234, Academic Press.

of organelles and are connected to the sieve tube elements by way of numerous plasmodesmata. A symplasmic mode of phloem loading is structurally indicated by high numbers of branched plasmodesmal links between leaf mesophyll, bundle sheath cells, and the companion cell and sieve element association. The passage of photosynthates occurs through the bundle sheath cell into a companion cell, with a final transfer into the sieve element. An absence or reduced frequency of plasmodesmata between minor vein cells is structural evidence of probable apoplastic loading. However, whether loading is primarily apoplastic, symplastic, or a combination of both remains unknown. It is likely that more than one mode of loading occurs in different taxa. Parenchyma cells are widely believed to function as "intermediary cells" in the transfer (or loading) of photosynthates between mesophyll parenchyma and sieve elements. No structural specializations are evident in phloem cells associated with sugar unloading.

In some families of flowering plants, the sheath cells around the minor leaf veins are modified as phloem and xylem **transfer cells.** These cells are characterized by unlignified wall ingrowths that either surround the entire cell (e.g., in phloem transfer cells) or are restricted to that portion of the wall adjacent to a vascular element (as in xylem transfer cells). Plasmodesmata usually are present only between adjacent transfer cells. The plasma membrane follows the outline of the ingrowths, amplifying its surface area and conferring on these cells their special functions in solute transport. The specialized wall and membrane structure is an adaptation facilitating a more efficient exchange of solutes across the plasma membrane. Transfer cells are infrequent in the leaves of monocotyledons, while in dicotyledons they appear to be very common or even universal in some families (e.g., Asteraceae) and rare or absent in others (e.g., Caryophyllaceae). Vascular transfer cells appear to be more common in herbaceous plants than arborescent ones and occur only rarely in roots.

## KRANZ ANATOMY AND PHOTOSYNTHESIS

It has been known for several years that the leaves of a taxonomically limited number of monocotyledonous and dicotyledonous families (all Amaranthaceae and some Chenopodiaceae, Gramineae, Euphorbiaceae, and Portulacaceae) are distinguished by the presence of enlarged vascular bundle sheath cells around the minor veins. Although numerous variations exist, the conspicuous one- or two-layered sheath (or "Kranz") is composed of large, green, thick-walled cells that contain specialized starch-rich chloroplasts with few or no internal granal membranes. Walls of the sheath cells may be modified as well. In some plants, the plastids are centrifugally or centripetally distributed within the sheath cell. The vascular bundles are flanked by one or two layers of green mesophyll cells that often extend radially from the bundle sheath. Numerous plasmodesmata exist between mesophyll and bundle sheath cells. There usually are reduced volumes of green mesophyll cells between adjacent vascular bundles, usually forming two layers, rather than the usual condition of many mesophyll cells between vascular strands. In

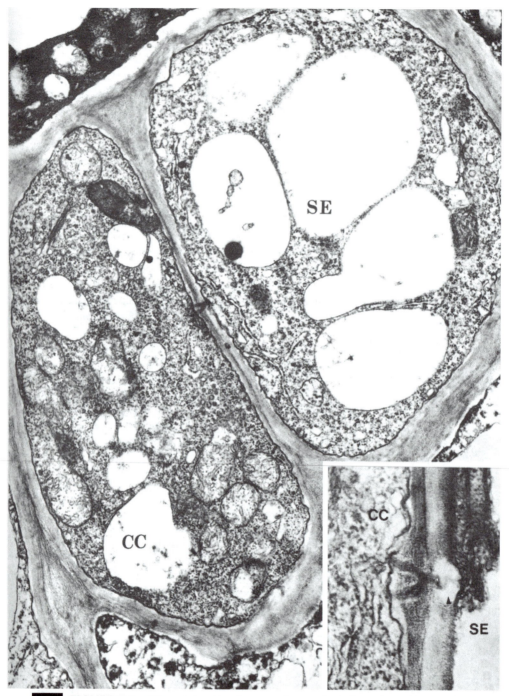

**FIGURE 7.6**   Differentiating sieve element and companion cell complex in *Hieracium floribundum*. A late stage in the ontogeny of a sieve element (SE) and adjacent companion cell (CC). The cytoplasm is becoming progressively less dense in the sieve element. A branched plasmodesma is present in the walls between the adjacent cells, ×26,500. (*Inset*) A branched plasmodesma between a mature sieve element (SE) and its adjacent companion cell (CC). A callose deposit (arrow) appears to constrict the single plasmodesmatal trunk in the wall of the sieve element. ×52,000. Reprinted with permission from Peterson and Yeung (1975), *Can. J. Bot.* **53**, 2745–2758.

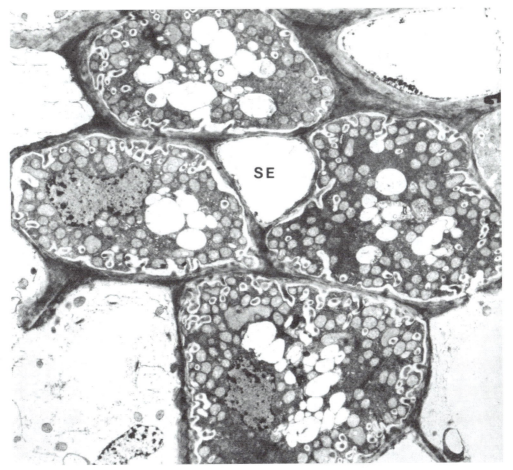

**FIGURE 7.7** Minor vein structure in the rhizome of *Hieracium floribundum*. A protophloem sieve element (SE) is surrounded by parenchyma cells, which have formed wall ingrowths. Walls not adjacent to the sieve element have the best developed wall ingrowths. Numerous mitochondria in a dense cytoplasm are characteristic of these transfer cells. ×5170. Reprinted with permission from Peterson and Yeung (1975), *Can. J. Bot.* **53**, 2745–2758.

some plants, each mesophyll cell is in contact with a bundle sheath cell. Plants showing this type of structure display a very regular and sequential depletion and synthesis of starch in leaf cell types. Haberlandt called this **Kranz leaf anatomy.**

All energy used by living organisms depends on the complex process of oxygenic photosynthesis, which is carried out by green plants. Energy-rich organic molecules in the form of sugar are synthesized from low-energy atmospheric $CO_2$ through a fundamental series of chemical reactions that utilize the energy of absorbed sunlight. The mechanisms of photosynthesis involve the acquisition of energy and the fixation of carbon dioxide. Photosynthesis consists of two interdependent series of reactions, the photochemical light reactions and the metabolic dark reactions; the former are dependent on light and the latter, on temperature. The light reactions represent the first stage of photosynthesis and convert light energy into

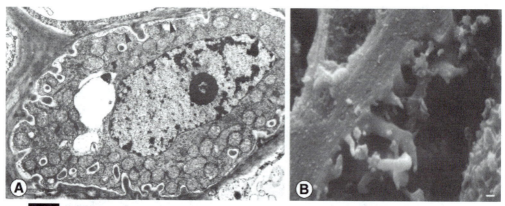

**FIGURE 7.8**   Transfer cells. (A) TEM photomicrograph of a transfer cell from *Hieracium floribundum* with numerous wall ingrowths, many mitochondria, and a large nucleus with prominent nucleolus. The transfer cell is adjacent to a sieve element. Also evident are numerous membranous structures (arrows) between the plasmalemma and the cell wall. × 10,000. (B) SEM view of transfer cell from corn, *Zea mays*. Note wall ingrowth that provides much greater plasma membrane surface area to facilitate apoplastic movement of solutes. A, reprinted with permission of Peterson and Yeung (1975), *Can. J. Bot.* **53,** 2745–2758. B, courtesy of A. Jones.

chemical energy. The dark reactions use the chemical energy products of the light reactions to convert carbon from carbon dioxide to simple sugars. The essential dark reactions involve the combining of $CO_2$ with a five-carbon sugar in a series of reactions called the Calvin–Benson cycle. This reaction yields an unstable intermediate that breaks down into two molecules of a three-carbon acid. Plants that photosynthesize in this manner are called $C_3$ plants. An enzyme known as rubisco catalyzes the formation of organic molecules from $CO_2$. In plants that utilize the $C_3$ photosynthetic pathway, rubisco is found in all the photosynthetic cells of the leaf.

A small number of angiosperms of tropical and temperate distributions have evolved a different $CO_2$ fixation system. In this case, the first stable fixation product is a four-carbon acid; therefore, these plants are called $C_4$ plants. The highly efficient $C_4$ photosynthetic system is correlated with Kranz leaf anatomy and requires the metabolic cooperation of leaf mesophyll and bundle sheath cells; it can utilize higher light intensities to fix $CO_2$. In the mature leaves of plants that utilize the more complex $C_4$ pathway, the enzyme rubisco is restricted to the specialized bundle sheath cells. The carboxylation phase of the $C_4$ pathway occurs in mesophyl cells and produces $C_4$ acids, which then rapidly diffuse to neighboring bundle sheath cells. In the bundle sheath cells, decarboxylation of the $C_4$ acids releases $CO_2$ for refixation by rubisco. These reactions cause $CO_2$ to be concentrated in the bundle sheath cells in the vicinity of rubisco. Various sheath cell wall modifications are thought to prevent $CO_2$ from escaping. In $C_4$ plants, there is a spatial separation of initial $CO_2$ assimilation in the mesophyll and the Calvin–Benson cycle in the bundle sheath. This increases photosynthetic efficiency and reduces metabolically wasteful photorespiration.

This structural condition also can occur in the internodal region, as in the sedge *Eleocharis vivipara*. This sedge is of particular interest because it alters

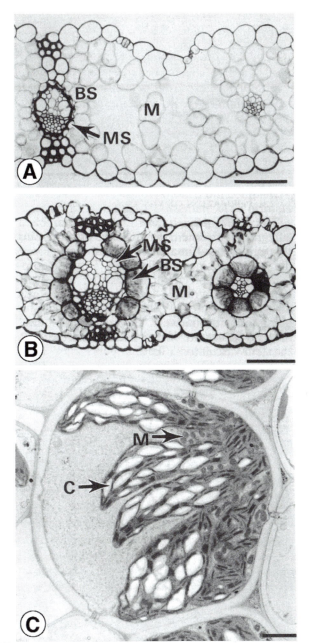

**FIGURE 7.9**    Leaf cross sections of C$_3$ and C$_4$ grasses. (A) *Bromus tectorum,* a C$_3$ plant. (B) *Zea mays,* a C$_4$ plant. (C) Ultrastructure of bundle sheath cell of *Panicum effusum,* a C$_4$ plant. Note centripetal position of chloroplasts and mitochondria. Scale bar = 50 μm in A and B; scale bar = 2 μm in C. Abbreviations: BS, bundle sheath; C, chloroplast; M in A and B, mesophyll; M in C, mitochondrion; MS, mestome sheath. Reprinted with permission from Nelson and Dengler (1992). *Int. J. Plant Sci.* **153,** S93–S105. Published by the University of Chicago Press.

its photosynthetic physiology depending on the environment. It utilizes $C_4$ photosynthesis when growing terrestrially but changes to $C_3$ photosynthesis when growing under submerged conditions. The mature internodal region of terrestrial plants possesses typical Kranz structure with well-developed bundle sheath cells, whereas bundle sheath cells are absent or poorly developed in submerged individuals. The correlation between this structural condition and a more efficient form of photosynthesis has only been substantiated during the past few decades and now represents one of the most striking examples of a close relationship between anatomy and function.

## XYLEM STRUCTURE AND WATER MOVEMENT

The xylem, or woody plant tissue, serves three important functions: (1) upward, long-distance transport of water from roots to the aerial portions of the plant; (2) support for the crown by providing mechanical strength and rigidity to the axis with lignin-reinforced secondary walls; and (3) storage reservoir for excessive metabolic products. From an economic standpoint, a knowledge of the functioning of the water transport system in plants is essential to understanding and combating many fungus-caused vascular wilt diseases. Attention has already been given to the fact that the movement of water in plants takes place in many vertically arranged, dead cells that consist only of cell walls and a central cavity or lumen. Physiologically, protoplasm is a hindrance to the rapid ascent of water and dissolved ions. It is remarkable, therefore, that the principal cell types involved in this process follow a developmental pathway that leaves them devoid of living contents at maturity. Most explanations of the ascent of water depend on this unusual structural property.

The long-distance transport of water from roots to leaves and other plant parts occurs within the xylem in a series of highly specialized cells called tracheary elements. Among these cell types is the **tracheid,** a long tapering imperforate cell with lignified secondary wall thickenings. As noted earlier, in all tracheary elements the protoplast disappears at maturity. Individual gymnosperm tracheids range in mean length from about 2 mm in *Juniperus* to about 6 or 7 mm in *Sequoia sempervirens* and some species of *Araucaria* and *Agathis*. Tracheid diameters are often 40 to 60 µm or more in *Sequoia*. Tracheids occur in vertical files with much overlapping so that the characteristic bordered pits in walls of adjacent elements are opposite each other. Water flows axially from cell to cell through the pits and thus across the pit membranes. This cell type is typically the only water-conducting element in lower vascular plants and gymnosperms.

During the course of evolution, the tracheid gave rise to two other cell types that individually have assumed the tracheid's functions and have largely replaced the tracheid in the flowering plants. One of these, the **vessel element,** was derived through modifications usually involving shortening and widening of the cell and, most importantly, the loss or perforation of the end walls. The latter gives many vessel elements an open tubular form through which fluids pass. Vessel elements are arranged end to end in an axial series, often with little

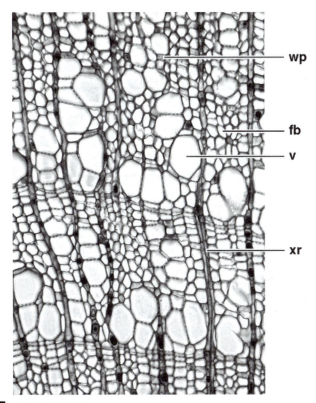

**FIGURE 7.10**   Transverse section of wood of *Tilia americana* (Tiliaceae) showing two growth ring boundaries. Abbreviations: fb, wood fiber; v, vessel element; wp, wood axial parenchyma; xr, xylem ray.

or no overlapping, and form an open continuous pipelike structure called the **vessel.** In axial water movement, water enters an individual vessel element at the lower end and flows through the lumen before exiting the open perforation plate at the opposite end. Vessels do not extend in parallel throughout the xylem but occasionally change direction and cross or connect with neighboring vessels, therefore spreading water tangentially during its ascent in the stem. Vessels never end in isolation but begin and end in vessel element clusters. Among species for which total vessel length has been measured, vessel length varies from less than one meter to many meters. Estimates of the minimum length of the longest of these vessels include 60 cm for *Acer* and 3 m for *Fraxinus.* Generally, wide vessels are longer than narrow ones. Some oak earlywood vessels can increase up to 11 m in length, and some vines have vessels that are 8 m or more in total length. Like the tracheid, vessel element walls are lignified and provided with interruptions called pits in various shapes and arrangements that are important in water conduction. The fact that hormonal auxin flow controls vessel diameter and density has been confirmed by recent studies demonstrating that low-level streams of auxin control the differentiation of wide earlywood vessels in ring-porous trees. Moderate or high auxin levels applied to ring-porous trees at the time of bud break limit the size of the earlywood vessels and result in diffuse porous wood.

**FIGURE 7.11**    SEM of isolated, barrel-shaped vessel element of *Fraxinus americana* (Oleaceae). ×500. Courtesy of E. A. Wheeler.

A second evolutionary derivative of the tracheid is the wood fiber in which all the supportive features of the tracheid have been intensified. The vessel element and wood fiber thus are complementary; together they perform the dual functions of the original tracheid. Although the exact function of living xylem cells has not been completely clarified, it is evident that vessel-associated parenchyma cells serve an important role in the long distance transport of sap.

A characteristic feature of tracheary elements is the presence of structurally complicated bordered pits on the lateral walls. Bordered pits provide wall openings between the cavities of two cells and are designed to facilitate the rapid movement of liquids through woody tissues. Entire vessels terminate in pairs or clusters of tracheary elements because water moves laterally from one vessel to the next through the intervessel pits. The bordered pits of adjacent cells are positioned exactly opposite to one another, typically occurring as pit pairs. In conifers as well as a few dicotyledons, tracheid to tracheid pit pairs have a perforated dividing **pit membrane** that has a central circular thickened region called the **torus** that divides the pit cavity into two halves. During pit pair differentiation, the enzymatic removal of noncellulosic matrix components from the unlignified pit membrane leaves a radiating microfibrillar network called the **margo**. Intervessel pit membranes of hardwoods typically lack a distinct torus, although exceptions to this generalization occur. The permeable membrane and torus normally occupy a median position, and water moves readily between cells by passing easily and rapidly through the perforations in the pit membrane. Pits in this condition are said to be open. When the pressure in the two adjacent tracheids becomes unequal, as when

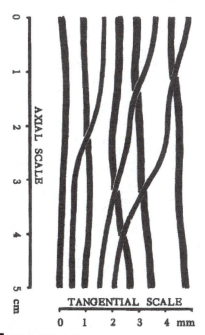

**FIGURE 7.12**  Diagrammatic representation of vessel distribution in *Fraxinus americana* (Oleaceae) in tangential view. This illustration is based on tracings of 20 individual vessels over an axial distance of 5 cm. Reprinted with permission of Zimmermann and Tomlinson (1967), *IAWA Bull.* **1**, 2–6.

air enters a conducting cell prior to heartwood formation or at sites of injury, the membrane and torus are forced to one side, sealing off the pit opening or aperture in a valvelike action. When a cell dries out, the pit membrane also is drawn away from its central position, perhaps by the surface tension of the last droplet of water in the pit, and becomes attached to one or other of the interior surfaces. When the pit membrane is laterally displaced to one side so as to block one of the apertures, the pit is referred to as being closed or **aspirated**. Once pits are aspirated, the wood becomes rather impermeable to fluids. The percentage of aspirated pits increases with increasing distance into heartwood and away from the functional sapwood. Authors have stressed the functional significance of pit borders in providing wall strength and supporting the displaced membrane, enabling the pit pair to retain a large membrane surface area for the movement of liquids in the nonaspirated condition. The pit border thus may help to prevent membrane rupture during membrane displacement.

The path of water and ion flow through the plant begins at the root hair zone of actively growing roots. Most of the water is absorbed by the plant through delicate root hairs that extend from the epidermal cells. Water uptake decreases with root age. Water and dissolved solutes readily move along a radial pathway between cortical parenchyma cells in the intercellular spaces. The innermost layer of the root cortex becomes differentiated as the **endodermis**. The radial and transverse cell walls of the endodermal cells become impregnated with a "diffusion-proof" substance called **suberin**. This modified

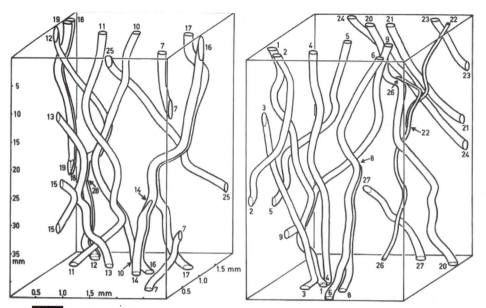

**FIGURE 7.13**    Course of vessels in the tangential and radial direction in a piece of wood of *Cedrela fissilis* (Meliaceae). Individual vessels are arbitrarily separated into two blocks so that they can be seen more clearly. Vessels are numbered at the point where they leave the block. Numbers with arrows indicate vessels terminating within the block. Vessel-to-vessel movement of water must be visualized through pits in areas where vessels run in pairs or groups. Note that the drawing is axially foreshortened about 10 times. Reprinted with permission from Zimmermann (1971), "Trees, Structure and Function," Springer-Verlag New York, Inc.

region of the wall is termed the **Casparian strip,** named after the German botanist J. X. Robert Caspary who first described it. Because there are no intercellular spaces between the cells of the endodermis and the suberized deposits of the Casparian strip inhibit the passage of water through the radial walls, substances entering or leaving the stele must pass through the tangential wall and living protoplasts of the endodermal cells. Therefore, soil solution is forced to move across selectively permeable membranes. The endodermis functions as a differentially permeable layer surrounding the stele. The rate of development of suberized endodermal walls is variable, although endodermal cells directly opposite the protoxylem may remain thin walled and without an extensive suberin layer. These cells are called **passage cells** and the movement of water and other materials between the cortex and stele is probably largely restricted to this region. Inside the endodermis, water must pass through the cell walls of the pericycle region of the root meristematic zone. Water finally enters the stele and moves into the principal conductors of water, the xylem tracheary elements.

The fact that water moves within the xylem is well known. The actual pathway of water movement within the stem is less understood, although determining which vessels are functional in transport can be followed by tracing the flow of dye that has been injected into the wood. In ring-porous species, water has been shown to move primarily through the largest vessels of the current year's growth increment, although not always in a strictly lin-

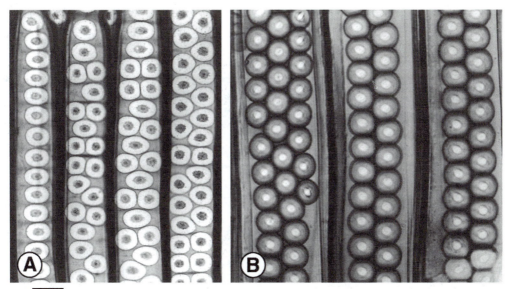

**FIGURE 7.14**  Circular bordered pits from conifer wood. (A) Radial longitudinal view of tracheids from *Cedrus deodara* showing uniseriate and biseriate pit arrangement. Note variation in size and number of intertracheary pits. (B) Radial longitudinal section of stem wood tracheids of *Agathis australis* showing biseriate and triseriate arrangement of pits through which water moves from tracheid to tracheid. Both courtesy of the Bailey-Wetmore Wood Collection, Harvard University.

ear direction. Conifers and diffuse porous dicotyledonous woods often possess a more complicated spiral grain structure. Wood "grain" refers to the arrangement and direction of axial orientation of xylem elements in relation to the long axis of the plant. In these more complex cases, water follows the helical path of the grain. The angle of the spiral grain may change from year to year, so the helical path of the water becomes altered in different growth increments. Even in the simplest xylem constructions, the three-dimensional distribution of vessels has not been well characterized. It is clear, however, that some vessels extend obliquely through the xylem. When two vessels come in contact, they remain together over a variable distance, allowing for lateral water exchange across the pit pairs on the side walls.

## Water Conductance

**Conductance** refers to the ability of capillary tubes or tracheary elements to transport water from one part of the plant body to another. Various theories have been proposed to explain the long-distance transport of water through the enclosed, nonliving tracheary elements of a plant. The **transpiration cohesion tension theory,** or simply the **cohesion tension theory,** has gained wide acceptance and is based on and affected by several factors, including the diameter and length of the conduit and the high degree of mutual cohesion of water molecules. A significant force is required to overcome the cohesive attraction between water molecules and break the column apart. Water molecules also show a great deal of adhesion to the conducting cell wall. This attraction of

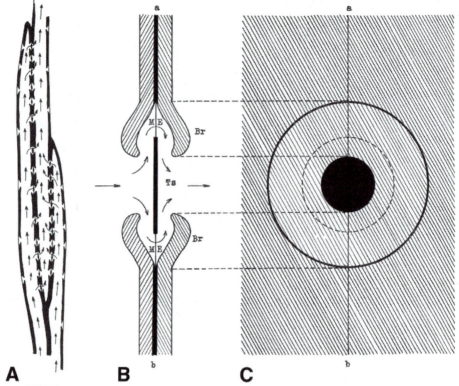

**A**          **B**          **C**

**FIGURE 7.15**  Bordered pit structure and water flow through adjacent tracheids. (A) Longitudinal section through radial walls of the overlapping ends of three adjacent tracheids. Note the concentration of bordered pits, which facilitate the flow of the ascending currents of water. (B) Section of bordered pit pair cut through (C) at a-b. The pit aperture is surrounded or bordered by the embossed portion of the secondary wall. The pit membrane and torus occupy a median position. Arrows indicate pathway of water flow through pit pair. (C) Section of radial wall of a tracheid showing a bordered pit in surface view. Abbreviations: Br, pit border; ME, pit membrane; Ts, torus. Reprinted from the *Forestry Quarterly* (Volume 11, number 1, page 13) published by the Society of American Foresters, 5400 Grosvenor Lane, Bethesda, MD 20814–2198. Not for further reproduction.

water molecules for the inside wall of conducting cells creates resistance to liquid flow. Perforation plates also create resistance to fluid flow, although this effect is small compared with the resistance of the walls. Calculations have shown that simple perforation plates provide less resistance than scalariform plates, however, because simple plates are arranged closer together along the vessel column, resistance values overlap with those of scalariform plates. The greater resistance to water flow in tracheids, as compared to vessel elements, arises from their smaller diameters and the absence of specialized perforation plates.

The upward movement of water requires a motive force that can exert a pull on the water column. This pull is created by capillary forces produced by the loss of water from the wet cell walls of leaves and other parts of the shoot. The process of water loss at the top of the plant is called transpiration and is viewed as the principal factor that results in a reduction in the diffusion pres-

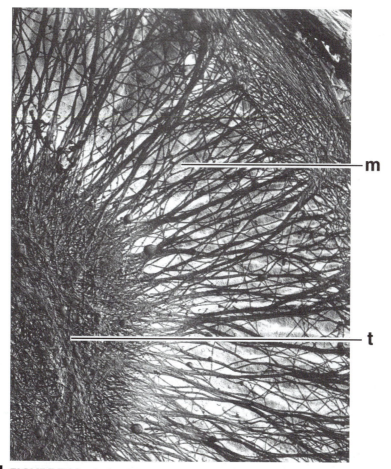

**FIGURE 7.16** Surface SEM view of part of softwood bordered pit membrane. (Cypress, *Taxodium*). The radial network of microfibrils extending outward comprise the margo (m) that support the thickened, matlike central torus (t).

sure of water in the leaf cells. As a consequence of transpiration, water moves from cell to cell until it leaves the cells adjacent to the tracheary elements. Ultimately, water is transported out of the xylem because of the cohesiveness of water columns. The rate of water conduction is largely determined by the rate of transpiration.

All available information indicates that the upward pull on the cohering molecules places the continuous columns of water within the tracheary elements under considerable tension (negative pressure) for long periods. If internal tensions within the conducting stream reach sufficiently high levels, the water columns can be punctured or broken and snap apart. If a water column is broken, the formation of a partial vacuum as a result of the separation of its parts can fill the conduit lumen and permanently render the conducting element nonfunctional. The actual rupture of the water column that is under tension is called a **cavitation.** The threshold at which a water column breaks depends less on the tensile strength of water and more on the size of the vessel

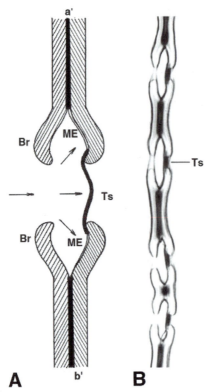

**FIGURE 7.17**    Bordered pit structure and function. (A) Section of bordered pit showing valvelike action of the torus. (B) Sectional view of bordered pits of *Cedrus* showing aspirated pit membrane tori pressed over the pit apertures in the bordered pit secondary wall of the left-hand cell. × ca. 800. Abbreviations: a′ to b′, compound middle lamella; Br, pit border; ME pit membrane; Ts, torus. A, reprinted from *Forestry Quarterly* (Volume 11, number 1, pages 12–20) published by the Society of American Foresters, 5400 Grosvenor Lane, Bethesda, MD 20814–2198. Not for further reproduction. B, courtesy of the Bailey-Wetmore Wood Collection, Harvard University.

elements, which is specific to species and organ. As a result, plants and plant parts may vary in their resistance to water stress. Bordered pits play a major role in determining the spread of cavitations and the resulting cavitation pressure of the xylem. Their permeability to an air–water interface is critical. They function well up to a certain pressure at which point they fail. In conifers where a torus is present, evidence suggests that they fail following the torus slipping from its aspirated sealing position, allowing air to leak through the large pores of the margo and through to the next tracheid. In angiosperms, the air leaks through pores in the pit membrane, or in some groups, the membrane may even tear or rip.

Although positive osmotic root pressures can potentially refill cavitated vessel elements in woody plants, vessel refilling by this method has only been documented in the grapevine *(Vitis)*. High tensions, freezing, or cosmic radiation can also introduce air or other gases into the water column where it accumulates in expanding bubbles known as **embolisms.** Embolism is the consequence of cavitation and refers to a gas-filled conduit. Embolism occurs as

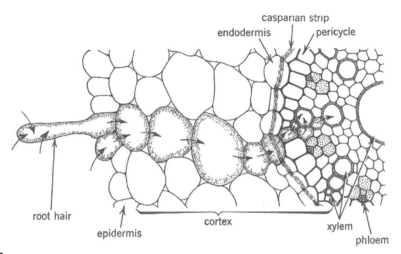

**FIGURE 7.18**   Part of transection of wheat root (*Triticum*), illustrating the kinds of cells that may be traversed by water and salts absorbed from the soil before they reach the tracheary elements of the xylem. Arrows indicate direction of movement through a selected series of cells. × 330. Reprinted with permission from Esau (1965), "Plant Anatomy," 2nd ed., John Wiley & Sons.

gases diffuse into the vapor void. Embolism also can occur when the vascular system is physically damaged allowing the bulk flow of air into the xylem. Gas embolisms can additionally occur spontaneously and are known to form normally in transpiring plants. The percentage of embolized vessels may vary at different times of the day. Freezing-induced embolisms within the xylem of trunks and branches can break the water column and thus represent a particularly serious problem for plants. Air embolisms are unable to pass axially from one tracheid to another because air bubbles do not traverse pit membranes. Perforated vessel elements, on the other hand, provide no such barrier, and disabling embolisms can rapidly spread through the open perforation plates. The spread of embolisms within a conducting element, and from vessel element to element through the perforation plates, causes the entire vessel to become disfunctional. A particularly puzzling matter is how air or other gases can be prevented from getting into the individual and collective water conduits and accumulating in bubbles that can cause a permanent blockage in the column. Also, plants need to deal with xylem embolisms after they have formed. The cohesion theory of water transport assumes that embolisms form only rarely and, once formed, become isolated from nonembolized columns because the high negative pressures in the columns would prevent them from being refilled. It has been suggested that the retention of narrow tracheids in dicotyledonous wood represents a means of protecting the conductive system when vessels become partially or totally nonfunctional. The closure of stomates results in the regulation of transpiration that may prevent excessive cavitation events in some instances. The grouping or aggregation of vessels in woods without conductive tracheids has also been cited as a way of adding conductive safety by providing alternate pathways for water movement in the event one or more vessels in a group are rendered nonfunctional. Among woody dicotyledons, xylem disruption can be overcome by the addition of a

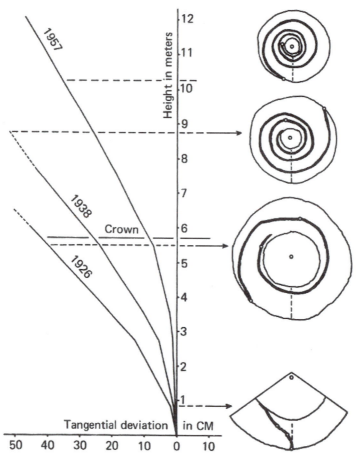

**FIGURE 7.19**     The path of water movement in stems. At the right are transverse sections through the stem of a 97-year-old *Abies concolor* at 0.84, 5.49, 8.74, and 10.24 m in height. Dye had been injected into the xylem through a radial bore hole at the base of the trees. This hole, vertically projected into the sections, is shown as a dashed line. The dye marks in the sections are indicated by a bold line. At the left, the course of dye movement, indicating the axial orientation of tracheids, is shown diagrammatically on the unfolded layers of growth rings. The changes of pitch of spiral grain are obvious from the comparison of the growth rings of 1926, 1938, and 1957. Reprinted with permission from Zimmermann (1971), "Trees, Structure and Function," Springer-Verlag, New York, Inc.

new increment of xylem during the next growing season. Monocotyledons, on the other hand, face potentially serious safety problems because damaged tracheary elements cannot be replaced. As a result, monocotyledons are protected against the effects of detrimental embolisms by having the ability to refill embolized conducting columns and, therefore, reestablish the continuity of the water column within the vessel.

## Hydraulic Segmentation and Safety

The site where two organs are connected is known to possess specific anatomical features, usually related to conduction safety. Anatomical modifications

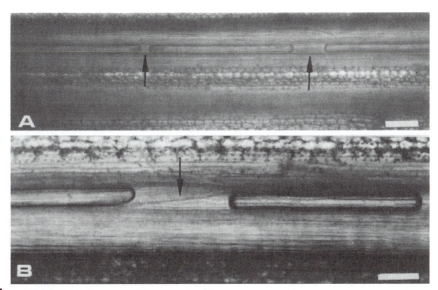

**FIGURE 7.20** Longitudinal sections (100–150 μm in thickness) of petioles of the palm *Rhapis excelsa* previously showing resistance above base level and xylem embolisms. (A) Intact length of vessel with bubbles arranged in series, one in each vessel element (scalariform perforation plates marked by arrows, bar = 200 μm). (B) Meniscus at scalariform perforation plate. Bar = 100 μm. Reprinted from Sperry (1985), *IAWA Bull n.s.* **6**, 283–292.

occur at the union of stem and leaf, at branch junctions, and at the site of root and stem connection. From the standpoint of conduction, the plant body shows segmentation, and marked hydraulic constriction occurs at a number of junctions as a result of the presence of narrow vessels or tracheids. In some grasses, the xylem tissue in the root and shoot junction of the plant axis shows such anatomical segmentation. The root and shoot junctions in rye, wheat, and barley, for example, appear to be constructed as **hydraulic safety zones** in which the root vessels are separated from stem vessels by the presumably less vulnerable, imperforate tracheids. Other cereals, however, such as maize, sorghum, and oats, have "unsafe" systems in which continuous strands of vessels extend from the roots to the shoot. It has been suggested that the hydraulic segmentation of the root vessel system from the shoot and vessel system may represent an important adaptation to the extremes of dry and cold environments.

The presumed tension on the water column coupled with the adhesion of water molecules to the cellulosic side walls of tracheary cells causes the wall of the tracheary element to be susceptible to collapse if it is not adequately reinforced. Woody plants are thus faced with complex anatomical, physiological, and cost–benefit problems in relation to environmental stresses and growth habit. On the one hand, the efficient long-distance transport of large volumes of water to the leafy crown of the plant is essential to a plant's survival. On the other hand, protecting the conducting elements from cavitation events and the mechanical support of the plant axis also is a necessity. These often competing demands have resulted in different structural solutions over the course of evolution.

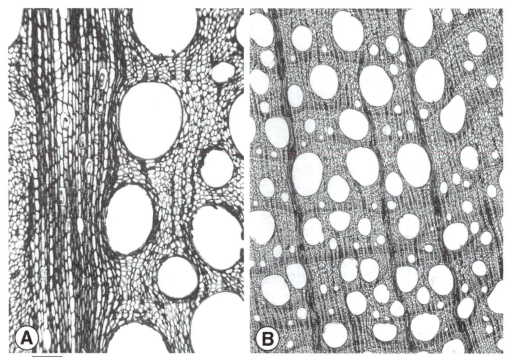

**FIGURE 7.21**    Anatomy of woody vines. (A) Transverse section of wood of the liana *Davilla rugosa* (Dilleniaceae). Note large, solitary vessels, ×55. (B) Transverse section of wood of *Actinidia chinensis* (Actinidiaceae) showing combination of large and small vessels, ×ca. 35.

We see this tradeoff at work in woody vines. Vines (lianas) are perennial plants with narrow stems relative to the leaf surface area supplied; they typically twine or climb to great heights and thus do not rely on their own axis strength for support. A characteristic anatomical feature of vines is the presence of exceptionally wide vessel elements (up to 500 µm) that may remain conductive from two to several years and are often present in combination with various types of tracheids. Lianas generally have greater vessel density than closely related tree species. This anatomical construction allows narrow stemmed vines to supply high volumes of water per unit time to a large leaf biomass. However, because wide vessels are often susceptible to failure under conditions of water stress, the hydraulic architecture of vines is one that appears to maximize conductive capability at the apparent expense of safety to the conducting system and to the strength of the axis. Within a transverse section of stem, the wider vessels are often more vulnerable than the smaller ones because the wider vessels have more permeable pit membranes.

## Water Flow

Water flow through the lumen of dead vessel elements is often compared to the movement of liquid through a bundle of perfectly cylindrical pipes. Such comparisons apply the equation known as the **Hagen–Poiseuille Law,** a 19th-century hydraulic engineering principle that can be used to model xylem conductance. The flow of water through capillaries is found to be proportional to the fourth power of the conducting element radius. For xylem tissue, vessel element diameters can be measured and the summation of their diameters to the fourth power can be calculated to estimate relative water conductance. The Hagen–Poiseuille Law is significant because it emphasizes the influence of a few large-diameter vessels on the overall hydraulic conductivity of stems.

Water flows in wider vessels more freely than in narrower pores, although, as we have seen, such wide vessels are at a potentially greater risk of cavitation, especially in freeze and thaw cycles. For example, one very wide vessel that is four times the median diameter of other vessels in a stem or bundle will theoretically transport an amount of water equal to the total transported by 256 other vessels of a median or small size class ($4^4$). Put in another way, in order to conduct an equal volume of water at a given pressure gradient, a maple tree with comparatively narrow pores requires approximately 7000 times more vessels per transection than does a ring porous oak tree with very wide earlywood vessels. However, the measured upper limit of conductance rarely reaches theoretical limit because vessels are usually not perfectly circular in transection and have a nonlinear shape. In addition, the presence of lateral wall pits and perforation plates influences water flow, although experimental studies of water flow through xylem vessels suggest that simple perforation plates create only insignificant resistance to flow. It is less clear what effect scalariform perforation plates may have on water flow. Some data show that scalariform perforation plates add about a 6 to 8% increase in resistance to flow per unit length. Small errors in measuring diameters or in accounting for all vessels in a stem section also are magnified tremendously when calculations are made. Nevertheless, this principle does point out the value of a few large-diameter vessels when maximum water transport is required rather than maximal mechanical strength.

At the same time that water is moving upward through the axial pathway, it also is circulating horizontally across the wood from tracheary cell to tracheary cell as well as through the wood rays. The tracheids of the earlywood of the first-formed growth rings of conifers generally not only tend to be much shorter and narrower than the tracheids of the outermost growth rings of large mature stems but also are provided with smaller and less numerous bordered pits. Thus, the area of tenuous pit membranes through which water and other liquids can pass most rapidly from tracheid to tracheid varies more or less markedly in different parts of a stem or root. Wood rays are aggregations of living parenchyma cells that extend radially between the vertical tracheary elements from the center of the stem outward to the cambium. They are active in storing carbohydrates and in transporting water over short distances.

Where ray cells and the water-conducting tracheary elements meet, they communicate with each other by means of pit pairs in their side walls. In conifers, the pitting formed in a radial section by the walls of a ray cell and those of an axial tracheid is termed **cross field pitting**. In dicots, **ray vessel pitting** occurs between a ray cell and a vessel member. The pitting that occurs in these situations is commonly composed of **half-bordered pit pairs** involving a simple pit and a bordered pit. The tracheary element side pit is bordered,

**FIGURE 7.22**  Radial section of wood of *Sequoia* showing pitting formed by the intersecting walls of longitudinal tracheids with ray parenchyma cells (cross field pitting). Pits enable the ray cells to secure water from the vertical water-conducting tracheids. ×400. Courtesy of the Bailey-Wetmore Wood Collection, Harvard University.

whereas the ray parenchyma side pit is simple. A major function of wood rays is to remove water from the tracheary cells and transport it along a horizontal pathway. Having moved across the pit pair, the movement of water within the rays is usually comparatively slow because the ray parenchyma cells contain a protoplast that hinders rapid water transport. Because a ray is composed of a large number of small cells, water also must pass across many pitted cross walls. The rays of some conifers possess specialized **ray tracheids** that open to the vertically oriented tracheids by means of small bordered pits and presumably provide a more efficient means of water movement. Given the small proportion of wood they represent, however, it is unclear how much they can contribute to water conduction.

## ANATOMICAL RESPONSES TO MINERAL DEFICIENCY

Among the many requirements for normal plant growth and development is an adequate supply of the common elements. Experiments have increased our

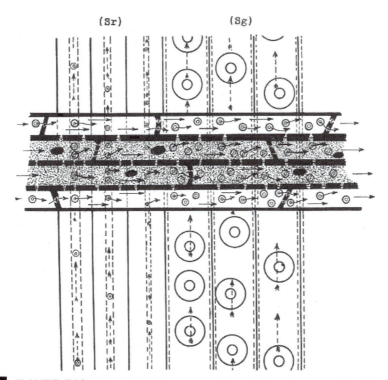

**FIGURE 7.23** Flow of liquid between conifer tracheids and ray parenchyma. Longitudinal radial section through wood, showing ray crossing vertical water-conducting tracheids. The water ascending through the underlying summerwood and springwood cells passes into the ray cell through the small pits, which are shown in surface view. The two rows of cells in the center of the ray are the ray parenchyma. The marginal cells without contents are the ray tracheids or horizontal water-conducting elements. Abbreviations: (Sr), tracheids of summerwood; (Sg) tracheids of springwood. Reprinted from Bailey (1915).

appreciation for the importance of proper nutrient balances for optimum plant growth. A number of "major" elements are necessary for growth. These include phosphorus, potassium, nitrogen, calcium, iron, sulfur, and magnesium in addition to various "minor" (or trace) elements such as boron, copper, manganese, molybdenum, chlorine, and zinc. Plants lacking one or more of the major or minor mineral salts usually develop characteristic visible **mineral deficiency symptoms,** such as stunted growth, general chlorosis or yellowing due to reduced chlorophyll content, necrosis or death of tissues, and anthocyanin formation (resulting in red color). In stems, symptoms such as the development of slender, woody stems in otherwise herbaceous plants, and a reduced flower and seed set are common.

To record these changes, carefully controlled experiments have been carried out that employ solution culture techniques that deprive plants of one or more of the essential elements. These experiments have not only revealed external changes in plants but also clearly established that mineral deficiencies can have severe effects on internal structure as well. Understanding these changes is necessary in order to provide a complete diagnosis of the symptoms associated with a particular deficiency and is of practical importance for the fields of agriculture, horticulture, plant nutrition, plant breeding, and pathology. Diagnosis is complicated, however, by the fact that anatomical symptoms vary with respect to the time of appearance and sequence in which organs are affected, the age and species of plant studied, the environment, and the severity of the deficiency.

Soil nitrogen availability provides an outstanding example of the affects of essential elements on plant structure. Plants growing in nitrogen-deficient habitats tend to be characterized by leaves that show classical xeromorphic syndromes. That is, foliage has a dense covering of trichomes, a thick cuticle, a high proportion of vascular tissue, multiple palisade layers, abundant sclerenchyma, and small thick-walled cells. In contrast, nitrogen levels are consistently high in thin, predominantly mesomorphic leaves. Calcium deficiency results, on the other hand, in necrosis of apical meristems or in the complete differentiation of meristematic cells. Because calcium is a constituent of the middle lamella, loss of cell wall integrity occurs. The percentage of axis cross-sectional area occupied by the vascular tissue, cortex, and pith has also been shown to be positively correlated with the concentration of calcium. Low levels of potassium (K+) have been related to a decrease in leaf cell size along with a compaction of cells per unit area. These features are associated with decreased leaf thickness, degradation of chloroplasts, and abnormal leaf cell ultrastructure.

A deficiency in one or more of the trace elements (e.g., copper, zinc, or boron) produces overall plant dwarfing as well as specific changes in leaf structure. Leaves often undergo involution, change color, and develop necrosis. Mesophyll cells frequently separate and ultimately undergo lysis. Plants lacking boron have been described as showing dramatic changes in the meristematic tissues of the root and shoot apices and vascular cambium, with meristematic activity noticeably reduced. Anatomical injury from boron deficiency can begin as early as three or four days after plants have been placed in a boron-deficient nutrient solution. At this time plants still appear normal and healthy. By the time individual plants show visible external changes, marked internal injury has usually developed. Such results point to the extreme sensitivity of plants to nutritional deficiencies. Vascular differentiation has been found to be altered or even discontinued. Necrosis is reported to extend from the cambium into the phloem and xylem. Abnormalities in stomatal and epidermal cells of leaves also have been described. Detailed studies of the anatomical responses of tomato, turnip, and cotton to variations in boron nutrition have verified earlier observations of cambial cells enlarging radially and becoming necrotic in some areas. Hypertrophy, necrosis, and hyperplasia extend into the phloem and, in extreme cases, inward into the xylem parenchyma. Less dramatic responses to boron deficiency can include more secondary growth with smaller vessels, changes in quantities of tanniniferous substances, and changes in the amount and degree of maturation of fibers and collenchyma. Boron deficiency in corn leads to changes in the mesophyll region of the leaf and injury to the small veins. Cells often fail to differentiate and the leaf collapses. Some areas of the leaf develop hypertrophy of cells, especially on the lower epidermis.

Although many experiments have been undertaken on the nutrition of plants, the results have not always been conclusive and uniform. Some anatomical symptoms associated with mineral deficiencies are known to be highly variable and related to environmental changes. Furthermore, the relationship between internal and external symptoms cannot always be directly correlated.

## SUMMARY

In maintaining themselves as living systems, plants carry out numerous complex processes that constitute their physiology. Many of these physiological processes are closely related to plant structure. Included among these basic activities are photosynthesis, respiration, metabolism, nutrition, and translocation and mobilization of water and photosynthetic solutes. Xylem and phloem occur in close spatial relationship and form a continuous integrated, long-distance transport system throughout the plant body. The rapid movement of organic solutes takes place through the phloem. The principal functional units of the phloem tissue of flowering plants include the major conducting conduit called the sieve tube element, along with its associated companion cells and a variety of integrating phloem parenchyma cells. The great degree of continuity of protoplasts (brought about through the development of sieve area pores) and modifications to the protoplast are features that facilitate the transport of photosynthesis products from cell to cell. Because companion cells and phloem parenchyma are nucleate at maturity, it is assumed that they also are actively involved in the translocation process. Movement of sugars involves the loading of assimilates into sieve tubes at sites of production, the subsequent long-distance transport of sugars between the green source organs and the sites of sugar consumption or storage, and lastly the removal or unloading of the translocates in the cells of the sink organs. The mechanism of phloem translocation is complex. The plant maintains a descending solute concentration gradient and a turgor pressure gradient within the phloem from the base (source) to the roots (sink). Consequently, a mass flow of solution occurs through the sieve tubes. Sieve tube elements are ideally structured for the long-distance transport of these assimilates. The process by which the major products of photosynthesis are selectively and actively delivered to the sieve elements is known as phloem loading. Sugars are subsequently released during phloem unloading.

The structure of the minor veins plays an important role in the process of loading sugars in the leaves. A symplastic mode of loading is indicated by high numbers of plasmodesmata. In some families, the sheath cells around the minor leaf veins are modified as phloem and xylem transfer cells. These cells are characterized by unlignified wall ingrowths. Leaves of some monocotyledons and dicotyledons possess enlarged vascular bundle sheath cells around the minor veins. This leaf structure is called Kranz leaf anatomy. It is associated with a more efficient mode of photosynthesis, $C_4$ photosynthesis.

The long-distance transport of water from roots to leaves occurs within the xylem in a series of highly specialized cells called tracheary elements. A characteristic feature of tracheary elements is the presence of structurally complicated bordered pits on the lateral walls. Bordered pits provide wall openings between the cavities of two cells and are designed to facilitate the rapid movement of liquids. The path of water flow begins in the root, where it moves along a radial pathway, crossing the endodermis and entering the tracheary cells. In the wood, water moves primarily through the current year's growth increment, often along a spiral pathway. Conductance refers to the ability of capillary tubes or tracheary elements to transport water from one part of the plant to another. One widely accepted theory of transport is the transpiration–cohesion–tension theory. It is based on and affected by several factors, including the diameter and length of the conduit and the degree of mutual cohesion of water molecules. Breaks in the water column are called cavitations and represent serious problems for plants. The flow of water through the lumen of dead vessel elements follows the Hagen–Poiseuille Law: the flow of water through capillaries is proportional to the fourth power of the conducting element radius. At the same time that water is moving upward through the axial pathway, it also is circulating horizontally across the axis from tracheary cell to tracheary cell. This movement of water also involves its passage through pit pairs.

Plants lacking one or more of the major or minor minerals develop characteristic mineral deficiency symptoms. Many of these deficiencies can have severe effects on the plant's internal structure. Alterations in meristem structure and function, vascular differentiation, leaf structure, and cell ultrastructure as well as loss of cell wall integrity can occur.

## ADDITIONAL READING

1. Armstrong, W. (1972). A re-examination of the functional significance of aerenchyma. *Physiologia*, Pl. 27, 173–177.
2. Bailey, I. W. (1913). The preservative treatment of wood. II. The structure of the pit membrane in the tracheids of conifers and their relation to the penetration of gases, liquids, and finely divided solids into green and seasoned wood. *For. Quart.* 11, 12–20.
3. Bailey, I. W. (1915). The effect of the structure of wood upon its permeability. No. 1—The tracheids of coniferous timbers. *Am. Railway Engineering Assoc. Bull.* 174, 1–19.
4. Bailey, I. W. (1958). The structure of tracheids in relation to the movement of liquids, suspensions, and undissolved gases. *In* "The Physiology of Forest Trees" (K. V. Thimann, Ed.), pp. 71–82. Ronald Press, New York.
5. Behnke, H. D., and Sjölund, R. D. (Eds.). (1990). "Sieve Elements: Comparative Structure, Induction and Development." Springer-Verlag, New York.
6. Bollard, E. G. (1960). Transport in the xylem. *Annu. Rev. Plant Physiol.* 11, 141–166.
7. Brouwer, R. (1965). Water movement across the root. *Soc. Exp. Biol. Symp.* 29, 131–149.
8. Brown, W. V. (1975). Variation in anatomy, associations, and origins of Kranz tissue. *Am. J. Bot.* 62, 395–402.
9. Canny, M. J. (1995). A new theory for the ascent of sap. Cohesion supported by tissue pressure. *Ann. Bot.* 75, 343–357.
10. Canny, M. J. (1997). Vessel contents during transpiration-embolisms and refilling. *Am. J. Bot.* 85, 1225–1230.
11. Canny, M. J. (1998a). Transporting water in plants. *Am. Sci.* 82, 152–159.
12. Canny, M. J. (1998b). Applications of the compensating pressure theory of water tansport. *Am. J. Bot.* 85, 897–909.
13. Coleman, G., and Coleman, W. J. (1990). How plants make oxygen. *Sci. Am.* 262, 50–67.

14. Ehleringer, J. R., Sage, R. F., Flanagan, L. B., and Pearcy, R. W. (1991). Climate change and the evolution of $C_4$ photosynthesis. *Trends Ecol. Evol.* **6**, 95–99.

15. Ellerby, D. J., and Ennos, A. R. (1998). Resistance to fluid flow of model xylem vessels with simple and scalariform perforation plates. *J. Exp. Bot.* **49**, 979–985.

16. Esau, K. (1965). "Plant Anatomy," 2nd ed. John Wiley & Sons, New York, London.

17. Esau, K. (1966). Explorations of the food conducting system in plants. *Am. Sci.* **54**, 141–157.

18. Esau, K. (1969). "The Phloem," Encyclopedia of Plant Anatomy, Vol. 5, Pt. 2. Gebrüder Borntraeger, Berlin.

19. Evert, R. F. (1984). Comparative Structure of Phloem. *In* "Contemporary Problems in Plant Anatomy" (R. A. White and W. C. Dickison, Eds.), pp. 145–234. Academic Press, San Diego.

20. Feild, T. S., Zwieniecki, M. A., Donoghue, M. J., and Holbrook, M. (1998). Stomatal plugs of *Drimys winteri* (Winteraceae) protect leaves from mist but not drought. *Proc. Natl. Acad. Sci.* (USA) **95**, 14256–14259.

21. Gamalei, Y. V. (1989). Structure and function of leaf minor veins in trees and herbs. *Trees* **3**, 96–110.

22. Gibson, A. C. (1996). "Structure-Function Relations of Warm Desert Plants." Springer-Verlag, Berlin.

23. Gregory, S. C., and Petty, J. A. (1973). Valve action of bordered pits in conifers. *J. Exp. Bot.* **24**, 763–767.

24. Greulach, V. A. (1973). "Plant Function and Structure." MacMillan Publishing Co., New York.

25. Gunning, B. E. S. (1976). The role of plasmodesmata in short distance transport to and from the phloem." *In* "Intercellular Communication in Plants: Studies on Plasmodesmata" (B. E. S. Gunning and A. W. Robards, Eds.), pp. 203–227. Springer-Verlag, Berlin.

26. Gunning, B. E. S., and Pate, J. S. (1974). Transfer cells. *In* "Dynamic Aspects of Plant Ultrastructure" (A. W. Robards, Ed.), pp. 441–480. McGraw-Hill, London.

27. Gunning, B. E. S., Pate, J. S., and Briarty, L. G. (1968). Specialized "transfer cells" in minor veins of leaves and their possible significance in phloem translocation. *J. Cell Biol.* **37**, C7–C12.

28. Gunning, B. E. S., Pate, J. S., and Green, L. W. (1970). Transfer cells in the vascular system of stems: Taxonomy, association with nodes, and structure. *Protoplasma* **71**, 147–171.

29. Haberlandt, G. (1965). "Physiological Plant Anatomy" (M. Drummond, Transl. from the fourth German edition), reprint edition. Stechert Hafner, New York.

30. Hargrave, K. R., Kolb, K. J., Ewers, F. W., and Davis, S. D. (1994). Conduit diameter and drought-induced embolism in *Salvia mellifera* Greene (Labiatae). *New Phytol.* **126**, 695–705.

31. Jarvis, P. G., and Slatyer, R. O. (1970). The role of the mesophyll cell wall in leaf transpiration. *Planta* **90**, 303–322.

32. Kozlowski, T. T., Hughes, J. F., and Leyton, L. (1967). Dye movement in gymnosperms in relation to tracheid alignment. *Forestry* **40**, 209–227.

33. Laetsch, W. M. (1974). The $C_4$ syndrome: A structural analysis. *Annu. Rev. Plant Physiol.* **25**, 27–52.

34. Leyton, L., and Armitage, I. P. (1968). Cuticle structure and water relations of the needles of *Pinus radiata* (D. Don). *New Phytol.* **67**, 31–38.

35. Markhart III, A. H., and Smit, B. (1990). Measurement of root hydraulic conductance. *HortScience* **25**, 282–287.

36. Milburn, J. A. (1991). Cavitation and embolisms in xylem conduits. *In* "Physiology of Trees" (A. S. Raghavendra, Ed.), pp. 163–174. Wiley, New York.

37. Nelson, T., and Dengler, N. G. (1992). Photosynthetic tissue differentiation in $C_4$ plants. *Int. J. Plant Sci.* **153**, S93–S105.

38. Noskowiak, A. S. (1963). Spiral grain in trees. A review. *For. Prod. J.* **13**, 26–275.

39. Owston, P. W., Smith, J. L., and Halverson, H. G. (1972). Seasonal water movement in tree stems. *For. Sci.* **18**, 266–272.

40. Pate, J. S., and Gunning, B. E. S. (1969). Vascular transfer cells in angiosperm leaves. A taxonomic and morphological survey. *Protoplasma* **68**, 135–156.

41. Pate, J. S., and Gunning, B. E. S. (1972). Transfer cells. *Annu. Rev. Plant Physiol.* **23**, 173–196.

42. Peterson, R. L., and Yeung, E. C. (1975). Ontogeny of phloem transfer cells in *Hieracium floribundum*. *Can. J. Bot.* **53**, 2745–2758.

43. Salleo, S., Lo Gullo, M. A., and Siracusano, L. (1984). Distribution of vessel ends in stems of some diffuse and ring-porous trees: The nodal regions as "safety zones" of the water conducting system. *Ann. Bot.* **54**, 543–552.

44. Scholander, P. F. (1958). The rise of sap in lianas. *In* "The Physiology of Forest Trees" (K. V. Thimann, Ed.), pp. 3–17. Ronald Press, New York.

45. Schulz, A. (1998). Phloem, structure related to function. *In* "Progress in Botany" (H.-D. Behnke, K. Esser, J. W. Kadereit, U. Lüttge, and M. Runge, Eds.), Vol. 59, pp. 429–475. Springer-Verlag, Berlin, Heidelberg.

46. Siau, J. F. (1984). "Transport Processes in Wood," Springer Series in Wood Science. Springer-Verlag, Berlin, Heidelberg.

47. Smith, W. K., Vogelmann, T. C., DeLucia, E. H., Bell, D. T., and Shepherd, K. A. (1997). Leaf form and photosynthesis. *BioScience* **47**, 785–793.

48. Sperry, J. S. (1985). Xylem embolism in the palm *Rhapis excelsa. IAWA Bull. n. s.* **6**, 283–292.

49. Sperry, J. S. (1995). Limitations on stem water transport and their consequences. *In* "Plant Stems: Physiology and Functional Morphology." (B. L. Gartner, Ed.), pp. 105–124. Academic Press, San Diego.

50. Sperry, J. S., and Tyree, M. T. (1988). Mechanism of water stress-induced xylem embolism. *Plant Physiol.* **88**, 581–587.

51. Sperry, J. S., Donnelly, J. R., and Tyree, M. T. (1988b). Seasonal occurrence of xylem embolism in sugar maple *(Acer saccharum). Am. J. Bot.* **75**, 1212–1218.

52. Sperry, J. S., Holbrook, N. M., Zimmermann, M. H., and Tyree, M. T. (1987). Spring filling of xylem vessels in wild grapevine. *Plant Physiol.* **83**, 414–417.

53. Torrey, J. G., and Clarkson, D. T. (Eds.) (1975). "The Development and Function of Roots," Third Cabot Symposium. Academic Press, London, New York.

54. Tyree, M. T., and Dixon, M. A. (1986). Water stress-induced cavitation and embolism in some woody plants. *Physiol. Plant.* **66**, 397–405.

55. Tyree, M. T., and Ewers, F. W. (1991). The hydraulic architecture of trees and other woody plants. *New Phytol.* **119**, 345–360.

56. Tyree, M. T., and Sperry, J. S. (1989). Vulnerability of xylem to cavitation and embolism. *Annu. Rev. Plant Physiol. Plant Mol. Biol.* **40**, 19–38.

57. Tyree, M. T., Davis, S. D., and Cochard, H. (1994). Biophysical perspectives of xylem evolution: is there a tradeoff of hydraulic efficiency for vulnerability to dysfunction. *IAWA J.* **15**, 335–360.

58. Van Bel, A. J. E. (1993a). Strategies of phloem loading. *Annu. Rev. Plant Physiol. Plant Mol. Biol.* **44**, 243–282.

59. Van Bel, A. J. E. (1993b). The transport phloem. Specifics of its functioning. *In* "Progress in Botany" (H.-D. Behnke, U. Lüttge, K. Esser, J. W. Kadereit, and M. Runge, Eds.), Vol. 54, pp. 134–150. Springer-Verlag, Berlin, Heidelberg.

60. Van Bel, A. J. E. (1996). Carbohydrate processing in the mesophyll trajectory in symplastic and apoplasmic phloem loading. *In* "Progress in Botany" (H.-D. Behnke, U. Lüttge, K. Esser, J. W. Kadereit, and M. Runge, Eds.), Vol. 57, pp. 140–167. Springer-Verlag, Berlin, Heidelberg.

61. Van Bel, A. J. E., and Kempers, R. (1997). The pore/plasmoderm unit; key element in the interplay between sieve element and companion cell. *In* "Progress in Botany" (H.-D. Behnke, U. Lüttge, K. Esser, J. Kadereit, and M. Runge, Eds.), Vol. 58, pp. 278–291. Springer-Verlag, Berlin, Heidelberg.

62. Varney, G. T., McCully, M. E., and Canny, M. J. (1993). Sites of entry of water into the symplast of maize roots. *New Phytol.* **125**, 733–741.

63. Williams, W. T., and Barber, D. A. (1961). The functional significance of aerenchyma in plants. *Symp. Soc. Exp. Biol.* **15**, 132–144.

64. Yeung, E. C., and Peterson, R. L. (1972). Xylem transfer cells in the rosette plant *Hieracium floribundum. Planta* **107**, 183–188.

65. Yeung, E. C., and Peterson, R. L. (1974). Ontogeny of xylem transfer cells in *Hieracium floribundum. Protoplasma* **80**, 155–174.

66. Yeung, E. C., and Peterson, R. L. (1975). Fine structure during ontogeny of xylem transfer cells in the rhizome of *Hieracium floribundum. Can. J. Bot.* **53**, 432–438.

67. Zimmermann, M. H. (1961). Movement of organic substances in trees. *Science* **133**, 73–79.

68. Zimmermann, M. H. (1971). Transport in the xylem. *In* "Trees. Structure and Function." (M. H. Zimmermann and C. L. Brown, Eds.), pp. 169–220. Springer-Verlag, New York, Heidelberg.

69. Zimmermann, M. H. (1973). The monocotyledons: Their evolution and comparative biology. IV. Transport problems in arborescent monocotyledons. *Q. Rev. Biol.* **48**, 314–321.

70. Zimmermann, M. H. (1982a). Piping water to the treetops. *Nat. Hist.* **91**, 6–13.

71. Zimmermann, M. H. (1982b). Functional xylem anatomy of angiosperm trees. *In* "New Perspectives in Wood Anatomy" (P. Baas, Ed.), pp. 59–70. Martinus Nijhoff/Dr. W. Junk Publ., The Hague, Boston, London.

72. Zimmermann, M. H. (1983). "Xylem Structure and the Ascent of Sap." Springer-Verlag, Berlin.

73. Zimmerman, M. H., and Brown, C. L. (1971). "Trees, Structure and Function." Springer-Verlag, Berlin, New York.

74. Zimmermann, M. H., and Jeje, A. A. (1981). Vessel-length distribution in stems of some American woody plants. *Can. J. Bot.* **50**, 1882–1892.

75. Zimmermann, M. H., and Milburn, J. A., (Eds.). (1975). "Transport in Plants. I. Phloem Transport." Springer-Verlag, Berlin.

76. Zimmermann, M. H., and Tomlinson, P. D. (1967). A method for the analysis of the course of vessels in wood. *IAWA Bull.* **1**, 2–6.

# 8

# ECOLOGICAL ANATOMY

Ecology is the study of an organism's relationship with its environment. As a result of the study of ecology, the life histories of plants and animals can be interpreted and questions relating to why any given species is successful in a particular type of habitat are addressed. The distributions and survival of plants are thought of in terms of both time and space and depend on growth and reproduction, all of which have a physiological and structural basis. A clear understanding of plant ecology, therefore, is based upon the effect of environmental factors—edaphic, biotic, and climatic—and upon the physiology and anatomy of the plant, considered individually or as a member of a community.

Having evolved from aquatic ancestors, plants have subsequently migrated over the entire surface of the earth, inhabiting tropical, arctic, desert, and alpine regions. Some species have adopted an epiphytic lifestyle; others have become parasites. During the course of this radiation, some angiosperms have independently returned to an aquatic habitat in either fresh or salt water. Anatomy can provide important clues relating to these major paths of adaptive radiation within defined groups of plants. This chapter focuses on a few examples of the anatomical strategies and adaptive responses that enable plants to survive and function in different environmental conditions.

## HABITAT AND PLANT STRUCTURE

An **adaptation** is any aspect of a plant that promotes its welfare in the environment it inhabits. Adaptations can take the form of external morphological

**295**

modifications, histological changes in tissues and cells, or physiological specializations. Any plant that is able to survive and reproduce in its environment is adapted to some degree to that environment. Adaptations are heritable and thus are the result of evolutionary change. Plants, of course, do not adapt themselves; rather, they become adapted as a result of natural selection and heritable variation. Adaptations can be specialized in the sense that they are adaptations to a particular subset of features within the environmental heterogeneity, or adaptations can be generalized if they represent specializations to a broad range of environmental features. Adaptations enable plants to survive in extreme conditions of intense sun or shade, cold or heat, physiologically wet or dry microhabitats, and soil mineral deficiences. Adaptations to a particular habitat or environmental condition directly influence conduction of food and water, rates of transpiration, temperature of tissues, and the effects of wind and humidity. In many cases, the functional significance of a particular structural feature is not known or can only be suspected. Adaptations also protect plants from insect damage and from grazing by large animals and may permit abscission or dormancy of an organ under unfavorable conditions. Structural adaptations of leaves and stems also can be related to the digestibility characteristics of plant tissues; they therefore can have economic implications.

Plants traditionally have been loosely lumped into broad categories on the basis of their possessing common syndromes of more or less similar modifications in presumed response to a known environmental factor or combination of factors. For example, plants that require abundant soil, water, and a relatively humid atmosphere and that live in regions of average or optimal water availability are termed **mesophytes.** The majority of familiar plants in temperate zones exhibit mesophytic structure. Many species, on the other hand, survive under extremes of water supply or other adverse conditions. Those plants known as **hydrophytes** grow on the surface of water or are submerged at various depths in soil covered by water or in soil that is usually saturated with water. These plants have no need to conserve water but possess adaptations that enhance light absorption and provide for a more efficient exchange and movement of gases within the plant and with the surrounding environment. Many more species live in regions where the supply of available water is deficient or fluctuates widely on a seasonal basis. These plants are called **xerophytes.** They inhabit dry regions such as deserts or semideserts, temperate and tropical scrub or savannah communities, and other dry places. These species possess xeromorphic adaptations, and represent a highly diversified and structurally complex group that exhibits adaptations, which enable them to survive and reproduce in environments characterized by generally low preciptation and atmospheric conditions that promote rapid water loss. Natural selection in this group has resulted in adaptations that are involved primarily with support, water transport and storage, and prevention of water loss. These adaptations result in most xeric plants losing comparatively little water through transpiration as compared to mesic- and hydric-adapted species. This collection of taxa includes a systematically varied group that have often converged to resemble one another in vegetative structure, as a result of the influences of the environment.

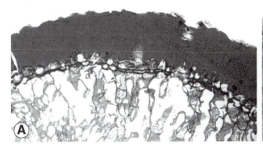
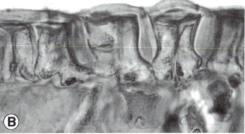

**FIGURE 8.1** Cuticle structure in cactus stems. (A) Transverse section of stem of *Ariocarpus* showing extensive development of cuticle. (B) Transverse section of stem of *Escontria chiotilla* showing cuticle extending between epidermal cells. Courtesy of S. Cornejo Loza and T. Terrazas-Salgado.

Some plants exhibit quite obvious structural adaptations, which help maintain the species in a particular adverse habitat. Other adaptations are less obvious, such as those involving diversity in cell wall structure or cell shape, and can only be appreciated following careful microscopic examination. Accompanying the differences in structure, there may also be an increased production of essential oils, resins, alkaloids, and other compounds that enable a plant to grow in particular habitats. Botanists have frequently interpreted these ecological categories to be the direct result of a single environmental factor. Although water deficit and high temperatures are the most obvious stresses that a land plant faces, numerous environmental factors affect a plant. There is typically an interplay among environmental and internal influences, and one factor may be altered by the effect of another. Not all xerophytic or hydrophytic adaptations occur together in the same individual. Generally some combination of features is observed within a plant. Accordingly, the practice of relating specific adaptations to a single environmental variable may be convenient but is not always justified.

Investigators have approached the study of **ecological plant anatomy** by using three general types of procedures: (1) the "floristic approach," in which the investigator examines and characterizes an entire assemblage of unrelated species occupying a single habitat; (2) the "systematic approach," in which a group of related species that show adaptive radiation into varied habitats along geographical clines is studied; and (3) the "experimental approach," in which controlled experiments are performed, usually of a single species.

## ECOLOGICAL LEAF ANATOMY

The variations in plant structure that are commonly affected by environmental factors are particularly strongly expressed in the morphology and anatomy of leaves. The leaf, in fact, has often been considered the most anatomically variable organ of the plant, and leaf adaptations have historically been used as indicators of environmental conditions. Smooth leaves with entire margins and apical drip tips are more common in lowland tropical rain forests. Leaf areoles of tropical and xeric climate plants tend to be smaller and with fewer freely ending veinlets than those of plants from cooler areas with adequate rainfall.

The ecological significance of leaf surface sculpturing, or cuticular ornamentation of the outer epidermal surface, is evidenced by the fact that woody species from moist tropical or subtropical environments tend to posses less conspicuous surface sculpturing. Herbaceous plants with thin, ephemerous leaves are reported to typically possess elaborate sculpture. As a rule, succulents have thick cuticles that are heavily encrusted by waxes and that show prominent ornamentation of epicuticular waxes. Hydrophytes have little or no surface sculpture. Stomata typically occur on the lower leaf surface. Because the upper epidermal surface of a leaf is exposed directly to sun, the temperature on the upper surface tends to be significantly higher than that of the lower surface. Because high leaf temperatures cause increased transpiration rates, stomata positioned on the lower surface and away from direct sunlight reduce water loss. Wax deposits and highly pubescent leaf surfaces can also reflect sunlight and decrease transpiration rates. These adaptations demonstrate the often close relationship between anatomy and the efficiency of physiological processes. It also must be remembered that the shoot apex comprises the growing tip of the stem and represents the site of new leaf production. Because the shoot apex is very sensitive to environmental stress, particularly water loss, the ability of the growing apex to withstand drought is critical to a plant's survival.

## Sun and Shade Leaves

The level of illumination that a leaf receives during development is perhaps the single most influential environmental factor affecting mature leaf structure. Structurally, leaves vary with respect to both the light intensity and light quality to which they are exposed during development. Within the same individual, leaves produced in bright light conditions (**sun leaves**) tend to be smaller and thicker and to have increased mesophyll tissue per unit area and higher density of stomata, veins, and chlorophyll when compared to leaves exposed to shade (**shade leaves**). The differences in internal leaf structure are related to blade thickness and the amount and distribution of palisade and spongy mesophyll tissue. These structural variations are associated with regulating light and $CO_2$ profiles within leaves and maximizing photosynthetic efficiencies. Sun leaves fix carbon faster than do shade leaves. Low light can also cause a reduction in mean vessel density. The differences in anatomy and physiology between sun and shade leaves are often reflected in species differences in shade tolerance, successional status, and drought aversion.

In sun leaves, the mesophyll zone usually contains an increased number of compact palisade layers, and each successive layer has greater palisade cell elongation. The foliar cuticle is often thicker than in shade leaves, and hairs, if present, are abundant. In these leaves, the well-developed cuticle may reflect light rays and protect the underlying tissues from excessive radiation. In some species, including tropical trees, for unexplained reasons foliar vein density increases with increasing height of the leaf insertion point, an indication that leaves adapt to microclimatic changes. Leaf areole size also may be larger near the bottom of an individual as compared with the upper, more illuminated leaf areoles. Mechanical tissues also are more strongly developed in some sun-grown leaves.

It has been suggested that the thickened cell walls of full sun leaves serve to resist the mechanical action of wind and prevent these leaves from bending. A number of factors are known to affect stomatal frequency in leaves, including water and sunlight. Sun leaves typically possess more numerous stomata than shade leaves and, as a result, transpire more abundantly and perhaps benefit from a cooling effect. The extreme sensitivity of plants to sunlight is evidenced by the fact that developing leaves of cowpea *(Vigna sinensis)* subjected to a single day of shade will have a decreased stomatal frequency, whereas plants placed in full sun for a day will have an increased stomatal number.

Leaves that develop under conditions of low irradiance are thinner, as, for example, those found on understory species or those lower down on an

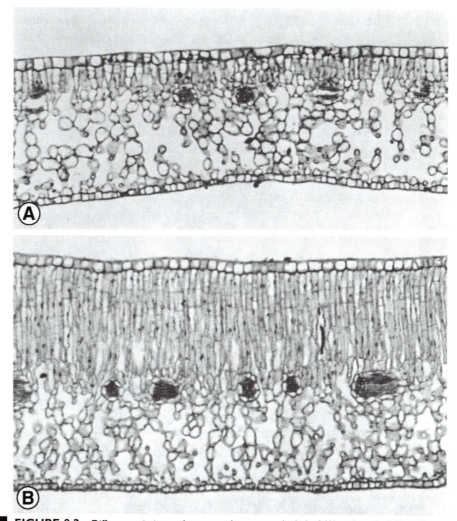

**FIGURE 8.2** Differences in internal structure between a shade leaf (A) and sun leaf (B) of *Prunus caroliniana*. Note extensive development of palisade mesophyll in the sun leaf form. Reprinted with permission, from Brown (1971), Springer-Verlag New York, Inc.

individual plant. They have a mesophyll region composed of poorly defined palisade cells and abundant intercellular spaces. Shade leaves also can possess special epidermal cells on the lower surface that are frequently larger than neighboring cells, are conspicuously lens shaped, and contain many chloroplasts. These cells are assumed to function in the capture of light. Gymnosperm leaves show similar responses to sun and shape conditions. The leaves of shade-grown conifers are thinner, have a reduced palisade mesophyll, reduced stomatal frequencies, and a higher chlorophyll content than sun-grown leaves.

The environmentally induced anatomical variation that occurs during leaf development is known to have significant consequences for photosynthesis. The development of a more structurally specialized palisade mesophyll in sun leaves is positively correlated with photosynthetic capacity. In some species, a thicker lamina promotes more efficient water use and lower transpiration rates under conditions of high radiation. The addition of columnar cells to the palisade region in sun leaves appears to facilitate the exchange of carbon dioxide between mesophyll cells and intercellular air spaces and plays an important role in the distribution of light within the leaf. Because photosynthesis depends on the balance between internal concentrations of both light and carbon dioxide, increased leaf thickness and the development of palisade layers directly influence this balance and optimize the rate of whole leaf photosynthesis. An increase in thickness of the upper epidermis and cuticle may have a protective function under high light conditions. Increased epidermal layers perhaps serve to increase leaf reflectance and protect underlying photosynthetic cells from excessive irradiance. In the process, leaf temperatures are maintained at optimal levels for physiological processes.

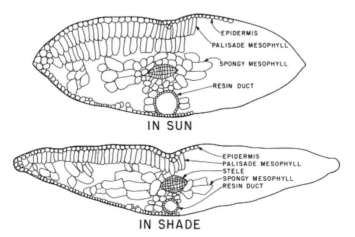

IN SUN

IN SHADE

**FIGURE 8.3**   Transverse sections of typical hemlock needles developed in full sunlight and in the shade of a dense Douglas fir canopy. Reprinted from *Forest Science* (**23**(2): 195–203) published by the Society of American Foresters, 5400 Grosvenor Lane, Bethesda, MD 20814-2198. Not for further reproduction.

## Xeromorphic Leaves

In addition to physiological differences, xerophytic plants typically possess one or more striking morphological and anatomical modifications to water stress, many of these are associated with leaf structure. It is important to emphasize, however, that xeromorphic foliar structure is sometimes correlated with the absence of certain nutrients in the soil. Low soil nitrogen levels can result in increased lignification of the leaf epidermis and nonveinal sclerenchyma. High nitrogen levels can cause a decrease in the amount of wood formed. Plants that live in habitats where the supply of available water is deficient have been classified into broadly defined categories on the basis of the different adaptive strategies they have developed:

1. *Drought-escaping species* are plants with a compressed growth cycle that complete vegetative growth and reproduction within the short period of time when conditions are favorable.

2. *Drought-evading species* are plants that are able to reduce water loss, or compensate for water loss, by possessing some specialized structural feature or set of features, such as an extensive root system.

3. *Drought-enduring species* are plants that are able to survive even when water uptake cannot take place or is severely reduced. This group includes species with a number of highly divergent specializations such as temporary leaf loss, changes in leaf angle and the rolling or folding of leaf blades, or permanent adaptations such as succulence or leaves reduced to spines. The leaves of some xerophytic grasses contain enlarged epidermal cells that store water and occur in dispersed longitudinal rows. These cells are called **bulliform cells** and under conditions of water deficit they lose turgor and thus constrict in upon themselves, causing the grass lamina to fold or roll inward edge to edge. Bulliform cells may occur only on the upper surface or both upper and lower surfaces. As a result, the entire upper leaf surface forms a protective barrier to air movement that limits water loss. Bulliform cells are recognizable by the cell wall ridges that are present when the cells are constricted.

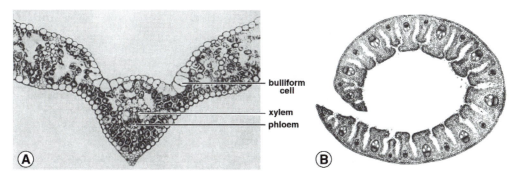

**FIGURE 8.4**    Transverse sections of grass leaves. (A) *Poa* leaf showing enlarged bulliform cells in the upper epidermis. (B) Beachgrass (*Ammophila breviligulata*) leaf showing curled condition following water loss.

The effects of environmental factors on leaf morphology and anatomy have produced a wide variety of adaptations in different taxa. Some features, such as reduction in leaf size, increased thickness of the outer walls of epidermal cells, increased thickness of cuticle, increased trichome density, reduction in stomatal pore area, and stomata recession below the blade surface (or confined to pits or grooves on the undersurface of the blade), probably contribute to a reduction in the rate of loss of absorbed water. Other characteristics, such as an increase in mechanical cells and wall lignification, increased succulence and water storage capacity, and the accumulation of mucilage, probably help the plant to tolerate dehydration and deal with water stress.

## Ericoid Leaves

A reduction in leaf size has occurred in diverse plant families. The progressive reduction in leaf area has resulted in the production of small leaves of very different form and shape. In a few cases, the ultimate reduction in leaf evolution has formed needlelike or scalelike projections and even resulted in the loss of leaves. The formation of a short, linear leaf form is associated with a reduction in leaf surface. This form is known as **ericoid,** named after its occurrence in many genera of the Ericaceae and related families. These leaves display extreme ecological specialization and often have stomatal grooves that extend the entire length of the lower surface. Trichomes may completely fill the grooves or be located only along the rims. In many species, the lamina has prominent inrolled margins or the leaf has become cylindrical in outline. Leaves of this general type also can exhibit parallel changes in internal anatomy, such as a general foliar sclerification.

## Sclerophyllous Leaves

In dry environments the leaves of some dicotyledons have retained a broad lamina and become increasingly leathery. Xerophytes that have increased cutinization and lignification of the leaves are commonly called coriaceous or **sclerophyllous.** This leaf type tends to be thick, composed of many highly lignified cells with thickened and strongly cutinized outer epidermal walls. Venation is typically dense, and individual vascular bundles are increased in size. Trichome density is often accentuated, and in extreme cases the trichomes are lignified. One or more layers of heavily lignified cells are sometimes distributed immediately below the epidermis to form a **hypodermis.** The increase in sclerified cells is correlated with a well-developed and lignified bundle sheath around the veins, in addition to the presence of diverse forms of sclereids. Because turgor pressures cannot always remain at high levels within tissues of xerophytes, extensive sclerenchyma development is required for support. There is frequently extensive development of palisade parenchyma in the mesophyll and a corresponding decrease in spongy mesophyll and intercellular spaces. The mesophyll in some sclerophyllous leaves is **isobilateral;** that is, palisade parenchyma occurs on both sides of the blade. In some species the stomata are sunken in pits on the lower surface of the blade, with cuticular rims overarching the depression. In this way, the open-

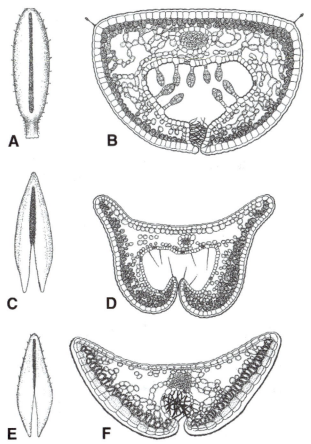

**FIGURE 8.5**  Leaf morphology and anatomy of Ericales. (A) *Empetrum nigrum,* leaf. × 8. (B) *Empetrum nigrum,* transverse section of leaf. × 114.  (C) *Cassiope tetragona,* entire leaf. × 9. (D) *Cassiope tetragona,* transverse section of leaf. × 9. (E) *Calluna vulgaris, entire leaf.* × 12.  (F) *Calluna vulgaris,* transverse section of leaf. × 114. Reprinted with permission from Hagerup (1953), *Phytomorphology* 3, 459–464.

ing between the guard cells is protected from contact with the outside air and consequently from air of low humidity. This anatomical type is not always correlated with aridity, however, as evidenced by the occurrence of the same complex of features in many woody bog plants and plants growing in nutrient-poor soils. The common occurrence of sclerophylly in many monocotyledons and dicotyledons has been related to long-lived leaves, which in turn is an adaptation to nutrient limitation. These features may also render leaves unpalatable to animals.

## Succulent Leaves

Another large group of xerophytes possesses fleshy or succulent leaves, stems, or roots. As in other xerophytes, epidermal cells are frequently somewhat thickened and heavily cutinized. Succulent trees in Baja California are reported to store up to six times more water per volume than co-occurring nonsucculent

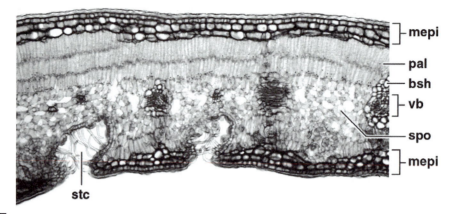

**FIGURE 8.6**   Transverse section of leaf of *Nerium oleander* (Apocynaceae). Abbreviations: bsh, vascular bundle sheath; mepi, multiple epidermis; pal, palisade mesophyll; spo, spongy mesophyll; stc, stomatal crypt lined with trichomes; vb, vascular bundle.

trees. Succulents are adapted to deliver huge amounts of storage water and in these plants mucilaginous substances are prominent. Colorless storage cells are usually large and often thin-walled. Frequently the walls are reinforced to prevent collapse when turgor is reduced and the leaves generally contain little or no sclerenchyma. Species of the widely distributed family Crassulaceae illustrate this condition particularly well.

A further important adaptation found mainly, but not exclusively, in succulent plants of the family Crassulaceae is a variant type of photosynthetic carbon dioxide fixation known as Crassulacean acid metabolism (CAM). Although this pathway is not especially efficient, CAM represents a metabolic sequence that promotes photosynthesis in very hot and arid environments. Some succulents such as the Cactaceae initially fix carbon dioxide into oxaloacetate, in a manner similar to the more common photosynthetic pathway. Unlike most plants, however, this occurs only at night and is associated with open stomates in leaves or stems. Stomata normally open at dawn and close in the evening. The reversal of this mechanism in succulents represents a selective adaptation that conserves water during the period of prolonged exposure to daytime heat and high transpiration rates. Oxaloacetate is subsequently processed, and the later-formed compounds provide a source of internally generated carbon dioxide during the daytime. The result is that the malic acid content of photosynthesizing cells increases at night and decreases during the day. This can result in no net carbon dioxide fixation during hours of peak irradiance. Not all succulent plants engage in this type of photosynthesis. Interestingly, some species that are conventional $C_3$ plants undergo CAM following periods of water stress and can be regarded as facultative with regard to their photosynthetic physiology.

## Poikilohydric Plants

A limited number of flowering plants fall into a category known as **poikilohydric** plants; that is, their body structure and function vary dramatically with

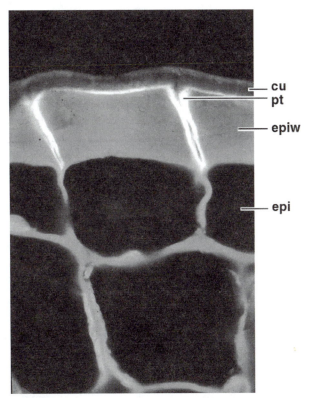

**FIGURE 8.7**  Transverse section of leaf epidermal cells of *Agave* (Agavaceae).  Note thick cuticle and conspicuously thickened outer wall of epidermal cells.  Abbreviations: cu, cuticle; epi, epidermal cell; epiw, outer wall of epidermal cell; pt, pectin.

environmental water availability. The cells, tissues, and organs of these plants are able to remain viable following cycles of extreme dehydration and rehydration. These plants are commonly referred to as **resurrection plants.** In addition to complex physiological adaptations, they possess a number of distinctive structural features. When dessicated, the leaves of resurrection plants shrink in size and often curl up. This is usually accompanied by severe wrinkling of the leaf epidermis. A close union between the plasma membrane and cell wall results in the folding of the entire cell upon desiccation, but plasmodesmatal connections are still maintained. In some resurrection plants the cell walls of tracheary elements have spiral thickenings. Leaf shrinkage is associated with the contractibility of the xylem elements, which have been shown to fold like an accordion during drying. The collapsibility of the helically stiffened xylem cells is considered to represent an adaptation to extreme environmental changes during the life of the plant. Most poikilohydric monocotyledons lose their leaf chlorophyll during prolonged dehydration. The loss of chlorophyll is correlated with the degradation or rearrangement of thylakoid membranes in chloroplasts. Upon rehydration, thylakoid membranes reform and chlorophyll and photosynthesis recover.

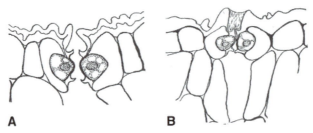

**FIGURE 8.8**  Stomatal structure in leaves of xerophytes. (A) Transverse section of stoma of a fleshy leaf of *Aloë*. Note thick cuticle and sunken position of guard cells. (B) Transverse section of stoma of a leaf of *Yucca filamentosa*. Note sunken guard cells.

## Halophytic Leaves

Halophytic plants, or **halophytes,** are those species able to tolerate saline environments. This adaptation is sometimes considered a special type of xerophytism because of the nature of the environments they inhabit. Halophytic species often show a diversity of structural and physiological adaptations that include succulent leaves with specialized "colorless" storage cells or "window cells" that increase the plants capacity for salt accumulation. Some species have secretory cells or tissues, enlarged tracheids

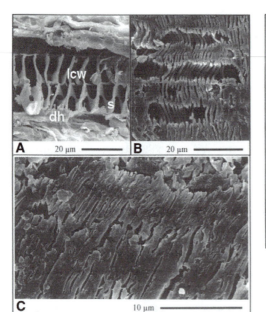

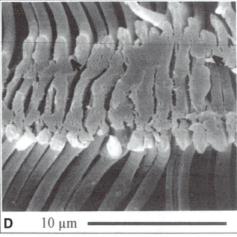

**FIGURE 8.9**  SEM of longitudinal sections of foliar vascular bundles of the Namibian endemic *Chamaegigas intrepidus* (Scrophulariaceae). This poikilohydric plant grows in ephemeral rock pools on granite outcrops. (A) Hydrated leaf. (B) Dehydrated leaf. (C) Air-dried leaf with extremely densely packed internal spiral bands. (D) Vessel of desiccated leaf showing thin longitudinal cell walls that are folded like an accordion. Abbreviations: dh, double helical sculpture; lcw, longitudinal cell wall; s, spiral band. Arrows, space between spiral bands. Reproduced with permission of Schiller *et al.* (1999), *Flora* **194,** 97–102.

terminating vein endings, structures such as salt glands and salt hairs, and a hypodermis. Halophytic plants concentrate salt in their tissues, moving it from the soil into and through the plant. The leaves of the salt-tolerant maritime mangrove (e.g., *Avicennia germinans*) contain large, tightly packed, multiple rows of specialized hypodermal cells that play a probable role in salt accumulation and storage. Because salt cannot be allowed to accumulate without limit, many halophytes (including mangroves) have special surface glands and hairs to remove NaCl from the underlying mesophyll cells and then actively secrete it on the leaf surface. The **salt glands** of *Avicennia* are formed in shallow pits on the upper leaf surface, whereas on the lower surface they are not sunken. Individual glands consist of two to four highly vacuolated basal cells, a cutinized stalk cell, and a minimum of eight terminal cells that are covered by a thin, perforated cuticle that is separated from the cell wall apically. All gland cells are interconnected by plasmodesmata. Salt is deposited in the subcuticular cavity of the gland and then ultimately reaches the leaf surface through the cuticular pores. The structurally complex salt glands of the facultative halophyte *Limonium gmelini* (Plumbaginaceae) are composed of 16 glandular cells overlying 4 collecting cells that are connected to mesophyll cells by plasmodesmata. A layer of cutin surrounds the collecting cells except at pore interruptions.

Other halophytes such as *Atriplex* (Chenopodiaceae) have **salt hairs** on the surface. These specialized hairs are composed of an enlarged terminal bladder cell that contains a large salt-accumulating and salt-storing central vacuole. The bladder cell is attached to the leaf epidermis by a stalk cell, through which the salt is actively transported. Eventually these salt-containing cells burst and crystalline salt is deposited on the epidermal surface. The surface salt provides a reflective coating that shades the leaf from direct sunlight and also decreases the palatability of the plant to herbivores. The sharp salt crystals also may function as a mechanical deterrent to herbivory. In the halophyte *Aster tripolium*, stomata close in response to high Na+ concentrations. This response prevents excessive accumulation of Na+ within the shoot by controlling transpiration rates.

## Alpine Plants

As a rule, alpine plants are exposed to bright light and high winds. Although the internal structure of alpine plant leaves varies with the species, they often show one or more adaptive features classically regarded as xeric specializations. The leaves of a number of montane species are covered with abundant hairs. A dense hair covering on leaves and stems is believed to serve some combination of the following functions: (1) reduce light absorption (increase light reflectance) during periods of high temperature, bright light, or drought; (2) reduce the diffusion of gases across the leaf and air interface; and (3) reduce predation by insects and larger herbivores. Leaf pubescence also can function to regulate internal leaf temperatures and avoid potentially lethal high temperatures. Leaves of alpine plants also often possess a medium to thick cuticle and thickened epidermal cell walls. The mesophyll tends to be compact and often contains multiple palisade layers.

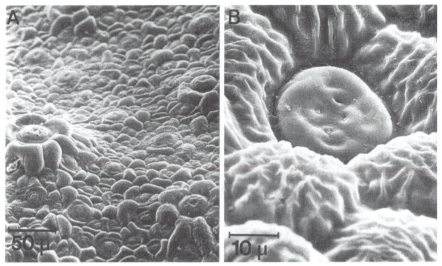

**FIGURE 8.10**   Salt glands of *Limonium sinuatum* (Plumbaginaceae). (A) SEM of leaf surface showing salt glands.  (B) High magnification of individual salt gland.  The gland develops from a single epidermal initial and at maturity consists of 20 cells.  Reprinted with permission from Wiehe and Breckle (1990), *Bot. Acta* **103**, 107–110, Copyright © Georg Thieme Verlag, Stuttgart, New York.

## Epiphytic Plants

One of the more interesting examples of structure and function relationships is found among some species of the monocot family Bromeliaceae. All members of the family possess trichomes to some extent; however, the leaf surfaces of the epiphytic subfamily Tillandsioideae are covered with a dense layer of elaborate **absorbing trichomes.** These plants have a xeromorphic structure, and the trichomes facilitate the rapid movement of water and minerals into the shoot. As a result of the trichomes, certain species are able to survive with no roots at all. Individual hairs are peltate structures composed of a multicelled disc of dead cells subtended by a multicellular stalk of living cells. The stalk penetrates the leaf epidermis and functions as a conduit to the mesophyll. The water-absorbing trichomes in bromeliads function in an interesting one-way action. They open to let water in when it rains but close to retain water when it dries out. The roots of these plants, when present, also possess a special covering tissue called **velamen** (discussed in the section headed "Aerial Roots").

## Hydromorphic Leaves

The secondary transition from a terrestrial to an aquatic environment has occurred independently within different groups of angiosperms. Extant water plants show all stages of this transition with emergent, floating, or permanently submerged leaves. All these hydrophytic leaf types vary considerably in response to changes in habitat, although all are characterized by marked increases in **aerenchyma,** parenchyma tissue with particularly large intercellu-

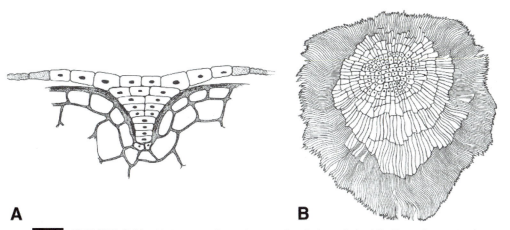

**A**                                                                      **B**

**FIGURE 8.11**   Trichome stalks and caps of tank bromeliads. (A) *Neoregelia,* sectional view. (B) *Neoregelia,* surface view of trichome cap. Reprinted with permission from Benzing (1970), *Bot. Gaz.* **131,** 23–31.

lar spaces. Whereas emergent upright aquatic leaves are otherwise similar in structure to those of nonaquatics, floating and submerged leaves possess a number of distinctive features that serve such functions as gas exchange and flotation. Free-floating plants such as the duckweed *(Lemna)* possess leaves modified to trap air and prevent waterlogging by having large, specialized trichomes and a thick cuticle.

The structural characteristics in aquatic plants are mainly related to a reduction of the protecting, supporting, and conducting tissues and the presence of air chambers in tissues of the leaf, stem, and root. Epidermal cells typically possess chloroplasts that increase light absorption and photosynthesis. Anatomical adaptations represent responses to excessive water content, which implies a decreased oxygen supply. The anatomy of leaves of the pondweed *Potamogeton* illustrates this syndrome; the leaves possess weakly developed supporting and conducting tissues and prominent intercellular air spaces. The stems of submerged aquatics like *Potamogeton* also contain large gas lacunae that provide pathways for the transport of oxygen to the roots. Although such pathways are continuous down the stem, they are interrupted by thin, perforated diaphragms. Experimental work has shown that these structures act to control leaks and provide most of the resistance in the mass flow of gases but contribute little resistance to diffusive gas transport.

With the notable exception of the genus *Brasenia* (Cabombaceae), likely stages in tissue and cell reduction have not been well documented among aquatic plants. Cuticle on the epidermis is generally absent or is extremely thin. The epidermis is not primarily protective in this case but is composed of epidermal cells that are modified into transfer cells in such a way that they facilitate the absorption of gases and nutrients directly from the water. Chloroplasts occur in the epidermal cells, although they decrease in abundance with increasing water depth and with decreasing light. Large air passages filled with gases are common in hydrophytes of this type. In addition to

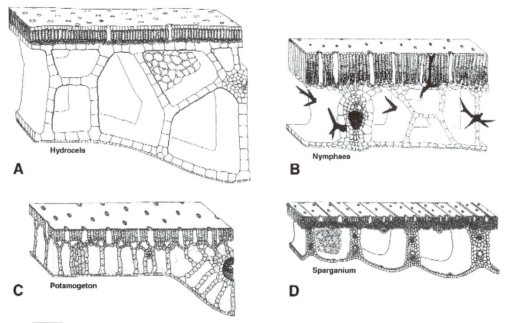

**FIGURE 8.12**    Leaf anatomy of floating leaves. (A) *Hydrocleis nymphoides* (Butomaceae). × 45.    (B) *Nymphaea tuberosa* (Nymphaeaceae).    Note astrosclereids in mesophyll. × 55.    (C) *Potamogeton nodosus* (Potamogetonaceae).    Note stomata on upper surface. × 28.    (D) *Sparganium fluctuans* (Sparganiaceae). × 60. Reprinted with permission from Kaul (1976). *Aquatic Bot.* **2,** 215–234.

providing very efficient mechanical stabilization to organs, the oxygen given off in photosynthesis as well as that from the air is stored and transported in these spaces, to be used again in respiration. The carbon dioxide from respiration also is held and used in photosynthesis. The function of air chambers in buoying some floating leaves is equally clear. Sclerenchyma is generally reduced or absent, although sclerenchymatous cells are conspicuous in the leaves of some marine angiosperms. Stomata are completely lacking in permanently submerged leaves. In floating leaves the lamina has retained a dorsiventral structure, and functional stomata are restricted to the adaxial (upper) surface of the lamina.

A number of unrelated aquatic plants develop leaves of two distinct morphologies and anatomies on the same stem—one form of leaf submerged and the other form aerial. This condition is known as **heterophylly** or leaf dimorphism. Shoot apices of submerged and aerial axes appear similar. Leaves originating from an aerial shoot apex are usually undissected, with an entire, lobed, or toothed blade, with stomata, and with a bifacial mesophyll with a palisade zone. Leaves arising from submerged apices are highly dissected, are essentially devoid of stomata and cuticle, have an undifferentiated mesophyll that often lacks palisade cells, and have reduced venation patterns. Among species with whorled phyllotaxy, the aerial stems typically have fewer leaves in each whorl as compared with the submerged segment of stem. In these cases, increase in the number of dissected leaves in a whorl increases total leaf surface area and the total assimilating area. Experiments with the mermaid

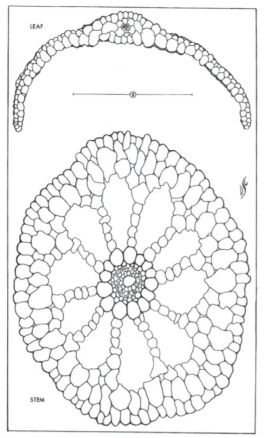

**FIGURE 8.13**   Anatomy of the aquatic plant *Najas flexilis* (Najadaceae).  Note reduced condition of vascular tissues. Reprinted with permission from Ogden (1974), Anatomical Patterns of Some Aquatic Vascular Plants of New York, New York State Museum.

weed *Proserpinaca palustris* have demonstrated that conditions of the photoperiod and other environmental factors can influence the form of leaf produced at the apex. We know from experiment that the developmental determination of heterphyllous leaf types involves the action of the hormone abscisic acid (ABA). When exposed to ABA, pondweeds form floating or aerial leaves instead of submerged leaves. Determination in these cases is a gradual process that is not finalized until late in development, when the leaf is almost fully expanded. Growing leaf primordia placed in different environmental regimes (aerial and submerged) at different stages of development will become either entirely one leaf type or the other, or they may show intermediate features, depending upon the stage of leaf development at which the shift in environment was made. If the transfer is made at an early stage, the leaf will resemble the leaf type characteristic of the new environment. If the shift is made later in development, the leaf will display intermediate features. A summary of selected anatomical differences between leaves of hydrophytes and xerophytes is provided in Table 8.1.

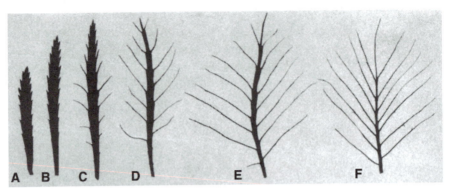

**FIGURE 8.14**    Leaf shape in *Proserpinaca palustris* (Haloragaceae). (A–F) Series of transitional leaves formed on shoots subjected to abrupt environmental change. (A, B) Aerial shoots. (E, F) Submerged shoots. Reprinted with permission from Schmidt and Millington (1968), *Bull. Torrey Bot. Club* **95**, 264–286.

## ECOLOGICAL WOOD ANATOMY

Although leaves have historically been the principal focus of ecological anatomical investigation, recent years have witnessed a renewed awareness among structural botanists of the profound influence of climatic variables on wood structure, or what is often referred to as the **hydraulic architecture** of woody trees. The innovative research of the late Harvard University professor

**TABLE 8.1    Common Anatomical Differences Between Leaves of Hydrophytes and Xerophytes**

|  | **HYDROPHYTES** | **XEROPHYTES** |
|---|---|---|
| Morphology | Often marked heterophylly; blade dissected or ribbon-like | Blade often reduced in size or absent |
| Texture | Thin | Leathery or succulent |
| Cuticle | Thin or absent | Thick |
| Epidermis | Thin-walled | Thick-walled |
| Stomata | Reduced in number or absent; restricted to upper surface | Sunken; sometimes in depressions or grooves |
| Mesophyll | Aerenchymatous | Compact; strongly developed palisade |
| Water storage tissue | Absent | Frequently present with associated mucilage |
| Sclerenchyma | Reduced or absent | Often well-developed; either fibers or sclereids; accompanying veins or beneath epidermis |
| Vascular tissue | Reduced in amount | Well-developed; dense |
| Other features | Epidermis often with high concentrations of chloroplasts; transverse septa | Capacity for folding or curling |

Martin H. Zimmermann and two contemporary plant anatomists, the American Sherwin Carlquist and the Dutch botanist Pieter Baas, ranks as significant in this field. Much of the structural diversity encountered within the secondary xylem of woody plants has a functional and adaptive explanation and can be directly related to plant habit as well as to varying atmospheric conditions and soil moisture availability. The hydraulic design of trees, shrubs, and vines, although less studied than that of leaves, clearly has an immense influence on the overall movement of water within the plant body and is perhaps the single most important factor in determining plant size, the vulnerability of stems to periods of drought, the water storage capacity of tissues, as well as the geographic distribution of woody species. Virtually all aspects of xylem structure can be influenced in either a qualitative or a quantitative manner by ecological conditions. **Ecological wood anatomy** refers to the study of correlations between the ecological and floristic preferences of taxa, and various wood anatomical characters that relate to function.

## Growth Rings

Perhaps the most widely recognized example of the influence of climate on xylem structure and function is found among some north temperate tree species, which form conspicuous growth layers or rings. Growth rings vary in thickness from year to year because of the effects of the environment. When trees reach a period of maximum growth, there are usually two alternating phases of rapid and slow cambial activity. The two phases proceed independently and can usually be correlated with different environmental conditions, such as periods of high and low rainfall. Growth rings are visible because there is seasonal variation in the abundance and character of certain kinds of cells formed within the growth layer. With slower growth at the end of the growing season, the conducting cells and fibers are smaller and often thicker walled (**latewood** or **summerwood**). The first-formed wood of each year (**earlywood** or **springwood**) is composed of much wider and thin-walled elements with proportionally large lumens that form a distinct line of larger cells adjacent to the smaller cells of the previous summerwood. The excess photosynthate that is generated in springtime is utilized in the production of new vegetative shoots and leaves and not incorporated into tracheary wall thickening. Toward the end of the summer, available photosynthates are moved into forming thickened tracheary cell walls. Woods in which the vessels of the earlywood are distinctly wider than those of the latewood of the same ring (and form a well-defined zone or ring in cross-sectional appearance) are termed **ring-porous** (*Quercus, Ulmus, Fraxinus, Castanea*). Interestingly, wood anatomists have observed that ring porosity does not typically occur in the temperate region woods of the Southern Hemisphere, although it is quite common in those of the Northern Hemisphere. This asymmetry has not been explained.

Because vessel diameter appears to be functionally related to the volume of water being conducted through the xylem (Chapter 7), differences in pore diameter in ring-porous trees reflect differences in the amount of water transported at various times throughout the growing season. They document the

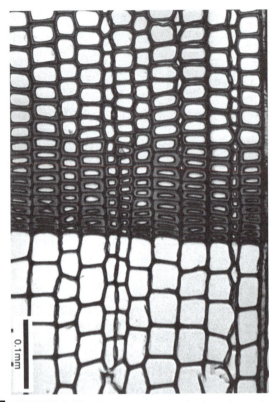

**FIGURE 8.15**  Transverse section of softwood annual ring transition showing thick-walled late-wood and thin-walled earlywood tracheary cells.  Courtesy of E. S. Wheeler and North Carolina State University.

fact that ring-porous woods are adapted to temperate seasonal climates. During the early part of the season when growth rates are high and soil water is abundant, the wide vessels of the earlywood of the current year are capable of handling a large volume of water flow per unit time and account for nearly all conduction within the stem. During the latter part of the season, as water availability decreases and associated stresses and tensions within the water columns increase, the narrow vessels of the latewood are advantageous and transport an increasing percentage of water. In woods that possess both vessel elements and tracheids, the tracheids presumably provide safety to the system against disabling air embolisms and cavitation. Ring porosity, therefore, al-lows for adequate water conduction throughout the entire growing season by combining these two adaptive strategies. The first maximizes the efficiency of water conduction before the initiation of leaves during the early part of the growing season; the second ensures adequate conduction as conditions of water stress increase later in the year. Ring-porous species, such as elm and chestnut, that operate a highly effective water-conducting system with high water flow efficiency in the outer ring of wide vessels are at high risk, however, because diseases affecting only a few large vessels can be very destructive.

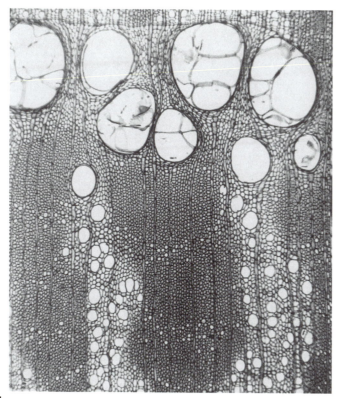

**FIGURE 8.16** Transverse section of the ring-porous wood of *Quercus alba* (Fagaceae). The wide earlywood vessels are filled with thin-walled tyloses. Courtesy of the Bailey-Wetmore Wood Collection, Harvard University.

## Hydraulic Architecture

Vessel element structure and distribution patterns provide a clear example of the significance of ecology on the hydraulic architecture of dicotyledonous trees and shrubs. Statistical analysis reveals a number of trends in specialization related to vessel element morphology and abundance that are directly related to changes in plant habit as well as latitude and altitude of provenance. Figure 8.17A illustrates the positive correlation between vessel frequency and increasing latitude in the genus *Ilex*. A positive correlation also is found between vessel element length and latitude. Figure 8.17B demonstrates a negative correlation between vessel diameter and increasing latitude. Although the functional significance of this correlation is not clear, it is evident that tropical species of *Ilex* possess longer and wider vessel elements than the temperate representatives of the same species. Similar correlations are known for many other genera. Statistically significant correlations also exist between latitude and other wood anatomical features, such as perforation plate bar number, imperforate tracheary element length, and ray size and frequency.

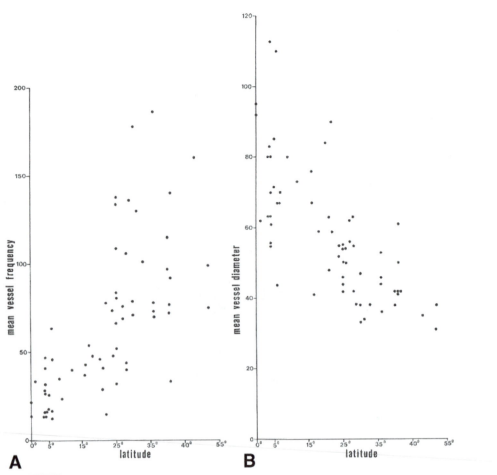

**FIGURE 8.17** Correlations between vessel frequency, latitude, and diameter in 64 species of *Ilex* (Aquifoliaceae). (A) Vessel frequency and latitude. (B) Vessel diameter and latitude. Reprinted with permission from Baas (1973), *Blumea* **21**, 193–258.

Habit-related anatomical modification may be superimposed upon phylogenetic advancement. Many trends are the result of complex interrelationships between ecology and plant morphology. In many diverse groups of dicotyledons, features such as vessel element length, tangential diameter, wall thickness, and frequency are known to be under the strong control of one ecological factor, water availability. It has been hypothesized that comparatively long and wide structurally primitive vessel elements are of adaptive value (or are not disadvantageous) only in mesic habitats, where soil moisture is not limiting and foliar transpiration is fairly slow and uniform owing to a constantly high relative humidity. Shorter and narrower but more numerous vessel elements per unit area, with simple perforation plates, have evolved under drier conditions of both low soil moisture and atmospheric humidity. The narrowness of vessels in classically "xeromorphic" dicotyledonous taxa is generally correlated with an increased number of vessels per unit area of xylem transection. These and other anatomical trends can occur among different woody dicot taxa along ecological gradients.

Although these correlations have been made by relating wood structure and plant habitat, their functional significance is not always immediately clear. For example, xylem cavitation, caused by the seeding of air into the vascular system, is a major potential problem for plants. Xylem cavitation occurs at species-specific xylem pressures and is correlated with drought tolerance. That is, xeric species cavitate at lower xylem pressure than mesic species. Air seeding occurs at the pit membranes and xeric species cavitate at lower pressures because they have pit membranes that are less permeable to air seeding. Cavitation pressure and xylem conduit diameter are only weakly correlated. Conifers with small volume and diameter tracheids are not as a group more resistant to cavitation than angiosperms with large vessels. Ring-porous trees are not as a group more vulnerable to cavitation than diffuse-porous trees. The reason for the weak correlation is that cavitation does not depend on conduit size, but on pit membrane structure. Pit membrane structure is more important in cavitation events than conduit size itself.

Some authors have advocated calculating a "vulnerability index" of wood anatomy (VULN), by dividing mean vessel diameter by the mean number of vessels per square millimeter. The lowest indices of vulnerability appear in xerophytic taxa, with narrow but very numerous vessels per unit area. Theoretically, this xylem formulation ("high vessel redundancy") is much safer physiologically than one that consists of wider and fewer vessels per unit area because it will restrict air embolisms to a smaller and more localized portion of the transpiration stream in the event of disruptively high negative pressures, freezing, or other types of injury to the xylem vessels. It also is possible to calculate an index of wood anatomical mesomorphy (MESO), in which the vulnerability index is multiplied by the mean length of the vessel elements.

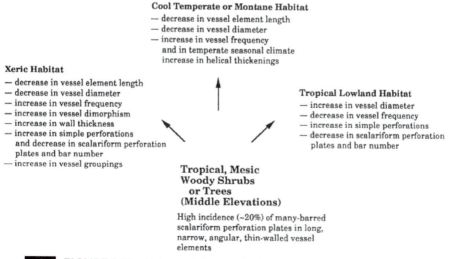

**Cool Temperate or Montane Habitat**
— decrease in vessel element length
— decrease in vessel diameter
— increase in vessel frequency
   and in temperate seasonal climate
   increase in helical thickenings

**Xeric Habitat**
— decrease in vessel element length
— decrease in vessel diameter
— increase in vessel frequency
— increase in vessel dimorphism
— increase in wall thickness
— increase in simple perforations
   and decrease in scalariform perforation
   plates and bar number
— increase in vessel groupings

**Tropical Lowland Habitat**
— increase in vessel diameter
— decrease in vessel frequency
— increase in simple perforations
— decrease in scalariform perforation
   plates and bar number

**Tropical, Mesic
Woody Shrubs
or Trees
(Middle Elevations)**

High incidence (~20%) of many-barred scalariform perforation plates in long, narrow, angular, thin-walled vessel elements

**FIGURE 8.18** Major ecophyletic trends of vessel element specialization. Reprinted with permission from Wheeler and Baas (1991). *IAWA Bull. n.s.* **12,** 275–332 (adapted from Dickison).

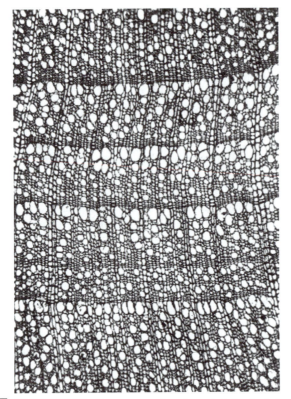

**FIGURE 8.19**   Transverse section of wood of the shrub *Crossosoma bigelowii* (Crossosomataceae) that inhabits arid regions.  Note high frequency of narrow vessels. × 55.

Plants that are considered mesophytic on ecological and macromorphological grounds typically exhibit wood mesomorphy indices greater than 200, whereas xerophytic taxa rarely possess indices of xylem anatomical mesomorphy in excess of 75. Table 8.2 provides a comparison of these quantitative features of vessel element morphology along with calculations of wood vulnerability and anatomical mesomorphy for the world flora in general and for the western Australian flora in particular.

### Adaptive Radiation in Dilleniaceae

The Dilleniaceae are a family of 11 genera distributed pantropically and ranging in habit from trees, shrubs, subshrubs, and rosette trees to woody climbers. The family is an excellent example of adaptive radiation and structural diversification at both the generic and specific levels of comparison, from mesic habitats toward both wetter and more xeric conditions. Morphological and anatomical diversity is particularly striking in the largest and most primitive genus *Hibbertia*.

Some mesic species of *Hibbertia* are shrubs up to 3 m tall. The mesomorphic xylem of these plants contains comparatively few angular and thin-walled vessels, with scalariform perforations containing 30 bars or more and

▮▮▮ **TABLE 8.2  Wood Anatomy of Western Australian Florulas Compared to Categories from Other Areas Showing Indices of Vulnerability (V) and Mesomorphy (M)**

| FLORA | VESSEL DIAMETER (µm) | VESSELS (PER mm²) | VESSEL ELEMENT LENGTH (µm) | V | M |
|---|---|---|---|---|---|
| World flora | | | | | |
| Mesic primitive woods | 109 | 47 | 1385 | 2.29 | 3172 |
| Rosette trees | 79 | 31 | 412 | 2.25 | 1051 |
| Vines and lianas | 157 | 19 | 334 | 8.22 | 2745 |
| Annuals | 61 | 162 | 186 | 0.38 | 71 |
| Desert shrubs | 29 | 353 | 218 | 0.08 | 17 |
| Stem succulents | 72 | 64 | 259 | 1.33 | 344 |
| Arctic shrubs | 27 | 559 | 245 | 0.10 | 25 |
| Western Australian flora | | | | | |
| Karri understory shrubs | 46 | 74 | 385 | 0.62 | 239 |
| Coastal shrubs | 41 | 118 | 349 | 0.34 | 119 |
| Bog shrubs | 37 | 195 | 361 | 0.19 | 69 |
| Sand heath shrubs | 32 | 40 | 307 | 0.14 | 43 |
| Desert shrubs | 37 | 192 | 217 | 0.19 | 41 |

Reprinted with permission from Carlquist (1977), *Am. J. Bot.* **64**, 887–896.

among the longest tracheary elements within the genus. Their leaves lack obvious drought-resistant features. Major trends of specialization within the genus have been toward reduction in overall plant stature, resulting in small, semiprostrate or prostrate, thin-stemmed shrubs. Decreased leaf sizes also have accompanied migration into drier microhabitats and both plant and leaf size reductions are correlated with xeromorphic xylem anatomical specializations. The wood anatomy of many xeric hibbertias is characterized by growth rings, shorter and thicker walled tracheary elements, and more numerous and narrower vessel elements with a reduced number of bars per scalariform perforation plate.

Closely correlated with the changes in plant habit and wood structure that have resulted from extremes in the habitat, leaves of the Australian hibbertias exhibit marked trends in form, texture, thickness, and venation. Progressive reduction in leaf size among all xeric and numerous semixeric species has resulted in small leaves of diverse and often needlelike form. Leaf size reduction has been accompanied by the frequent acquisition of such drought-resistant, ericoid features as thick cuticle, woolly vestiture, revolute margins with thick-walled epidermal cells, hypodermis, and stomata confined to abaxial grooves extending the length of the leaf. The vegetative morphological and wood anatomical characteristics of *Hibbertia* and other xeric taxa have generally evolved as a single adaptive unit in response to environmental selection. These correlated features, sometimes termed **adaptive character syndromes,** appear as multiple facets of the plant hydrovascular system and

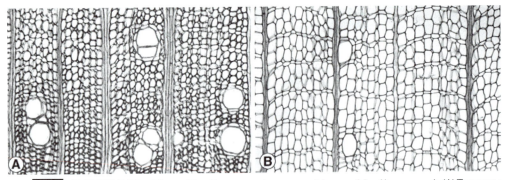

**FIGURE 8.20**   Effects of flooding on wood structure of *Annona glabra* (Annonaceae). (A) Transverse section of wood from tree flooded four months per year. (B) Transverse section of wood from permanently flooded tree.  Courtesy of L. Espinosa Yáñez and T. Terrazas-Salgado.

presumably evolved as structural and functional units in response to selective environmental pressures.

## ROOTS

### Xeromorphic Roots

Like leaves and stems, roots also have become structurally adapted to different habitat conditions and plant lifestyles. The root systems of certain xerophytes are widely spreading and shallow for optimal water absorption. A short tuberized root is common in some xeric taxa and there is an absence of root hairs under drought conditions. Xeromorphic roots also can be succulent because they contain large water storage cells in the cortex. Some plants from dry regions produce roots that develop a thick, sclerified, and exfoliating bark that cracks and splits off in large sheets. Cactaceae and other succulents conserve water during prolonged dry periods by having few, if any, lateral branch

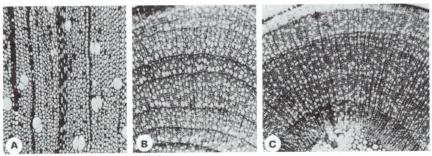

**FIGURE 8.21**   Wood anatomy of *Hibbertia* (Dilleniaceae). (A) *H. lucens*, transverse section showing thin-walled vessels and fibers. (B) *H. exutiacies*, transverse section showing irregularly shaped growth rings and numerous pores. (C) *H. uncinata*, transverse section showing growth rings and numerous, narrow pores.  Reprinted with permission from Rury and Dickison (1984), Academic Press, San Diego.

**FIGURE 8.22** Transverse section of leaf of the xerophytic Australian species. *Hibbertia pungens* (Dilleniaceae). Note abaxial stomatal grooves that are lined with trichomes.

roots; the entire root system may be covered by layers of thick bark. As the soil water potential increases, the dormant root system quickly forms new lateral roots, termed **rain-induced roots,** that are able to absorb large quantities of water. Some plants that grow in nutrient-poor soils form dense clusters of determinate lateral rootlets. These are sometimes called proteoid roots, after their occurrence in some Proteaceae where they are believed to be involved in phosphate acquisition. Specialized cluster roots also occur in members of the Casuarinaceae, Myricaceae, Betulaceae, and Moraceae.

## Hydromorphic Roots

The roots of aquatic plants range from well developed like those of terrestrial species through various degrees of specialization, including root dimorphism with respect to the presence or absence of root hairs. Because mechanical support and conducting efficiency are no longer necessary, vascular systems are to be ontogenetically and phylogenetically reduced and narrow, with weakly lignified tracheary elements. This reduction in vascular tissue tends to be correlated with the degree of specialization of the plant. Emergent aquatics possess the least reduction in the quantity of conducting tissues, whereas submerged plants have the most reduced. In cases of extreme reduction, the stele can be only 50 µm in diameter and composed of very few sieve tube elements and tracheary elements that are thin-walled and probably vestigial. In hydromorphic roots, the sieve tube elements are commonly distributed in an unusual pericyclic manner. Prominent cortical air spaces are formed by the breakdown or separation of alternate plates of cells and are continuous with those in the stem and leaves. In plants with a well-developed lacunose ground tissue, there is a ready diffusion of oxygen from the stems and leaves down to the roots that are embedded in an oxygen-deficient substrate. Root growth, therefore, is not normally oxygen limited. In rare instances, carbon dioxide is known to be absorbed by the roots and the gas moves through the air spaces to the leaves where it is photosynthetically fixed. Aquatic plants lacking abundant air space systems in their vegetative organs normally grow in rapidly moving streams and near waterfalls where oxygen appears not to be a limiting factor. The Podostemaceae are one of these families. The air chambers in aquatic plants are regularly traversed by septa or **diaphragms** that are solid or porous and sometimes constructed of stellate parenchyma. Diaphragms are thought to serve three functions: (1) to provide strength and stabilization to organs and prevent collapse of the air chambers, (2) to restrict the movement of water when the organ is damaged and internal flooding occurs, and (3) to provide a lateral transport pathway across the cortex.

## Flooded and Drought-Stressed Roots

Roots also can undergo major adaptive changes during periods of transient drought stress and flooding. Plants temporarily deprived of water show a pattern of root cell death from the epidermis inward, a process that is arrested at the endodermis. In *Allium cepa,* the Casparian strip and suberin lamellae of the endodermis have been shown to resist water movement from the stele of the root to the dry soil, allowing internal layers to survive for up to 24 weeks in a medium containing no free water.

A common anatomical response to root zone flooding is the development of aerenchymatous tissue in the root cortex and the suberization of both exodermal and endodermal cells. Root growth in flooded saline conditions is an especially widespread agricultural problem. Some species respond to excess soil salinity by developing a number of adaptive mechanisms that allow them to maintain some growth under these conditions. Salinity stress is known to induce a reduction in root growth rate, along with reduced lateral root development, changes in the timing of cell maturation, and increased cellular vacuolation and other cytological alterations.

## Aerial Roots

**Aerial roots** arise from some other aerial organ and function as supporting and protective organs; they also play important roles in translocation and gas exchange. Aerial roots occur in many unrelated plant families and reach their greatest development among tropical species. Mangrove species and members of the genus *Pandanus* are noted for aerial root formation. The aerial roots of woody vines are supporting structures and may grow downward and make contact with the soil. Aerial roots also are characteristic of many epiphytes, such as orchids. Mature orchid roots are composed of three major tissue regions: a central vascular stele, a cortex containing an inner endodermal layer and outer thick-walled exodermis, and an external covering known as the velamen. The specialized velamen tissue arises from the undifferentiated protoderm and at maturity consists of multiple layers of dead cells. The velamen has been assumed to be involved in the uptake, accumulation, and movement of water and salts under the conditions of low mineral and moisture availability that exists in the forest canopy. It appears, however, that the velamen actually performs a number of functions, including adhesion of the root system to its substratum.

## FROST HARDINESS

The effects of temperature on the growth and survival of plants play a major role in determining the natural distribution patterns of many wild and culti-

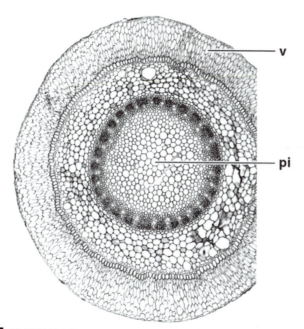

**FIGURE 8.23**   Transverse section of polyarch aerial root of the orchid *Dendrobium radicum* showing outer velamen (v) and inner pith (pi).

vated plants. Understanding the morphological and anatomical bases of frost hardiness could be very useful. Clarifying the overall phenomenon of frost resistance and identifying the basic physiological processes that underlie it can facilitate selection of economically valuable frost-hardy cultivars. There have been numerous studies of the relationship between frost-hardiness and leaf anatomy in several species of potato *(Solanum)*. Among potatoes, both frost-hardy (frost-killing temperature –4.0° C or colder) and frost-nonhardy species occur. No consistent relationship has been found between various gross morphological features and frost hardiness. Foliar anatomy, however, provides a number of significant characteristics that are differentially distributed among species in temperate regions or at high elevations in the Andes, where frost is a major problem. The anatomical characters of hardy species are probably the result of adaptation to environmental stress resulting from low temperature, and in some cases water stress as well.

Of particular note, all hardy taxa of *Solanum* have been found to possess two very distinct palisade layers, whereas all except three of the nonhardy species have one palisade layer in the leaf mesophyll (Table 8.3). In addition to this defining feature, the stomatal index on the upper leaf surface is closely related to frost hardiness. Hardy species have a stomatal index that is three times greater than the nonhardy plants. No significant differences in cell size and intercellular space have been observed among the two categories of plants, despite the fact that some previous investigators have suggested that small cell size, thick cell walls, and low stomatal frequencies were features generally associated with frost hardiness in plants. Identifying other leaf anatomical characters associated with extreme cold in other plants will be useful in selecting

**TABLE 8.3.   Leaf Anatomical Differences Between Hardy and Nonhardy**
*Solanum* **Species**[a]

|  | NONHARDY SPECIES | HARDY SPECIES |
|---|---|---|
| Frost-killing temperature (°C) | −3.5 or warmer | −4.0 or colder |
| Number of palisade layers | 1 | 2 |
| Palisade layer thickness (%) (palisade/palisade + spongy) | 51.2 ± 8.5 | 62.8 ± 8.9 |
| Leaf thickness (µm) | 223.9 ± 50.6 | 280.9 ± 56.4 |
| Stomatal index, upper surface (%) | 8.8 ± 6.2 | 26.7 ± 8.4 |
| Palisade cell dimensions (µm)[b] |  |  |
| Width | 14.4 ± 3.9 | 16.5 ± 3.5 |
| Length | 101.7 ± 36.6 | 99.9 ± 18.9 |
| Epidermal cell (no./0.1, 225 mm² leaf area) | 152.3 ± 147.7 | 103.7 ± 81.1 |

[a]The values are averages for 21 species.
[b]Only cells in the top palisade layer were measured.
Reprinted with permission from Palta and Li (1979), *Crop Sci.* **19**, 665–671.

cultivars and genetically engineering economically important species for improved cold tolerance.

Woody plants differ greatly in their vulnerability to damage by xylem cavitation resulting from freezing and thawing. Cavitation by freeze and thaw cycles occurs because of the persistence of air bubbles in the xylem sap caused by the degassing of the xylem during freezing. In contrast to cavitation caused by tension during drought, the vulnerability of a woody plant to cavitation by freezing is highly correlated with conduit diameter. Conifers and angiosperms with narrow vessels (mean diameter below 30 µm) are essentially resistant to cavitation by freezing and thawing under normal conditions. Species with mean diameter tracheary elements below 30 µm are resistant; species with mean diameters above 40 µm can be completely disabled by a single freeze and thaw cycle. Larger conduits form large air bubbles during freezing, which persist longer and require less xylem tension to expand when the xylem thaws and tensions redevelop.

## ANATOMY AND POLLUTION

Cities, industrial and agricultural operations, and individuals all dump toxic compounds whose many detrimental effects on plants have yet to be thoroughly documented. Pollutants represent a diverse assortment of materials that are added to the biosphere in quantities so great that they adversely affect the functioning of plants, animals, and humans. Research on the effects of pollution on plants has become more common, and understanding the anatomical bases of pollution-caused diseases is an essential part of this undertaking. Pollutants enter the biosphere as liquids, gases, fine particles (particulates), solids, and radioactive substances. They do serious damage to plants in the form of toxic gases, ozone, acid rain, insecticides, herbicides, and different types of radiation. In the case of liquids and gases, the various pollutants either adhere to the plant surface or enter the leaf through the cuticle or stomata, where physiological and structural responses are elicited. The relationship between visible injury and internal anatomy is not always evident. Plants that have endured long exposure or have received high concentrations of various chemicals often show visible and sometimes characteristic symptoms, depending upon the age of the individuals. In some instances, examination of anatomical changes in leaves and stems exposed to pollution can document the onset of total plant decline earlier and more accurately than is possible on the basis of external morphology and appearance. To understand the serious consequences of environmental contaminants on plants, it is necessary to clarify a few of the varied and sometimes subtle responses of plants to different classes of pollutants. Understanding these effects has proven difficult, due in part to many complex interactions. Nevertheless, quantitative and qualitative histological and cellular changes often represent one aspect of the total plant response. A consideration of the deleterious and often permanent changes that different categories of pollutants have on plant structure is essential to solving the major problems related to plants and pollution.

The incidence and severity of leaf injury to pollution is related to a number of interrelated variables. Knowledge of the following categories of data are required in order to fully clarify the effects of pollution on plants:

1. Age of tissue and plant.
2. Foliar wettability (the attraction of liquid to the leaf surface).
3. Concentration of the pollutant.
4. Length of exposure to the pollutant.
5. Amount and chemical nature of any pollutant entering the leaf by diffusing through the cuticle or through the stomata.
6. Cuticular characteristics and the deposition of epicuticular waxes.
7. Leaf pubescence.
8. Amount of intercellular space (such as the substomatal chamber).
9. Cell wall thickness.
10. Movement and assimilation of pollutant within the leaf.
11. Physiological factors affecting cellular sensitivity.
12. Environmental factors affecting cellular sensitivity, such as relative humidity, temperature, and light intensity.

## Anatomy and Plant Resistance

Leaf morphology can theoretically exert an effect on leaf resistance to harmful substances by altering the depth of the boundary layer. The potential role of pubescence in regulating pollution uptake and in determining the responses of plants to air pollution has been suggested by studies that show considerably higher trichome frequencies and individual hair lengths in healthy plants growing in assumed high-pollution areas, as compared to the same species from low-pollution sites. This response, however, is contradicted by reports showing that certain plants with highly pubescent leaves are more affected by air pollution factors than those with a smooth leaf surface. Results from a number of studies have supported the hypothesis that leaf surface wax can be a major factor in controlling laminar resistance.

Because stomata are believed to represent a major site of pollutant penetration in some plants, stomatal structure, frequency, and distribution have been assumed to be significant variables affecting plant sensitivity and overall leaf resistance. Smog artificially applied in darkness, for example, is much less effective in producing injury than the same concentration applied in light when stomata are open. However, attempts to demonstrate a general pattern of relationship between stomatal resistance and the regulation of pollutant absorption have been inconclusive. The anatomical and physiological bases of plant phytotoxin resistance or sensitivity are not yet well understood.

## Effects of Pollution on Wood

Atmospheric pollution has become a major cause of tree disease in many regions of the world. Although the relationship between airborne pollutants and forest decline is complex, studies of the anatomical characteristics of wood from pollution-diseased trees indicate that tree ring width patterns often can be used to measure the effects of pollution on tree growth over long time scales. In view of the vast economic importance of wood, it is of considerable interest to know if the wood from diseased trees shows any chemical, physical, or technical differences when compared with that from healthy trees. Does the wood from diseased trees show any reduction in overall quality? As a general rule, softwoods such as fir and spruce have been shown to be more sensitive to air pollutants than hardwoods such as beech and oak.

The most common structural alterations associated with pollution in forest communities are an overall suppression of tree growth and a reduction in the xylem growth ring width as the result of a reduction in cambial activity. Pollution-induced growth suppression also can result in the formation of discontinuous rings, as well as in the occurrence of missing rings. In many cases, it is possible to document the initiation of tree stress responses by the careful study of growth rings. Changes in the growth ring pattern typically become visible before foliar symptoms are evident in the crown of the tree. However, because abrupt changes in growth rate also can be a reflection of mechanical injury to the stem, roots, or crown, of destruction of the foliage by insects, fungi, or persistent drought, and of changes in light conditions, the analysis of growth rings is not a simple process. In the narrow rings of diseased trees, the proportion of earlywood and latewood may either remain similar to healthy trees or be altered to different degrees. No influence of gaseous pollutants on the xylem sap velocities of either oak, beech, or Douglas fir has been demonstrated.

Although changes in growth may not always be associated with structural changes in individual cells, such anatomical alterations have been reported in some trees. Among the cell characteristics that are symptomatic of disease in

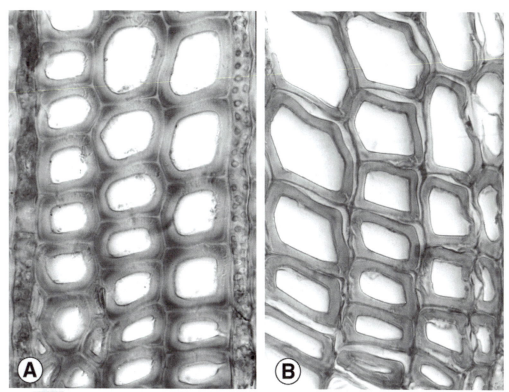

**FIGURE 8.24** Transverse sections of wood of *Abies religiosa* (Pinaceae) from nonpolluted (A) and polluted areas (B) of Mexico. Atmospheric pollution resulted in thinner walled tracheids. Courtesy of Bernal Salazar Sergio and Teresa Terrazas.

plants are reduced tracheary element length and reduced wall thickness. Electron micrographs have not revealed any alterations of the ultrastructure of cell walls or the pit structure in diseased conifers. Although environmental pollutants can result in a serious reduction in wood production in some forest areas, the chemical composition of cell walls and the physical and mechanical strength of wood from diseased trees appear to be similar to those of healthy individuals, and thus there is no overall loss in wood quality.

## Gaseous Air Pollutants

Destructive atmospheric phytotoxins such as sulfur dioxide ($SO_2$), hydrogen fluoride (HF), and nitrogen dioxide ($NO_2$) are emitted by automobiles, by other internal combustion engines, and by various factories and refineries. The effects of these pollutants upon leaf structure can be dramatic and varied and often result in significant reductions in growth biomass and in acute and chronic tissue damage. Native vegetation or plants experimentally fumigated with hydrogen flouride or sulfur dioxide typically show shrinkage and eventual collapse of leaf mesophyll cells. Sulfur dioxide is injurious to plant tissues in concentrations as low as 0.3 to 0.5 ppm. Gaseous air pollutants enter the

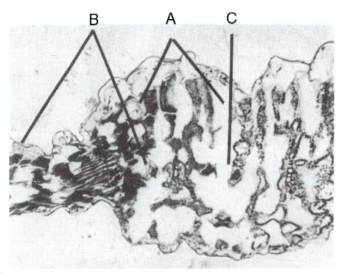

**FIGURE 8.25**  Transverse section of leaf of *Petunia hybrida* showing varying zones of damage from 2 ppm $SO_2$ fumigation (0.5 ppm benomyl drench). (A) Early stage of palisade breakdown. (B) Advanced stage of mesophyll damage. (C) Unaffected region of mesophyll. × 100. Reprinted with permission from Uhring (1978), *J. Am. Soc. Hort. Sci.* **103**, 23–27.

leaf primarily through the stomata. Within the mesophyll, the gas goes into solution on the wet mesophyll cell walls and forms toxic ions or acids. Some pollutants cause a decrease or total loss of cell membrane semipermeability in the vicinity of the stomata. This injury results in the leakage of cell sap from mesophyll cells into the intercellular spaces, eventually appearing as an oiliness on the surface of the affected leaf. Any redistribution or distortion of chloroplasts results in a reduction in net photosynthesis, which can be followed by the collapse of both lower and upper epidermal cells. Peroxyacyl nitrates specifically cause spongy mesophyll cells to collapse and be replaced by air pockets. This tissue damage produces a symptom known as leaf glazing on either the lower leaf surface or both surfaces of the lamina in some plants. The resulting damage is usually visible as areas of chlorosis or yellowing on the lamina. In addition to foliar chlorosis, phytotoxic air pollutants commonly increase the rate of leaf abscission. High levels of sulfur dioxide may cause tissue between the veins and on leaf edges to die, although vascular tissues can show little apparent injury when compared to other leaf tissues.

### Ozone Injury to Plants

Ozone ($O_3$) is produced naturally, but the release of nitrogen dioxide from automobile and other internal combustion engine exhausts, and its subsequent combination with oxygen in sunlight, results in excessive levels of ozone in the atmosphere. Ozone is a major component of smog and is known to cause serious injury to plants even after only short exposure at very low concentrations of 0.1 to 0.3 ppm. Young, expanding leaves are the most susceptible to pollution injury. Ozone enters the leaf through stomata, and leaves sensitive to ozone reportedly show a rapid and progressive series of structural changes, ending in cellular collapse. Palisade parenchyma cells are especially affected following the disruption of the cell membrane. In addition to these responses, ozone stress in leaves can produce increased hair density and thickened epidermal cell walls. An alteration of organelle fine structure and metabolic malfunction is a further consequence of this damage. Lipid droplets and tanninlike depositions can increase in the mesophyll vascular bundle sheath cells of some leaves. The effects of ozone are visible as mottled, stippled, or bleached patterns on the upper leaf surface. A premature loss of leaves has been used as an indication of ozone stress in some conifer species, as well as in some dicots such as citrus and grapes.

## Acid Rain

Industrial pollution in the form of sulfur and nitrogen oxides undergoes a chemical transformation into strong acids upon contact with atmospheric moisture, producing wet or frozen acidic precipitation over vast areas of the earth, often far removed from the source of emission. Rain or snow is considered acid if it has a pH below 5.6 and contains large quantities of sulfuric and nitric acids. Exposure of leaves to acid rain produces a number of histological alterations and injuries in the form of surface lesions, smaller mesophyll cells, fewer intercellular spaces, and the collapse of epidermal, palisade, and spongy mesophyll cells. A diverse collection of plants has reacted to simulated acid precipitation by forming foliar galls. These develop through a combination of cell enlargement (**hypertrophy**) and cell proliferation (**hyperplasia**) and become raised above the leaf surface. In plants that have been subjected to simulated acid rain events, most lesions develop near trichome and hydathode bases, near stomata, and along veins. In contrast to gaseous pollutants, which gain entry into the leaf primarily through the stomata, liquid acid rain appears to enter through the cuticle. In this sequence, therefore, cellular damage is most often initiated at the epidermis but then develops internally with more pollutant exposure. The extent of leaf damage may be a function of the area of the leaf in contact with the rainwater and the rate of absorption of pollutants per unit area.

## Carbon Dioxide

In view of the potentially serious consequences of a global rise in atmospheric $CO_2$ levels because of the industrial revolution, attention has been directed at the effects of increased $CO_2$ levels on plant growth rates and leaf anatomy. $CO_2$ enrichment studies have shown leaves of different species to undergo changes in stomatal density and strong morphological modification, particularly in leaf shape. Experiments also have revealed significant increases in leaf thickness under enriched $CO_2$ conditions (Table 8.4). In some plants, the increase in lamina thickness can be correlated with the formation of an additional palisade layer. This response has been observed in various species of dicotyledons with $C_3$ photosynthesis and in the needles of pine. The formation of significantly thicker leaves in response to high $CO_2$ levels has not been recorded in corn,

**TABLE 8.4. Effect of $CO_2$ Concentration on Total Leaf Thickness and on Epidermis, Palisade, and Spongy Parenchyma Layers in Soybean Leaves**

| $CO_2$ TREATMENT (ppm) | LEAF THICKNESS ($\mu$m) | | | |
|---|---|---|---|---|
| | Total | Epidermis | Palisade | Spongy |
| 340A[a] | 183b[b] | 26a | 82c | 75a |
| 340 | 161b | 20b | 75c | 66a |
| 520 | 218a | 23ab | 99b | 96a |
| 718 | 214a | 25a | 97b | 92a |
| 910 | 211a | 20b | 115a | 76a |
| LSD 0.05 | 15 | 3 | 9 | 56 |

[a] The value 340A is ambient plot (no chamber); other values are from within chambers. Values for $CO_2$ are daytime means.
[b] Entries followed by the same letter within a column are not different at the 5% significance level. Reprinted with permission from Thomas and Harvey (1983), *Bot. Gaz.* **144**, 303–309. Published by the University of Chicago Press.

even though it is a $C_4$ plant. Increased $CO_2$ levels also result in decreases in stomatal frequencies. This response can be documented by growing plants under controlled conditions in which $CO_2$ levels are manipulated.

An understanding of the potential effects of increasing $CO_2$ levels on wood structure is important in order to predict changes in wood quality in the future. It has been hypothesized that elevated atmospheric $CO_2$ levels along with warm temperatures and adequate rainfall should stimulate radial stem growth and produce wider growth rings. Results of $CO_2$ enrichment studies on wood structure in *Pinus* document some random changes in tracheary element length and tangential diameter. No changes in wood density have been found, and in many cases, no effects of $CO_2$ enrichment on wood growth rings have been reported. Overall, elevated $CO_2$ levels have been found to have no significant effect on conifer tracheid dimensions. Some data suggest that wood density increases in response to higher $CO_2$ levels in hardwoods such as sweetgum, although further study is needed. Scientists have predicted that increased atmospheric $CO_2$ levels will cause changes in world temperature and moisture availability, factors known to produce both quantitative and qualitative changes in wood structure.

## Ionizing Radiation

It is well established that structural modifications in plant tissues also can be brought about through the effects of ionizing radiation. An increase in the level of background radiation by the release of different categories of radioactive materials is known to produce abnormalities of form as well as anatomical changes in different parts of the plant. Radioactive substances emit three classes of ionizing radiation: alpha radiation with comparatively little penetrating power, beta radiation with an intermediate degree of penetration, and short wave length gamma radiation that is similar to X-rays and that has considerable penetrating power. Gamma rays are emitted from radioactive substances such as cobalt-60 and cesium-137. Radioactive materials represent an important potential contaminant of the environment and, like chemicals, are known to cause serious structural injury to many higher plants. Extremely short wave length ultraviolet light rays also present a danger to plants.

The anatomical responses of a particular plant vary with radiation dose rates, age of the plant, radiosensitivity of the plant, and the specific tissue receiving radiation. It has long been recognized that the meristematic cells forming the shoot apex are extremely sensitive to radiation and readily undergo cytological and genetical alterations. Shoot apical meristems are known to show differential radiosensitivity, thus cells comprising the various meristematic zones respond differently. For a number of plant species, the most radiosensitive region of both the apical and axillary meristems is the second tunica layer ($L_2$ layer). One type of radiation injury to this part of the plant is the induction of a chimera, a combination of tissues of different genetic constitution in the same part of the plant.

A variety of responses have been reported in plants exposed to ionizing radiation. These include the loss of apical dominance and abnormal branching patterns. Leaves of irradiated plants typically exhibit marked and variable changes in color, form, and texture. The size of the lamina is frequently reduced, along with an increase in pubescence. Leaves often show an increased rate of abscission, irregular blade structure including tissue necrosis, and distorted venation. Modifications can be observed in the arrangement and size of palisade cells, in the enlargement of the spongy mesophyll cells, and in an increase in intercellular space as compared with control individuals. Some reports have described an elongation of palisade cells following radiation, whereas others described wider and shorter palisade cells. Distortion in the vascular tissue, most notably the xylem, also has been described. Increases and decreases in the amount of vascular tissue have been noted. Following exposure to acute gamma radiation, the foliage of *Sequoia* has been described as gnarled and swollen, with abnormal mesophyll and vascular tissue. There also can be an extensive development of transfusion tissue and an abnormal pattern of resin duct formation. The anatomical changes induced by exposure to ionizing radiation are accompanied by physiological changes and the total plant response represents a balance between this two activities.

## Ultraviolet B Radiation

Ultraviolet (UV) light rays are the extremely short wave length component of sunlight, lying just beyond the end of the visible spectrum. Plants are continuously exposed to varying levels of UV-B radiation that has damaging effects on their physiology and structure. As a result, plants have developed damage repair mechanisms and a number of whole plant responses to this stress. Leaves subjected to high levels of UV-B undergo structural changes as well as (1) damage to DNA to proteins and membranes, (2) alterations in transpiration and photosynthesis, and (3) striking changes in growth. These structural changes include curling to reduce the leaf area exposed to UV. Epidermal cells can either collapse or increase in size and become thicker walled, along with undergoing a decrease in numbers of stomata and an increase in the thickness of the entire lamina. The damaged mesophyll is characterized by a marked elongation of palisade cells. However, an exceptionally thick-walled and lignified epidermis can weaken the severity of these UV-B effects on the underlying mesophyll cells. This is sometimes accompanied by a redistribution of chlorophyll away from the adaxial surface. Morphological changes such as shorter internodes, decreased plant height, increased axillary branching, and changes in leaf shape also can occur.

## Insecticides and Herbicides

The chemical control of insects and weeds is generally referred to as control by insecticides, pesticides, and herbicides. Chemical poisons may be applied as solutions, suspensions, emulsions, dusts, and aerosols. In addition to being toxic to insects, these poisons may have long-lasting residual effects on other animals and on the general vegetation. Chlorinated compounds such as DDT (dichlorodiphenyltrichloroethane) have especially long-lasting destructive effects on the environment. To control weeds that cannot be controlled through normal agricultural practices, herbicides can often be used with success. The improper or careless application of these chemicals, however, can produce serious injury to crops as well as to other species.

Herbicides are available under a wide variety of trade names, in various concentrations, and in dry and liquid form. 2, 4–D (2, 4 dichlorophenoxyacetic acid) is commonly used to control weeds. The application of 2, 4–D is known to produce structural, physiological, and biochemical abnormalities in susceptible plants, and ultimately causes death. Treated leaves from susceptible plants tend to exhibit increases in thickness, formation of strongly vacuolate parenchymatous replacement tissue, and reduced or distorted vascular bundles. Castor bean *(Ricinus communis)*, is one plant in which the histological responses to 2, 4–D have been examined in detail. Castor bean is an important oil seed crop in India, as well as a potential weed. The type and extent of response is related to the method of herbicide application and the age of the plant at the time of application. Seeds and seedlings demonstrate the highest degree of susceptibility to 2, 4–D. When this chemical is applied to seedlings, leaves do not expand normally and meristematic activity is induced in the interfascicular region of the hypocotyl, petiole, and stem. The formation of new tissues results in swelling of the organ, and this frequently ruptures the outer tissues. Pith cells show marked enlargement. The major anatomical responses of *Ricinus* to 2, 4–D are an increased activity of the interfascicular cambium, lack of differentiation of phloem elements, and a rupturing of tissues. Treated plants possess more compact leaf mesophyll with more abundant chloroplasts as compared to controls.

Studies of a number of different plants have shown that growth-regulating substances such as cytokinin, auxin, and auxinlike substances have a strong influence on the anatomical responses of plants. When sprayed with growth-regulating substances, leaves typically show an increase in leaf blade thickness, increased production of secondary vascular tissues in the stem, stimulation of the endodermal or pericyclic regions of the stem to become meristematic, and a general proliferation of vascular parenchyma and pith cells. Stomatal number is sometimes decreased compared to control plants, and cellular ultrastructure also can be altered.

## SUMMARY

Ecology is the study of an organism's relationship with its environment. Plants have evolved many different anatomical strategies and adaptive responses to various environmental conditions that enable them to survive and function under different conditions. Adaptations enable plants to survive in extreme conditions of intense sun or shade, cold or heat, physiologically wet or dry microhabitats, and soil mineral deficiences. Plants that live in regions of average or optimum water availability are termed mesophytes. Hydrophytes grow on the surface of water or submerged at various depths, in soil covered by water, or in soil that is usually saturated with water. Species that live in regions where the supply of available water is deficient or fluctuates widely on a seasonal basis are called xerophytes. The variations in plant structure that are commonly affected by environmental factors are particularly strongly expressed in the morphology and anatomy of leaves. Within the same individual, leaves produced in bright light conditions (sun leaves) tend to be smaller and thicker and have increased mesophyll tissue per unit area as compared with leaves exposed to shade (shade leaves). Xeromorphic leaves have developed different adaptive strategies. These divergent specializations include temporary leaf loss, changes in leaf angle, and the rolling or folding of leaf blades, or permanent adaptations such as succulence or leaves reduced to spines or needles. Some features, such as reduction in leaf size, increased thickness of the outer walls of epidermal cells, increased thickness of cuticle, increased trichome density, reduction in stomatal pore area, and stomata recessed below the blade surface or confined to pits or grooves on the undersurface of the blade, are features thought to contribute to a reduction in the rate of water loss. Other characteristics such as an increase in mechanical cells and wall lignification, increased water storage capacity, and the accumulation of mucilage probably help the plant tolerate dehydration and deal with water stress. Halophytic plants are those able to tolerate saline environments. Halophytic species typically have salt hairs or salt glands that remove salt from underlying tissue. Hydrophytic leaf types vary considerably in response to change in habitat, although all are characterized by marked increases in aerenchyma, (i.e., parenchyma tissue with particularly large intercellular spaces).

Ecological wood anatomy refers to the study of correlations between the ecological and floristic preferences of taxa and various wood anatomical characters in relation to function. Growth rings are a widely recognized example to the influence of climate on xylem structure. Vessel element structure and distribution patterns provide clear examples of the significance of ecology on the hydraulic architecture of trees and shrubs. Some species show correlations in wood structure and leaf anatomy in response to environmental extremes.

Roots also have become structurally adapted to different habitat conditions and plant life styles. Rain-induced roots, hydromorphic roots, flooded and drought-stressed roots, and aerial roots are all adaptations to changing conditions.

Plants that have had long exposure to or received high concentration of various chemical pollutants often show visible injury and changes in anatomy. Modifications in the arrangement, structure, and size of cells can also occur following exposure to ionizing radiation, insecticide and herbicides, and acid rain. Leaf morphology and anatomy can exert an effect on leaf resistance to harmful substances. The most common structural alteration associated with pollution in forest communities is an overall suppression of tree growth and a reduction in xylem growth ring width as a result of a reduction in cambial activity.

## ADDITIONAL READING

1. Abrams, M. D. (1990). Adaptations and responses to drought in *Quercus* species of North America. *Tree Physiol.* **7,** 227–238.
2. Arnold, D. H., and Mauseth, J. D. (1999). Effects of environmental factors on development of wood. *Am. J. Bot.* **86,** 367–371.
3. Ashton, P. M. S., and Berlyn, G. P. (1992). Leaf adaptations of some *Shorea* species to sun and shade. *New Phytol.* **121,** 587–596.
4. Ashton, P. M. S., and Berlyn, G. P. (1994). A comparison of leaf physiology and anatomy of *Quercus* (section of *Erythrobalanus*-Fagaceae) species in different light environments. *Am. J. Bot.* **81,** 589–597.
5. Baas, P. (1973). The wood anatomical range in *Ilex* (Aquifoliaceae) and its ecological and phylogenetic significance. *Blumea* **21,** 193–258.
6. Baas, P. (1986). Ecological patterns in xylem anatomy. *In* "On the Economy of Plant Form and Function"(T. J. Givnish, Ed.), pp. 327–352. Cambridge University Press, Cambridge.
7. Baas, P. (1990). Ecological trends in the wood anatomy and their biological significance. *In* "Anatomy of European Woods" (F. H. Schweingruber, Ed.), pp. 739–765. P. Haupt, Bern, Stuttgart.
8. Baas, P., and Bauch, J. (Eds.) (1986). "The Effects of Environmental Pollution on Wood Structure and Quality." International Association of Wood Anatomists, Leiden.
9. Baas, P., and Miller, R. B. (Eds.) (1985). "Functional and Ecological Wood Anatomy." Proceedings of the Martin H. Zimmermann Memorial Symposium. *IAWA Bull n. s.* **6,** 279–397
10. Baker, E. A., and Bukovac, M. J. (1971). Characterization of the components of plant cuticles in relation to the penetration of 2, 4–D. *Ann. Appl. Biol.* **67,** 243–253.
11. Benzing, D. H. (1970). Foliar permeability and the absorption of minerals and organic nitrogen by certain tank bromeliads. *Bot. Gaz.* **131,** 23–31.
12. Benzing, D. H., and Burt, K. M. (1970). Foliar permeability among twenty species of the Bromeliaceae. *Bull. Torrey Bot. Club* **97,** 269–279.
13. Bobrov, R. A. (1955). The leaf structure of *Poa annua* with observation on its smog sensitivity in Los Angeles County. *Am. J. Bot.* **42,** 467–474.
14. Brown, C. L. (1980). Growth and form. *In* "Trees, Structure and Function" (M. H. Zimmermann and C. L. Brown, Eds.), pp. 125–167. Springer-Verlag, New York, Heidelberg, Berlin.
15. Caldwell, M. M. (1971). Solar UV irradiation and the growth and development of higher plants. *In* "Photophysiology" (A. C. Giese, Ed.), Vol. 6, pp. 131–177. Academic Press, New York.
16. Carlquist, S. (1957). The genus *Fitchia* (Compositae). *Univ. Calif. Publ. Bot.* **29,** 1–144.

17. Carlquist, S. (1975). "Ecological Strategies of Xylem Evolution." University of California Press, Berkeley.

18. Carlquist, S. (1977). Ecological factors in wood evolution: a floristic approach. *Am. J. Bot.* **64**, 887–896.

19. Carlquist, S. (1988). "Comparative Wood Anatomy: Systematic, Ecological, and Evolutionary Aspects of Dicotyledon Wood." Springer-Verlag, Berlin, New York.

20. Cen, Y.-P., and Bornman, J. F. (1990). The response of bean plants to UV-B radiation under different irradiances of background visible light. *J. Exp. Bot.* **41**, 1489–1495.

21. Cline, M. G., and Salisburg, F. B. (1966). Effects of ultraviolet radiation on the leaves of higher plants. *Rad. Bot.* **6**, 151–163.

22. Davison, A. W., and Barnes, J. D. (1998). Effects of ozone on wild plants. *New Phytol.* **139**, 135–151.

23. DeLucia, E. H., Day, T. A., and Vogelmann, T. C. (1992). Ultraviolet-B and visible light penetration into needles of two species of subalpine conifers during foliar development. *Plant Cell Environ.* **15**, 921–929.

24. Donaldson, L. A., Hollinger, D., Middleton, T. M., and Souter, E. D. (1987). Effect of $CO_2$ enrichment on wood structure in *Pinus radiata* D. Don. *IAWA Bull. n. s.* **8**, 285–289.

25. Ehleringer, J., and Björkman, O. (1978). Pubescence and leaf spectral characteristics in a desert shrub, *Encelia farinosa*. *Oecologia* **36**, 151–162.

26. Ehleringer, J. R., and Mooney, H. A. (1978). Leaf hairs: Effects on physiological activity and adaptive value to a desert shrub. *Oecologia* **37**, 183–200.

27. Evans, L. S., Gmur, N. F., and Da Costa, F. (1977). Leaf surface and histological perturbations of leaves of *Phaseolus vulgaris* and *Helianthus annus* after exposure to simulated acid rain. *Am. J. Bot.* **64**, 903–913.

28. Fahn, A., and Cutler, D. F. (1992). "Xerophytes." Gebrüder Borntraeger, Stuttgart.

29. Ferenbaugh, R. W. (1976). Effects of simulated acid rain on *Phaseolus vulgaris* L. (Fabaceae). *Am. J. Bot.* **63**, 283–288.

30. Gaff, D. F. (1989). Responses of desiccation tolerant "resurrection" plants to water stress. *In* "Structural and Functional Responses to Environmental Stress: Water Shortage" (H. Kreeb, H. Richter, and T. M. Hinckley, Eds.), pp. 255–268. SPB Academic Publ., The Hague.

31. Gibson, A. C. (1983). Anatomy of photosynthetic old stems of nonsucculent dicotyledons from North American deserts. *Bot. Gaz.* **144**, 347–362.

32. Gibson, A. C. (1996). "Structure-Function Relations of Warm Desert Plants." Springer-Verlag, Berlin.

33. Gouret, E., Rohr, R. and Chamel, A. (1993). Ultrastructure and chemical composition of some isolated plant cuticles in relation to their permeability to the herbicide, diuron. *New Phytol.* **124**, 423–431.

34. Gunckel, J. E. (1957). Symposium on the effects of ionizing radiation on plants. IV. The effects of ioning radiation on plants: Morphological effects. *Quart. Rev. Biol.* **32**, 46–56.

35. Hagerup, O. (1953). The morphology and systematics of the leaves in Ericales. *Phytomorphology* **3**, 459–464.

36. Hättenschwiler, S., Schweingruber, F. H., and Körner, Ch. (1996). Tree ring responses to elevated $CO_2$ and increased N deposition in *Picea abies*. *Plant Cell Environ.* **19**, 1369–1378.

37. Iqbal, M., and Ghouse, A. K. M. (1985). Impact of climatic variation on the structure and activity of vascular cambium in *Prosopis spicigera*. *Flora* **177**, 147–156.

38. Jackson, L. W. R. (1967). Effect of shade on leaf structure of deciduous tree species. *Ecology* **48**, 498–499.

39. James, S. A., Smith, W. K., and Vogelmann, T. C. (1999). Ontogenetic differences in mesophyll structure and chlorophyll distribution in *Eucalyptus globulus* ssp. *globulus* (Myrtaceae). *Am. J. Bot.* **86**, 198–207.

40. Janes, M. A. K., Gaba, V., and Greenberg, B. M. (1998). Higher plants and UV-B radiation: Balancing damage, repair and acclimation. *Trends Plant Sci.* **3**, 131–135.

41. Jetter, R., Riederer, M., and Lendzian, K. (1996). The effects of dry $O_3$, $SO_2$ and $NO_2$ on reconstituted epicuticular wax tubules. *New Phytol.* **133**, 207–216.

42. Johnson, E. D. (1926). A comparison of the juvenile and adult leaves of *Eucalyptus globulus*. *New Phytol.* **25**, 202–212.

43. Kangasjarvi, J., Talvinen, J., Utriainen, M., and Karjalainen, R. (1994). Plant defense systems induced by ozone. *Plant Cell Environ.* **17**, 783–794.

44. Kapoor, M. L., Joshi, B. C., and Natarajan, A. T. (1965). Effect of chronic gamma radiation on epidermal hairs of some varieties of wheat and barley. *Rad. Bot.* **5**, 265–269.

45. Kaul, R. B. (1971). Diaphragms and aerenchyma in *Scripus validues*. *Am. J. Bot.* **58**, 808–816.

46. Kaul, R. B. (1972). Adaptive leaf architecture in emergent and floating *Sparganium*. *Am. J. Bot.* **59**, 270–278.

47. Kaul, R. B. (1973). Development of foliar diaphragms in *Sparganium eurycarpum*. *Am. J. Bot.* **60**, 944–949.

48. Kaul, R. B. (1976). Anatomical observations on floating leaves. *Aquatic Bot.* **2**, 215–234.

49. Kozlowski, T. T. (1979). "Tree Growth and Environmental Stresses." University of Washington Press, Seattle.

50. Lamoreaux, R. J., and Chaney, W. R. (1977). Growth and water movement in silver maple seedlings affected by cadmium. *J. Environ. Qual.* **6**, 201–205.

51. Levering, C. A., and Thomson, W. W. (1971). The ultrastructure of the salt gland of *Spartina foliosa*. *Planta* **97**, 183–196.

52. Lewis, T. E. (Ed.) (1995). "Tree Rings as Indicators of Ecosystem Health." CRC Press, Boca Raton, Florida.

53. Liphschitz, N., Adiva-Shomer-Ilan, Eschel, A., and Waisel, Y. (1974). Salt glands on leaves of Rhodes grass *(Chloris gayana* Kth). *Ann. Bot.* **38**, 459–462.

54. Lortie, C. J., and Aarssen, L. W. (1996). The specialization hypothesis for phenotypic plasticity in plants. *Int. J. Plant. Sci.* **157**, 484–487.

55. Nobel, P. S. (1976). Photosynthetic rates of sun versus shade leaves of *Hyptis emoryi* Torr. *Plant Physiol.* **58**, 218–223.

56. Noguchi, Y. (1935). Modification of leaf structure by x-rays. *Plant Physiol.* **10**, 753–762.

57. Ogden, E. C. (1974). Anatomical patterns of some aquatic vascular plants of New York. Bulletin 424. New York State Museum. Albany, New York.

58. Pääkkönen, E., Günthardt-Goerg, M. S., and Holopainen, T. (1998). Responses of leaf processes in a sensitive birch *(Betula pendula* Roth) clone to ozone combined with drought. *Ann. Bot.* **82**, 49–59.

59. Pallardy, S. G. (1989). Hydraulic architecture and conductivity: An overview. *In* "Structural and Functional Responses to Environmental Stresses: Water Shortage" (K. H. Kreeb, H. Richter, and T. M. Hinckley, Eds.), pp. 3–19. SPB Academic Publishing, The Hague.

60. Palta, J. P, and Li, P. H. (1979). Frost-hardiness in relation to leaf anatomy and natural distribution of several *Solanum* species. *Crop Sci.* **19**, 665–671.

61. Parker, J. (1952). Desiccation in conifer leaves: Anatomical changes and determination of the lethal leve. *Bot. Gaz.* **114**, 189–198.

62. Pyykkö, M. (1966). The leaf anatomy of East Patagonian xeromorphic plants. *Ann. Bot. Fenn.* **3**, 453–622.

63. Ram Mohan, H. Y., and Satsangi, A. (1963). Histomorphological responses of *Ricinus communis* L. to 2, 4–D. *Phytomorphology* **13**, 267–284.

64. Reich, P. B., Uhl, C., Walters, M. B., and Ellsworth, D. S. (1991). Leaf lifespan as a determinant of leaf structure and function among 23 Amazonian tree species. *Oecologia* **86**, 16–24.

65. Rury, P. M., and Dickison, W. C. (1984). Structural correlations among wood, leaves and plant habit. *In* "Contemporary Problems in Plant Anatomy." (R. A White and W. C. Dickison, Eds.), pp. 495–540. Academic Press, San Diego.

66. Sayre, J. D. (1920). The relation of hairy leaf covering to the resistance of leaves to transpiration. *Ohio J. Sci.* **20**, 55–86.

67. Schiller, P., Wolf, R., and Hartung, W. (1999). A scanning electron microscopical study of hydrated and desiccated submerged leaves of the aquatic resurrection plant *Chamaegigas intrepidus*. *Flora* **194**, 97–102.

68. Schmidt, B. L., and Millington, W. F. (1968). Regulation of leaf shape in *Proserpinaca palustris*. *Bull. Torrey Bot. Club* **95**, 264–286.

69. Schneider, E. L., and Carlquist, S. (1996). Conductive tissue in *Ceratophyllum demersum* (Ceratophyllaceae). *Sida* **17**, 437–443.

70. Schoch, P-G., Zinsou, C., and Sibit, M. (1980). Dependence of the stomatal index on environmental factors during stomatal differentiation in leaves of *Vigna sinensis* L. *J. Exp. Bot.* **31**, 1211–1216.

71. Sharma, G. K., and Butler, J. (1975). Environmental pollution: Leaf cuticular patterns in *Trifolium pratense* L. *Ann. Bot.* **39**, 1087–1090.
72. Sherwin, H. W., Pammenter, N. W., February, E. D., Willigen, C. Vander, and Farrant, M. (1998). Xylem hydraulic characters, water relations and wood anatomy of the resurrection plant *Myrothamnus flabellifolius* Welw. *Ann. Bot.* **81**, 567–575.
73. Shields, L. M. (1950). Leaf xeromorphy as related to physiological and structural influences. *Bot. Rev.* **16**, 399–447.
74. Shimony, C., and Fahn, A. (1968). Light and electron-microscopical studies on the structure of salt glands of *Tamarix aphylla* L. *Bot. J. Linnean Soc.* **60**, 283–288.
75. Solberg, R. A., and Adams, D. F. (1956). Histological responses of some plant leaves to hydrogen fluoride and sulfur dioxide. *Am. J. Bot.* **43**, 755–760.
76. Stover, E. L. (1944). Varying structure of conifer leaves in different habitats. *Bot. Gaz.* **106**, 12–25.
77. Stover, E. L. (1951). "An Introduction to the Anatomy of Seed Plants." D. C. Heath and Co., Boston.
78. Swiecki, T. J., Endress, A. G., and Taylor, O. C. (1982). The role of surface wax in susceptibility of plants to air pollutant injury. *Can. J. Bot.* **60**, 316–319.
79. Taylor, G. E., Jr. (1978). Plant and leaf resistance to gaseous air pollution stress. *New Phytol.* **80**, 523–534.
80. Thomas, M. D. (1981). Gas damage to plants. *Annu. Rev. Plant Physiol.* **2**, 243–254.
81. Thomas, J. F., and Harvey, C. N. (1983). Leaf anatomy of four species grown under continuous $CO_2$ enrichment. *Bot. Gaz.* **144**, 303–309.
82. Thomson, W. W., Dugger, W. M. Jr., and Palmer, R. L. (1966). Effects of ozone on the fine structure of the palisade parenchyma cells of bean leaves. *Can. J. Bot.* **44**, 1677–1682.
83. Tomlinson, P. B. (1986). "The Botany of Mangroves." Cambridge University Press, Cambridge.
84. Tucker, G. F., and Emmingham, W. H. (1977). Morphological changes in leaves of residual western hemlock after clear and shelterwood cutting. *For. Sci.* **23**, 195–203.
85. Uhring, J. (1978). Leaf anatomy of *Petunia* in relation to pollution damage. *J. Am. Soc. Hort. Sci.* **103**, 23–27.
86. Very, A. A., Robinson, M. F., Mansfield, T. A., and Sanders, D. (1998). Guard cell cation channels are involved in Na+-induced stomatal closure in a halophyte. *Plant J.* **14**, 509–521.
87. von Willert, D. J. (1992). "Life Strategies of Succulents in Deserts with Special Reference to the Namib Desert." Cambridge University Press, New York.
88. Wali, Y. A., and Abdel-Salam, A. S. (1969). Histological studies of gamma irradiation artichoke heads. *Am. Soc. Hort. Sci.* **94**, 507–509.
89. Wheeler, E. A., and Baas, P. (1991). A survey of the fossil record for dicotyledonous wood and its significance for evolutionary and ecological wood anatomy. *IAWA Bull. n. s.* **12**, 275–332.
90. Wiehe, W., and Breckle, S. W. (1990). Die Ontogenese der Salzdrüsen von *Limonium* (Plumbaginaceae). *Bot. Acta* **103**, 107–110.
91. Witztum, A. (1978). Mesophyll intrusion induced by far-UV damage to leaves of *Zebrina pendula*. *Bot. Gaz.* **139**, 53–55.
92. Woodward, F. I. (1987). Stomatal numbers are sensitive to increases in $CO_2$ from pre-industrial levels. *Nature* **327**, 617–618.
93. Ziegler, H., and Lüttge, U. (1966). Die Salzdrüsen von *Limonium vulgare*. I. Die Feinstruktur. *Planta* **70**, 193–206.

# III

# ECONOMIC AND APPLIED PLANT ANATOMY

# 9

# GENETICS AND PLANT BREEDING

Genetics, the science of heredity, deals with the study of variation, its transmission in inheritance, and its expression in development. The basic unit of inheritance is the **gene,** which is transmitted essentially unchanged from cell to cell and from generation to generation. The visible **characteristics** determined by the genes on the chromosomes of the parents are passed on to offspring in general but occur according to definite ratios or percentages, based on tendencies for dominance, recessiveness, and other gene characteristics. Our concept of the gene has undergone many transformations during this century, culminating in the contemporary view that genes are composed of sequences of nucleotide triplets that encode the information to build proteins. **Mutations** are usually stable changes in this genetic material. They provide the primary means of understanding gene structure and function in plants by disrupting normal developmental programs and subsequently affecting mature structure and physiology. Numerous mutations have been characterized and mapped, and mutations that control many aspects of plant development and anatomy have been identified and the genes sequenced. To understand the effect of mutations on plant organization and development, a complete knowledge of the morphology and structure of normal plants is required to distinguish between the normal and the mutated case.

A great deal of modern research combines genetics and molecular biology to characterize the structure of genes and their targets and to clarify the processes by which genes control unit characters, that is, how genes are expressed and regulated. Recent discoveries in plant science have illustrated the potential of modern techniques for the production of genetically engineered plants that are resistent to pathogens, herbicides, and adverse environmental conditions.

## MOLECULAR GENETIC ANALYSIS

Molecular genetic analysis of the structure and expression of structural genes in plants has changed significantly in recent years. Like other multicellular organisms, flowering plants contain many enzymes, each one usually produced in different tissues and in different cell types. The identification of shoot- and root-expressed genes and the location of specific gene products within the plant is beginning to clarify basic questions of development. Once genes that function in the expression of anatomical traits have been identified, their sequences and expression patterns can be compared in attempts to learn how genotypes have been altered during evolution. Each gene-directed protein also carries in its amino acid sequence some information pertaining to its evolutionary history and origin, thus providing clues to the evolutionary history of that plant. The results of such studies are not only of theoretical importance but also of enormous practical significance because they hold potential for the development of useful new plant varieties. At the present time genetics and plant anatomy are handicapped by inadequate information relating to how specific genes exert their influence on anatomical characters. In some sense, the study of genetics has changed from the analysis of externally expressed character states (phenotypic variation) to the study of the cellular basis of genetics and finally to a consideration of the molecular basis of gene expression and regulation. In this progression, plant geneticists cannot afford to neglect the knowledge of other disciplines such as anatomy, in their effort to understand the complex mechanisms and results of gene action.

## ANATOMY AS AN INDICATOR OF HYBRIDIZATION

**Hybridization** occurs when two different naturally occurring plant populations show a breakdown in reproductive isolation and an exchange of genes between diverse individuals takes place. Such cross breeding results in increased variation in which parental characteristics are recombined in new individuals called **hybrids.** Two closely related species (or rarely, two genera) may undergo only occasional crossing, thus maintaining the identities of the two parental species but increasing variability by allowing some gene exchange. Successive backcrossing of the naturally occurring hybrids to either or both of the parental species results in further movement of genes from one species to another. This gene migration from one species into another is termed **introgression** and the process by which it occurs is called **introgressive hybridization.** The diverse group of individuals that is produced by this event is referred to as a **hybrid swarm.**

For the most part, morphological and cytological data have been the primary bases for the determination of hybrids. In addition to these traditional sources of evidence, however, there is anatomical evidence for natural hybridity, although the apparent usefulness of anatomy in hybrid analysis is not fully appreciated and so it has not been extensively studied. The role of leaf anatomy as an aid in the identification and characterization of plant hybrids can be illustrated by the mint genus *Salvia*. *Salvia apiana* and *S. mellifera* provide a classic example of introgressive hybridization between two closely related species. Thorough analysis has uncovered a diverse set of quantitative characters that are of value in the identification of the hybrids between these species. These characters relate to leaf margins, venation pattern, aspects of sectioned leaves, and epidermal characteristics. In most instances the hybrids are more or less intermediate in quantitative features, although statistical analysis has shown that only some features are reliable for determining the hybrid nature of putative hybrid plants. In the case of *Salvia,* no statistically significant differences can be found between the two parent species with respect to depth of sinuses on leaf margins, thickness of leaves when measured through the veins, abundance of glandular trichomes, or abundance of stomata on abaxial surfaces of leaves. Only the length of margin recurvature, abundance of nonglandular trichomes on upper leaf surfaces, abundance of trichomes on lower leaf surfaces, and abundance of stomata on upper surfaces of leaves are consistently useful in separating hybrids and both parental species. Clearly, quantitative anatomical data are required when dealing with putative hybrids.

The Aloineae are a well-defined tribe of the monocotyledonous family Liliaceae that offers additional insight into the inheritance of leaf anatomical characters. Within this complex, evidence has been gathered for the genetic regulation of leaf surface characters resulting from intrageneric and intergeneric crosses. Leaves of *Aloë* possess sunken stomata with apertures overarched on either side by a cuticular rim. Each stomata is surrounded by four subsidiary cells. Epidermal cells have micropapillae on the outer surfaces and wax is present in small flakes or particles. In this complex, there is compelling evidence that some epidermal features such as stomatal position, size, and surface papillation are under direct genetic control and are only marginally affected by environmental variables. Epidermal features in *Aloë* are species specific, and hybrid plants with normal chromosome complements typically possess an intermediate surface morphology that can only be recognized microscopically. This is true for both bispecific and bigeneric hybrids. Experimental crosses involving diploid *Aloë rauhii* Reynolds plants with tetraploid *A. dawei* Berger individuals produce triploid hybrids. Although an intermediate condition results with respect to stomata size, the hybrids show a dosage effect and more closely resemble *A. dawei,* the tetraploid parent, than *A. rauhii,* the diploid parent. Chromosome aberrations such as duplications, deletions, and chromosome loss also have a direct effect on the expression of leaf surface characteristics in this genus.

Although data are very limited, a careful consideration of wood and bark anatomy has shown that the internal characters of woody hybrids, like most external morphological features, tend to be intermediate in their expression. In most instances, such as hybrids within the genus *Eucalyptus,* the range of variation in quantitative and qualitative wood anatomical characters either is not significantly different from either parent or is intermediate between the two parents. This can be seen from data on vessel element size and distribution and the physical and mechanical properties of wood.

Cytogenetics, that is, variation in chromosome number and size, also can be correlated with aspects of anatomy. Most systems of classification now

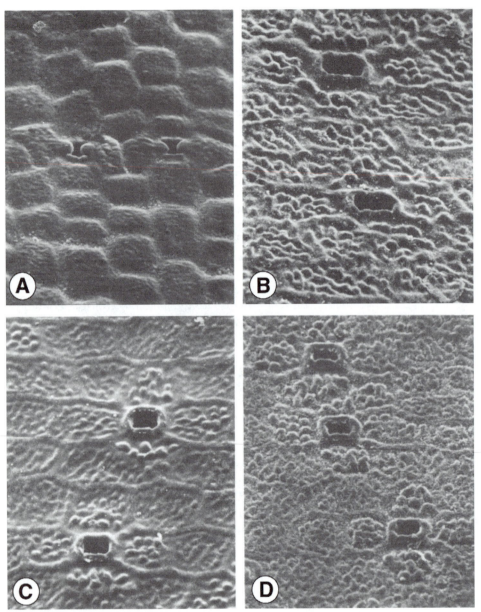

**FIGURE 9.1**   Influence of chromosome variation on the organization of the leaf epidermis in a hybrid *Aloë* (Liliaceae). (A) *A. rauhii.* (B) *A. dawei.* (C, D) Normal hybrids. All × 300. Reprinted with permission of Brandham and Cutler (1978), *Bot. J. Linnean. Soc.* **77,** 1–16.

divide the grass family (Gramineae) into tribes, forming a system that is partially based on karyotypic differences and leaf anatomical features. Research on the grasses has shown them to be a group of plants with an evolutionary history that is dominated by the phenomena of hybridization and by increases in chromosome number through doubling (polyploidy). Grass leaves possess a variety of taxonomically useful foliar characteristics that reflect differences in epidermal cell type, composition, and arrangement, as well as differences in mesophyll histologies. Various leaf anatomical conditions can be used to circumscribe grass taxa up to the level of tribe.

## GENE EXPRESSION AND ANATOMICAL TRAITS

A large number of mutants have been identified. They affect a wide range of morphological and anatomical traits, ranging from embryogenesis to the adult plant. Some genes display organ and tissue specificity, whereas others are only expressed in certain cell types at particular stages of development, for example, genes that are expressed in the procambium and reveal the early stages of the developmental program controlling vascular differentiation. Some mutants function to disrupt the normal spatial patterning of stomata and guard cell–mother cell development. One mutant, termed *Too Many Mouths* (*tmm*), is associated with the formation of stomata in groups, where normally there would only be one. The *tmm* gene controls the developmental pathway by which protodermal cells form stomata. The expression of a mutant gene or gene families can result in the loss of an organ or cell, or it may alter the morphology or function of a target structure to various degrees.

It has been estimated that hundreds of genes are involved in the many aspects of embryo development, including pattern formation and meristem function. The structural and functional integrity of the shoot apical meristem is disrupted by a recessive mutation in the *WUSCHEL* (*wus*) gene. Plants showing the *wus* mutant fail to properly organize a shoot meristem in the embryo. Defective shoot meristems terminate activity prematurely and have aberrant flat apices. One embryo pattern mutant, called *shoot meristemless,* results in the absence of epicotylar shoot development. Mutants displaying the meristemless gene were originally believed to be totally devoid of shoot apical meristem activity. It is now known that these plants form cotyledons during embryology and thus have a functioning apical meristem during early development, but this nascent meristem undergoes early abortion. Other genes control shoot apical meristem development and growth and include mutants with an abnormal apical meristem zonation organization and unusual patterns of leaf primordia production. Additional pattern mutants that produce seedlings devoid of either hypocotyl, root, or cotyledons are known. Numerous mutants have been isolated from *Arabidopsis thaliana*. These mutants directly control root development such as epidermal cell formation, root hair development, and directed root growth. *Arabidopsis* is an especially useful species for study because it possesses a small genome.

## THE WILTED GENE

*Zea* (corn) has been the subject of intensive genetic research. In fact, more is known about the genetics of this species than any other plant. One interesting study carried out nearly 40 years ago identified a gene from maize that directly controls a physiologically important anatomical feature and developmental program. The gene of interest is the recessive mutant named *wilted* (*wi*). Corn plants that are homozygous recessive (*wi/wi*) for this gene pair show symptoms of chronic water deficiency early in development during periods of moderate to high transpiration rates. Anatomical observation has shown that *wilted* plants possess many vascular bundles with abnormal tracheary element development. Furthermore, the presence of abnormal vascular bundles is correlated with plant development because the basal internodes on the stem have numerous normal bundles, whereas the youngest internodal regions contain mostly or entirely abnormal bundles. As a result, leaves display the *wilted* condition in an upward succession from the lowermost leaf.

In normal corn vascular bundles the primary xylem differentiates centrifugally and a pair of large metaxylem vessel elements develop laterally to

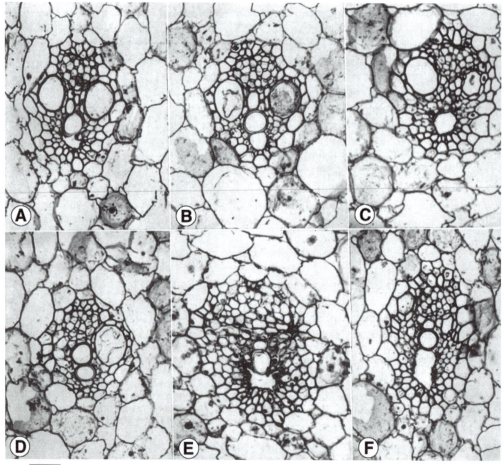

**FIGURE 9.2**  Vascular bundles of a *wilted* corn stem showing various effects of the gene *wilted (wi)*. (A) All metaxylem matured normally. (B) Metaxylem tardy in maturation. (C) One metaxylem vessel element matured normally and one tardily. (D) One metaxylem vessel element absent. (E) Both metaxylem vessels obliterated during maturation. (F) Only one or no metaxylem elements differentiated. All × 250. Reprinted with permission from Postlethwait and Nelson (1957), *Am. J. Bot.* **44,** 628–633.

other tracheary cells and adjacent to a prominent protoxylem lacuna. In contrast, the abnormal bundles in *wilted* individuals undergo delayed metaxylem differentiation; this is combined with different mature anatomies in which one or both of the metaxylem elements are degenerate, destroyed during ontogeny, or missing altogether. The primary effect of the *wilted* gene, therefore, is to control an anatomical feature directly related to water conduction. Plants showing the trait have leaves that are temporarily or permanently wilted. The phloem is unaffected. Classic genetical analysis of controlled crosses involving *wilted* have demonstrated that the *wilted* condition is regulated by a simple recessive gene that follows a Mendelian pattern of inheritance. That this is the case is indicated by the fact that all plants produced from selfed (pollinated from its own pollen or a closely related like plant) *wilted* individuals are also *wilted*. The progeny resulting from crosses between normal and *wilted* individuals are all normal, whereas backcrosses between the genotypes (*Wi/Wi* × *wi/wi*) × *wi/wi* give progeny with a genetic ratio of 50% *wilted* and 50% normal, the expected outcome if the mutant were recessive.

## TRICHOME DEVELOPMENT

Genetic analysis also has demonstrated the role of genes in regulating trichome development and distribution on the plant. In *Arabidopsis*, trichomes are unicellular and exist as branched outgrowths of the epidermis. Three separate categories of mutations at 21 different loci are known to affect trichome development. One group results in nearly complete absence of trichomes. Two genes from this group, *glabrous* (*GL1*) and *transparent testa glabra* (*ttg*), are essential for the initiation of trichome development and have been the subject of intensive study. Recessive mutations in either gene prevent the initiation of most trichomes, presumably because they result in a loss of function. These same two genes also regulate root hair development. All characterized mutant alleles of the *GL1* locus, a gene that encodes a member of the *Myb* family of transcriptional regulators, affect only trichome development. A second group of trichome mutants results in a reduction in both the number and degree of trichome branching. A final class of mutants causes a distorted and abnormal enlargement of trichomes and is represented by the *distorted* (*dis*1 and *dis*2) genes.

## CORN LEAF DEVELOPMENT

Gene families controlling leaf development and maintenance of the shoot apex as a formative region also have been characterized, whereas still others regulate the transition from a vegetative to a reproductive meristem. Molecular genetic analysis of maize leaf development has provided an opportunity to study the control of pattern formation, cell division, and cellular differentiation in plants. The study of leaf mutations has led to the identification of genes that apparently play an important role in the indeterminate growth of the shoot, in altering leaf cell fates, and in the spatial control of cell division. Developmental genetics has provided evidence supporting the view that the apical meristem contains recognizable zones that remain organized and formative and whose integration is closely related to the ability to initiate leaf primordia.

The developing grass leaf is composed of two parts, a **blade** domain, the thin, flattened distal region of the leaf, and the **sheath** domain, the basal, non-expanded leaf stalk enclosing the stem. Each leaf also contains a small, scale-like, epidermally derived flap of tissue called the **ligule** on the adaxial surface between the sheath and the blade. At the base of the blade at its union with the sheath, a pair of small projections termed **auricles** are formed from opposite sides of the sheath and, at maturity, partially extend around the stem. A number of studies have focused on the molecular genetics of corn leaf initiation and determination. We now know that the sheath, blade, and auricle are anatomically identifiable in such features as venation, internal histology, and the surface characteristics of epidermal cell shape and trichome pattern. Current studies are directed at understanding how such leaves are initiated and become differentiated into domains with divergent anatomies, including the genetic basis of leaf determination and how leaf features become fixed during development.

In corn, different dominant mutations are known to regulate determination patterns within the leaf that result in the formation of altered patterns of blade, sheath, and auricle development, and mature histology. One mutation that has received extensive study is called the *Knotted-1 (Kn*1) mutation. This mutation is known to produce alterations in the vein system and change patterns of cell division and differentiation in all layers of the corn leaf. Dominant *Knotted-1 (Kn*1) mutations cause cells in the vicinity of the lateral veins of the blade to acquire features more typical of the sheath and auricle, and to sporadically form fingerlike outgrowths of tissue called knots. The *Kn*1 gene has been isolated and characterized, and the protein it encodes has been identified. Genetic analysis has demonstrated that the *Kn*1 gene product is normally expressed in the shoot apex, where it may promote indeterminate growth. In mutant plants, the gene is also expressed ectopically in developing lateral veins of the leaf, and its ectopic expression in the developing blade seemingly results in the acquisition of vein sheath characteristics. It has been hypothesized that *Kn*1 delays the progression of cells in the blade to their final fates, causing them to respond to sheath-determining signals they would not normally perceive.

## CELL WALL GENES

It has been estimated that hundreds or even thousands of genes are linked to the growth and dynamics of plant cell walls at different stages of plant development and under both normal or wound-stressed conditions. As was discussed in Chapter 1, cell wall development involves the synthesis, interaction, and deposition of numerous components, all originating by the combined action of specific enzymes and genes. The genetic approach to the study of plant cell walls can be illustrated by a mutant of the widely studied plant *Arabidopsis,* in which the polysaccharide composition of the wall is altered.

Mutant plants have a dwarfed stature, and their cell walls are weakened. It has been found that these characteristics are caused by the plants' inability to synthesize a wall sugar that is a component of the pectic and hemicellulosic wall polysaccharides. Although the anatomy of plant organs is not changed, this allelic mutation affects the mechanical properties of the primary wall. Cell wall studies also have been directed at the structure and expression of genes coding for cell wall structural proteins and a large number of wall proteins have been characterized. Some of these cell wall genes are known to be developmentally regulated.

## FLORAL DEVELOPMENT

Although not included within the realm of anatomy, there is an extensively studied group of plant genes whose products are preferentially expressed during floral development in a taxonomically diverse collection of species. Most angiosperms form perfect, complete flowers with four concentric whorls of floral parts, sepals, petals, stamens, and carpels. These arise in precisely defined numbers and positions on the floral meristem. Genetic studies have identified mutations in which this normal floral development is altered. Different general classes of genes have been recognized: (1) **homeotic** (organ identity) **genes** whose products specify the radial sequence (or order) of development of the different categories of floral primordia, (2) **cadastral genes** that control the spatial pattern of development and thus determine the amount of area of floral meristem that contributes to the formation of a particular organ, (3) **meristic genes** that affect organ numbers, and (4) **meristem identity genes** that specify whether a meristem will remain indeterminate or become a determinate floral meristem or be transformed into an inflorescence. At the present time, little is known regarding the pathways that determine how this genetic information actually regulates organ formation.

Mutations in homeotic genes produce morphological variants in which the type of organ that arises during development varies from the normal condition. As a result, various atypical combinations or sequences of floral parts can occur, or one or more categories of parts are arrested during early ontogeny. For example, perianth parts can occur in positions normally occupied by stamens and carpels, or flowers can display homeotic conversions of perianth parts into fertile stamens and carpels. Experimental work has shown that different classes of homeotic genes interact to determine the nature of organs that will arise at any given location on the meristem.

## ANATOMY, HEREDITY, AND PLANT BREEDING

The yield, appearance, quality, and other attributes of plants, including internal structural properties, are profoundly influenced both by the environment and by heredity. In some instances, humans have found it advantageous to artificially adjust the hereditary characteristics of species in an effort to fit plants to the environment so that they will produce not only a more desired product but also a better quality product that is more useful or more conveniently harvested. A significant component of breeding experimentation directly or indirectly involves anatomical characters. Identifying desirable anatomical traits is a prerequisite to the breeding of improved varieties, however, no amount of variety can overcome the detrimental influence of environmental factors such as a low supply of available nutrients, poor cultural practices, or the damaging effects of bad weather. Likewise, a poor or unadapted variety cannot make the most of the advantages presented by a favorable productive environment. This statement is valid whether one is referring to agricultural crop plants or trees that have been bred to secure maximum yield or improve wood quality for human needs.

The breeding of new varieties starts with the selection of a plant having particularly favorable genetic variation. A desirable trait is one that in some way benefits the desired final product. To produce new individuals possessing this feature, plant breeders undertake controlled crossbreeding followed by the careful selection of progeny that show the desired trait and the rejection of those that do not. Characteristics of a good agricultural crop would be uniformity of size of plants, strong root systems, resistance to disease and insects, high yield, and adaptation. Long-term breeding programs in silviculture (cultivation of trees), on the other hand, might be more interested in improving aspects of wood quality.

## WOOD QUALITY

Overall wood quality is determined from the combined effects of several interrelated and variable wood properties that are positively and negatively correlated. It is often difficult to select specific wood properties to be investigated genetically because there may be conflicting interests influencing the direction of a commercial breeding program. Sawmill operators are most concerned with branch properties and stem straightness. At the pulpmill the elimination of short tracheid length and maintenance of low wood density frequently take precedence. Many wood properties are under different degrees of genetic control, and these can be altered to some extent through genetic experimentation and intensive selection. In the wood processing industry, for example, improved tear resistance of papers, hardness of furniture woods and stiffness of construction lumber are desirable qualities that are directly related to cell wall characteristics and cell size. Some properties can be manipulated during processing but others are stable features that can only be altered by a focused tree-breeding program. Important heritable wood properties that are related to variation in softwood and hardwood stems are listed in Table 9.1. A number of these same features also are important in formulating a breeding program for fiber crops.

Wood specific gravity and density are important in determining wood quality (Chapter 15). Both vary widely across species and even across individuals. As will be discussed at greater length in Chapter 15, specific

███████ **TABLE 9.1    Heritable Wood Properties Related to Stem Variation**

| Stem properties | Weight-related properties |
|---|---|
| Width of annual ring | Basic density |
| Proportions of juvenile and mature wood | Dry matter content |
| Proportions of early- and latewood | **Branch-related properties** |
| Amount of reaction wood | Knot content |
| Amount of heartwood | Spike knot frequency |
| Amount of spiral grain | |
| | **Chemical properties** |
| **Tracheid properties** | Content of extractives |
| Cell wall thickness | Content of lignin |
| Tracheid diameter | Content of cellulose |
| Diameter of lumen | Content of hemicellulose |
| Tracheid length | |
| Microfibril angle | |

Reprinted with modification and permission from Stahl and Ericson (1991), "Genetics of Scots Pine," Elsevier, Amsterdam.

gravity refers to the ratio of the weight of a given volume of air-dried wood to the weight of an equal volume of water at its greatest density (4°C). When calculating specific gravity, it is always necessary to indicate the moisture content of the sample at the time its volume was measured. By wood density we mean the weight of wood expressed per cubic volume. Both concepts are related to the amount of cell wall material that has been deposited. Summerwood, therefore, is denser and has a higher specific gravity than springwood. Specific gravity should not be thought of as a single wood character but as a complex of features, each of which is under strong genetic control. The effects of variation in each of these features on specific gravity are illustrated here.

Along with cell length and lignin content, both specific gravity and density have a significant impact on the durability, strength, machinability, acoustic properties, and paper yield among other characteristics of the final product. Wood density is particularly important. Its genetic manipulation can benefit both the structural lumber and pulping industries. At different times the objective of a breeding program can be to either increase or decrease density. In the production of certain high-quality papers, the most desired wood is one having low density because the collapse of thin-walled fibers results in good paper printability. Because density is correlated with hardness and strength, woods containing thick-walled cells are very dense in addition to being hard and strong. Unfortunately, trees that undergo rapid radial growth tend to form wood with a low percentage of summerwood and heartwood and therefore a low basic density, all of which result in low-quality sawn lumber as well as low pulp yield.

Attention has recently been directed at cell wall microfibril angle, as one structural feature that has the potential of being altered for significant economic gain. Cells near the pith of a tree have high microfibril angles that decline gradually toward the cambium. Microfibril angle has two major influences on wood properties. Most importantly, as microfibril angle declines, cell wall stiffness increases. Longitudinal wood shrinkage increases with microfibril angle but in a highly nonlinear manner. Stiffness of wood increases fivefold as the microfibril angle decreases from pith to cambium. Because microfibril angle cannot be manipulated during various wood processing procedures, the selection of trees having a below average microfibril angle would be of potential benefit to the wood construction industry.

As a tree grows, wood properties vary between the stem apex and base and across the radial diameter of the stem from the center out. **Juvenile wood** is formed in the upper part of the older tree and is located internally near the pith. Juvenile wood is successively covered by **mature wood**. These two classes of wood are formed at different stages of a plant's development and vary with regard to several quantitative and qualitative features. The relative proportion of the two wood types within a tree is of great importance in determining wood quality. Juvenile wood is correlated with short fibers, low cellulose content, prevalence of spiral grain, and high longitudinal shrinkage and low strength and is undesirable for most solid wood construction needs. Spiral grain causes lumber to have reduced strength and stiffness and a tendency to twist when dried. The utilization of juvenile wood as pulpwood is also to be avoided. Properties such as density, tracheid and fiber length, strength, cell wall thickness, percentage of latewood, and higher transverse shrinkage all increase in mature wood. Because these properties are related to tree growth characteristics, juvenile and mature wood can be artificially selected for in the desired direction to meet market requirements. The increased growth rate that has been selected for in most breeding programs leads to a corresponding increase in juvenile wood content. Most tree-breeding programs focus on increasing wood production, while the improvement of wood properties is given secondary priority.

## SUMMARY

Genetics, the science of heredity, deals with the study of variation, its transmission in inheritance, and its expression in development. Gene mutations provide the primary means of understanding gene structure and function in plants. The genes controlling developmental traits can be identified by isolating mutations that alter or block that particular type of development. In some cases, mutations affecting development have been found to occur in structural genes, encoding enzymes that are necessary for a specific function that characterizes a particular developmental pathway. A large number of mutants have been identified that affect a wide range of anatomical traits, ranging from embryogenesis to the adult plant. Some genes display organ and tissue specificity. Numerous mutants have been isolated that directly control root and shoot development, meristem behavior, epidermal cell formation, root hair and trichome development, and vascular tissue differentiation. The recessive mutant named *wilted* is of special interest. Corn plants that are homozygous recessive for this gene pair show symptoms of chronic water deficiency early in development. Anatomical observation has shown that *wilted* plants possess many vascular bundles with abnormal tracheary element development. The abnormal bundles in *wilted* individuals undergo delayed metaxylem differentiation that is combined with different mature anatomies in which one or both of the metaxylem elements are degenerate or destroyed. Another mutation of corn is called *Knotted-1*. This mutation is known to produce alterations in the vein system and change patterns of cell division and differentiation in all layers of the corn leaf.

Hybridization occurs when two different naturally occurring plant populations show a breakdown in reproductive isolation and an exchange of genes between diverse individuals takes place. In most instances the hybrids are more or less intermediate in quantitative features. The Aloineae are a well-defined tribe of the monocotyledonous family Liliaceae that provides insight into the inheritance of leaf anatomical characters. Within this complex, evidence has been gathered for the genetic regulation of leaf surface characters resulting from intrageneric and intergeneric crosses. In this complex, there is compelling evidence that some epidermal features such as stomatal position, size, and surface papillation are under direct genetic control. Epidermal features in *Aloë* are species specific, and hybrid plants with normal chromosome complements typically possess an intermediate surface structure that can be recognized microscopically.

Wood quality is determined by the combined effects of several interrelated and variable wood properties. Heritable wood properties include the amount of juvenile and mature wood, grain pattern, individual cell size and wall characteristics, and cell wall chemical properties. Some wood properties can be manipulated during processing but others are stable features that can only be altered by a focused tree-breeding program.

## ADDITIONAL READING

1. Aeschbacher, R. A., Schiefelbein, J. W., and Benfey, P. N. (1994). The genetic and molecular basis of root development. *Annu. Rev. Plant Physiol., Plant Mol. Biol.* 45, 25–45.
2. Baas, P. (1978). Inheritance of foliar and nodal characters in some *Ilex* hybrids. *Bot. J. Linnean Soc.* 77, 41–52.
3. Barton, M. K., and Poethig, R. S. (1993). Formation of the shoot apical meristem in *Arabidopsis thaliana*: An analysis of development in the wild type and shoot meristemless mutant. *Development* 119, 823–831.
4. Brandham, P. E., and Cutler, D. F. (1978). Influence of chromosome variation on the organization of the leaf epidermis in a hybrid *Aloë* (Liliaceae). *Bot. J. Linnean Soc.* 77, 1–16.
5. Cannon, W. A. (1909). Studies in heredity as illustrated by the trichomes of species and hybrids of *Juglans, Oenothera, Papaver,* and *Solanum. Carnegie Inst. Wash. Publ.* 117, 1–67.

6.  Cousins, S. M. (1933). The comparative anatomy of the stem of *Betula pumila, Betula lenta* and the hybrid *Betula jackii. J. Arn. Arb.* **14**, 351–355.

7.  Cutler, D. F. (1972). Leaf anatomy of certain *Aloë* and *Gasteria* species and their hybrids. *In* "Research Trends in Plant Anatomy" (A. K. M. Ghouse and Yunus Md., Eds.), pp. 103–122. Tata McGraw Hill, New Delhi.

8.  Cutler, D. F., and Brandham, P. E. (1977). Experimental evidence for the genetic control of leaf surface characters in hybrid Aloineae. *Kew Bull.* **32**, 23–32.

9.  Dolan, L. (1996). Pattern in the root epidermis: An interplay of diffusible signals and cellular geometry. *Ann. Bot.* **77**, 547–553.

10. Dolan, L. (1997). SCARECROW: Specifying asymmetric cell divisions throughout development. *Trends Plant Sci.* **2**, 1–2.

11. Gavalas, N., Bosabalidis, A. M., and Kokkini, S. (1998). Comparative study of leaf anatomy and essential oils of the hybrid *Mentha x villoso-nervata* and its parental species *M. longifolia* and *M. spicata. Isr. J. Plant Sci.* **46**, 27–33.

12. Giertych, M., and Matyas, C. (Eds.) (1991). "Genetics of Scots Pine." Elsevier, Amsterdam, Oxford, New York, Tokyo.

13. Hall, L. N., and Langdale, J. A. (1996). Molecular genetics of cellular differentiation in leaves. *New Phytol.* **132**, 533–553.

14. Langdale, J. A., and Kidner, C. A. (1994). Bundle sheath defective, a mutation that disrupts cellular differentiation in maize leaves. *Development* **120**, 673–681.

15. Larkin, J. C., Marks, M. D., Nadeau, J., and Sack, F. (1997). Epidermal cell fate and patterning in leaves. *Plant Cell* **9**, 1109–1120.

16. Larkin, J. C., Oppenheimer, D. G., Lloyd, A. M., Paparozzi, E. T. and Marks, M. D. (1994). Roles of the GLABROUS1 and TRANSPARENT TESTA GLABRA genes in *Arabidopsis* trichome development. *Plant Cell* **6**, 1065–1076.

17. Lemieux, B. (1996). Molecular genetics of epicuticular wax biosynthesis. *Trends Plant Sci.* **1**, 312–318.

18. Mayer, U., Torres-Ruiz, R. A., Berleth, T., Miséra, S., and Jürgens, G. (1991). Mutations affecting body organization in the *Arabidopsis* embryo. *Nature* **353**, 402–407.

19. Medford, J. I., Elmer, J. S., and Klee, H. J. (1991). Molecular cloning and characterization of genes expressed in shoot apical meristems. *Plant Cell* **3**, 359–370.

20. Nelson, T., and Dengler, N. G. (1992). Photosynthetic tissue differentiation in $C_4$ plants. *Int. J. Plant Sci.* **153**, S93–S105.

21. Nelson, T., and Langdale, J. A. (1992). Developmental genetics of $C_4$ photosynthesis. *Annu. Rev. Plant Physiol., Plant Mol. Biol.* **43**, 25–47.

22. Postlethwait, S. N., and Nelson, O. E., Jr. (1957). A chronically wilted mutant of maize. *Am. J. Bot.* **44**, 628–633.

23. Pri-Hadash, A., Hareven, D., and Lifschitz, E. (1992). A meristem-related gene from tomato encodes a dUTRase: Analysis of expression in vegetative and floral meristems. *Plant Cell* **4**, 149–159.

24. Pryor, L. D., Chattaway, M. M., and Kloot, N. H. (1956). The inheritance of wood and bark characters in *Eucalyptus. Aust. J. Bot.* **4**, 216–239.

25. Pyke, K. (1994). *Arabidopsis*—Its use in the genetic and molecular analysis of plant morphogenesis. *New Phytol.* **128**, 19–37.

26. Reiter, Wolf-Dieter, Chapple, C. C. S., and Somerville, C. R. (1993). Altered growth and cell walls in a fucose-deficient mutant of *Arabidopsis. Science* **261**, 1032–1035.

27. Roffer-Turner, M., and Napp-Zinn, K. (1979). Investigation on leaf structure in several genotypes of *Arabidopsis thaliana* (L.) Heynh. *Arabidopsis Inform. Serv.* **16**, 94–98.

28. Smith, L. G., and Hake, S. (1994). Molecular genetic approaches to leaf development: Knotted and beyond. *Can. J. Bot.* **72**, 617–625.

29. Stahl, E. G., and Ericson, B. (1991). Inheritance of wood properties. *In* "Genetics of Scots Pine" (M. Giertych and C. Mátyás, Eds.), pp. 231–241. Elsevier, Amsterdam.

30. Uchino, A., Sentoku, N., Nemoto, K., Ishii, R., Samejima, M., and Matsuoka, M. (1998). $C_4$-type gene expression is not directly dependent on Kranz anatomy in an amphibious sedge *Eleocharis vivipara* Link. *Plant J.* **14**, 565–572.

31. Verma, D. P. S. (Ed.) (1993). "Control of Plant Gene Expression." CRC Press, Boca Raton, Florida.

32. Walker, J. C. F., and Butterfield, B. G. (1996). The importance of microfibril angle for the processing industries. *N. Z. For.* **40**, 34–40.

33. Webb, A-A., and Carlquist, S. (1964). Leaf anatomy as an indicator of *Salvia apiana-mellifera* introgression. *Aliso* **5**, 437–449.

34. Wissenbach, M., Überlacker, B., Vogt, F., Becker, D., Salamini, F., and Rohde, W. (1993). Myb genes from *Hordeum vulgare*: Tissue-specific expression of chimeric Myb promotor/Gus genes in transgenic tobacco. *Plant J.* **4**, 411–422.

35. Yearbook of Agriculture. U.S. Department of Agriculture. (1937). U.S. Government Printing Office, Washington, DC.

36. Zobel, B., and Talbert, J. (1984). "Applied Forest Tree Improvement." John Wiley & Sons, New York.

37. Zobel, B. J., and Buitjtenen, J. P. van (1989). "Wood Variation. Its Causes and Control." Springer-Verlag, Berlin.

# 10
## DEFENSE MECHANISMS AND STRUCTURAL RESPONSES OF PLANTS TO DISEASES, PESTS, AND MECHANICAL INJURY

Like all living things, plants are continuously subjected to a wide variety of potential disease-causing agents and to the stresses caused by mechanical injury. Generally speaking, the causes of disease fall into two broad groups, parasitic and nonparasitic. An organism that lives in or on another individual for part or all of its life, securing its food directly from the tissues of the host organism but not benefiting the host in any way, is referred to as a **parasite.** Many but not all parasites are pathogenic, that is, disease-producing organisms. **Disease** refers to any disturbance in functioning and growth that causes a lower operating efficiency or a breakdown in the plant's metabolism. Diseases that afflict plants generally result from microbial infections, invasions of the body by pathogenic viruses, bacteria, fungi, or other microorganisms. Diseases also can result from infestations of insects and nematode worms. The reaction of the affected plant to the cause of the disturbance produces a spectrum of visible symptoms and consequences that are commonly recognized as the effects of disease, such as wilting, cell death and dieback, stunting, and discoloration. Many changes in cellular structure and metabolism occur when plants are infected by parasites. Some of these changes are thought to be closely related to the ability of the plant body to defend itself.

Vertebrate animals possess a sophisticated immune system that provides the body with a means of resisting infection. In the apparent absence of such a system, plants must rely on various other defense mechanisms. Among the multiple defenses that occur in plant cells and tissues in response to stress and invasion by pathogens and that are implicated in disease resistance and susceptibility are genetic and biochemical defenses; other defense mechanisms are structural in nature. Plants also can be subjected to mechanical injury by chewing animals or by forces in the physical environment that create open wounds. These structural changes often are accompanied by chemical alterations in the host tissue that are thought to be inhibitory or toxic to pathogens.

Not all animal and microbial invasions in plants are lethal. In some woody species the larvae of insect cambium miners injure the cambial initials so that irregular patches of abnormal parenchymatous tissue are produced and become embedded in the wood, visible as a longitudinal streak on the surface. These scattered patches of wound parenchyma, termed **pith flecks,** are composed of cells that are irregular in size, shape, and arrangement. In some woods, pith flecks are common, but the plant does not appear to be obviously diseased. Many complicated insect and plant relationships produce an overgrowth in the host tissue in which the larvae develop. These localized overgrowths are termed **galls** and are caused by an abnormal increase in the number or size of small groups of cells. The shape and structure of galls may be diagnostic of an insect species, and they can be restricted to a specific part of a plant, for example, leaves, stems, or roots. Although they represent a deviation from the normal condition, galls usually do not noticeably harm the plant.

This chapter describes a few of the various anatomical responses of higher plants to wounding or infection and includes an analysis of how plants have evolved growth forms to cope with changing stress patterns in natural environments. Many defense responses in plants are genetically controlled, resulting in efforts to breed more anatomically decay- and disease-resistant individuals. An understanding of the structural bases of these responses is important in horticultural and floricultural practices.

## THE NATURE OF PLANT DEFENSES

Plant defenses can be passive, in the sense that they are preexisting barriers to infection. Examples of passive, **preinfectional structural defenses** are the cell walls and outer surface covering of the plant body. The multilayered cuticle of young stems and foliage, consisting of an insoluble polyester called cutin, provides a very effective barrier to invasion by pathogenic microbes. Some fungi produce extracellular enzymes capable of hydrolyzing cutin. The enzymes involved in disrupting this polymer barrier are called **cutinases.** Although some fungi have become adapted to penetrate the cuticular layer and epidermal cell wall, the cuticle is commonly breached only through openings such as stomata or wounds. Resin-secreting systems composed of vertical and horizontal resin ducts also act as defense barriers. In some plants, trichomes and thorns function to minimize invasion of specific pathogens. The intact outer periderm or bark of woody plants provides an equally successful barrier to infection that is not readily crossed. Norway spruce clones that are resistant to the bark beetle–vectored fungal pathogen *Ceratocystis polonica* possess axial secondary phloem parenchyma cells containing antifungal polyphenolic vacuolar deposits. Some woody species are more resistant to infection as young individuals when the bark is smooth and unbroken and become more susceptible to disease as older plants when the periderm becomes cracked.

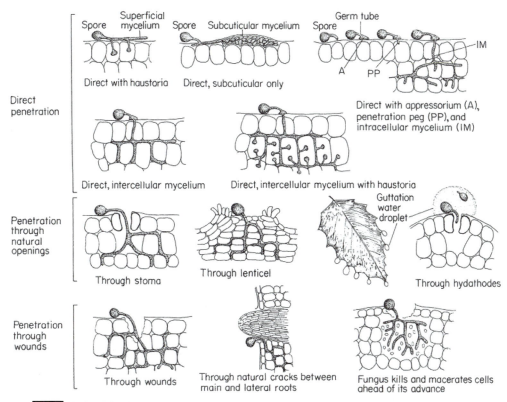

**FIGURE 10.1**   Methods of penetration and invasion of leaves by fungi. Reprinted with permission from Agrios (1997), "Plant Pathology," 4th ed., Academic Press.

Active mechanisms of resistance, referred to as **postinfectional host responses,** are changes induced in cells and tissues in direct response to injury or the activities of harmful invaders. Infection of plant tissues results in transient or permanent long term changes in cell or tissue structure. These changes are produced directly by the parasite or result from a plant defense response. A plant's response to infection or wounding can be a nonspecific, general response to damage or a very parasite-specific response, resulting in a predictable and well-defined set of cellular changes. Most active mechanisms of host resistance attempt to limit damage or infection to the smallest possible volume of cells. Research has led to the identification and characterization of specific signaling molecules that elicit defensive responses that often are some distance removed from the site of wounding.

## RESPONSES OF PLANTS TO WOUNDING AND INVASION BY MICROORGANISMS

The principal anatomical response of plants to the formation of wounds and infection is to wall off (wall out) the injured or diseased region in an effort to contain the damage to a small area within a single tissue. This adaptation restricts the movement of pathogens in and around the infected region. The

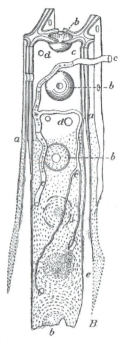

**FIGURE 10.2**   Fungal hyphae in pine tracheids. Abbreviations: a, cell wall of tracheids; b, bordered pits; c, hyphal filament of fungus; d, holes in cell wall produced by fungal hyphae, which gradually dissolve the walls as shown at e. Reproduced from Roth (1895), Bull. 10, U.S. Department of Agriculture, Division of Forestry.

walling off process (termed **compartmentalization**) occurs in the primary tissues of stem and leaves, as well as in the secondary xylem and bark of stems and roots. The compartmentalization of infection is a continuous process and an individual tree may have many isolated areas of decay. The isolation process proceeds by forming increasingly more formidable barriers and is often followed by the regeneration of new tissues. Plants have a capacity for regenerating lost or damaged organs and tissues and together these activities constitute a nonspecific response to injury. Trees and other plants survive only as long as they are able to successfully wall off the spread of infection to small tissue areas and then regenerate new healthy tissues.

Tissues are modified in a number of different ways to form structural and chemical barriers. Among the varied defense mechanisms that plants have developed is the ability to alter cell wall structure. The activation of genes coding for cell wall proteins following wounding or pathogenic infection results in the rapid polymerization of proline-rich proteins that form a tightly bound wall to render the wall indigestible by invading pathogens. Lignification of walls is another common response of plant tissues to wounding or infection, thus providing resistance and forming a barrier. Cell wall lignification may even take place in advance of a spreading fungal infection, suggesting that a chemical induction mechanism is present. Infection also may stimulate the suberization of walls as a resistance response, causing senescence and death of cells and checking the further movement of disease. The addition of suberin to the walls of

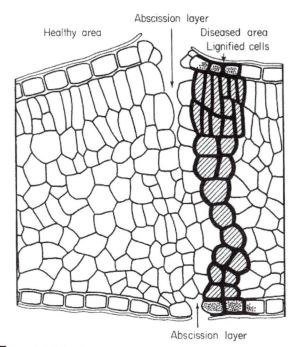

**FIGURE 10.3**  Formation of an abscission layer around a diseased spot of a *Prunus* leaf. Reprinted with permission from Agrios (1997), "Plant Pathology," 4th ed., Academic Press.

wounded tissues and injured cells seals off exposed surfaces and helps replace the damaged original tissue layers. Like lignin, suberized walls have the potential of limiting the size of an infected area and appear to play an important role in the compartmentalization of decay. Suberized axial parenchyma can develop within the secondary xylem of trees and form a barrier zone separating uninfected from infected wood. The relative rates of growth of the pathogen, as compared to the rate of suberization and cork formation by the host, often determine the fate of the infection and therefore of the host.

The carbohydrate callose is formed in a variety of living cells, apparently in response to injury. Callose is especially evident in sieve tube elements following physical damage of the phloem. Sieve tube elements are very sensitive to injury, so that when a cell is cut the solute quickly flows toward the wound site. The combination of phloem protein plugging and callose deposition rapidly seals off the injured area. As callose accumulates over the sieve pores and eventually covers the entire sieve plate, it provides protection against the loss of translocating sugar. Because phloem protein and callose appear to form instantaneously with injury or pressure release in the sieve tube element, it has proven exceedingly difficult to prepare phloem for microscopic observation without inducing these defense mechanisms. It has been suggested that the plugging of sieve elements with sufficient wound callose to occlude the pores represents an important evolutionary adaptation that insures the functioning of the phloem by preventing the loss of excessive amounts of assimilates.

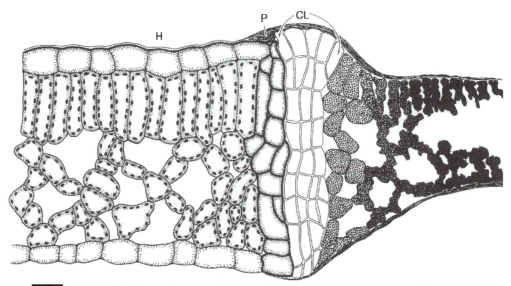

**FIGURE 10.4**   Formation of a cork layer (CL) between infected and healthy (H) areas of leaf, (P), phellogen. Reprinted with permission from Cunningham (1928), *Phytopathology* **18**, 717–751.

Protective barriers also may form as a result of meristematic activity immediately adjacent to a wound or infection site. In these cases, meristematic cells are usually initiated from parenchymatous cells and form a mass of **protective callus** by cell division and enlargement. The ability of plants to form new tissues largely results from the proliferative capabilities of parenchyma cells. The walls of callus cells typically become filled with secondary metabolites such as suberin, lignin, or phenolics. A new vascular and cork cambium often regenerates within the wound tissue, forming a new permanent barrier zone.

The stems of some trees bear scars that reflect the continuous struggle between pathogens and wound cork. This is seen in Aspen trees (*Populus tremuloides*), which are frequently invaded by a fungus that enters the trunk through open wounds and lenticels. The fungal mycelium grows in the outer cortex and stimulates the plant to form a subdermal phellogen to wall off the invasion. However, hyphae frequently penetrate the newly formed cork cells and stimulate the formation of another more internally positioned periderm. Successive layers of wound periderm thus are created. They push the more peripheral tissues outward, rupturing the original periderm and causing the rough bark surface that is characteristic of some trees. In this manner, the bark of trees is transformed from smooth to rough.

## Abscission

During normal growth, perennial plants regularly shed organs or parts, thus forming exposed wounds on the stem. Leaves, branches, flowers, and fruits all may be lost. The loss or injury of parts results in exposed wounds, which are liable to infection for a period of time immediately following abscission if they are not sealed or isolated. The complex phenomenon of organ shedding involves hormonal and anatomical interactions that form the sequence of events known as **abscission.** This process results in the separation of cells along a narrow band at a predetermined location at the base of the organ to be shed. The hydrolytic enzymes involved in cell wall softening and breakdown within the site of organ detachment are known as **cellulases** and, in some plants, such as *Impatiens* and *Citrus*, show increased activity when abscission is initiated. The gaseous plant growth regulator ethylene is known to promote abscission, whereas the hormone auxin retards the separation process.

The events of defoliation provide a classic example of wound response in which an open wound becomes sealed with a protective covering. Although details of the abscission are somewhat variable among plants, a narrow, transverse plane of cells forms at the base of the petiole (or petiolule) known as the **abscission zone.** The abscission zone can be recognized by cells that are smaller than in the adjacent petiole. This represents a region of weakness and is composed of two histologic regions. Cells of the **separation layer** (or layers) are directly involved in the weakening process. They develop an abnormal wall chemistry and swelling that results in their easily being pulled apart along the pectin-rich middle lamella. Electron microscope observations show that breakdown of the wall is not restricted to the middle lamella but involves adjacent areas of the primary wall. SEM micrographs show masses of intact cells on the fracture surface and have transformed the previously held view that the cells in the separation layer collapsed, dissolved, or broke apart during the abscission process. New observations show the separation of intact turgid cells. Following the separation of cortical cells, the xylem tissue stretches and ruptures. Separation of the petiole from the stem starts at a locus above the vascular tissue and spreads radially across the petiole. The fracture line

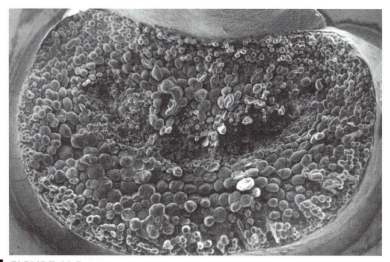

**FIGURE 10.5** SEM of leaf fracture surface in *Impatiens sultani*. Note the absence of broken cells and the ring of enlarged parenchyma cells, which is raised above the general cortex. Courtesy of R. Sexton.

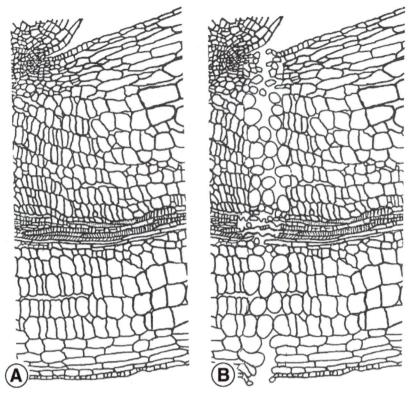

**FIGURE 10.6**   Diagrams of the process of leaf abscission in *Impatiens*. (A) Tracing of a micrograph of a longitudinal section of the petiole base. Note the smaller cells of the abscission zone toward the left-hand side. (B) Diagram of abscission process. Two or three rows of cells toward the distal end of the abscission zone separate from one another, round up, and increase in volume. This expansion stretches and snaps the xylem tracheary cells bridging the layer and facilitates the separation of collenchyma and parenchyma. Reproduced with permission of Sexton et al. (1984). Copyright © American Society of Plant Physiologists.

extends around cells leaving intact cells with active cytoplasms on the scar surface. It has been proposed that a localized accumulation of proteins in the exposed cells of the wound temporarily protects the underlying tissue from disease-causing pathogens. Later in the abscission process layers of suberin or lignin are deposited in the cells on the stem side of the break to form a **protective layer** over the exposed surface. A periderm is subsequently initiated beneath the protective layer and becomes continuous with the periderm layers of the stem. This highly regulated process of organ detachment indicates that the abscission zone contains specialized cells that respond to developmental signals in a highly specific manner.

## Tissue Regeneration

In some herbaceous dicotyledons, severe wounding of the stem may sever vascular bundles. Vascular regeneration can reestablish the continuity of the transport system by means of a complex developmental process involving

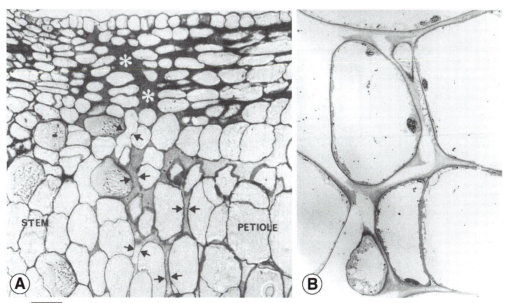

**FIGURE 10.7** Leaf abscission in *Impatiens*. (A) PAS-stained longitudinal section through the petiole base 16 hr after induction of abscission. Note the separation of cells along the fracture line as it passes through the collenchyma (starred) and parenchyma of the upper cortex (arrows). × 200. (B) Electron micrograph of the separation layer cells from a weakened leaf abscission zone. Note that the cell walls have degraded, allowing the cells to separate. The inner layer of the wall remains intact, and the cytoplasm is apparently normal. A, reproduced with permission of Sexton *et al.* (1984). Copyright © American Society of Plant Physiologists. B, courtesy of R. Sexton.

structural and physiological differentiation of parenchyma into xylem and phloem elements. Depending upon the severity of the wound, vascular elements reform in a few to several days and bridge the wound site to reconnect with intact vascular tissue above and below. Phloem cells appear first and arise from the division and differentiation of ground parenchyma. Wound-induced tracheary cells form somewhat later from the direct transformation of vacuolated parenchyma cells. The course of wound phloem regeneration in various dicotyledonous stems is illustrated in Figure 10.9. Following the formation of regenerated xylem and phloem, a cambium is sometimes formed and secondary vascular tissues can accumulate in the wound area. The induction and control of vascular regeneration involves the action of hormones such as auxin and cytokinin, although the mechanism of their action is only partially understood.

Large wounds that occur in stems as a result of natural branch loss or from pruning of limbs are eventually covered over by new tissues produced largely by activities of the vascular and cork cambia. The manner and speed with which new secondary tissues are formed, however, depend in large part upon which underlying tissues are exposed following wound formation and under what conditions the healing proceeds. The manner of wound response can be different at different times of the year and at different locations on shrubs and trees. An important function of the vascular cambium is the formation of callus, or **wound tissue,** that forms around the periphery of the severed branch. Wound

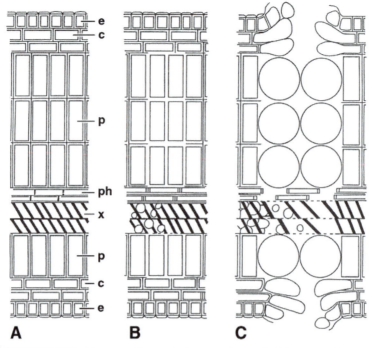

**FIGURE 10.8**    Diagrams of longitudinal sections showing leaf abscission in *Impatiens*. (A) Separation layer zone before abscission begins. (B) Early stage of abscission showing breakdown of the middle lamella in the cortical parenchyma. Tyloses develop in the xylem tracheary elements, and the phloem sieve plates become callosed over. (C) The final stage is characterized by middle lamella breakdown and expansion of cortical parenchyma cells. Finally only the xylem vessels bridge the separation zone, and these rupture along the line of the most distally located tyloses. Abbreviations: c, collenchyma; e, epidermis; p, cortical parenchyma; ph, phloem; x, xylem. Courtesy of R. Sexton.

callus is also parenchymatous tissue, originating by division of parenchyma cells that dedifferentiate, enlarge, and become meristematic.

If a small section of bark is removed from a tree so that the vascular cambium is attached to the bark, a surface of outer xylem is exposed. If the exposed xylem surface is kept moist, a uniform wound callus will develop across the entire exposed surface. The callus cells in this case originate primarily from proliferating xylem ray and axial parenchyma. After a period of time, the more peripheral layers of callus will become suberized. A new cork cambium differentiates adjacent to the preexisting phellogen in the surrounding bark. A new vascular cambium also will form beneath the phellogen as a continuous extension of the original cambium, differentiating across the callus centripetally from the outer margins of the wound. Following a period of early irregular divisions, both cambia produce secondary tissues of normal structure and appearance. As a general rule, phloem parenchyma cells and immediate cambial derivatives are readily capable of undergoing further differentiation and forming callus. Parenchyma cells of the secondary xylem, in contrast, lose the ability to proliferate callus tissue in cases of deep-seated wounds. Because older xylem parenchyma cells will not divide, when a limb

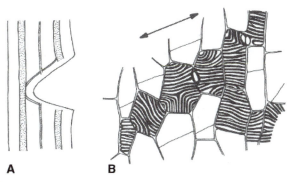

**FIGURE 10.9**  Regeneration of vascular tissue in a *Coleus* stem after vascular bundles were severed. (A) Diagrammatic longitudinal section through internode 8 days after wounding showing normal vascular bundles (dotted) and oblique regeneration connections between them across the outer pith around wound gap. (B) Drawing showing the short regenerated tracheary elements that differentiated from parenchyma cells. The arrow indicates the direction of the differentiation. Reprinted with permission from Sinnott and Bloch (1945), *Am. J. Bot.* **32**, 151–156.

is pruned, the wound tissue typically develops along the margins of the scar and then grows inward on all sides until the entire cut surface is covered. The wound callus that forms around the scar is initially of uniform size and concentric in outline. In later years, under the influence of growth regulators, the scar usually becomes spindle shaped as a result of the unequal growth of callus on the lateral flanks of the wound. Callus cells become suberized, lignified, or filled with phenolic substances.

Knowledge of these paths of regeneration of new cambia on wounded tree or shrub axes has practical applications in the cultivation of woody plants. When pruning a limb, the cut should be made so that stubs are not left on the main stem. The collar at the base of the branch should be left intact in order that the cambial zone surrounding the wound can assist in closing the surface rapidly. If a surface wound is wrapped or covered by lanolin or polyethylene film, a new bark will generally regenerate within the same growing season. But if wounds are left uncovered or the exposed cells are painted or sprayed with toxic substances, the parenchyma cells will be damaged and callus formation will proceed more slowly.

Vines are perennial plants with stems that twine or climb and do not grow upright without the support of other plants. Many lianas are woody and have been found to have stems that attain considerable length. The stem anatomy of lianas is characterized by many different anatomical types (see Chapter 4). These "anomalous" structural patterns include the presence of wide unlignified rays, much xylem parenchyma, included phloem, lobed xylem, disjunct cambia, supernumerary cambia, and multiple vascular cylinders. The functional significance of these anatomical types has largely been unexplained. However, the histological consequences of experimentally induced injuries to woody vines, as might occur by physical rubbing or animal gnawing, have recently been documented. It has been shown that these structural anomalies of vine stems participate in extensive callus formation and in the rapid healing of damaged vascular tissues before the leafy crown dies. Such anomalies

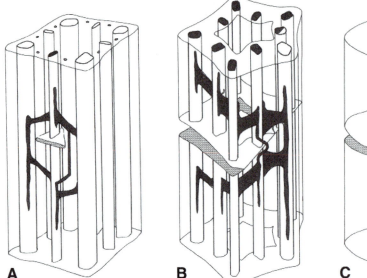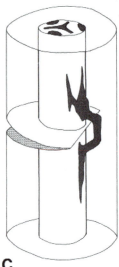

**A**                              **B**                              **C**

**FIGURE 10.10**   The course of wound phloem in *Coleus* (A) and *Cucurbita* (B) stem internodes and *Pisum* (C) roots where single bundles and the entire stele were severed. Reprinted with permission from Kollmann and Schulze (1993), Phloem regeneration. "Progress in Botany" Vol. 54, pp. 63–78. Copyright © Springer-Verlag.

also permit an orderly longitudinal splitting along defined rupture planes that limit vascular dysfunction. They may have adaptive significance for lianas because they limit injury and promote healing.

## Grafting

Grafting is a process whereby two plants are joined so that they become anatomically and physiologically united and grow as a single individual. Grafts can occur naturally but are more commonly prepared in the vegetative propagation of commercially valuable plants such as decorative shrubs and fruit and nut trees. Grafting is possible only in plants that have a degree of compatibility and a vascular cambium. There are many different methods of commercial grafting, but all depend upon the induction of wound responses by the vascular cambia of the two individuals being joined. In the preparation of a stem, graft segments of stems are cut in a manner so that the vascular cambia of the two segments can be placed in contact. The upper or top part that is transferred is known as the **scion** and the lower portion is the **rootstock** (or stock).

A number of developmental stages can be recognized in the formation of a graft union. The early stage begins within a few days and is characterized by the death of cell layers at the graft interface and the generation of parenchymatous wound callus that fills the gap between the two graft components. The callus is formed by all living undamaged cells at the graft site, including ray parenchyma, phloem parenchyma, and cambial initials and their

**FIGURE 10.11** Healing of a severed lateral branch in hickory (*Carya tomentosa*). Wound callus has formed and become spindle shaped due to the more rapid growth of the callus tissue along the vertical sides of the wound.

immediate derivatives. The callus eventually merges to form a single tissue in regions adjacent to the cambia. Tracheary cells arise within the callus from the transdifferentiation of parenchyma cells. This is followed by the differentiation of callus parenchyma to form new cambial initials and the subsequent union of the newly formed cambium with the original cambia in both stock and scion. The final stage in graft formation is the production of new secondary vascular tissues by the newly differentiated cambium, thus permitting the interconnection of scion and stock and the passage of nutrients and water between the two components of the graft and so forming a single individual.

## CYTOLOGICAL REACTIONS TO INVASION BY PARASITES

When plants are attacked by pathogenic microorganisms, many changes can be seen within the individual cells being penetrated or in cells in the immediate vicinity of an invasion site. Plant cells respond to attempts at penetration at a specific location in a number of ways. At the cellular level plants have the ability to alter cell ultrastructure and functioning to form localized cytological barriers to fungal invasion. By rapidly accumulating a mass of cytoplasm at the site of a fungal hyphal entry, a "cytoplasmic aggregate" can form within a single cell. Such aggregates contain a normal complement of cellular organelles and are believed to function by producing a cellular response to penetration.

Deposition of heterogeneous materials by a cytoplasmic aggregate between the host cell plasma membrane and cell wall at the site of fungal invasion is known as a **papilla.** Papillae are thought to confer resistance to the cells possessing them. Chemically papillae are composed of either callose, lignin or other phenolic derivatives, cellulose, or silicon. It is possible that papillae function as a mechanical barrier to fungal penetration or inhibit penetration of fungal hyphae in other ways. Fungal penetration of the host appears to provide the necessary stimulus for their formation, but the exact role of papillae in relation to disease resistance remains unclear. Ultrastructural studies show dramatic changes in the cytoplasmic membrane system of cells invaded by fungal hyphae; however, the significance of these changes also is unknown.

Another cytological phenomenon, known as the halo effect, has been observed. A halo is sometimes observed as a locally modified region of the host cell wall around a penetration site or at sites of potential penetration. The mechanism of halo induction is unknown and its nature is poorly understood. Haloes appear to be regions of localized deposition of cuticular lipid and opaline silica. One possible function might be to reduce water loss from the host cell at the penetration site. The halo region also is reported to be lignified or suberized, or show callose deposition between the cell wall and the plasma membrane. Very little is known about how the cell secretes and deposits these substances. Some reports have indicated halo development may represent a form of cellular resistance to pathogen penetration.

Nematodes are parasitic (or free living) wormlike pests whose feeding activities inflict major damage on vegetables and crops such as potatoes and soybeans. Nematodes enter the tips of growing roots and become confined to one or a few cells. Following entry they induce dramatic changes in these cells in order to support the prolonged removal of nutrients by the parasite. Many of the complex cell modifications induced by nematodes are unique among parasites. Some nematodes form a **syncytium** (a multinucleated cytoplasm) within the stele by a combination of cell wall digestion and cell hypertrophy. Root knot nematodes form root galls associated with parenchyma and cortical cell enlargement.

One of the most dramatic examples of plant responses to invasion by parasites is the highly specialized cellular adaptations called **giant cells** that are associated with nematode feeding. Nematodes inject substances into the cells around its head, and this initiates a process that creates giant feeding cells. Giant cells are highly specialized adaptations that divert water and nutrients from the growing plant and that function as nutrient sinks from which soluble assimilates are ingested by the worm and replenished by the plant. Giant cells are distinguished by having many (50 to 70) enlarged, highly lobed nuclei and a high DNA content. These cells possess a higher than normal number of organelles and increased cellular metabolism. Giant cells also develop the specialized wall ingrowths typical of transfer cells. It is hypothesized that their high rate of metabolic activity stimulates mobilization of photosynthates from shoots to roots and then to the giant cells, where the solutes are ingested by the feeding nematodes. Giant cells and nematodes are totally dependent on one another. If either is removed, the other dies.

## STRUCTURAL BASES OF RESISTANCE

**Resistance** refers to the ability of a plant to withstand, oppose, lessen, or overcome the attack of fungal or bacterial pathogens and insect pests. **Susceptibility** refers to the inability of a plant to defend itself against an organism or to overcome the effects of invasion by a pathogenic organism. Resistance and susceptibility are relative in the sense that plants may possess varying degrees of resistance and susceptibility to different organisms and at different times of development. Different types of resistance exist in plants, although most species appear to depend on some combination of structural and physiological characters and the interactions between them to prevent successful invasion or attack by parasites or harmful insects. The realization that plant resistance can be influenced by morphology and anatomy implies that resistance can be improved by careful breeding programs. An increase in resistance to one pest, however, may result in an increased susceptibility to another.

Although the effectiveness of structural factors in conferring resistance to plants has been a matter of some debate, numerous studies indicate that resistance in some plants is due at least in part to anatomical factors. In these instances, anatomical characters can obstruct or completely restrict the entry, penetration, feeding, or movement of pests. An outline of the structural features that can potentially increase plant resistance follows:

## I.   REMOTE FACTORS (e.g., COLOR, SHAPE, SIZE)

## II.  CLOSE RANGE OR CONTACT FACTORS

    A.  Thickening of cell walls
    B.  Stem characteristics (e.g., solidness, lignification)
    C.  Foliage morphology

        1.  Foliar pubescence (e.g., trichomes glandular or nonglandular, bristles, spines)
        2.  Surface waxes

    D.  Incrustation of minerals in cuticles
    E.  Anatomical adaptation of organs

        1.  Denticles on the midribs of leaves
        2.  Increases in the number of vascular bundles in the leaf and stem
        3.  Degree of lignification of cell walls
        4.  Relative amount of sclerenchymatous tissue.

Certain potato varieties with resistance to the fungus *Pythium debaryanum* have a higher crude fiber content than more susceptible plants. The higher fiber content is attributed to an increase in secondary wall deposition.

In this case, resistance to infection is apparently due to the resistance of the cell wall to mechanical puncture by the invading fungal hyphae. Some fruits, such as plums, become more susceptible to rot when they begin to ripen. This is believed to be due to the softening of the middle lamellae between the cells, which allows the fungus to force its way through the tissue more rapidly.

## Trichomes

There are a number of examples in which resistance to insect attack (through feeding or oviposition) has been attributed to the effects of various types of foliar trichomes and their abundance. Trichomes can influence the searching behavior of an insect, physically entrap or impale an insect, severely hinder the movement of an adult insect or larva on the plant, or elevate eggs and nymphs from the leaf surface where they are exposed to parasites, predators, or pathogens. One of the most commonly cited examples of the influence of foliar pubescence on entomophagous species is the effect of the hooked trichomes of field beans, *Phaseolus vulgaris,* on the movement of potato leafhopper (*Empoasca fabae*) nymphs. The specialized hairs of field beans have been shown to act as physical barriers that significantly influence the movement of adult leafhoppers. Densely pubescent varieties of soybeans also are known to be highly resistant to the potato leafhopper. Protective secretions produced by the enlarged terminal heads of glandular trichomes can act as a major plant defensive mechanism as well. The glandular hairs of some *Solanum* species, for example, reduce the degree of infestation of young Colorado potato beetle larvae, and the glandular trichomes of tobacco limit the damage of the tobacco hornworm. Products produced by trichomes of the genus *Nama* (Hydrophyllaceae) contain insect antijuvenile and juvenile hormone activity that substantially affects the growth, development, and physiology of insects and represents yet another line of defense against herbivorous pests.

## Laticifers

In the discussion of cell types known as laticifers presented in Chapter 2, we noted that the viscous, milky secretion that exudes from these specialized cells sometimes serves a defensive function upon drying by clogging the mouthparts of insects attempting to feed on any latex-producing plant. Experimental investigations carried out around the turn of the century showed that if latex-forming plants were drained of latex, they could be made palatable to insects. Unaltered plants, on the other hand, did not show insect damage. Latex is specifically known to be a deterrent to feeding by ants, beetles, and caterpillars. Because the laticiferous system of many plants is located in association with the phloem, incisions made across the major leaf veins drain the plant of latex. Interestingly, recent observations have shown that certain insects have naturally evolved the behavior of leaf vein cutting. By cutting the leaf veins at a location proximal to the site of feeding prior to feeding, the latex is drained, rendering the plant defenseless against the pest.

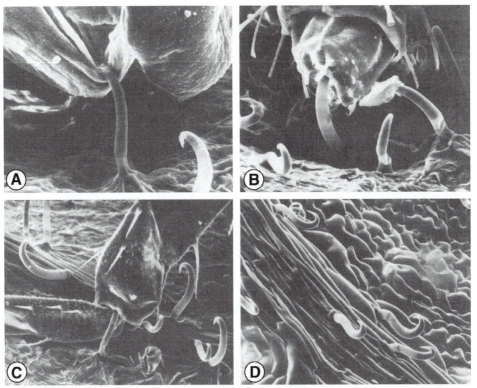

**FIGURE 10.12** Hooked trichomes as a physical plant barrier in *Phaseolus vulgaris* to the potato leafhopper. (A) Trichome inserted in abdomen of a leafhopper nymph. × 700. (B) Trichome embedded in posterior of abdomen. × 700. (C) Trichome embedded in membranous tissue between leg segment. × 350. (D) Procumbent hooked trichomes of the lima bean cultivar Henderson Bush. × 350. Reproduced with permission from Pillemer, E. A. and W. M. Tingey. (1976). *Science* **193**, 482–484. Copyright © 1967, American Association for the Advancement of Science.

## Dutch Elm Disease and Tyloses

Aspects of plant structure are sometimes correlated with the degree of resistance to certain diseases, not only between individuals but also to different times in the growing season. Dutch elm disease provides an example of one of these correlations. Dutch elm disease, which attacks species of the genus *Ulmus,* is one of the most devastating tree diseases in North America and Europe. The disease was introduced into the North American continent and is caused by an ascomycetous fungus, which after introduction into a functional vessel element, grows, moves rapidly, and produces spores (conidia) within the xylem vessels. The fungus is transmitted from tree to tree by one principal vector, the elm bark beetle. The fungus grows within the xylem sap and most fungal spores are transported throughout the tree within the transpiration stream of the wide vessels of the most recently formed earlywood. Elm trees show the greatest susceptibility to disease at the time of formation of the earlywood component of an annual ring. Living cells of the xylem such as parenchyma cells and cambial cells also can be affected structurally by the pathogen.

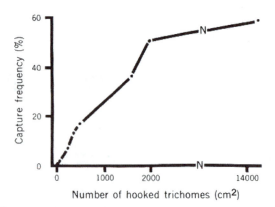

Number of hooked trichomes (cm²)

**FIGURE 10.13**   Relationship between hooked trichome density and capture frequency of leafhoppers on field beans. Reproduced with permission from Pillemer, E. A. and W. M. Tingey. (1976). Hooked trichomes: a physical plant barrier to a major agricultural pest. *Science* 193, 482–484. Copyright © 1967, American Association for the Advancement of Science.

Within the vessels, the fungus causes changes in the pH of the xylem sap and produces metabolites such as cell wall degrading enzymes, growth substances, and toxins. All these products interact to cause wilting and the eventual death of the tree. An infected region of wood is characterized by a dark discoloration of the newly formed growth ring resulting from the oxidation and polymerization of phenolic compounds in the xylem fluid and axial and ray parenchyma. Gums also become deposited in the lumina of the vessels. Because the spread of the fungus throughout the tree is accomplished in large part by the movement of spores in the xylem transpiration stream and the tangential spread of the fungus occurs from vessel to vessel, anatomical differences in the vascular systems of individual trees are potentially of considerable significance in conferring resistance and limiting the spread of the fungus.

A slow lateral and vertical spread of discolored and diseased wood is correlated with a xylem structure having a higher percentage of solitary pores and shorter as well as comparatively narrower vessel elements. Less resistant trees tend to possess vessels that are closely grouped and longer on average than those of more resistant individuals. It is thought that fungal hyphae enter vessel elements through pits on the lateral walls of adjacent vessel members. The more intervessel pits that are present between grouped vessels, the greater and presumably more rapid is the rate of spread of the pathogen. Solitary pores that are narrow and short indicate a high degree of compartmentalization of the xylem, potentially resulting in localization of the colonizing pathogen to a small area of the wood. The decreased susceptibility of trees during the period of latewood formation may be caused by the small diameter of the vessels in this region of the growth ring. Resistance is further affected by the accumulation of gums, gels, or phenols and the formation of abundant tyloses that apparently assist in restricting pathogen movement.

**Tyloses** have been reported to be a possible mechanism of resistance to certain diseases in a number of woody plants. They develop as cytoplasmic outgrowths from an adjacent ray or axial parenchyma cell through a pit in a vessel

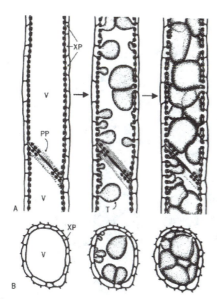

**FIGURE 10.14** Development of tyloses in xylem vessels. Longitudinal (A) and transverse section (B) views of healthy vessels (left) and of vessels with tyloses. Vessels at right are completely clogged with tyloses. Abbreviations: PP, perforation plate; V, xylem vessel element; XP, xylem parenchyma cell; T, tylosis. Reprinted with permission from Agrios (1997), "Plant Pathology," 4th ed., Academic Press.

element wall, partially or completely blocking the vessel lumen. Tyloses form when pressures and other influences between the vessel element and parenchyma cell cause ballooning of the protoplast into the nonliving tracheary element. They are normal in some woods, but they also can develop in response to wounding, invasion by fungus pathogens, or virus infection. Tyloses are generally more abundant in heartwood than sapwood and within the heartwood tyloses originate almost exclusively from the ray cells and not the axial parenchyma. Data suggest that there is a significant difference in tylosis formation between susceptible and resistant clones of elms after inoculation with the pathogen (Fig 10.16). It is possible that the rapid formation of tyloses in advance of the pathogen effectively seals off the vessels and confines the disease. In other species, such as red maple (*Acer rubrum*), the ability to plug vessels rapidly after wounding appears to be a major factor in limiting the spread of infection. Grapevines infected with virus diseases also show a number of anatomical changes in the wood, bark, and leaf mesophyll. For example, changes induced by Pierce's disease include gum formation and an excessive development of tyloses in the vessels, both of which appear to be the result of the virus.

## Heartwood

Among the most dramatic changes that occur during the transformation of sapwood into heartwood is the accumulation of a wide diversity of extractive substances including tannins, essential oils, gums, aromatic polyphenols, and salts of organic acids. These materials become deposited in heartwood cell lumens and cell walls of some species and produce a dark-colored xylem with

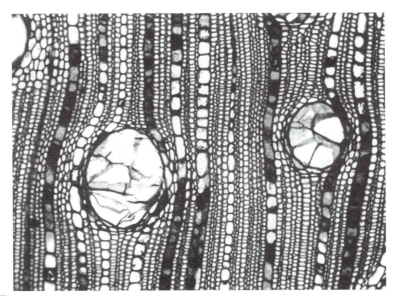

**FIGURE 10.15**   Transverse section of wood of *Anacardium* (Anacaradiaceae) showing tyloses in vessels. Note vasicentric axial parenchyma around vessels. × 180. Courtesy of E. A. Wheeler.

an often distinctive odor, that provides a well-known resistance to decay as well as being a repellent to moths and other insects. On occasion, wound-induced discolorations, often initiated by the death of branches, are superimposed on normal heartwood and are variously referred to as "false heartwood," "wound heartwood," and "redheart." Storage chests, cabinets, and wardrobes constructed of heartwood lumber of eastern United States red cedar ( *Juniperus virginiana*) give particularly good protection for woolen clothes against damage from moths.

## STRUCTURAL ASPECTS OF VIRUS MOVEMENT IN PLANTS

Viruses are obligate parasites; that is, they must obtain their food directly from the tissues of other living organisms. Viruses have a simple structure compared to other organisms, but they are responsible for many destructive plant diseases. Most viruses that cause plant disease are transmitted by insects, principally those that have sucking mouth parts, such as aphids and leafhoppers. The insect penetrates the host with its stylet and seeks a particular tissue. In the course of feeding, the virus is transmitted to the plant. Certain viruses are tissue specific. There exist three classes of virus plant tissue association: (1) viruses of general distribution throughout the fundamental and vascular tissues, (2) viruses restricted to the phloem tissue, and (3) viruses restricted to the xylem tissue. Because the parenchyma cells are the only living elements in the xylem, the virus is thought to multiply in those cells. Most viruses use the phloem mode of transport. The anatomic effects of disease-producing viruses are also either tissue nonspecific or tissue specific.

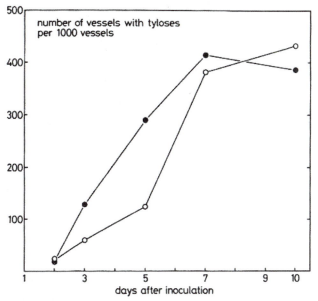

**FIGURE 10.16**  Number of vessels of *Ulmus* showing tyloses per 1000 vessels in transverse section: o, susceptible trees to disease; •, resistant trees to disease. Courtesy of D. M. Elgersma.

Tissue-nonspecific viruses can occur in all tissues containing living parenchyma cells and produce a general necrosis of cells. Tissue-specific viruses, such as those restricted to the phloem, are confined to this tissue and result in phloem cell degeneration and death.

Understanding virus movement through and out of the vascular tissue requires detailed knowledge of the anatomy of phloem and xylem systems, especially the ontogeny and structure of minor veins and surrounding sheath cells. The transport of the virus to the vascular tissue involves cell-to-cell movement into and between mesophyll cells, mesophyll cells to bundle sheath cells, bundle sheath cells to vascular parenchyma and companion cells, and finally entry into sieve elements or tracheary cells. Viral movement within the plant occurs in two ways: cell-to-cell movement through plasmodesmata and long-distance movement in the sieve tubes of the phloem. The interrelated mechanisms of movement may be different in these two pathways. Cell-to-cell movement of viruses takes place through plasmodesmata—the strands of cytoplasm that pass through a pore in the cell wall and join the protoplasts of two adjacent cells.

Plant viruses appear to be capable of modifying plasmodesmatal structure, although the complex basis for these changes is not well understood. Most plant viruses are either globular (with a diameter of 18–80 nm), helical, or filamentous rods (rigid or flexuous with diameters ranging from 10–25 nm and lengths up to 2.5 µm). Because viruses are generally larger than the plasmodesmatal "microchannels," each virus must alter the molecular size exclusion limits of the plasmodesmata in order to pass through them. Some viruses genetically encode for a "movement protein" that is essential for virus plasmodesmatal interactions and the subsequent cell-to-cell spread of the virus.

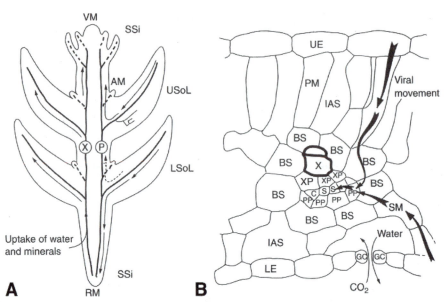

**FIGURE 10.17** Virus entry into the plant, and long-distance transport pathways throughout the whole plant and between cells. (A) The vascular system of the plant. After the virus has gained entry into the sieve element, its final destination will be governed by the specific sink tissues being fed by this particular phloem network. (B) Transverse section of a source leaf, in the region of a minor vein, illustrating the cell-to-cell pathway by which the virus can move all the way from the epidermis into the sieve elements. Abbreviations: VM, vegetative meristem; SSi, strong sink; AM, axillary meristem; USoL, upper source leaves; LSoL, lower source leaves; RM, root meristem; UE, upper epidermis; PM, palisade mesophyll; IAS, intercellular air space; BS, bundle sheath; XP, xylem parenchyma; C, companion cell; PP, phloem parenchyma; SM, spongy mesophyll; GC, guard cell; LE, lower epidermis. Reprinted from *Trends in Plant Science*, Vol. 1, Gilbertson, E. L., and Lucas, W. J. How do viruses traffic on the "vascular highway"?, pp. 260–268, Copyright © with permission of Elsevier Science.

Under the influence of a movement protein, the diameter of the plasmodesmatal canal increases. Furthermore, some plant viruses have the ability to induce infected cells to form new and sometimes transient cytoplasmic tubules across existing cell walls, thereby gaining entrance to neighboring cells.

Velocities and the direction of movement support the view that phloem-limited viruses travel over long distances within the sieve elements, although exactly how viruses enter and exist in mature phloem cells remains largely unanswered. Viruses introduced into the conducting tissue move rapidly within the translocation stream, whereas viruses introduced into epidermal or fundamental cells travel relatively slowly until they reach the vascular tissue. Velocities and direction of movement support the view that phloem-limited viruses travel with the food in the sieve elements. A more complete understanding of the structure and functioning of the plasmodesmata interconnecting phloem cells will be necessary to further clarify virus movement.

## SUMMARY

Plants are continuously subjected to a wide variety of potential disease-causing agents and the stresses caused by mechanical injury. Disease refers to any disturbance in functioning or growth that causes a lower efficiency or breakdown in the plant's metabolism. The reaction of plants to disease is varied. Multiple defenses occur in plants in response to stress and the invasion of pathogens. Some of these are structural and are often accompanied by chemical alterations in cells and tissues. Plant defenses can be preinfectional or postinfectional responses. Defenses can be passive in the sense that they are preformed barriers to infection or active mechanisms of resistance, formed in direct response to injury or the activities of harmful invaders. The cuticle provides a very effective passive defense. The principal active anatomical response of plants to wounding and infection is to wall off the injured or diseased region. The walling off process is termed compartmentalization. Lignification of cell walls and callose deposition is another common response of plant tissues to injury. Protective barriers can also result from meristematic activity at the wound site. In these cases, parenchymatous protective callus is formed.

The complex process of organ shedding involves hormonal and anatomical interactions that form the sequence of events known as abscission. During leaf abscission a narrow, transverse zone of cells form at the base of the petiole known as the abscission zone. This represents a region of structural weakness and is composed of the separation layer and protective layer. Severe wounding of the stem can sever vascular bundles and result in vascular regeneration that reestablishes vascular continuity. Grafting is a process whereby two plants are joined so that they become anatomically and physiologically united. The union of stem graft segments involves formation of parenchymatous wound callus and the activity of a newly formed vascular cambium. The final stage in graft formation is the production of new secondary vascular tissue, thus permitting the interconnection of graft segments.

Resistance refers to the ability of a plant to withstand, oppose, lessen, or overcome the attacks of fungal or bacterial pathogens and insect pests. Susceptibility refers to the inability of a plant to defend itself against an organism or to overcome the effects of invasion by a pathogenic organism. Structural features that can potentially increase plant resistance include thickening and lignification of cell walls and increasing the amount of sclerenchymatous tissue, in addition to foliar pubescence, surface waxes, and mineral accumulation. Laticifers also can serve a defensive function by clogging the mouthparts of insects attempting to feed on a latex-producing plant. Resistance is further effected by the accumulation of gums, gels, or phenols and the formation of abundant tyloses that apparently assist in restricting pathogen movement. In elm trees, tylosis formation is correlated with increased resistance to Dutch elm disease.

Understanding virus movement through and out of the vascular tissue requires detailed knowledge of the anatomy of phloem and xylem systems. The transport of virus to the vascular tissue involves cell-to-cell movement into and between mesophyll cells, mesophyll cells to bundle sheath cells, bundle sheath cells to vascular parenchyma and companion cells, and finally entry into sieve elements or tracheary cells. Viral movement within the plant occurs in two ways: cell-to-cell movement through plasmodesmata and long-distance movement in the sieve tubes of the phloem. Plant viruses are capable of modifying plasmodesmatal structure, although the complex basis for these changes is not well understood.

## ADDITIONAL READING

1. Addicott, F. T. (1982). "Abscission." Univ. of California Press, Berkeley, California.
2. Addicott, F. T. (1991). Abscission: Shedding of parts. *In* "Physiology of Trees" (A. S. Raghavendra, Ed.), pp. 273–300. Wiley, New York.
3. Agarwal, R. A. (1969). Morphological characteristics of sugarcane and insect resistance. *Ent. Exp. Appl.* **12**, 767–776.
4. Agrios, G. N. (1997). "Plant Pathology," 4th ed. Academic Press, San Diego.
5. Binder, B. F. (1995). Trichomes of *Nama* (Hydrophyllaceae) that produce insect-active compounds. *Aliso* **14**, 35–39.
6. Blanchette, R. A., and Biggs, A. R. (Eds). (1992). "Defense Mechanisms of Woody Plants Against Fungi," Springer Series in Wood Science. Springer-Verlag, Berlin, Heidelberg.
7. Borger, G. A. (1973). Development and shedding of bark. *In* "Shedding of Plant Parts" (T. T. Kozlowski, Ed.), pp. 205–236. Academic Press, New York.
8. Cunningham, H. S. (1928). A study of the histologic changes induced in leaves by certain leaf-spotting fungi. *Phytopathology* **18**, 717–751.
9. Dussourd, D. E., and Eisner, T. (1987). Vein-cutting behavior: Insect counterploy to the latex defense of plants. *Science* **237**, 898–901.
10. Ecale, C. L., and Backus, E. A. (1995). Time course of anatomical changes to stem vascular tissues of Alfalfa, *Medicago sativa,* from probing injury by the potato leafhopper, *Empoasca fabae. Can. J. Bot.* **73**, 288–298.
11. Elgersma, D. M. (1982). Resistance mechanisms of elms to Dutch elm disease. *In* "Resistance to Diseases and Pests in Forest Trees" (H. M. Heybroek, B. R. Stephen, and K. von Weissenberg, Eds.), pp. 143–152. Centre for Agri. Publ. and Doc., Wageningen.
12. Esau, K. (1948). Some anatomic aspects of plant virus disease problems. II. *Bot. Rev.* **14**, 413–449.
13. Esau, K. (1961). "Plants, Viruses, and Insects." Harvard University Press, Cambridge, Massachusetts.
14. Fiori, B. J., and Lamb, R. C. (1982). Histological method for determining resistance of *Pyrus ussuriensis* X. *P. communis* hybrids against the pear psylla. *J. Econ. Ent.* **75**, 91–93.
15. Fisher, J. B. (1981). Wound healing by exposed secondary xylem in *Adansonia digitata* (Bombacaceae). *IAWA Bull.* **2**, 193–199.
16. Fosket, D. E., and Roberts, L. W. (1964). Induction of wound-vessel differentiation in isolated *Coleus* stem segments in vitro. *Am. J. Bot.* **51**, 19–25.
17. Francheschi, V. R., Krekling, T., Berryman, A. A., and Christiansen, E. (1998). Specialized phloem prenchyma cells in Norway spruce (Pinaceae) bark are an important site of defense reactions. *Am. J. Bot.* **85**, 601–615.
18. Gawadi, A. G., and Avery, G. S. (1950). Leaf abscission and the so-called "abscission layer." *Am. J. Bot.* **37**, 171–180.
19. Gibson, R. W. (1971). Glandular hairs providing resistance to aphids in certain wild potato species. *Ann. Appl. Biol.* **68**, 113–119.
20. Gibson, R. W. (1974). Aphid-trapping glandular hairs on hybrids of *Solanum tuberosum* and *S. berthaultii. Potato Res.* **17**, 152–154.
21. Gibson, R. W. (1976). Glandular hairs of *Solanum polyadenium* lessen damage by the Colorado potato beetle. *Ann. Appl. Biol.* **82**, 147–150.
22. Gilbertson, E. L., and Lucas, W. J. (1996). How do viruses traffic on the "vascular highway"? *Trends Plant Sci.* **1**, 260–268.
23. Gonzalez-Carranza, Z. H., Lozoya-Gloria, E., and Roberts, J. A. (1998). Recent developments in abscission: Shedding light on the shedding process. *Trends Plant Sci.* **3**, 10–14.
24. Hampson, M. C., and Sinclair, W. A. (1973). Xylem dysfunction in peach caused by *Cytospora leucostoma. Phytopathology* **63**, 676–681.
25. Head, G. C. (1973). Shedding of roots. *In* "Shedding of Plant Parts" (T. T. Kozlowski, Ed.), pp. 237–293. Academic Press, New York.
26. Heybroek, H. M., Stephan, B. R., and von Weissenberg, K. (Eds.) (1982). "Resistance to Diseases and Pests in Forest Trees." Proceedings of the Third International Workshop on the Genetics of Host–Parasite Interactions in Forestry. Centre for Agricultural Publishing and Documentation, Waginingen, the Netherlands.
27. Howe, W. L. (1949). Factors affecting the resistance of certain cucurbits to the squash borer. *J. Econ. Ent.* **42**, 321–326.

28. Ikeda, T., and Kiyohara, T. (1995). Water relations, xylem embolism and histological features of *Pinus thunbergii* inoculated with virulent or avirulent pine wood nematode, *Bursaphelenchus xylophilus* L. *Exp. Bot.* **46**, 441–449.

29. Johnson, B. (1956). The influence on aphids of the glandular hairs on tomato plants. *Plant Path.* **5**, 179–202.

30. Kollmann, R., and Schulz, A. (1993). Phloem regeneration. *In* "Progress in Botany" (H. D. Behnke, U. Lüttge, K. Esser, J. W. Kadereit, and M. Runge, Eds.), Vol. 54, pp. 63–78. Springer-Verlag, Berlin, Heidelberg.

31. Kozlowski, T. T. (1969). Tree physiology and forest pests. *J. For.* **69**, 118–122.

32. Kozlowski, T. T. (1973). Extent and significance of shedding of plant parts. *In* "Shedding of Plant Parts" (T. T. Kozlowski, Ed.), pp. 1–44. Academic Press, New York.

33. Küster, E. (1925). "Pathologische Pflanzenanatomie." Gustav Fischer, Jena.

34. Levin, D. A. (1973). The role of trichomes in plant defense. *Quart. Rev. Biol.* **48**, 3–15.

35. Lipetz, J. (1970). Wound healing in higher plants. *Int. Rev. Cytol.* **27**, 1–28.

36. Mares, M. A., Ojeda, R. A., Borghi, C. E., Giannoni, S. M., Diaz, G. B., and Braun, J. K. (1997). How desert rodents overcome halophytic plant defenses. *Bioscience* **47**, 699–704.

37. Melching, J. B., and Sinclair, W. A. (1975). Hydraulic conductivity of stem internodes related to resistance of American elms to *Ceratocyctis ulmi*. *Phytopathology* **65**, 645–647.

38. Neely, D. E. (1970). Healing of wounds on trees. *J. Am. Soc. Hortic. Sci.* **95**, 536–540.

39. Nelson, R. S., and van Bel, A. J. E. (1998). The mystery of virus trafficking into, through and out of vascular tissue. *In* "Progress in Botany" (H. D. Behnke, K. Esser, J. W. Kadereit, U. Lüttge, and M. Runge, Eds.), Vol. 59, pp. 476–533. Springer-Verlag, Berlin, Heidelberg, New York.

40. Newbanks, D., Bosch, A., and Zimmermann, M. H. (1983). Evidence for xylem dysfunction by embolism in Dutch elm disease. *Phytopathology* **73**, 1060–1063.

41. Noel, A. R. A. (1968). Callus formation and differentiation at an exposed cambial surface. *Ann. Bot.* **32**, 347–359.

42. Norris, D. M., and Kogan, M. (1980). Biochemical and morphological bases of resistance. *In* "Breeding Plants Resistant to Insects" (F. G. Maxwell, and P. R. Jennings, Eds.), pp. 23–62. John Wiley & Sons, New York.

43. Obrycki, J. J. (1986). The influence of foliar pubescence on entomophagous species. *In* "Interactions of Plant Resistance and Parasitoids and Predators of Insects" (D. J. Boethel and R. D. Eikenbarry, Eds.), pp. 61–83. Ellis Horwood Limited, Chichester.

44. Patton, R. F., and Johnson, D. W. (1970). Mode of penetration of needles of eastern white pine by *Cronartium ribicola*. *Phytopathology* **60**, 977–982.

45. Pillemer, E. A., and Tingey, W. M. (1976). Hooked trichomes: A physical plant barrier to a major agricultural pest. *Science* **193**, 482–484.

46. Roth, F. (1895). "Timber: An Elementary Discussion of the Characteristics and Properties of Wood." Bull. No. 10, U. S. Department of Agriculture Division of Forestry, Washington, DC.

47. Sexton, R. (1976). Some ultrastructural observations on the nature of foliar abscission in *Impatiens sultani*. *Planta* (Berl.) **128**, 49–58.

48. Sexton, R. (1994). Abscission. *In* "Handbook of Plant and Crop Physiology" (M. Pessarakli, Eds.), pp. 497–525. Marcel Dekker, New York, Basel, Hong Kong.

49. Sexton, R., and Redshaw, A. J. (1981). The role of cell expansion in the abscission of *Impatiens sultani* leaves. *Ann. Bot.* **48**, 745–756.

50. Sexton, R., Burdon, J. N., Reid, J. S. G., Durbin, M. L., and Lewis, L. N. (1984). Cell wall breakdown and abscission. *In* "Structure, Function, and Biosynthesis of Plant Cell Walls" (W. M. Dugger and S. Bartnicki-Garcia, Eds.), pp. 195–221. American Society of Plant Physiologists, Rockville, Maryland.

51. Shain, L. (1995). Stem defense against pathogens. *In* "Plant Stems: Physiology and Functional Morphology" (B. L. Gartner, Ed), pp. 383–406. Academic Press, San Diego.

52. Shigo, A. L. (1985). Compartmentalization of decay in trees. *Sci. Am.* **252**, 96–103.

53. Sinnott, E. W., and Bloch, R. (1945). The cytoplasmic basis of intercellular patterns in vascular differentiation. *Am. J. Bot.* **32**, 151–156.

54. Soe, K. (1959). Anatomical studies of bark regeneration following scarring. *J. Arn. Arb.* **40**, 260–267.

55. Struckmeyer, B. E., Beckman, C. E., Kuntz, J. E., and Riker, A. J. (1954). Plugging of vessels by tyloses and gums in wilting oaks. *Phytopathology* **44**, 148–153.

56. Wood, R. K. S. (Ed.). (1982). "Active Defense Mechanisms in Plants." Plenum Press, New York, London.

# ▌▌ HERBS, SPICES, AND DRUGS

No discussion of the applied aspects of plant anatomy would be complete without emphasizing the anatomical basis of commercially valuable and therapeutic chemicals that occur in herbs, spices, and drugs. These chemicals are secreted in a wide variety of wild and cultivated plants, although their exact role in the vital activities of the plant is often imperfectly understood. Many plant secretions are believed to serve as animal attractants, as in the case of pleasant or unpleasant floral fragrances that function to attract pollinators. Secretions can also serve as food rewards in the form of nectar, oils, and food bodies. Various chemicals commonly provide the plant with its only means of defense by discouraging the visitation of harmful insects and other herbivores. Human beings have utilized these products for thousands of years as condiments, medicines, fragrances, dyes, and ornaments. Herbs, spices, and medicinal plants have served important traditional and ritualistic functions for many people throughout the world.

The distinction between herbs and spices is not easily made. Generally speaking, an herb is derived from the leaves and flowers of herbaceous plants and tends to be mild in taste. Spices, on the other hand, are usually derived from the stems, seeds, and fruit of woody plants and are strong tasting. Drugs are chemicals that possess medicinal value, are psychoactive, or are poisonous to some degree. Although modern chemistry and the pharmaceutical industry have produced a wide array of beneficial synthetic drugs, those of plant origin have been known and used by humans for millennia. Even in our era of synthetic products, botanically based drugs continue to be sought after and refined by modern science.

The discipline that deals with the biological, biochemical, and economic aspects of naturally occurring drugs and their constituents is called **pharma-cognosy.** Pharmacognosy involves the collection, selection, identification, evaluation, preservation, and commerce of drugs, including those of vegetable origin. Useful drugs of plant origin can be derived or extracted from tissues of the root, rhizome or stem, bark, leaves, flowers, fruits, and seeds. During the 19th century and the early part of the 20th century, advances in pharma-cognosy depended upon the correct identification of drug plants and exact descriptions of drug plant structure. This need was filled by the publication of detailed histological descriptions of drug-yielding plants. Even today, during the processing of drug-producing plant parts, raw materials continue to be routinely subjected to microscopic observation. This is especially important when dealing with powdered and sectioned plant tissues. In addition to macroscopic examination and the use of various chemical tests, microscopic evaluation is essential for the correct identification of powdered plant mater-ial and the detection and quantitative analysis of adulterants. The contamina-tion of drugs by the addition of foreign organic matter not only results in a product of inferior quality but also can lead to an increased rate of spoilage and deterioration. Illegal adulteration also can occur by the complete substi-tution of one plant tissue for another. To maintain official standards for drug purity, the analyst must be an experienced microscopist and possess a good knowledge of the anatomy of drug plants. Until modern times, plant anatomy was widely recognized as a major course of study in the curricula of pharma-ceutical institutions throughout the world. When a plant part is powdered, few if any morphological characters are available for use in drug identifica-tion. During this process, cells and tissues become variously disassociated and broken into a **maceration.** As a result, the preparation must be characterized using the evidence provided by individual cells, that is, tracheary and fibrous elements, sclereids, leaf epidermal cells, trichomes, crystal or silica bodies, and starch grains. By law, each commercially important drug of vegetable origin has an official definition that includes, in some instances, a description con-sisting of the details of the internal structure of the plant part constituting the source of the drug, as microscopically viewed in both sectioned and powdered forms. We will examine some of the interrelationships between anatomy and the production and evaluation of plant secretory products, but to appreciate this relationship, it is necessary to clarify the salient anatomical features of secretory structures.

## CLASSIFICATION OF SECRETORY STRUCTURES

The secretory products referred to as herbs, spices, and drugs are secreted by the plant in a variety of ways and in the form of a host of complex organic substances, such as volatile oils (essential oils), liquid or amorphous deposi-tions called gums, resins, latex, mucilage, mineral salts, and various chemicals such as alkaloids and glycosides. In many cases, various substances occur together and are secreted in specialized structures or cells termed **secretory structures** and **secretory cells.**

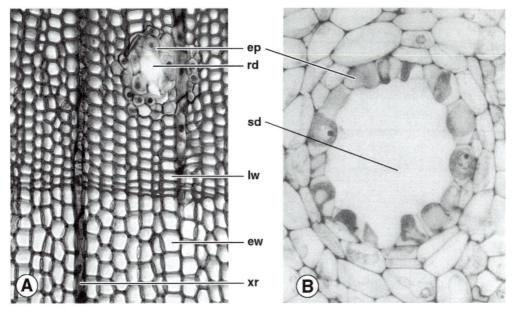

**FIGURE 11.1**   Structure of the resin duct and secretory duct. (A) Transverse section of wood of Scotch pine, *Pinus sylvestris,* in vicinity of a growth ring boundary showing resin duct. (B) Transverse section of secretory duct(sd) in *Rhus toxicodendron* (Anacardiaceae). × 680. Abbreviations: ep, epithelial cell; ew, earlywood tracheid; lw, latewood tracheid; rd, resin duct; xr, xylem ray. B, courtesy of P. G. Mahlberg.

Not only do secretory structures produce a variety of products, but they also occur in many diverse forms and locations in the plant body. Intergradation in structure is common. This diversity is illustrated by the essential oils that are formed in secretory cavities or glandular trichomes and that convey distinct odor and taste to the plant part possessing them. The fruits of orange and lemon trees are well-known examples of tissues containing secretory idioblasts filled with aromatic oils. Many different taxa possess elongated secretory canals, ducts, or cavities that become lined with specialized secreting **epithelial cells.** Most plant anatomists separate secretory structures on a topographic basis, recognizing externally and internally positioned types. Examples of these two categories and the source of a few herbs, spices, and drugs produced by each category are provided later. This account has been adapted from the work of Rudolf Schmid, and in most cases definitions follow the recommendations of Katherine Esau.

## External Secretory Structures

**Trichomes or Papillae.** Trichomes are outgrowths from the epidermis. Trichomes vary in size and complexity and include hairs, scales, and other structures and may be glandular or stinging types. **Glandular trichomes** (or glands) have a unicellular or multicellular head composed of secretory cells, usually borne on a stalk of nonglandular cells. These structures are found on rose petals, saffron (*Crocus sativus*) stigmas and styles, hops (*Humulus lupulus*) female inflorescence bracts, marijuana (*Cannabis sativa*) leaves and flowers, mint and oregano (family Labiatae) leaves, and vanilla (*Vanilla planifolia*) fruits. Insectivorous plants that trap insects on the viscid leaf surface, such as *Pinguicula* (Lentibulariaceae), have glandular trichomes that secrete nectar, mucilages, or digestive juices.

**Colleters.** A colleter is a multicellular appendage or a trichome that produces a sticky secretion. They are found on buds and leaves of many woody species.

**Nectaries.** Nectaries are multicellular glandular structures that secrete a liquid containing organic substances including sugar. They occur in flowers (**floral nectaries**) and on vegetative plant parts (**extrafloral nectaries**).

**Hydathodes.** Hydathodes are structural modifications of vascular and ground tissues, usually in a leaf, that permit the release of water through a pore in the epidermis. Hydathodes may be secretory in function.

**Stigmas.** A stigma is a region of the carpel, in most taxa at the apex of the style, that serves as a surface upon which the pollen germinates.

## Internal Secretory Structures

**Secretory Cells (Oil Cells).** Secretory cells are cells specialized with regard to secretion or excretion of one or more, often organic, substances. These structures are found in bay (*Laurus nobilis*) leaves, cinnamon (*Cinnamomum culilaban*) bark, citronella (*Cymbopogon nardus*) leaves, nutmeg/mace (*Myristica fragrans*) seeds, black pepper (*Piper nigrum*) fruits, pepper (paprika, cayenne, chili) (*Capsicum annuum* var. *annuum*) fruits, star anise (*Illicium verum*) fruits, and wintergreen (*Gaultheria procumbens*) leaves.

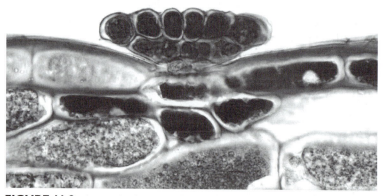

**FIGURE 11.2**   Longitudinal section of a glandular trichome of *Pinguicula* (Lentibulariaceae).

**Secretory Cavities.**  Secretory cavities are glands lined by specialized glandular cells and containing a large secretion-filled space. These structures occur in allspice (*Pimenta dioica*) fruits, citrus (*Citrus* spp.) fruit peels, and clove (*Syzygium aromaticum*) flower buds.

**Secretory Ducts (Canals).**  Secretory ducts are a category of secretory structures that includes resin ducts and gum ducts. Commonly this term refers to elongated ducts containing a secretion derived from the epithelial cells lining the duct. These structures are found in chamomile (*Anthemis* spp.) flowers, myrrh (*Commiphora abyssinica*) stems, tarragon (*Artemisia dracunculus*) leaves, parsley (*Petroselinum crispum*) leaves, and fennel (*Foeniculum vulgare*) fruits.

**Laticifers.**  The term *laticifer* refers to a cell or cell series containing a characteristic fluid called latex as found in opium (*Papaver somniferum*) fruit.

## ANATOMY OF SECRETORY STRUCTURES

The production of oils and resins in plants is generally associated with either the occurrence of external secretory structures known as glandular trichomes or internal oil and resin cells and ducts. **Glandular trichomes** vary widely in structure with species, and more than one structural type can be present on a given plant. Both quantitative and qualitative differences in oil content between different trichome types of the same plant can occur. Included among the structural variation in glandular trichomes are capitate stalked or stalkless forms. Stalked glands possess a basal cell (or cells), a unicellular or multicellular stalk, and a unicellular or multicellular spherical secretory head or disc. Sessile glands have only a unicellular or multicellular head that accumulates the secretory product. A variety of trichome types occur on flowers of *Cannabis*. In addition to various nonglandular hair types, three types of glandular hairs are found on *Cannabis*. These have been described as bulbous, capitate sessile, and capitate stalked and are believed to be the primary site of synthesis of the important narcotic drug tetrahydrocannabinol (THC). The capitate sessile forms are attached directly to the organ surface whereas the bulbous glands and capitate stalked types have short or tall stalks. The drug occurs in the resinous substance secreted by the plant. The most abundant resin production occurs in flowering carpellate plants growing in hot, dry climates. *Marijuana* is a mixture of the resin plus assorted tissue fragments from the carpellate inflorescence. *Hashish* is the term that refers to the pure resin. It is of interest that another species from the same family, *Humulus lupulus* (hops), secretes a resinous substance whose components contribute to the making of beer. Large cup-shaped glands are formed on bracteoles, perianth parts, and stipular bracts and fill with a resinous secretion known as lupulin, which contributes taste and aroma to beer.

The ontogeny of oil- or resin-producing glandular trichomes commonly begins with the enlargement and periclinal or anticlinal division of a single protodermal or epidermal initial. Later divisions form the stalk cells and basal

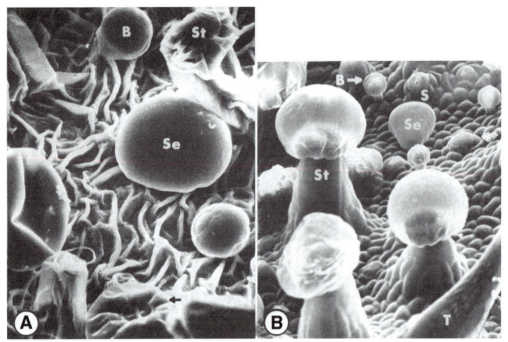

**FIGURE 11.3**   Glandular hairs on bracts of pistillate plants of *Cannabis* (Cannabaceae). (**A**) Portion of old bract with full-sized bulbous (B), capitate sessile (Se), and capitate stalked (St) glands. Note complete collapse of head of stalk gland in upper right and abscised head from gland stalk in lower center (arrow). ×700. (**B**) Portion of old bract bearing full-sized bulbous (B), capitate sessile (Se), capitate stalked (St), and nonglandular trichome (T). Note several stalked stomates. ×390. Reprinted with permission from Hammond and Mahlberg (1973), *Am. J. Bot.* **60,** 524–528.

cells in the epidermis. In some instances more than one initial is present, as in *Cannabis*. The early stage of ontogeny is followed by divisions of the stalk cell in either the periclinal or anticlinal planes to form additional stalk cells and the secretory cells, also known as disc cells. At maturity the unicellular or multicellular head contains an enlarged subcuticular space that is initiated by periclinal splitting of the outer wall of the disc cells. The secretory cavity becomes filled with essential oil or resin as secretion proceeds. This extracellular space begins to form in juvenile glands following the separation of the cuticle from the outermost walls of the head cells. The outer zone of the outer wall of the disc cells forms a subcuticular wall under the cuticle. The reservoir enlarges as the secretory substance is secreted into this cavity from the disc cells. Different cytoplasmic organelles are involved in the secretion of products in glands of various taxa. In some plants the glandular trichome secretory cells contain numerous Golgi apparati or endoplasmic reticulum that are reported to be active in secretion. The disc cells of *Cannabis* contain specialized plastids that synthesize secretions. The secretion products pass through the plasma membrane and the cell walls of the disc cells as they move into the secretory cavity. When the cuticle is ruptured, the secretion product is released.

Some plants, such as the familiar stinging nettle, have conspicuous **stinging hairs** that produce skin-irritating substances. The stinging hairs of nettle have the general construction of a fine capillary tube with an enlarged blad-

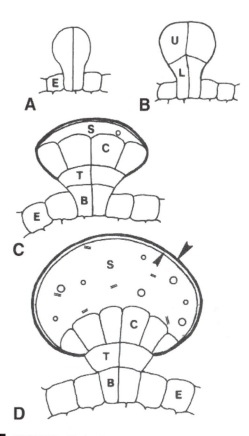

**FIGURE 11.4** Diagrammatic representation of glandular trichome development in *Cannabis*. (A) Enlarged glandular trichome initial in epidermis following anticlinal cell division; (B) Periclinal division separates an upper secretory cell tier from a lower tier. (C) Anticlinal divisions in upper tier form a tier of secretory disc cells, the outer wall of which splits tangentially to form the secretory cavity. A periclinal division in the lower tier forms the stipe and basal cells of gland embedded in epidermis. (D) Mature gland consists of an enlarged secretory cavity covered with a sheath consisting of cuticle and subcuticular wall. Secretions and other components are present in the secretory cavity. Abbreviations: B, basal cell; C, secretory disc cell; E, epidermal cell; L, lower tier cell; O, secretory vesicle; S, secretory cavity; T, stipe cell; U, upper secretory tier cell. Reprinted with permission of Kim and Mahlberg (1997a). *J. Plant Biol.* **40,** 61–66.

derlike base that is surrounded by raised epidermal cells. The tapering upper part of the hair is closed in the form of a small spherical tip. The apical bulb breaks off along a predetermined line when the hair makes contact with the skin, allowing the exposed needlelike point to penetrate the skin. As the hair bends under the pressure of contact, the pharmacologically active chemicals within the enlarged hair base are forced through the needlelike tube and into the wound, causing an itching and burning sensation. The active substances have yet to be identified with certainty.

## SECRETORY CELLS AND CAVITIES

Secretory cells constitute one type of internal secretory structure and are usually associated with wood rays or parenchyma or are isolated in the fibrous tissue.

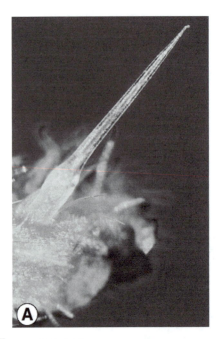
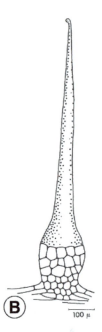

**FIGURE 11.5**   Stinging trichomes of nettle (*Urtica*, Urticaceae). (A) Stinging hair of *U. dioica*. (B) Multicellular stinging hair of *U. urens* showing bulbous base, shaft, and distal swollen tip.

The term *secretory cell* is applied to secretory idioblasts that show a considerable diversity of structure and that contain a variety of different oils, mucilages, and other substances. Secretory cells are almost always larger than surrounding elements and sometimes they resemble enlarged, densely staining parenchyma cells. These specialized cells can occur throughout the plant body and include the category called oil cells. **Oil cells**  can vary in content and are often visible as transparent dots in the leaf. Secretory oil cells are especially common in members of primitive, woody Magnolialian families, where they show a diversity in the type of oil and in ontogeny. Oil cells are generally spherical and appear to intergrade with mucilage cells with which they are morphologically very similar. A summary of the major structural and chemical differences between oil cells and mucilage cells as presently understood is provided later. It is to be especially noted that mature oil cells possess three-layered, suberized walls. Intermediate conditions exist and are accounted for in Table 11.1

Mucilage cells of some taxa contain crystals, especially bundles of needle-like crystals known as raphides. The common ornamental houseplant *Dieffenbachia* (dumbcane) and other members of the predominantly tropical arum family (Araceae) are pertinent in this respect because painful and prolonged paralysis of the mouth occurs if the plant parts are chewed, as might be done by a child. This severe reaction is caused by the combined action of chemicals with the ejection of numerous crystal raphides that become embedded in tissues of the mouth and throat.

Internal secretory spaces containing oils or resins are usually described as cavities or canals of indeterminate length. These structures arise from groups or

**TABLE 11.1  The Main Differences Between Oil Cells and Mucilage Cells**

| | OIL CELLS | MUCILAGE CELLS |
|---|---|---|
| Cell wall | Typically thee layered with a suberized layer sandwiched between cellulosic layers; often with a cellulosic cupule or wall protuberance (suberized layer and cupule apparently absent from a number of taxa) | Cellulosic; suberized layer and cupule typically absent |
| Site of oil or mucilage accumulation | Between cupule wall and plasmamembrane; this is preceded by accumulation of small droplets within cytoplasm; in a number of cases, the oil drop is formed entirely within the cytoplasm | Between cell wall and plasmamembrane (rarely in the vacuoles) |
| Main organelles involved in secretion | Plastids (sometimes endoplasmic reticulum) | Golgi vesicles (rarely plastids and plasmamembrane or endoplasmic secretion |
| Light microscopic appearance of secretion product | Homogeneous, "shiny" | Layered in the dehydrated state |
| Chemistry of secretion product | Terpenes, fats, flavonoid aglycones | Polysaccharides |

Reprinted with permission from Baas and Gregory (1985), *Isr. J. Bot.* **34**, 167–186.

rows of cells, which are similar to mucilage cells, in their ontogeny and structure. *Syzygium aromaticum* (Myrtaceae), the clove of commerce, is reported to be commercially the most important flower bud in the world. The aromatic, volatile oil of this plant is found within secretory cavities distributed throughout the flower and fruit. The comparisons between oil cells, cavities, and elongate ducts are not so easily made. Cavities and canals (or ducts) are intercellular spaces that become filled with secretory products. Cavities and ducts have been imprecisely defined and the variations and intergradations in form are not well known. It has long been thought that these categories of internal secretory structures arise in one of three ways. **Lysigenous cavities,** as often described in *Citrus* fruits, have been reported to originate by the disintegration and dissolution of glandular cells, with the secretion product contained within the enlarging cavity or filling the cavity upon cell breakdown. **Schizogenous secretory cavities,** in contrast, arise by the separation of cell walls of adjacent glandular cells along the middle lamella so as to form a cavity or canal in the center that is lined with a uniseriate or biseriate secreting epithelium. The oil cavities in tissues of *Eucalyptus* are of the schizogenous type. Schizogenously formed resin canals that ramify throughout the vegetative and reproductive tissues represent an outstanding anatomical feature of the Anacardiaceae and are the source of the very toxic skin irritant secreted by poison ivy and poison sumac. Schizogenous cavities and ducts are also common in Asteraceae and yield the aromatic constituent of tarragon (*Artemisia*). **Schizolysigenous secretory cavities** undergo an initial schizogenous type of development that is followed by a lysigenous breakdown of glandular epithelial cells.

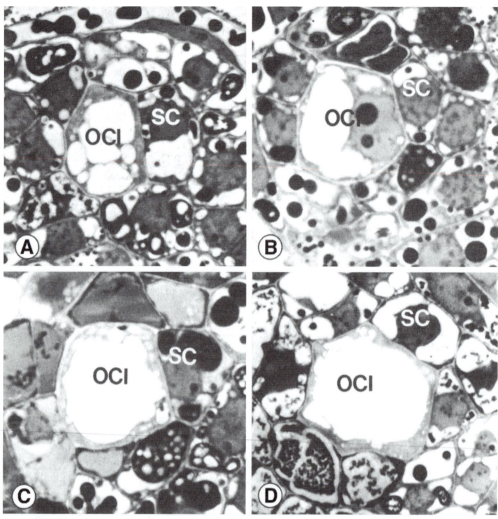

**FIGURE 11.6**   Oil cell development in the leaf of *Magnolia grandiflora* (Magnoliaceae). (A) The oil cell initial (OCI) and its sister cell (SC) at a young stage of development. (B) Expanded oil cell initial. (C) Continued oil cell expansion and vacuole coalescence. Cytoplasm is appressed against the wall. (D) Enlarged oil cell initial. All × 1387. Reprinted with permission from Postek and Tucker (1983), *Bot. Gaz.* **144,** 501–512. Published by the University of Chicago Press.

The oil-filled secretory cavities in the leaves and fruits of *Citrus*, such as lemon and orange, have traditionally been described as lysigenous in origin. Although there has been a long-held view that some secretory cavities arise in the lysigenous manner, the lysigenous gland concept has undergone a recent reevaluation. It has been demonstrated that the lysigenous appearance of *Citrus* secretory cavities may be the result of artifacts produced by the use of certain solutions in the fixation stage of tissue preparation. By eliminating hypotonic fixative solutions, the epithelial cells appear living and functional at gland maturity. As a result of these findings, the lysigenous category of secretory cavity development may not be a valid developmental concept.

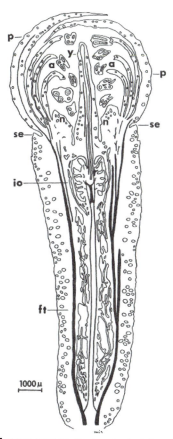

**FIGURE 11.7**  Longitudinal view of the flower of *Syzygium aromaticum* (Myrtaceae), the spice clove. The many circles in all the floral parts represent the secretory cavities containing the essential oil for which clove is prized. The heavy black lines represent vascular tissue. Abbreviations: a, androecium; ft, floral tube; n, nectary; io, inferior ovary; p, petal; se, sepal; y, style. Reprinted with permission from Schmid (1972), *Bot. Jahrb. Syst.* **92,** 433–489.

Members of the carrot family (Apiaceae or Umbelliferae) characteristically produce oils that can be extracted and utilized as flavorings. Depending upon the species, the leaves, roots, or fruits are used in various forms (fresh, dry, cooked). Umbell plants contain volatile compounds that are associated with prominent secretory canals or ducts ("oil vessels") that can be observed in fruit transection.

The drug opium is derived from the milky exudate, or latex, obtained from the capsular fruit of *Papaver somniferum* (poppy family Papaveraceae). **Latex** is produced by an internal system of branching and anastomosing **laticiferous tubes.** Latex is a heterogeneous secretory product containing alkaloids that is present in nearly all parts of the poppy plant; however, the immature capsular wall produces latex in abundance. Incisions made into the unripe fruit wall severs the latex tubes and the white latex exudes and hardens on the outer surface of the fruit. The air-dried latex is scraped off and processed. Laticifers in poppy are of the articulated type, meaning that the system forms from the union of individual cells by the perforation and fusion of cell end walls (see Chapter 2). In this way, rows of thin-walled latex tubes become interconnected by bridges that differentiate from parenchyma cells, creating an extensive network throughout the plant.

## ANATOMICAL AND PHARMACOGNOSTIC STUDIES ON HERB, SPICE, AND DRUG PLANTS

The microscopic evaluation of spices and drugs deals almost exclusively with the appearance of plant tissues in sectional view and powdered form. The histological characterization of these plants refers to the minute structure and arrangement of tissues and is necessary so that authentic tissue samples can be easily and properly identified through their microscopic characters.

### Cinnamon

True spice cinnamon is prepared from the bark of a tree (usually *Cinnamomum zeylanicum*) that is cultivated in India and Sri Lanka. Sticks of cinnamon are produced by gathering bark from fast growing coppiced young trees and cutting the strips transversely and longitudinally. Layers of the inner bark are dried and rolled into a quill. The bark has secretory oil cells that produce the volatile essence well known for its wide use as a spice and flavoring. Cinnamon bark is irregularly fissured. The middle zone contains groups of sclereids, and phloem fibers are abundant. The oil containing secretory cells are scattered with cortical and secondary phloem parenchyma. The powdered commercial cinnamon sold in grocery stores is almost always a blend of the barks of different species of cinnamon. The mixing of barks can either improve the aromatic quality of the product or reduce its quality. To assist in the identification of crude samples, keys to the identification of barks and their powders from anatomical characteristics of species of *Cinnamomum* are available.

### Ginger

Ginger is the washed and dried, decorticated rhizome of the monocotyledonous plant *Zingiber officinale,* a member of the family Zingiberaceae. Ginger is extensively cultivated throughout the tropics. The rhizome contains scattered vascular bundles, starch-filled ground parenchyma, and special secretive oil and resin cells. The yellowish oil cells contain a very aromatic, volatile oil that is referred to as a flavor. Other members of the Zingiberaceae also produce valuable secretion products such as the oils extracted from cardamon seeds. Prior to the establishment of strict drug standards, powdered ginger was commonly the subject of widespread adulteration. Included among the adulterants were starches, cereals, sawdust, pepper to increase pungency, and a variety of dangerous additives.

### Peppermint

Peppermint (*Mentha piperita*) is a perennial herb that grows to approximately 3 feet in height. Like most members of the mint family Labiatae (or

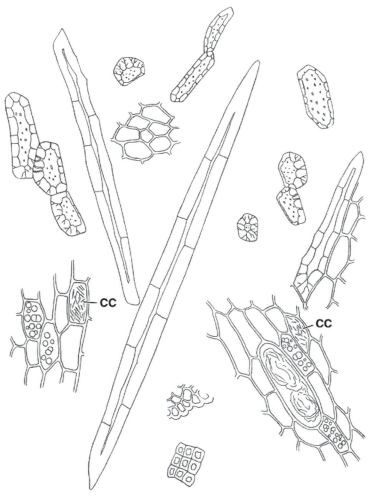

**FIGURE 11.8** Powdered cinnamon showing dispersed sclerenchymatous elements in the form of fibers and sclereids, parenchymatous cells and crystal-bearing (cc) and starch-containing cells.

Lamiaceae), it has a pungent mint smell. The herb is a source of the volatile oil of peppermint that is widely used in cooking and as a commercial flavoring for jelly, candies, medicines, toothpastes, and mouthwashes. The oil is derived from dried mature leaves and plant tops that are subjected to steam distillation. A native of Europe, cultivated plants are grown in the United States. Histological sections through the leaf lamina show an upper epidermis composed of large, clear epidermal cells and possess few or no stomata. The mesophyll consists of a single layer of columnar palisade cells and about five layers of irregularly shaped sponge cells. The lower epidermis is composed of small epidermal cells and numerous stomata. Both glandular and nonglandular trichomes are present. The nonglandular hairs are uniseriate, papillose, and one to eight celled. The glandular hairs have a one- or two-celled stalk and a one- to eight-celled secretory head. These hairs secrete the aromatic oil. In powdered form, fragments of epidermis are present with thin, wavy anticlinal walls. Both trichome types are evident as are fragments of leaf mesophyll and leaf veins. It is not uncommon for other species of mint to be used as adulterants.

## Cinchona

Not all herbs, spices, and drugs are secreted by distinct secretory structures; some are formed diffusely in the tissue systems of the body. For instance, coffee is derived from fruits of *Coffea arabica,* chocolate from fruits of *Theobroma cacao,* and licorice from tissues of roots and rhizomes of *Glycyrrhiza glabra. Cinchona,* the source of quinine that has been used to treat malaria, also falls into this category. The important alkaloid quinine is derived from *Cinchona* bark where it is formed primarily in the parenchyma cells of the middle bark zone. There are significant chemical and anatomical differences among the barks of several *Cinchona* species, and powdered *Cinchona* is routinely analyzed histologically as a method of quality control.

## Cascara sagrada

Cascara sagrada (*Rhamnus purshiana*) is a native American woody plant that was originally used by native Americans and early settlers of the Pacific northwest for its medicinal powers. The drug of commerce is obtained from tissues of the dried bark.

## Nirbisi

The botanical origin of all drug and medicinal plants used by different peoples throughout the world is not completely known. In such cases anatomical study can assist with a more exact knowledge of plant identities. Aconites (aconiti tuber) are a traditional source of crude drug in Nepal, and are used as an antidote to poison. The crude drug is obtained from tuberous roots and is referred to as Nirbisi or Nirmasi. The drug was long believed to be from the roots of *Delphinium denudatum,* a member of the Ranunculaceae. Recent histological examination of tuberous roots from Nepalese markets, however, has shown Nirbisi to be derived not from *Delphinium* but derived from roots of *Aconitum orochryseum,* also a member of the Ranunculaceae. Unlike *Delphinium,* the roots of *Aconitum* contain a continuous cambium that has a stellate appearance in transectional view. Also, the primary root possesses a few sclereids in the cortex, whereas the secondary root shows a total absence of sclereids.

## SUMMARY

Many plants produce secretions that are believed to serve as animal attractants or food rewards in the form of nectar, oils, and food bodies. Some chemicals provide plants with their only means of defense by discouraging the visitation of harmful insects and herbivores. Human beings use these products as condiments, medicines, fragrances, dyes, and ornaments. An herb is derived from the leaves and flowers of herbaceous plants and tends to be mild in taste. Spices are usually derived from stems, seeds, and fruits of woody plants and are strong tasting. Drugs are chemicals that possess medicinal value, are psychoactive, or are poisonous to some degree. Pharmacognosy is the discipline that deals with the biological, biochemical, and economic aspects of naturally occurring drugs and their constituents. Secretory products are composed of a variety of organic molecules and are secreted by plants in specialized secretory structures and secretory cells. Different types of secretory structures often produce chemically identical products. Many different plants possess elongated secretory canals, ducts, or cavities that become lined with specialized secreting epithelial cells. External secretory structures include glandular trichomes, colleters, glands, nectaries, and hydathodes. Glandular trichomes vary widely in structure but typically possess a basal cell, stalk, and unicellular or multicellular secretory head. At maturity, the head contains an enlarged subcuticular space that becomes filled with essential oil or resin as secretion proceeds. Stinging hairs represent another trichome type and produce skin-irritating substances. Internal secretory structures encompass oil cells, secretory cavities and canals, and laticifers. Schizogenous secretory cavities arise by the separation of cell walls of adjacent glandular cells along the middle lamella so as to form a central cavity or canal that is lined with secretory epithelium.

The phamacognostic study of herb, spice, and drug plants involves the microscopic evaluation of plant tissues in sectional and powdered form. The histological characterization of these plants refers to the minute structure and arrangement of cells and tissues and is necessary so that authentic tissue samples can be easily and properly identified.

## ADDITIONAL READING

1. Ascensao, L., Marques, N., and Pais, M. S. (1995). Glandular trichomes on vegetative and reproductive organs of *Leonotis leonurus* (Lamiaceae). *Ann. Bot.* **75**, 619–626.
2. Baas, P., and Gregory, M. (1985). A survey of oil cells in the dicotyledons with comments on their replacement by and joint occurrence with mucilage cells. *Isr. J. Bot.* **34**, 167–186.
3. Carr, D. J., and Carr, S. G. M. (1970). Oil glands and ducts in *Eucalyptus* l'Hérit. II. Development and structure of oil glands in the embryo. *Aust. J. Bot.* **18**, 191–212.
4. Carr, S. G. M., and Carr, D. J. (1969). Oil glands and ducts in *Eucalyptus* l'Hérit. I. The phloem and the pith. *Aust. J. Bot.* **17**, 471–513.
5. Claus, E. P., and Tyler, V. E. (1965). "Pharmacognosy," 5th ed. Lea and Febiger, Philadelphia.
6. Dayanandan, P., and Kaufman, P. B. (1976). Trichomes of *Cannabis sativa* L. (Cannabaceae). *Am. J. Bot.* **63**, 578–591.
7. Edlin, H. L. (1969). "Plants and Man. The Story of our Basic Food." The Natural History Press, New York, Garden City. (Published for the American Museum of Natural History.)
8. Fahn, A. (1979). "Secretory Tissues in Plants." Academic Press, London, New York.
9. Fahn, A. (1988a). Secretory tissues and factors influencing their development. *Phyton* **28**, 13–26.
10. Fahn, A. (1988b). Secretory tissues in vascular plants. *New Phytol.* **108**, 229–257.
11. Hammond, C. T., and Mahlberg, P. G. (1973). Morphology of glandular hairs of *Cannabis sativa* from scanning electron microscopy. *Am. J. Bot.* **60**, 524–528.
12. Hayward, H. E. (1938). "The Structure of Economic Plants." MacMillan, New York.
13. Jackson, B. P., and Snowdon, D. W. (1968). "Powdered Vegetable Drugs." J. A. Churchill, London.

14. Kim, E.-S., and Mahlberg, P. G. (1995). Glandular cuticle formation in *Cannabis* (Cannabaceae). *Am. J. Bot.* **82**, 1207–1214.

15. Kim, E.-S., and Mahlberg, P. G. (1997a). Cytochemical localization of cellulase in glandular trichomes of *Cannabis* (Cannabaceae). *J. Plant Biol.* (Korea) **40**, 61–66.

16. Kim, E.-S., and Mahlberg, P. G. (1997b). Immunochemical localization of tetrahydro-cannabinol (THC) in cryofixed glandular trichomes of *Cannabis* (Cannabaceae). *Am. J. Bot.* **84**, 336–342.

17. Kim, E.-S., and Mahlberg, P. G. (1997c). Plastid development in disc cells of glandular trichomes of *Cannabis* (Cannabaceae). *Mol. Cells* **7**, 352–359.

18. Mansfield, W. (1916). "Histology of Medicinal Plants." John Wiley & Sons, New York.

19. Morrison, J. C., and Polito, V. S. (1985). Gum duct development in almond fruit, *Prunus dulcis* (Mill.) D. A. Webb. *Bot. Gaz.* **146**, 15–25.

20. Parry, J. W. (1962). "Spices. Their Morphology, Histology and Chemistry." Chemical Publishing Company, New York.

21. Postek, M., and Tucker, S. C. (1983). Ontogeny and ultrastructure of secretory oil cells in *Magnolia grandiflora* L. *Bot. Gaz.* **144**, 501–512.

22. Schmid, R. (1972). Floral anatomy of Myrtaceae. I. *Syzygium. Bot.-Jahrb. Syst.* **92**, 433–489.

23. Schmid, R. (1992). "Diversity of Plants and Fungi." Burgess International, Edina, Minnesota.

24. Scott, D. H. (1886). On the occurrence of articulated laticiferous vessels in *Hevea. J. Linnean Soc. Bot.* **21**, 566–573.

25. Southoron, W. A. (1960). Complex particles in *Hevea* latex. *Nature* **188**, 165–166.

26. Thurston, E. L., and Lersten, N. R. (1969). The morphology and toxicology of plant stinging hairs. *Bot. Rev.* **35**, 393–412.

27. Tucker, S. C. (1976). Intrusive growth of secretory oil cells in *Saururus cernuus. Bot. Gaz.* **137**, 341–347.

28. Turner, G. W., Berry, A. M., and Gifford, E. M. (1998). Schizogenuos secretory cavities of *Citrus limon* (L.) Burm. F. and a reevaluation of the lysigenous concept. *Int. J. Plant Sci.* **159**, 75–88.

29. Vaughan, J. G. (1970). "The Structure and Utilization of Oil Seeds." Chapman and Hall, London.

30. Werker, E., Ravid, U., and Putievsky, E. (1985). Structure of glandular hairs and identification of the main components of their secreted material in some species of the Labiatae. *Isr. J. Bot.* **34**, 31–45.

31. West, W. C. (1969). Ontogeny of oil cells in the woody Ranales. *Bull. Torrey Bot. Club* **96**, 329–344.

32. Winton, A. L., Moeller, J., and Winton, K. B. (1916). "The Microscopy of Vegetable Foods." John Wiley, New York.

33. Youngken, H. W. (1926). "A Textbook of Pharmacognosy," 2nd ed. P. Bakiston's Son and Co., Philadelphia.

# 12

# FIBERS, FIBER PRODUCTS, AND FORAGE FIBER

Many species of plants yield natural fibers that are interwoven for the manufacture of a variety of valuable products including fabrics, cordage, paper, and brushes. Plant fibers also are incorporated into various composite materials in which a matrix phase is reinforced with fibers. **Fibers** lend a particular combination of strength and resilience to materials made with them and the production and improvement of fiber-based products cannot be done intelligently without consideration of total fiber structure through collaboration between anatomists, chemists, and others. To the botanist, fibers are slender, attenuated cells, many times longer than wide, with a ratio of width to length averaging 1:10 or 1:20, although extremes have been reported to reach 1:4000. Individual fiber cells typically have channellike lumens and thick, pitted, strongly lignified secondary walls that are tapered at the ends. At maturity, these cells usually lose their protoplasts and thus seemingly play no active role in plant metabolism, although there are notable exceptions to this generalization. The majority of fibers, therefore, are interpreted to be supporting cells that provide resistance to compression and flexing forces.

In commerce, the definition of a fiber is decidedly different from that of the botanist. Any cell type that is long, durable, has a high tensile strength (resistance to elongation stress), and is easily available is referred to commercially as a fiber. The long seed hairs of cotton (*Gossypium*) are thus regarded as fibers, and in fact constitute the most economically important fiber in the world. Besides their preeminence in the textile industry as the raw material for spinning thread and weaving cloth, cotton "fibers" are used in the manufacture of a number of other products. The long, silky seed hairs of the tropical

Kapok tree (*Ceiba pentandra,* Bombacaceae) provide another well-known example of a commercially important plant product that is not technically a fiber. In addition to individual fibers, aggregations of fibers or **fiber bundles** are used in the manufacture of different products. Fiber bundles are extracted from certain nonwoody plants and can have diameters between 100 and 1000 µm with a very long length. Each category of natural fiber has its own characteristic pliability, fineness, durability, and tensile strength that make it desirable for the creation of a particular product, or worthy of attention for other economic reasons. The determination of exactly what fiber properties are required for different end uses and how these can be improved represents an area of active research. The experimental breeding of fiber-producing plants is an important aspect of this research, carried out with the aim of increasing fiber yield, improving fiber quality, and increasing plant nutritional characteristics.

## EXTRAXYLARY FIBERS

Plant anatomists conveniently divide true fibers into two broad categories: (1) **xylary** (or wood) **fibers** and (2) **extraxylary** (or bast) **fibers.** Xylary fibers are those found in the secondary xylem and fall into one of a number of different types, based primarily upon pitting and wall characteristics. Extraxylary fibers are generally classified on the basis of their site of origin or location within the plant body. It is therefore customary to speak in terms of cortical fibers, pericyclic fibers, phloem fibers, and leaf fibers. The phloem fibers of certain dicotyledons and the leaf fibers of monocotyledons (called soft fibers and hard fibers, respectively) are in the greatest demand commercially. Monocot leaf fibers are generally superior for the production of coarse fabrics like burlap, heavy duty ropes, and rugs. The fibers of monocotyledons often enclose foliar vascular bundles or appear as strands associated with vascular bundles or as independent strands. In many taxa the entire fibrovascular bundle serves as the commercially important unit fiber or fiber bundle. It is extracted from the leaves by a crushing and scrapping process known as **decortication,** which removes the leaf tissue surrounding the fibrous cells. Several species of monocotyledons are cultivated for their strong, economically valuable leaf fibers. Among the more important of these plants are *Musa textilis* (Musaceae), a species closely related to the banana and known as Manila hemp; *Agave sisalana* (Agavaceae), known as sisal hemp; *Furcraea gigantea* (Amaryllidaceae), known as Mauritius hemp; *Phormium tenax* (Liliaceae) or New Zealand hemp; and *Sansevieria* (Liliaceae), known as Bowstring hemp. In *Agave,* massive fibrous strands accompany the vascular bundles. In *Sansevieria,* some fiber strands are associated with the vascular bundles, whereas others are independent.

Fibers positioned in the phloem or cortical regions of the stem of dicotyledons include a number of commercially important fibers that are very strong and flexible. These fibers are used in the manufacture of cloth, fine linens, ropes, and threads. The durable phloem fibers of hemp were once used widely for sail cloth. Among the more important of these plants are *Linum usitatissi-*

*mum* (Linaceae), known as flax; *Cannabis sativa* (Cannabaceae) or hemp; *Corchorus capsularis* (Tiliaceae) or jute; and *Boehmeria nivea* (Urticaceae), known as ramie. The fibers of these plants can be remarkably long and are among the longest cells known in angiosperms. For example, an individual hemp fiber can reach approximately 10 cm in length, and flax fibers can range from 2.5 mm to 1.2 cm. Long fibers have also been isolated from ramie, whereas some fiber bundles of jute stem reach 1.5 to 3.6 m in length. Mature stems of *Linum usitatissimum* contain a multiseriate zone of conspicuous fibers, situated just external to the secondary phloem, that are of interest commercially. Anatomically the thick-walled fibers of flax are of particular interest because the secondary walls are often completely unlignified. Cigarette papers often are made from flax fibers, because their nearly pure cellulosic wall burns evenly without taste or color. In this species the rate of lignification is an indicator of fiber quality. Lignin deposition in and around the fiber bundles decreases fiber quality and delays the harvesting and refining processes. Lignification results in an undesirable increase in the crispness and a decrease in flexibility, softness, and divisibility of the fiber bundles. Delaying the onset of lignification or decreasing the rate of lignification could delay the start of the harvesting process (thus increasing yield) without any decrease in fiber quality, and is thus on many research agendas.

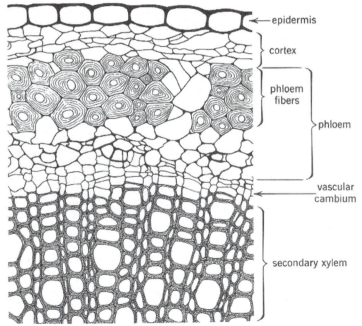

**FIGURE 12.1**   Transverse section of mature stem of flax, *Linum usitatissimum* (Linaceae), showing zone of thick-walled primary phloem fibers. ×320. From Esau (1965), "Plant Anatomy," 2nd ed. Reprinted by permission of John Wiley & Sons.

The stem of *Cannabis sativa* shows the presence of phloem fibers as well as cortical fibers. The phloem fibers that occur singly or in groups are smaller, shorter, and thinner walled than the larger and longer cortical fibrous elements. The location of stem fibers in two zones in the stem enables them to be traditionally harvested by a process known as retting, in which stems are thoroughly wetted and the soft tissues allowed to rot away. The fibers are then beaten or washed free and extracted as bundles or groups of fibers that are joined together. In the case of ramie, dissolution of the intercellular cementing substance that binds the bast fibers together is difficult and the individual fibers are characteristically "gummy" following treatment. If the retting process is not carried out properly, the fiber can be ruined or the quality lowered.

Indigenous peoples sometimes use the inner fibrous bark of some species in the manufacture of various products. Exceptionally strong and durable bast fibers are obtained by stripping, beating, and soaking the inner bark. The fibrous tissue is woven into a variety of items, including ropes, fishing nets, sacks, and clothing. Various peoples around the world also utilize the fibrous inner bark of different tree species to form a clothlike material known as bark cloth, bast cloth, or tapa. The manufacture of bark cloth generally involves stripping bark from the tree and separating the inner and outer tissue zones. The inner fibrous region is soaked and beaten to blend the best fibers together.

## XYLARY FIBERS AND PAPER MANUFACTURE

Wood is the predominant raw material in the manufacture of paper and other wood fiber products such as paperboard, fiber board, and insulation board. The mechanical and chemical conversion of wood into pulp for the manufacture of these products consumes vast quantities of timber. In a strict anatomical sense, xylary fibers are the elongated supporting cells of the ground tissue with thin or thick lignified walls and simple or distinctly bordered pits. In the manufacturing of wood products, however, the term *fiber* is a convenient commercial concept that includes a variety of wood cell types.

Paper is manufactured from the secondary xylem of both softwood (gymnosperm) and hardwood (dicotyledon) species, with each source providing a pulp with distinctive cellular characteristics. Wood pulp is composed of "fibers." The complex combination of characteristics of individual cells and their mode of incorporation into the paper web structure and interaction in the papermaking process determine the quality of the final product. Although paper products are largely composed of fibers, they also contain adhesives and strength additives. Pulp made from softwoods is rather homogeneous and is composed of long tracheids. Pulp derived from hardwood species, on the other hand, is heterogeneous and is composed of wide vessel elements as well as narrow, elongate, thick-walled cells that are true fibers in the strict histological sense. Knowledge of particular cell morphologies makes it possible to predict the resulting pulp characteristics. Softwoods generally produce coarse, strong pulps compared to hardwoods. Hardwood pulps typically result in finer, smoother papers that are used extensively in printing and writing grade stock. Some of the fundamental wood fiber properties that directly affect fiber product characteristics are fiber length, tensile strength, cross-sectional shape, wall thickness, elastic modulus (flexibility), as well as characteristics other than strength properties such as surface chemistry. However, it is often extremely difficult to isolate specific individual factors that control pulp and paper quality and to devise adequate methods for studying them under controlled experimental conditions. Even if one does so, it is difficult to determine their exact effects upon other variables in the complex.

Paper is made from cell-sized particles, continuously formed wet into thin sheets and dried over hot rolls. The initial step in the manufacturing process is the mechanical and chemical maceration of wood, resulting in the partial or total separation of its component cells to form individual fibers, fiber bundles, or fiber parts. The macerated xylem is commercially known as **wood pulp.** The process of chemical pulping removes lignin, and following further treatment that includes bleaching, beating, and refining, the pulp is transformed into paper sheets. There are numerous pulping procedures and countless different kinds and grades of paper and paperboard produced. Because pulp fibers are integrated into a web that forms paper, properties such as fiber type, structure and number, cell wall chemistry, and the manner in which they are joined, blended, and used with additives all have a significant effect on the final paper characteristics.

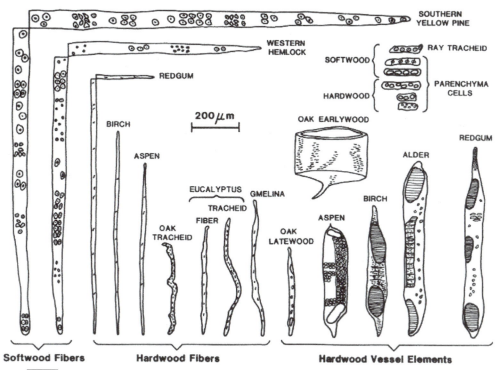

**FIGURE 12.2** Relative size of various softwood and hardwood fibers and hardwood vessel elements. From Heitmann, Jr. (1992). Used with permission of Canadian Pulp and Paper Association.

There are clearly many types and grades of paper and paper products, each one having specific characteristics that satisfy different use requirements. In some papers, for example, strength is the most important feature to be considered. In others, high tear resistance or folding endurance is paramount. Features such as smoothness or a high opacity are required for other products. All these properties depend upon the type of pulp utilized and the processing that it undergoes. Fiber structure, function, and wall chemistry thus have direct application to the production, utilization, and improvement of fiber-derived products. During the mid 1800s, paper manufacturers began replacing cotton and linen fibers with wood fiber. The wood required a chemical process to make the paper opaque and receptive to inks. To accomplish this, however, an acidic additive was built into the paper. Over time, the acid has caused these papers to become brittle and has made the paper documents unusable. Today, many libraries, archives, and record repositories are faced with collections that are deteriorating rapidly. To retard this process the acid in paper can sometimes be neutralized.

Among the important fiber characteristics that determine paper quality are cell length and wall thickness, although these features interact with other parameters. Most papermaking softwood fibers range in length from 3.0 to 5.0 mm. The more diverse hardwood fibers range between 0.7 and 2.5 mm. Fiber strength, length, and bonding within the web all interrelate to determine tear strength. The longer the average fiber length, the higher will be the tear resistance and folding strength of the final product in general. Fiber length has little effect on the mechanical properties of dry formed fiberboard, one category among numerous reconstituted products made from trees, but length does affect dimensional stability in the plane of a board. Long fibers promote board stability during water absorption or desorption. Long fibers also produce strong mats, which are needed for high-speed handling. Thick-walled fibers form paper with a low tensile strength but a high degree of resistance to tear. Such papers, however, tend to have a low folding endurance. Thick-walled fibers have the potential to form fewer interfiber bonds than thin-walled elements and this greatly affects paper characteristics.

Because wood density is directly related to wall thickness, lower density woods generally form a better papermaking raw material. Therefore, when softwoods are used as pulp, paper quality is directly related to the ratio of thick-walled latewood to thin-walled earlywood tracheids making up the pulp. For hardwood pulps the proportion of vessel elements to true fibers making up a wood also can affect pulp quantity and quality. A second influence on pulp properties in addition to cell wall thickness relates to whether the fiber lumen has collapsed during processing to form a ribbonlike fiber, or whether it remains open. Short, flexible, thin-walled fibers with collapsed lumens generally yield high-quality papers because they form dense, smooth, well-bonded webs with good tensile and fold strength.

After a fiber type is selected, pulped, and bleached, fiber structure and other characteristics are often further modified by beating and refining. This processing step is very important in determining the final strength of the product. The polysaccharide and phenolic polymers forming plant cell walls are very reactive, and the interfiber bonding of cellulose molecules makes a significant contribution to the paper's strength. Wall hemicelluloses also con-

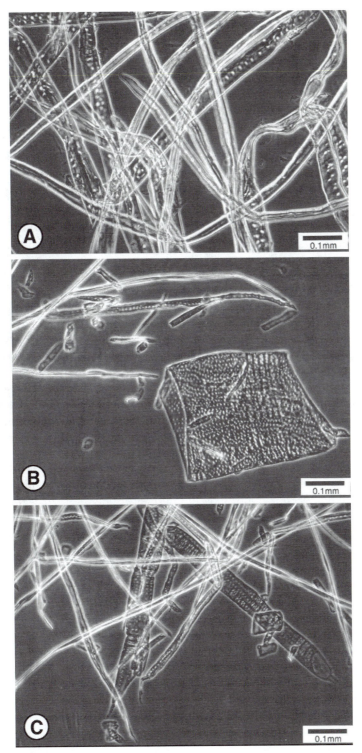

**FIGURE 12.3**   Wood pulp fibers. (A) Unrefined, bleached fibers from southern pine. (B) Unrefined mixed hardwood fibers. Note fiber size relative to softwood fibers and the presence of a large oak vessel element. (C) Mixed hardwood pulp. Courtesy of E. Wheeler and North Carolina State University.

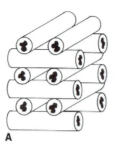

**FIGURE 12.4**   Fiber structure and paper construction. (A) Coarse, thick-walled fibers with uncollapsed lumens from latewood. (B) Thin-walled fibers from earlywood with collapsed lumens. These fibers are more flexible and form denser sheets, with better bonding and overall strength properties but with lower tear strength and opacity. From Heitmann, Jr. (1992). Used with permission of the Canadian Pulp and Paper Association.

tribute in a major way to forming fiber-to-fiber bonds in the papermaking process and thus are important when strong papers are desired. It is important to remove most of the lignin so as to decrease or eliminate fiber stiffness. To increase the potential for interfiber bonds to form, fibers are mechanically flattened and frayed (beaten) in order to damage the cell wall and thus to increase the area where bonding can occur. During the process of refining, fibers are usually cut and shortened, and the cell wall is altered by removal of the primary and $S_1$ layers. Overall, the numerous effects that refining has on fiber properties greatly influence the final properties of the pulp and paper.

The final step in controlling pulp properties involves the blending of different types of pulps and the addition of other substances (additives) to the mix. In some cases, a single-pulp fiber type may be used to form a finished product, but more commonly pulps are blended and additives are used to produce fine grades of paper.

## REACTION WOOD

In response to being naturally or experimentally displaced from a normal growth orientation, the main axis or lateral branches of actively growing stems are stimulated to form wood with an eccentric growth pattern that is composed of cells having atypical structure and chemical composition. This secondary xylem is known as **reaction wood** and it forms unilaterally along one side of the displaced axis in response to gravity. The amount of reaction wood formed by the vascular cambium is generally correlated with the degree of axis lean from the vertical upright position. It has been reported that trees are sometimes stimulated to form reaction wood in main stems deviating as

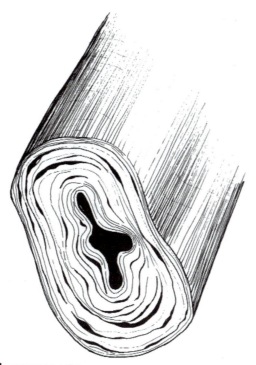

**FIGURE 12.5**   Refined chemical pulp fiber. Transverse section of fiber showing lamellar structure of wall that develops during refining. From Heitmann, Jr. (1992). Used with permission of the Canadian Pulp and Paper Association.

little as two degrees from the vertical. The subsequent longitudinal expansion or contraction of this tissue eventually results in the slow bending and gradual reorientation of the displaced previous year's growth increment back to its original direction. The development of reaction wood in stems and branches appears to represent a hormone (auxin)–mediated response. Reaction wood is not known to form in monocotyledonous stems. The reaction wood in conifers is known as **compression wood,** whereas reaction wood in dicotyledons is called **tension wood.**

## Compression Wood

The reaction wood of gymnosperms forms on the lower or compression side of the woody cylinder and is commonly known as compression wood, otherwise called redwood. It appears as eccentric layers of dense, reddish or yellowish brown wood and in transverse section is often identified by exceptionally wide growth layers that resemble successive layers of locally thickened regions of latewood. The reorientation of the axis is thought to occur by the expansion of this tissue. In response to changes in moisture content, reaction wood shrinks and swells more than normal wood does during the seasoning process. During drying, lumber with reaction wood becomes bowed and warped as a result of abnormal longitudinal shrinkage. A board with compression wood

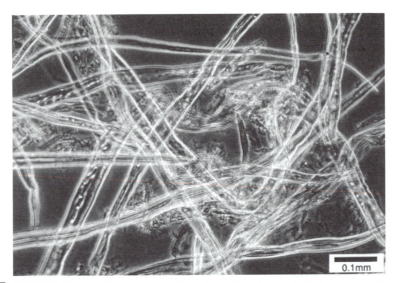

**FIGURE 12.6**   Refined softwood fibers. Note fiber damage, fibrillation, and debris as compared with unrefined fibers in Figure 12.3. Courtesy of North Carolina State University.

can shrink up to 6 or 7% more along the longitudinal axis than normal wood, although it usually has a higher specific gravity than normal wood. The strength of compression wood is inferior to that of normal wood. Lumber with substantial amounts of reaction tissue is of little or no use in the construction industry.

Anatomically, the tracheids of gymnospermous compression wood differ considerably in structure from those of normal wood. The tracheids are typically shorter, thicker walled, heavily lignified, and rounded in outline so that abundant intercellular spaces are present between the tracheids. The cell walls contain visible discontinuites or striations and the cellulosic microfibrils are oriented at a large angle (45°) relative to normal tracheids. It is this abnormal microfibril angle that results in increased longitudinal shrinkage. The lignin composition of compression wood is higher than that of normal wood, whereas its cellulosic content is lower. In commercially valuable softwoods, compression wood is a serious abnormality. Not only is it harder and denser than normal wood, but it also shows undesirable bowing and has poor nail holding capacity. The unusual chemistry of compression wood cells also reduces its desirability as pulp for the manufacture of paper.

### Tension Wood and Gelatinous Fibers

The reaction wood of arborescent dicotyledons forms along the upper (or tension) side of leaning stems and branches and is known as tension wood. A comparison of the macroscopic, microscopic, and ultrastructural features of tension wood and compression wood is presented in Table 12.1.

The tension wood of many, but not all, hardwoods is characterized by the presence of **gelatinous** (or **mucilaginous**) **fibers.** Some studies have reported that tension wood contains fewer and narrower vessels per unit area. Gelatinous

**FIGURE 12.7**  Transverse section of stem of *Pinus resinosa*. Photographed by transmitted light to show the relative opacity of dark compression wood zone compared with light normal wood zone. Reprinted with permission from Côté and Day (1965), *in* "Cellular Ultrastructure of Woody Plants," pp. 391–418, Syracuse University Press.

fibers possess a highly modified inner layer of the secondary wall, the gelatinous or G layer, that contracts excessively during drying. This property is the direct result of an altered cell wall ultrastructure and microfibril orientation. The G layer differentiates as an additional layer internal to the usual secondary wall, or it may replace one or more of the secondary wall layers. Gelatinous fibers have a higher cellulosic content but a lower lignin composition than normal fibers as a result of the chemical nature of their G layer. Unlike compression wood that is generally unusable for the production of quality pulp, tension wood can produce very satisfactory pulp if subjected to special treatment, although paper made from tension wood often has poorer strength properties. Because tension wood's gelatinous fibers contain higher than normal cellulose content and are usually deficient in bonding properties, the resulting papers are inferior in their folding, burst, and tensile strength properties. The reduced strength of paper made from tension wood fibers appears to be associated with

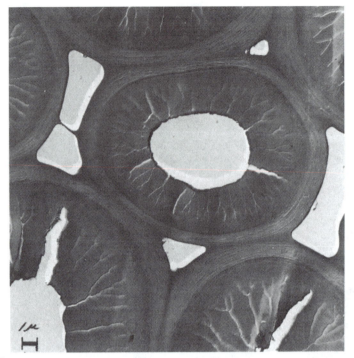

**FIGURE 12.8** Transverse section of compression wood in heartwood of eastern larch. Rounded tracheids, branched "checks," abnormally thick $S_1$ layer, and absence of an $S_3$ layer can be observed in this electron micrograph. Note the parent wall where it traverses the intercellular space at upper left. ×3200. Reprinted with permission from Côté and Day (1965), *in* "Cellular Ultrastructure of Woody Plants," pp. 391–418, Syracuse University Press.

the bulkiness contributed by the G layer (which hinders fiber collapse) and the lower hemicellulose content (which limits interfiber bonding), as well as with the presence of discontinuities (called slip planes) and minute compression failures.

The presence of tension wood in construction timber produces a woolly or fuzzy surface on portions of freshly cut boards. Gelatinous fibers are not easily sawn and extend as cell fragments from the board surface upon sawing and planing. Logs and processed lumber containing tension wood are weaker than normal wood and tend to split at the ends as a result of the release of internal stresses. Because tension wood can be intercalated between normal xylem in upright trunks and is not always easily detected, improved anatomical techniques to better and more easily identify tension wood would be commercially valuable.

## FORAGE "FIBER" AND ANIMAL NUTRITION

All classes of domestic and wild herbivores eat plants in some form, although they differ in the extent of their use of different plant species and plant parts. In whatever form plant parts are consumed, however, they must supply the animals with their nutrient requirements. The energy supply that most herbi-

**TABLE 12.1    Comparison of Tension Wood and Compression Wood Characteristics**

| | TENSION WOOD | COMPRESSION WOOD |
|---|---|---|
| Gross and physical characteristics; mechanical properties | Eccentricity of stem cross section: the "upper side" <br> Dry, dressed lumber: silvery sheen of tension wood zones in many species; darker than normal in certain tropical and Australian species <br> Green-sawn boards are woolly on surface <br> Longitudinal shrinkage may reach 1%+ <br> Particularly high tensile strength in dry tension wood; lower than normal in green condition | Eccentricity of stem cross section: "lower side" <br> Nonlustrous, "dead" appearance "Rotholz" darker than normal <br> Longitudinal shrinkage as high as 6–7% <br> Modulus of elasticity, impact strength, tensile strength: low for its density |
| Anatomical characteristics | Gelatinous (tension wood) fibers present though may be lacking in some species <br> Vessels reduced in size and number in tension wood zones | Rounded tracheids <br> Intercellular spaces <br> Transition pattern, springwood–summerwood, altered: more gradual than in normal wood |
| Microstructure | G layer present; convoluted or not <br> Slip planes and compression failures in tension wood fiber walls <br> G layer in secondary wall of gelatinous fibers in three types of arrangements: <br> $S_1 + S_2 + S_3 + G$ or $S_G$ <br> $S_1 + S_2 + G$ or $S_G$ <br> $S_1 + G$ or $S_G$ | Helical checks or cavities in $S_2$ <br> Slip planes and compression failures generally absent |
| Ultrastructure | Primary wall appears normal <br> $S_1$ may be thinner than normal <br> Microfibrillar orientation of G layer nearly parallels fiber axis; high parallelism within G layer | $S_3$ layer absent <br> $S_1$ may be thicker than normal <br> $S_2$ microfibrillar orientation approaches up to 45° <br> Ribs of cellulose parallel to direction of microfibril orientation; cellulose lamellae parallel with wall surface |
| Chemical composition | Lignification of tension wood fibers variable; unlignified or slightly lignified <br> Abnormally high cellulose content <br> Abnormally low lignin content <br> More galactan than normal <br> Less xylan than normal | "Extra" lignin deposited as a layer between $S_1$ and $S_2$ <br> Abnormally low cellulose content <br> Abnormally high lignin content <br> More galactan than normal <br> Less galactoglucomannan than normal |

Reprinted with permission from Côté and Day. (1965), *in* "Cellular Ultrastructure of Woody Plants," pp. 391-418, Syracuse University Press.

vores require can be obtained from plants, providing that a few conditions are met. These include not only an adequate supply of feed but also reasonable food quality, palatability, and digestibility. Plant cell digestion results from the action of enzymes produced by the animal as well as anaerobic bacteria and fungi found in the animal's gut. The consumption of a large amount of plant

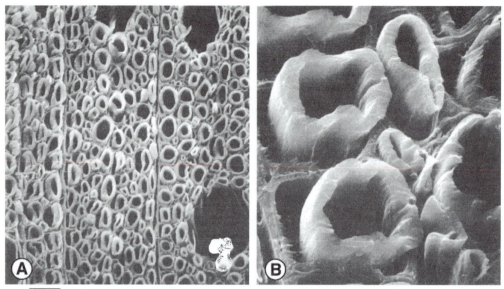

**FIGURE 12.9**  Tension wood with gelatinous fibers. (A) SEM of a cut, transverse surface of tension wood of *Populus tristis*. Note the presence of thick, loosely attached gelatinous layers in nearly every fiber. × 360. (B) Several gelatinous fibers of *Populus* showing the loose attachment of the G layer to the $S_2$. The thick G layer contributes to the bulk of the fiber. × 3000. Used with permission of Isebrands and Parham (1974), International Association of Wood Anatomists.

material of low digestibility (thistle) may not support an animal (horse) to the same extent as a much lower amount of more digestible vegetable matter (alfalfa). Ranchers and farmers, therefore, are very concerned with the quality, nutritional value, and digestibility of animal feeds, and thus have an interest in the improvement of commercial forage or "fiber" plants.

Studies of forage composition and feeding values suggest that increasing the ratio of leaves to stems is associated with increased digestibility. Increasing dry matter digestibility increases the amount of energy that is usable for animal production. As a rule, leafy feeds (alfalfa hay) are much more valuable for animal consumption than those with a high proportion of stems (timothy). For this reason, leaves are selectively grazed by animals. Typical figures on the digestible nutrients of alfalfa leaves (dry matter basis) may reach or surpass 65%, whereas alfalfa stems may fall to 46% or less. Plant structure is among the most important influences on the digestibility of forage tissues by rumen microorganisms and digestive enzymes. This has been demonstrated by numerous studies using both light and electron microscopy and by *in vitro* and *in vivo* experimentation. Variations in the site and type of lignification and the percentage of specific tissue types in different areas of the stem and leaf blade are especially significant in tissue degradation. The digestibility of some plant tissues also is related to structural changes that accompany plant aging or that are associated with environmental variables. The initial high digestibility of immature food samples generally declines with increased maturity, primarily because of increased cell lignification and an increased percentage of stem fraction in the samples. As digestion begins, leaf mesophyll and phloem are typically the first tissues to degrade, followed by epidermal cells and vascular bundle sheath cells. Sclerenchmya cell walls become partially degraded, whereas lignified vascular cells are essentially resistant to digestion. The efficient use of many forages as feeds for ruminants is limited by the low digestibility of the structural carbohydrates composing plant cell walls. The major detriment to digestibility is an increase in lignin accumulation in the vascular bundles of leaves and stems and in the interfascicular parenchyma of the stem. The concentrations and digestion characteristics of structural carbohydrates, as a result, influence the rate and extent of cell wall digestion. Because a decrease in the number of fibrous cells is commonly correlated with increased digestibility, livestock silage may be improved through genetic selection for decreased fiber concentration.

## Grass Leaf Digestion

It has been shown that total digestion is lower for grasses adapted to tropical regions than for species from temperate areas. This difference has been found to be partly related to the development of the stem and extreme lignification and partly related to marked differences in leaf anatomy, including the relative proportions of different tissue types, such as mesophyll, vascular bundles, and sheath cells. This difference in anatomy leads to a higher proportion of the less digestible types of tissue in tropical grasses. High temperatures increase the proportion of cell wall and decrease its digestibility in both leaf and stem. The leaf midrib in maize is poorly digested. Tropical grasses possess less of the easily digested mesophyll tissue and more of the hard to digest epidermis, vascular tissue, and sclerenchyma. Tropical grasses also have more vascular bundles in their stems and leaves than temperate grasses.

The grass genus *Panicum* is unusual because it contains species with leaf blades of markedly diverse anatomical structure associated with variations in the photosynthetic pathway (i.e., $C_3$, $C_4$, and intermediate $C_3/C_4$ types). Anatomical comparisons of the rate and degree of degradation of specific leaf tissues within different photosynthetic types and individual species of *Panicum* have shown that the digestibility of leaf tissue is high in $C_3$ species, but lower in $C_4$ and $C_3/C_4$ species. This is related to the fact that mesophyll cells are tightly packed together in tropical grasses as compared to temperate species. Epidermal cells are seen to be readily degraded in most plants. Parenchyma bundle sheaths are 90 to 100% digested both in $C_3$ and $C_3/C_4$ taxa, but those of many $C_4$ species are poorly degraded. The thick, suberized outer portion of the sheath cell wall is particularly resistant to degradation. The low degradation of the parenchyma bundle sheath appears to be a major influence on overall blade digestibility in $C_4$ species. As would be expected, vascular tissues are either not degraded or only partially digested. Results of *in vitro* digestion experiments of tissue types of *Panicum* species are presented in Table 12.2.

Research with *Panicum* suggests that an improvement in forage quality could result from plant breeding research with an aim to improve the digestibility of certain leaf tissues, such as the parenchyma bundle sheath cells of $C_4$ grass species, which have a major influence on limiting the digestibility of these

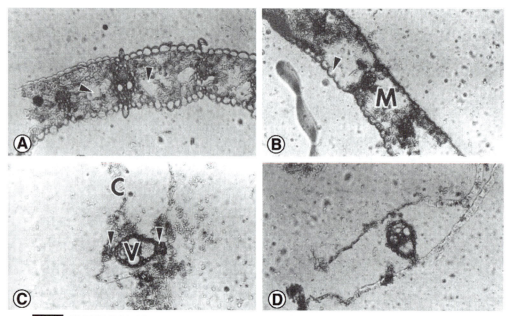

**FIGURE 12.10** Light micrographs of rumen microbial degradation of thin sections from orchard-grass, *Dactylis glomerata,* leaf blades. (A) Incubation in buffer 24 hr. Arrows indicate air space. (B) Incubation with rumen fluid 6 hr showing digestion of mesophyll (M) tissues and some epidermal cells (arrow). (C) Incubation with rumen fluid 17 hr showing residue of sclerenchyma cells (arrows), lignified vascular tissue (V), cuticle (C), and unidentifiable debris. (D) Incubation with rumen fluid 24 hr showing residue essentially the same as that after 17 hr. Reprinted with permission of Akin (1982), *Crop Sci.* **22,** 444–446.

plants. Modification of stem anatomy may also make tropical grasses more digestible by reducing the number of vascular bundles. Selection for improved varieties possessing less cutin has improved forage quality in some instances.

Another line of research is directed at altering the lignin and structural carbohydrate composition of forages in order to enhance fiber digestion by animals. Forage "fiber" is composed of a complex mixture of cell types, including parenchyma, xylem elements, and sclerenchyma; these differ considerably in cell wall composition and digestion characteristics. The corn (*Zea mays*) mutant termed brown midrib (*bm*) produces an altered lignin that makes the mutant plant more digestible than wild-type individuals. Mutant plants have an approximately 20% reduction in lignin content and a modified lignin composition. Mutants are characterized by the presence of a reddish brown pigment in leaf midribs and stem sclerenchyma. In maize, four distinct naturally occurring brown midrib mutants have been described. Brown midrib mutants of forages have been the focus of study over 30 years due to their potential agronomic importance.

## Legume Digestion

Some legume species (Leguminosae) result in economic loss to ranchers because ruminant animals may bloat following their ingestion. Forage pro-

**TABLE 12.2** Estimation of the *in vitro* Digestion of Tissue Types of *Panicum* Species at 6, 24, and 48 hr as Observed by Light Microscopy

| SPECIES | TIME OF DIGESTION (hr) | PERCENT TISSUE TYPE DIGESTED[a] | | | | |
| --- | --- | --- | --- | --- | --- | --- |
| | | Mesophyll | Parenchyma bundle sheath | Adaxial epidermis | Abaxial epidermis | Lignified vascular tissue |
| P. maximum (C₄) | 6 | 50 | 0 | 5 | 5 | 0 |
| | 24 | 100 | 65 | 100 | 100 | 10 |
| | 48 | 95 | 73 | 100 | 100 | 15 |
| P. antidotale (C₄) | 6 | 65 | 10 | 93 | 85 | 0 |
| | 24 | 100 | 85 | 100 | 100 | 0 |
| | 48 | 100 | 75 | 100 | 100 | 10 |
| P. virgatum (C₄) | 6 | 60 | 5 | 55 | 60 | 0 |
| | 24 | 100 | 75 | 100 | 100 | 10 |
| | 48 | 100 | 70 | 100 | 100 | 10 |
| P. decipiens (C₃/C₄) | 6 | 8 | 0 | 3 | 0 | 0 |
| | 24 | 60 | 90 | 75 | 100 | 3 |
| | 48 | 75 | 95 | 90 | 100 | 0 |
| P. laxum (C₃) | 6 | 3 | 8 | 0 | 0 | 0 |
| | 24 | 78 | 100 | 100 | 100 | 15 |
| | 48 | 83 | 100 | 100 | 100 | 15 |
| P. tricanthum (C₃) | 6 | 95 | 90 | 100 | 100 | 25 |
| | 24 | 98 | 100 | 100 | 100 | 30 |
| | 48 | 98 | 100 | 100 | 100 | 25 |

[a]Represents visual estimates based on replicated samples from two blades for each species.
Reprinted with permission from Akin, Wilson, and Windham (1983), *Crop Sci* **23**, 147–155.

teins are the major cause of legume pasture bloat in grazing animals and the greatest proportion of forage proteins are located in the mesophyll cells of the leaf. Bloat-safe legumes show less tissue damage and cell wall rupturing following mechanical disruption, as might result from chewing. Cell disruption caused by damaging whole leaves is influenced by mechanical tissue strength and cell-to-cell adhesion. Increasing the resistance to cell wall rupture or tissue damage reduces the bloat potential of some legumes.

Studies to identify bloat-safe characteristics in forage legumes and to develop methods for breeding bloat-safe traits have an important plant anatomi-

**TABLE 12.3  Characteristics Contributing to the Nonbloating Nature of Bloat-Safe Legumes**

| CICER MILK VETCH | SAINFOIN | BIRDSFOOT TREFOIL |
|---|---|---|
| Prominent reticulate venation pattern of secondary and tertiary veins[a,b] | Tannin sacs present[a] | Tannin sacs present[a] |
| Collenchyma and bundle sheath cell extensions attached to both epidermal layers[a,b] | Subepidermal cells adjacent to abaxial epidermis[b] | |
| Thick epidermal layer[b] | | |

[a]Hinders bacterial digestion.
[b]Increases mechanical strength.
Reprinted with permission from Lees, Howarth, and Goplen (1982).

cal basis. A comparison of the rates of digestion and mechanical strength of three bloat-causing legumes—alfalfa (*Medicago saliva*), white clover (*Trifolium repens*), and red clover (*Trifolium pratense*)—and three bloat-safe legumes—birdsfoot trefoil (*Lotus corniculatus*), cicer milk vetch (*Astragalus cicer*), and sainfoin (*Onobrychis uciifolia*)—has shown that leaf anatomy in some species does have an influence on the amount of mechanical damage and the rate of bacterial penetration sustained by the leaflets of legumes. Rumen bacteria initially digest bloat-causing legumes faster than bloat-safe plants. Anatomical characteristics that appear to restrict digestion within the leaflets are listed in Table 12.3.

Experimentation on the rates of digestion and on the mechanical strength of leaflets indicates that secondary and tertiary vein patterns and structure play an important role in leaves resisting microbial invasion. Research is under way to develop superior bloat-safe cultivars of alfalfa by altering leaf vein structure through selection and breeding.

## SUMMARY

Plant fibers are slender, attentuated cells, many times longer than they are wide, with thick, pitted, and usually strongly lignified walls. At maturity, fibers typically lose their protoplasts and thus play no role in metabolism. Fibers lend strength and resilience to materials made with them. Botanical fibers are divided into two broad categories: (1) xylary fibers and (2) extraxylary fibers. Xylary fibers are those found in the secondary xylem, whereas extraxylary fibers (also called bast fibers) are classified on the basis of their site of origin or location within the stem and leaf. Fibers positioned in the phloem or cortical regions of the stem and leaf fibers of monocotyledons include a number of commercially important cells that are very strong and flexible and that are used in the manufacture of various products.

Paper is manufactured from the wood of both softwood (gymnosperms) and hardwood (dicotyledons) species, with each source providing a pulp with distinctive cellular characteristics. Wood pulp is macerated tissue composed of a collection of cells called "fibers." The complex combination of characteristics of individual cells and how they are processed and incorporated into the paper web structure determines the quality of the final product. The process of pulping removes lignin from cell walls, and following further treatment that includes bleaching, beating, and refining, the pulp is transformed into paper sheets. Among the important fiber characteristics in determining paper quality are cell length and cell wall thickness.

In response to being naturally or experimentally displaced from their normal growth orientation, the main axis or lateral branches of actively growing stems are stimulated to form wood with an eccentric growth pattern that is composed of cells having both atypical structure and chemical composition. This secondary xylem is known as reaction wood. It forms unilaterally along one side of the displaced axis in response to gravity. The reaction wood of gymnosperms forms on the lower side of the woody cylinder and is known as compression wood. Tracheids of compression wood possess walls with visible discontinuities and abnormal microfibril angles. The reaction wood of arborescent dicotyledons forms along the upper side of leaning stems and is known as tension wood. The tension wood of many hardwoods is characterized by the presence of gelatinous fibers. Gelatinous fibers have a highly modified inner layer of the secondary wall that contracts excessively during drying and results in reduced strength properties.

The energy supply on which herbivores depend is obtained from their consumption of different plant species and plant parts. The consumption of large amounts of plant material of low digestibility may not support an animal to the same extent that a much smaller amount of highly digestible vegetable matter does. Variations in the degree of lignification and the percentage of fibrous tissue in stems and leaf blades are significant in determining tissue digestibility. The efficient use of many forages as animal feed is limited by the low digestibility of the structural carbohydrates comprising plant cell walls. Research with grasses suggests that an improvement in forage quality also could result from plant breeding with an aim to improve the digestibility of certain leaf tissues that now cause bloating in their wild-type form.

## ADDITIONAL READING

1. Akin, D. E. (1982). Section to slide technique for study of forage anatomy and digestion. *Crop Sci.* **22**, 444–446.
2. Akin, D. E., and Robinson, E. L. (1982). Structure of leaves and stems of arrowleaf and crimson clovers as related to *in vitro* digestibility. *Crop Sci.* **22**, 24–29.
3. Akin, D. E., Wilson, J. R., and Windham, W. R. (1983). Site and rate of tissue digestion in leaves of $C_3$, $C_4$, and $C_3/C_4$ intermediate *Panicum* species. *Crop Sci.* **23**, 147–155.
4. Ash, A. L. (1948). Hemp-production and utilization. *Econ. Bot.* **2**, 169.
5. Atal, C. K. (1961). Effect of gibberellin on the fibers of hemp. *Econ. Bot.* **15**, 133–139.
6. Berlyn, R. W. (1965). The effect of variations in the strength of the bond between bark and wood in mechanical barking. *Pulp Paper Res. Inst. Can. Res. Note* **54**.
7. Biermann, C. J. (1993). "Essentials of Pulping and Papermaking." Academic Press, San Diego.
8. Bolton, A. J. (1994). Natural fibers for plastic reinforcement. *Materials Tech.* **9**, 12–20.

9. Bolton, A. J. (1997). "Plant Fibers in Composite Materials: A Review of Technical Challenges and Opportunities." The Burgess-Lane Memorial Lectureship in Forestry, The Univ. of British Columbia, Vancouver.

10. Boyd, J. D. (1977). Basic cause of differentiation of tension wood and comparison wood. *Aust. For. Res.* **7**, 121–143.

11. Collings, T., and Milner, D. (1982). The identification of non-wood papermaking fibers: Pt. 3. *Paper Conservator* 7, 24–27.

12. Côté, W. A., and Day, A. C. (1965). Anatomy and ultrastructure of reaction wood. *In* "Cellular Ultrastructure of Woody Plants" (W. A. Côté, Jr., Ed.), pp. 391–418. Syracuse University Press, Syracuse, New York.

13. Dadswell, H. E., and Hillis, W. E. (1962). Wood. *In* "Wood Extractives and Their Significance to the Pulp and Paper Industries" (W. E. Hillis, Ed.), pp. 3–55. Academic Press, New York.

14. Esau, K. (1965). "Plant Anatomy." John Wiley & Sons, New York.

15. Fisher, J. B., and Mueller, R. J. (1983). Reaction anatomy and reorientation in leaning stems of balsa (*Ochroma*) and papaya (*Carica*). *Can. J. Bot.* **61**, 880–887.

16. Grabber, J. H., Jung, G. A., and Hill, R. R., Jr. (1991). Chemical composition of parenchyma and sclerenchyma cell walls isolated from orchardgrass and switchgrass. *Crop Sci.* **31**, 1058–1065.

17. Halpin, C., Holt, K., Chojecki, J., Oliver, D., Chabbert, B., Monties, B., Edwards, K., Barakate, A., and Foxon, G. A. (1998). Brown-midrib maize (*bml*)—A mutation affecting the cinnamyl alcohol dehydrogenase gene. *Plant J.* **14**, 545–553.

18. Heitmann, J. A. Jr. (1992). Pulp properties. *In* "Pulp and Paper Manufacture: Mill Electrical," Vol. 9, 3rd ed. (M. Kouris, Ed.), pp. 85–98. Joint Textbook Committee of the Paper Industry, Tappi, Atlanta.

19. Haygreen, J. G., and Bowyer, J. L. (1982). "Forest Products and Wood Science. An Introduction." The Iowa State University Press, Ames, Iowa.

20. Ilvessalo-Pfaffli, M.-S. (1995). "Fiber Atlas—Identification of Papermaking Fibers." Springer Series in Wood Science. Springer Verlag, Berlin, Heidelberg, New York.

21. Isebrands, -J. G., and Parham, R. A. (1974). Slip planes and minute compression failures in Kraft pulp from *Populus* tension wood. *IAWA Bull,* **1**, 16–23.

22. Isenberg, I. H. (1956). Papermaking fibers. *Econ. Bot.* **10**, 176–193.

23. Jung, H. G., and Allen, M. S. (1995). Characteristics of plant cell walls affecting intake and digestibility of forages by ruminants. *J. Animal Sci.* **73**, 2774–2790.

24. Jung, H. G., and Buxton, D. R. (1994). Forage quality variation among maize inbreds: Relationships of cell-wall composition and in vitro degradability of stem internodes. *J. Sci. Food Agri.* **66**, 313–322.

25. Kirby, R. H. (1963). "Vegetable Fibres. Botany, Cultivation, and Utilization." World Crops Books. Leonard Hill (Books) Limited, London.

26. Kundu, B. C. (1942). The anatomy of two Indian fibre plants, *Cannabis* and *Corchorus*, with special reference to the fibre distribution and development. *J. Indian Bot. Soc.* **21**, 93–128.

27. Kundu, B. C., and Sen, S. (1961). Origin and development of fibres in ramie (*Boehmeria nivea* Gaud.). *Proc. Natr. Inst. Sci. India*, B (Suppl.) **26**, 190–198.

28. Lees, G. L., Howarth, R. E., and Goplen, B. P. (1982). Morphological characteristics of leaves from some legume forages: Relation to digestion and mechanical strength. *Can. J.Bot.* **60**, 2126–2132.

29. Minson, D. J., and Wilson, J. R. (1980). Comparative digestibility of tropical and temperate forage—A contrast between grasses and legumes. *J. Aust. Inst. Agri. Sci.* **46**, 247–249.

30. Nguyen, H. T., Sleper, D. A., and Matches, A. G. (1982). Inheritance of forage quality and its relationship to leaf tensile strength in tall fescue. *Crop Sci.* **22**, 67–72.

31. Panshin, A. J., and de Zeeuw, C. (1980). "Textbook of Wood Technology: Structure, Identification, Properties, and Uses of the Commercial Woods of the United States and Canada." 4th ed. McGraw-Hill Book Co., New York.

32. Schery, R. W. (1961). "Plants for Man." Prentice-Hall, Englewood Cliffs, New Jersey.

33. Scurfield, G. (1973). Reaction wood: Its structure and function. Science **179**, 647–655.

34. Wardrop, A. B. (1965). The formation and function of reaction wood. *In* "Cellular Ultrastructure of Woody Plants" (W. A. Côté, Jr. Ed.), pp. 371–390, Syracuse University Press, Syracuse, New York.

35. Wardrop, A. B., and Dadswell, H. E. (1955). The structure and properties of tension wood. *Holzforschung* **9**, 97–103.
36. Wenham, M. W., and Cusick, F. (1975). The growth of secondary wood fibers. *New Phytol.* **74**, 247–271.
37. Wilson, B. F., and Archer, R. R. (1977). Reaction wood: Induction and mechanism of action. *Annu. Rev. Plant Physiol.* **28**, 23–43.
38. Wilson, J. R., and Minson, D. J. (1980). Prospects for improving the digestibility and intake of tropical grasses. *Trop. Grasslands* **14**, 253–259.

# 13

# FORENSIC SCIENCE AND ANIMAL FOOD HABITS

The use of plant anatomy in relation to historical interpretation and the law of evidence has a solid foundation. **Forensic botany,** or botanical evidence suitable for a law court, has proven to be useful in the following ways: (1) as scientific evidence for resolution of court cases involving property dispute, fraud, and matters of legal responsibility or obligation; (2) as scientific evidence in criminal cases offered to prove than an alleged crime has been committed; (3) as scientific evidence in murder, accidental death, or questioned death cases offered in connection with establishing the cause or time of death; and (4) as scientific evidence offered in connection with identifying the perpetrator of a crime or establishing the location of a crime.

In a law court, **evidence** is the information from which inferences may be drawn as the basis for proof of the truth or falsity of a disputed fact. Evidence is introduced to ascertain the probability that a certain event or series of events has occurred. Evidence takes many forms and can be obtained in many different ways. The two basic types of affirmative evidence are direct evidence and indirect circumstantial evidence. **Direct evidence** is information that, if true, directly resolves the issue without requiring any inference. **Indirect evidence,** or circumstantial evidence, is information which, if believed, can suggest the existence of the fact at issue. Thus, circumstantial evidence does not resolve the issue unless additional reasoning is used. As scientific knowledge grows and technology expands, techniques for the gathering of evidence have improved. Evidence that was formerly regarded as irrelevant is now treated as highly relevant (e.g., ballistic evidence, DNA evidence). In addition to the traditional types of courtroom evidence, information obtained from plants, including that resulting from the study of plant structure, has gained wide acceptance as **admissible forensic evidence** that can be used as either direct or indirect evidence. Plant material that can be used to aid in legal investigations may be obtained at the crime scene, from the clothing or body of a victim or suspect, or directly from the stomach contents of autopsied homicide victims. Victims and the individuals that commit crimes can inadvertently acquire plant parts on various areas of their body by falling on the ground, climbing trees, or roaming fields or wooded areas. Fruits and seeds that are adapted for animal transport can become attached to clothing or skin. Leaf or stem fragments can enter pockets or shoes. Unfortunately, most plant species are too common and widely distributed to be utilized in establishing an exact crime location. Some plants, however, are restricted to very specific habitats or community types and can provide conclusive evidence relating to the whereabouts and activity of the accused or the victim at a given period. Botanical evidence can sometimes also indicate whether the victim was moved subsequent to the commission of the crime. In breaking and entering cases, robberies, murders, and other types of crimes, wood fragments can become trace evidence. Doors, windows, walls, ceilings, or other wooden objects at the scene of the crime are sometimes broken or used as weapons. In the process, wood fragments become attached or linked to the perpetrators or victims. These fragments often can be identified microscopically and linked to standards from the crime scene.

Anatomical detective work also is an important component of obtaining information on the diets of wild animals. Evaluating such information is often an important first step in formulating comprehensive conservation programs or studying plant and animal interactions.

## PLANT ANATOMY AS FORENSIC EVIDENCE

The careful microanalysis of plant cells, tissues, or organ fragments at the light microscope and scanning electron microscope levels can provide compelling evidentiary arguments toward the resolution of many legal questions. In addition to surface examination, techniques involved in the preparation of plant tissues for microexamination, including thin sectioning, staining, and mounting of specimens on glass microscope slides, are utilized. Very small fragments require special handling and identification techniques. The information obtained is most often offered to the jury through expert testimony.

Food plants also provide a source of forensic evidence. A record of what a victim ate just before death can be obtained by identifying the plant cells present in stomach contents. Unlike animal cells, plant cells are characterized by the presence of a structurally complex outer cellulosic cell wall that is deposited during development and that is composed of largely indigestible materials. Plant cell walls differ in thickness and structure, and along with their overall shape and the relation of cells to one another, provide an important basis for identifying cell and tissue types and the organs and taxa from which they originated. Many plant cells remain largely undigested following their passage through the digestive tract, although the chewing process and the association of cells with digestive enzymes can significantly alter their appearance. Because food remains in the stomach for a limited period of time before moving on to the small intestine, plant cells retain many of their diagnostic features during this period. As a result, plant cells removed from the stomach contents of a victim can be recognized and identified many hours after death or poisoning has occurred. To aid in the identification of partially digested plant materials, a number of forensic laboratories have established collections of reference slides prepared by subjecting commonly eaten plant organs and tissues to various enzyme treatments to imitate digestion.

## PLANT CELL IDENTIFICATION

With experience, a student of plant anatomy can identify and classify plant cells recovered from a crime victim or suspect into the following categories: (1) ground or fundamental tissue composed of thin-walled, more or less isodiametric cells, sometimes with simple pits; (2) water-conducting tracheary tissue composed of thick-walled, elongate cells with characteristic spiral, scalariform, or pitted secondary wall deposits, and with end walls absent (vessel elements) or intact (tracheids); (3) mechanical tissue including very thick-walled, more or less isodiametric sclereids and very elongate fibrous cells with tapered and often forked ends and little wall pitting; and (4) dermal tissue composed of compactly arranged cells with little or no separation between cells that contain specialized stomatal pores and surface hairs (trichomes). In some cases the combination of different cell types can be used to identify the source of the tissue. It is virtually impossible, however, to correctly identify with absolute confidence the source of some cells and tissues. Many cell and tissue types integrate, and the features observed in macerated samples may not be distinctive for a particular taxon or organ. Nonetheless, because humans tend to consume a diet composed of relatively few different kinds of plants, it is often possible to identify with reasonable certainty the plant(s) or plant organs that an individual ate during a previous meal. In some cases, a particular restaurant that served the meal can even be identified.

Epidermal cells have thin or thick walls, have a striated or nonstriated surface, are pitted or nonpitted, and have straight or wavy lateral walls. If stomata are present, they occur in diagnostic arrangements and have a characteristic mature appearance. Foliar venation and associated vein sheath cells also can be distinctive. If surface hairs are present, they are uniseriate or multiseriate, unicellular or multicellular, branched or unbranched, and glandular or nonglandular. Distinctive sclereids (stone cells), starch grains, crystals of different forms and sizes, and pigment bodies provide additional distinguishing features. Onion epidermal cells, by way of example, have a recognizable elongate appearance with straight lateral walls. Chewed pear pulp contains scattered clusters of sclereids, and pineapple tissue has needlelike crystals called raphides. Citrus, beets, and spinach contain crystals of the druse type. Conspicuous surface cell features can be used to identify green beans and chili peppers. Very resistant silica bodies of diverse sizes and shapes allow us to definitively identify cereal grasses and bamboos. Seeds often have genus-specific epidermal cell pattern as well as distinctive sclereid composition.

The histological features of wood or wood fragments enable wood to be easily recognized as gymnospermous (softwood) or dicotyledonous (hard-

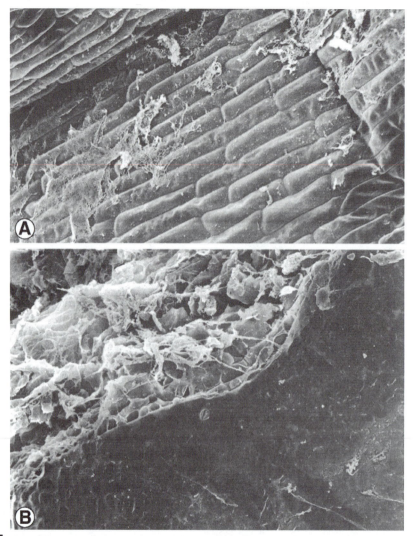

**FIGURE 13.1**   Anatomy of food plants. (A) SEM of chewed onion cells showing their characteristic elongate rectangular shape. (B) SEM of chewed cherry pulp (upper left) and peel (lower right). Both × 200. Reprinted with permission of Lane et al. (1990), *BioScience* **40,** 34–39. ©1990 American Institute of Biological Sciences.

wood), as well as frequently permitting their assignment to family and genus. Examination of even small slivers of wood can document the presence of growth rings, scalariform versus simple vessel element perforation plates, pitting types, fiber structure, and the occurrence of crystals or other inclusions. Woods of many genera, however, are not sufficiently different anatomically to permit specific identification. In cases involving small fragments of wood, it is important to examine the possibility that a physical match may exist, that is, the small wood piece might have come from a large one. The size, shape, color, and finish of the sample might be critical. An exact physical match can be very strong evidence and should never be overlooked. In larger samples, tool marks on the wood can prove to be very important.

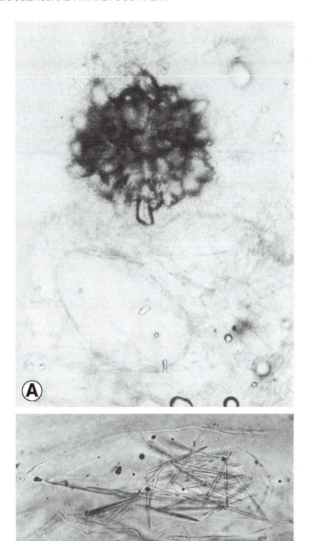

**FIGURE 13.2**  Anatomy of food plants. (A) Light micrograph of chewed pear pulp showing cluster of stone cells that are the source of the gritty texture of pears. × 150. (B) Light micrograph of chewed pineapple pulp, showing the needlelike raphides characteristic of the fruit. × 350. Courtesy of Dr. David O. Norris, University of Colorado. Reprinted with permission of Lane et al. (1990), *BioScience* **40**, 34–39. © 1990 American Institute of Biological Sciences.

## APPLICATION OF ANATOMICAL EVIDENCE IN CRIMINAL AND CIVIL LAW

The following abbreviated case histories provide examples of the application of plant anatomical evidence in criminal and civil law.

On the night of March 1, 1932, the infant son of Charles A. and Anne M. Lindbergh was snatched from his crib in a second floor room of the family home. Charles Lindbergh was an American hero, having been the first person to fly an airplane solo across the Atlantic Ocean. As a result, the kidnapping of his 21-month-old baby was a sensational news story. On May 12, 1932, the body of the young Lindbergh was found, and shortly thereafter, a man named Bruno R. Hauptmann was arrested and charged with murder. What followed in January of 1935 was one of the most celebrated murder trials in U.S. history. Hauptmann was a carpenter by trade and lived in the Bronx section of New York City. Among items that law enforcement officers were able to recover at the Lindbergh house on the night of the crime was a wooden ladder that was presumably used by the perpetrator of the crime to reach the second story window. The ladder was clearly a homemade structure that subsequently proved crucial in solving the case. In an effort to extract maximum information from the ladder, Arthur Koehler, head of the U.S. Forest Products Laboratory and an expert on wood anatomy, was asked to examine its wood. Koehler was initially sent some small splinters taken from the ladder but he subsequently asked if he could examine the entire ladder. Koehler used his knowledge of wood structure to show to the jurors' satisfaction that Hauptmann had built the ladder with his own tools using wood that was still

**FIGURE 13.3**    The ladder found at the scene of the Lindbergh kidnapping (A) and disassembled (B) for easy transport in a car. Courtesy of Regis B. Miller and the U. S. Forest Products Laboratory.

in his possession. After an extensive investigation, Koehler identified the different woods (Douglas fir, Ponderosa pine, birch, Southern yellow pine) used to construct the 19 ladder parts. By examining and measuring plane striation patterns on the wood surface and identifying unique saw tooth and chisel marks, he also demonstrated that different ladder sections had specific mill planer marks and had been planed, sawed, and chiseled by the tools found in Hauptmann's garage. Koehler discovered that some of the lumber used to build the ladder had come from a mill in South Carolina and been shipped to the lumber yard where Hauptmann worked. Finally, police noticed that a section of the attic floor in Hauptmann's house was missing. Wood anatomical study proved that one of the ladder rails had come from this piece of floor wood. Not only did old nail holes in the ladder wood fit perfectly into place over the attic floor timbers, but the pattern of growth rings in the ladder wood exactly matched those of the adjoining floor plank. This celebrated trial was the first publicized court case in which forensic botany was admitted as evidence. It played a major role in the trial's outcome, for Hauptmann was convicted of murder.

Sometimes an exact identification of plant materials is not necessary, as in a case involving explosives from Central America that was investigated by the U.S. Forest Products Laboratory. The wood filler or sawdust used in bombs confiscated in Central America was examined by experts at the Forest Products Laboratory to see if there was any evidence that the bombs had been made outside Central America. The reasoning was that wherever the bombs were made, local sawdust would have been used. Although the fragments were so small that they could not be sectioned, some were examined under the light microscope in an effort to find any woody species of north temperate origin. Although it was not possible to identify the exact species of any fragments, the investigators concluded that the wood filler was probably tropical in origin. This conclusion was possible because the cell fragments did not have spiral wall thickenings, a feature common in many north temperate species. In

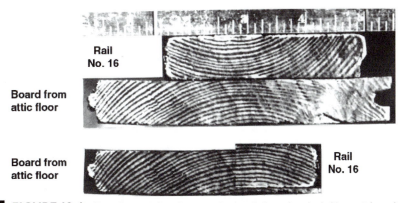

**FIGURE 13.4** Two photographs, when superimposed, show that the ladder upright and attic floorboard were formerly one and the same piece. Despite the fact that 1.4 in. of wood are missing between the two pieces, the growth rings show agreement as to curvature, number, variation in width and percentage of summerwood. Courtesy of Regis B. Miller and U.S. Forest Products Laboratory.

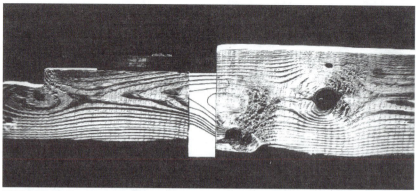

**FIGURE 13.5**   Matched halves of ladder rail 16 sawed from attic floorboard. Although a 35 mm (1.4 in.) piece of wood was missing, it was demonstrated that the two halves matched perfectly. The annual rings were identical in number, curvature, width, prominence, and grain pattern. Drawing depicts the missing piece. Courtesy of Regis B. Miller and the U.S. Forest Products Laboratory.

addition, almost all the fragments were of hardwood origin. (Most trees in the tropics are hardwoods; in north temperate regions, softwoods are common.) The only softwood fragment closely resembled a species of *Podocarpus*, one of the few softwoods confined to the tropics. A fragment from a monocot also was found (perhaps a palm), as well as several fragments that contained many prismatic crystals in the axial parenchyma and one fragment with small parenchyma cells or chambered cells. These features are common in some tropical species, but they are less frequent in north temperate species.

In the next criminal case, a young woman had been stabbed to death. It was believed that the victim had eaten her final meal at a fast food restaurant which had a very limited menu. Examination of the victim's stomach contents revealed the presence of cabbage, beans, and green peppers, as well as onions, tomato, and lettuce. Because three of these food plants were not available at the fast food establishment, investigators concluded that she must have eaten her last meal elsewhere and may have known her killer.

A final example involves the suspicious death of a young boy. Authorities suspected that the parents had drowned the child, although they claimed the death was accidental. In sworn statements, the parents said that just prior to the boy's accident they had fed him a snack of canned mixed fruit. Subsequent examination of stomach contents revealed no plant cells that could have come from either pears, peaches, cherries, grapes, or pineapple. This evidence thus exposed a major contradiction in the parents' sworn statements and thus raised doubt about the truthfulness of their story. The parents later confessed to the crime.

Plant anatomy has provided evidence to establish the cause of not only human deaths but also the deaths of animals. The author was recently asked to examine the stomach contents of a pony that had died suddenly and unexpectedly. To resolve the mystery and rule out the possibility of criminal activity, different investigative approaches were used. An analysis of the pony's hay and feed for mycotoxin proved negative. Examination of the ponies enclosure, however, revealed visual evidence of a chewed tree trunk of black locust

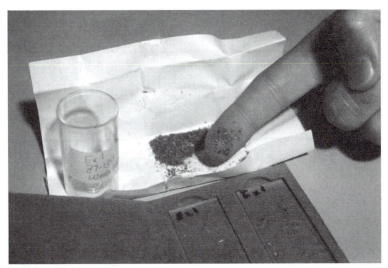

**FIGURE 13.6**   Sawdust from explosives confiscated in Central America. Courtesy of Regis B. Miller and the U.S. Forest Products Laboratory.

(*Robinia pseudoacacia*). Black locust is known to be toxic and thus it was hypothesized that this may have caused the poisoning. Examination of stomach contents showed, as would be expected, material dominated by grasses. Some of the maceration, however, was clearly not of grass origin. The sample included fibrous tissues containing abundant calcium oxalate crystals. Grasses do not contain such crystals. However, crystals of this type are known to be abundant in the bark of legumes, such as black locust. There also was evidence of dicotyledonous leaf material in the stomach contents. Although it was not possible to state with certainty that the tissue in the pony's stomach was derived from black locust, anatomical observation provided some evidence that this was the case.

The increased use of marijuana (*Cannabis sativa*) has led to court cases revolving around the botanical identity of small pieces of plant material. Anatomical information can be decisive in identifying the evidence on hand and, therefore, the fundamental question of whether a crime was committed. As part of the forensic identification procedure, tissue samples are routinely examined microscopically. The distinctive glandular and nonglandular trichomes occurring on the surface of the leaves, petioles, and inflorescence bracts are considered conclusive evidence for identification purposes. Furthermore, like other members of the Cannabaceae, *Cannabis sativa* contains cystoliths composed of calcium carbonate that will effervesce when dilute hydrochloric acid is added.

In urban areas, legal disputes can arise between property owners as a result of tree root damage to ground pipes and building foundations. The only feasible way of resolving the question of whose tree produced the offending roots is to prepare thin sections of the root wood for microscopic examination. It is known that tree root xylem can differ from stem xylem in certain details, so caution is required when making identifications.

**FIGURE 13.7**  Wood fragments from explosives viewed under a light microscope. Courtesy of Regis B. Miller and the U.S. Forest Products Laboratory.

Investigations at the Royal Botanic Gardens at Kew, England, report that they occasionally are asked to examine the contents of smoking cigars. In Great Britain it is illegal for certain small cigars to be made of any product other than tobacco (*Nicotiana*), including the cigar wrapping paper. *Nicotiana* is recognizable by the presence of a type of glandular trichome, but occasionally taxa other than tobacco can be identified.

The importation of commercial timbers necessitates their reliable identification in order to prevent fraud. There have been numerous cases of inferior woods being substituted for more valuable and expensive species, or included among shipments of commercially more valuable timber. Correct identification in such cases is especially important when a builder or architect has specified a particular wood for a purpose that cannot tolerate substitutes lacking critical properties (such as strength or earthquake resistance). Confusion can also arise because several species are sometimes sold under the same common name. Tropical mahogany is a favorite paneling wood and is used in the manufacture of cabinets or furniture and veneer. The name *mahogany,* however, is applied to a number of unrelated taxa belonging to different families. Genuine mahogany wood is derived from the genus *Swietenia* belonging to the family Meliaceae. Some woods called mahogany are botanically not true mahogany and lack its beautiful appearance. When cases of fraud are discovered in the commercial timber market, the testimony of an expert in wood anatomy is needed to verify the identification. In some instances there is no deliberate attempt to defraud because the wood has been misidentified in the field.

Many species of plants yield natural fibers used in the manufacture of fabrics and cordage. Once again, cases of fraud involving the substitution of less valuable fibers in the manufacture of ropes, twines, and textiles can be uncovered and prosecuted by using the evidence from microscopic examination. Cases of purposeful or accidental food and drug contamination and adulter-

**FIGURE 13.8**  Wood fragment of *Podocarpus* from explosives showing axial tracheids, ray cells, and cross-field pitting. Courtesy of Regis B. Miller and the U.S. Forest Products Laboratory.

ation are often detectable only by the anatomical method. Plant spices and seasonings are common in a variety of foods and the use of inferior products in place of more expensive seasonings (e.g., saffron) can often only be detected anatomically. An unknown foreign body in food that has been digested or that has caused injury may require identification. Not only is it important that human foodstuffs be quality controlled, but cases involving the contamination of animal feeds also require anatomical evidence.

## ANIMAL FOOD HABITS

Herbivorous animals are those that live by consuming plant materials. Obtaining diet and nutritional information on domestic and wild herbivores represents a major part of the research efforts of some state and federal agencies. Wildlife management personnel are especially interested in obtaining detailed food habit data and preferences for certain animals because the data relate to habitat requirements, population size and fluctuation, range, and complex interrelationships with other species. Evaluating such information is often an important first step in formulating comprehensive conservation programs.

Various methods are employed to determine the composition of domestic livestock diets as well as those of grazing wildlife and rodents. In recent years the discipline and techniques of plant anatomy have become increasingly important in these fields of research. Plant microtechnique and histology have now become widely established as accurate procedures for the determination of animal food habits. In the past, food sampling methods have been seriously hampered by the need for chemical tests and by the difficulty of obtaining specimens of ingested food materials large enough for standard macroscopic identification. By using histological analysis of herbivore diets, materials can be examined at any stage of digestion, both recently consumed and as fine particles obtained from stomach contents and fecal droppings. Fecal samples can be examined for traces of food remains during all seasons of the year and in all habitats. An additional advantage of this approach is that it is accessible to persons unfamiliar with traditional plant identification practices.

Surveys of this type have proven valuable in monitoring the availability, type, and quantity of particular herbaceous and woody plant species or the plant food classes consumed by particular animals at specific field locations, at various seasons of the year, and as the habitat changes over time. Becoming familiar with animal food habits and preferences has obvious implications for the development and use of habitats and the management and conservation of wildlife.

Wildlife biologists, ecologists, botanists, and environmental consultants wishing to utilize the microhistological method of plant identification must become adept at using a microscope, preparing slides, as well as understanding general anatomical descriptive terminology. The steps involved in tissue and slide preparation and in subsequent diet analysis are not provided here but are readily available in the literature. One general approach to microscopic identification has been to compare the appearance of research samples with drawings and photographs of plant tissue fragments included in various pictorial atlases and manuals listed in the Additional Reading. In most cases, the illustrations are selected to show the variations and detailed appearance of anatomical features of different potential food plants. As in all anatomical investigations, careful observation, attention to detail, and experience are necessary to identify unknown or in some cases uncommon plant species from isolated tissue fragments. Plant histological studies in these areas require the creation of a specially prepared reference collection for tissue identification. To make correct species identifications from research samples, the same careful microtechnical procedures need to be followed in preparing the reference collection as are applied in obtaining the research samples.

As a general rule, the epidermis and lignified vascular tissues have proven to be most abundant in research samples, and therefore most useful in plant identification. The epidermis possesses a number of important diagnostic characters that offer valuable clues for identification. Any one feature or combination of features may prove crucial for a correct identification; among these are size, shape and orientation of stomata, guard cells or subsidiary cells, structural peculiarities of epidermal cell walls, distinctive or specialized forms of trichomes, or, in the case of grasses, the occurrence and distribution of cork cells and silica bodies. Useful information also comes from the structure of xylem elements and the distribution of sclerenchyma.

Very little is known about bark structure and the preferential utilization of bark as food by various animals. Because bark anatomy has a major influence on the manner in which bark can be removed from a tree, bark fracture properties may be influential in the stripping and ingesting of selected barks by African elephants. It has been proposed that the ease with which bark is stripped from a tree, a direct result of bark structural properties, largely accounts for the species selected by these animals. According to this hypothe-

■■■ **TABLE 13.1 Extent of Debarking Compared to Fracture Properties, Method of Bark Removal by Elephants, and the Predominant Type of Sclerenchyma in the Secondary Phloem**

| SPECIES | EXTENT OF DEBARKING, AV+SD | BREAKAGE STRENGTH, AV+SD (kg) | BENDING STRENGTH | METHOD DEBARKING | TYPE OF SCLERENCHYMA |
|---|---|---|---|---|---|
| *Acacia albida* | 5 | 1.14 ± 0.38 | 4.00 ± 2.50 | SS | F(G) |
| *Xanthocercis zambesiaca* | 5 | <1 | 1.00 ± 0.00 | G | F(C) + S |
| *Schotia brachypetala* | 4 | 3.00 ± 1.41 | 3.00 ± 1.50 | SS | F(G) |
| *Acacia nigrescens* | 4 | 19.25 ± 3.54 | 60 + | LS | F(G) |
| *Acacia tortilis* | 3 | 27.64 ± 5.51 | 60 + | LS | F(G) |
| *Boscia albitrunca* | 3 | <1 | 1.38 ± 0.52 | G | F |
| *Colophospermum mopane* | 2 | 13.38 ± 1.92 | 60 + | SS | CCS |
| *Lonchocarpus capassa* | 2 | 6.00 ± 1.31 | 41.00 ± 5.50 | SS | F(G) |
| *Combretum hereroense* | 2 | 1.13 ± 0.35 | 1.25 ± 0.46 | B | S |
| *Croton megalobotrys* | 1 | <1 | 1.38 ± 0.52 | G | S |
| *Combretum imberbe* | 1 | <1 | 1.00 ± 1.50 | B | S |

Extent of debarking: 1 = slight damage (1–15%); 2 = mild damge (16–25%); 3 = moderate damage (26–35%); 4 = extensive damage (36–45%); 5 = severe damage (46–99%), where $(x - y\%)$ = mean percentage bark removed per species.

Bending strength: number of 90 bends before breaking of standard bark strip ($1 \times 2 \times 100$ mm). Method of bark removal: B = comes off in squares or blocks; G = tusk placed on bark and groove formed as tree is debarked; SS = removed as short strips; LS = removed as long strips.

Predominant type of sclerenchyma in secondary phloem (excluding dilatation tissue): CCS = lignified chambered crystalliferous strands; F = phloem fibers (lignified); F(C) = phloem fibers, exclusively cellulosic; F(G) = phloem fibers, exclusively or predominantly gelatinous; S = sclereids.

Reprinted with permission from Malan and van Wyk (1993), *IAWA J.* **14**, 173–185.

sis, tough, fibrous barks will be selected, whereas barks lacking fibers or having sclereids will be rejected. Preliminary studies have shown that the majority of tree species preferred by elephants have strong and pliable barks that are high in fiber. Exceptions to this rule occur, however, indicating that additional features are important in the selection process, perhaps chemical composition, moisture content, and nutritive value. The relationship between the extent of debarking and the type of bark sclerenchyma is compared in Table 13.1.

## SUMMARY

Forensic botany, or the use of botanical evidence in courts of law, has proven to be useful in both civil and criminal cases. Evidence obtained from plants, including structural information, has gained wide acceptance as either direct or indirect evidence. The careful microanalysis of plant cells, tissues, or organ fragments at both the light microscope and scanning electron microscope levels can provide compelling evidentiary arguments for resolving a variety of legal questions. Food plants also provide a source of forensic evidence. A record of what the victim ate just before death can be obtained by identifying the plant cells present in stomach contents. In some cases the exact restaurant that served the last meal can even be located. The histological features of wood or wood fragments enable wood to be readily identified by family and genus. One of the most famous cases involving the application of plant anatomical evidence in criminal law is the Lindbergh baby kidnapping of 1932. Wood anatomical evidence was crucial in solving the case and convicting the perpetrator. Anatomical evidence also has provided important information in other criminal cases, such as those involving the use of illegal drugs and cases of fraud involving the substitution of less valuable woods, fibers, and food products for those of higher quality.

Obtaining diet and nutritional information on domestic and wild herbivores represents another major application of plant anatomical research. Detailed food habit data and preferences are important as they relate to animal population size and fluctuation, range, and other complex interrelationships. Histological analysis of herbivore diet has the advantage that materials can be microscopically examined at any stage of digestion, both recently consumed and as fine particles obtained from stomach contents and fecal droppings. As an example of animal diet studies, it is possible to point to the relationship between bark structure and the preferential utilization of bark as food by various animals. Because bark anatomy has a major influence on the manner that bark can be removed from a tree, it has been suggested that bark fracture properties are influential in the stripping and ingesting of selected barks by African elephants.

## ADDITIONAL READING

1. Bates, D. M., Anderson, G. J., and Lee, R. D. (1997). Forensic botany: Trichome evidence. *J. Forensic Sci.* **42,** 380–386.
2. Baumgartner, L. L., and Martin, A. C. (1939). Plant histology as an aid in squirrel food-habit studies. *J. Wildlife Manag.* **3,** 266–268.
3. Bock, J. H., and Norris, D. O. (1997). Forensic botany: An under-utilized resource. *J. Forensic Sci.* **42,** 364–367.
4. Bock, J. H., and Norris, D. O. "A New Application of Plant Anatomy," Strategies for Success in Anatomy and Physiology and Life Science. No. 6. Benjamin/Cummings Publ. Co., Redwood City, California.
5. Bock, J. H., Lane, M. A., and Norris, D. O. (1988). "Identifying Plant Food Cells in Gastric Contents For Use in Forensic Investigations: A Laboratory Manual." U.S. Department of Justice, National Institute of Justice, Washington, DC.
6. Christensen, D. J. (1977). The ladder link. *Forest and People* Fourth Quarter, 8–11.
7. Dusi, J. L. (1949). Methods for the determination of food habits by plant microtechniques and histology and their application to cottontail rabbit food habits. *J. Wildlife Manag.* **13,** 295–298.
8. Graham, A. (1995). Symposium: Forensic botany: Plant sciences in the courts. *Am. J. Bot.* Supplement **82** (6), 104.
9. Graham, S. A. (1997). Anatomy of the Lindbergh kidnapping. *J. Forensic Sci.* **42,** 368–377.
10. Green, E. L., Blankenship, L. H., Cogar, V. F., and McMahon, T. (1985). "Wildlife Food Plants: A Microscopic View." Texas Agr. Exp. Station.
11. Haag, L. C. (1983). The investigation of the Lindbergh kidnapping case. *J. Forensic Sci.* **8,** 1044–1048.

12. Howard, G. S., and Samuel, M. J. (1979). "Atlas of Epidermal Plant Species Fragments Ingested by Grazing Animals." U.S. Department of Agriculture Tech. Bull. 1582. U.S. Dept. Agriculture, Washington, DC.

13. Koehler, A. (1937). Technique used in tracing the Lindbergh kidnapping ladder. *Am. J. Pol. Sci.* **27,** 712–724.

14. Lane, M. A., Anderson, L. C., Barkley, T. M., Bock, J. H., Gifford, E. M., Hall, D. W., Norris, D. O., Rost, T. L., and Stern, W. L. (1990). Forensic botany: Plants, perpetrators, pests, poisons, and pot. *BioScience* **40,** 34–39.

15. Lipscomb, B. L., and Diggs, G. M., Jr. (1998). The use of animal-dispersed seeds and fruits in forensic botany. *Sida* **18,** 335–346.

16. Malan, J. W., and van Wyk, A. E. (1993). Bark structure and preferential bark utilization by the African elephant. *IAWA J.* **14,** 173–185.

17. Mestel, R. (1993). Murder trial features tree's generic fingerprint. *New Scientist* **138,** 6.

18. Miller, R. B. (1994). Identification of wood fragments in trace elements. *In* "Proceedings of International Symposium on the Forensic Aspects of Trace Evidence," pp. 91–111. U.S. Department of Justice Federal Bureau of Investigation. FBI Academy, Quantico, Virginia.

19. Palenik, S. (1983). Microscopic trace evidence—The overlooked clue. *Microscope* **31,** 1–14.

20. Palmer, P. G. (1976). Grass cuticles: A new paleoecological tool for East African lake sediments. *Can. J. Bot.* **54,** 1725–1734.

21. Parkinson, S. T., and Fielding, W. L. (1930). "The Microscopic Examination of Cattle Foods." Headley Bros., London.

22. Stewart, D. R. M. (1967). Analysis of plant epidermis in faeces: A technique for studying the food preferences of grazing herbivorous mammals. *Aust. J. Biol. Sci.* **14,** 157–164.

23. Thornton, J. J., and Nakamura, G. R. (1972). The identification of marijuana. *J. Forensic Sci. Soc.* **12,** 461–519.

24. Vivra, M., and Holechek, J. L. (1980). Factors influencing microhistological analysis of herbivore diets. *J. Range Manag.* **33,** 371–374.

25. Winston, A. L. (1906). "The Microscopy of Vegetable Foods." John Wiley & Sons, New York.

26. Yoon, C. K. (1993). Botanical witness for the prosecution. *Science* **260,** 894–895.

# 14

## ARCHAEOLOGY, ANTHROPOLOGY, AND CLIMATOLOGY

Among the basic questions a student of archaeology and anthropology can ask are the following: Of what was a particular artifact made? When was the artifact created and used? From where did the material used in making the artifact come? If an artifact or building material is of botanical origin, these questions can often be answered most convincingly by subjecting them to anatomical examination. One form of evidence of past human activity is the recovery of plant materials such as seeds, fruits, stems, and wood that have been routinely utilized as food, in construction, as clothing, in religious or other ceremonies, or as fuel. The careful study of the anatomical record of these materials can be used to help reconstruct an accurate picture of the life of prehistoric peoples and to extend our knowledge of the otherwise undocumented climatic conditions under which they lived.

Some North American Indians constructed elaborate brushwood stake fences called fishweirs that were made from available woody plants and placed in a stream, river, or other body of water to catch fish. Microscopic investigation of the stakes and wattles from fishweirs can supply insight into the species ultilized in their construction, the time of year the weir was assembled and repaired, and the degree of degradation of the woody tissues. In one case, anthropologists determined that the weir was always built and repaired during the spring because none of the stakes exhibited a broad zone of differentiating xylem. Rather, the stakes had a completely differentiated outer layer of wood or incipient stages in the formation of a new annual increment of xylem. The accurate identification of the woody taxa used by ancient cultures for the construction of dwellings, utensils, and figure carvings can test existing theories about traditionally used woods and their relation to social and religious practices. The discovery of an unexpected wood sample at a particular site can have major implications for theories of long-distance migration of peoples and plants. The Center for Wood Anatomy Research of the United States Forest Products Research Laboratory in Madison, Wisconsin, maintains the world's largest research wood collection; over the last 80 years, its employees have identified many unusual wood samples, including specimens from such well-known archaeological sites as the La Brea tar pits in Los Angeles; Chaco Canyon; and the tombs of King Tut, King Midas, and King Herod.

## WOOD ANATOMY AND THE ENVIRONMENT

Preserved wood and charcoal samples are especially valuable in dating historical buildings and archaeological sites and for interpreting past climates. Trees are almost unique in forming absolutely datable annual growth increments that respond to and thus record environment changes. In previous chapters, we described the method by which trees increase in girth by adding a new layer of secondary xylem during each growing season. In temperate climates, this cambial cycle generally means that one additional increment of xylem is produced year by year and decade by decade over the entire surface of the plant axis. These successive layers of xylem within a tree trunk are called **growth rings** and can usually be readily seen on the transverse cut of a piece of wood. Because in a temperate climate one ring of wood is usually added to the tree each year, the rings are commonly referred to as annual rings. The base of a tree is its oldest region and the greatest number of growth rings are present at this level. Growth rings are visible and readily demarcated when ring boundaries are distinct. Successive rings are characterized by abrupt structural changes at the margins of two adjacent layers. These observable changes include alterations in cell wall thickness and cell radial diameter. Cells comprising the first-formed part of a growth ring (early or springwood) are less dense, larger, and thinner walled, whereas cells composing the later-formed part of the ring (late or summerwood) are more dense, smaller, and generally considerably thicker walled. The combination of a light-colored earlywood zone and dark-colored latewood zone forms a band that represents one growth ring. The abrupt transition between the latewood of the previous growth increment and the earlywood of the next growth increment forms a well-defined zone or ring.

Tree rings can tell many stories. Fire, lightning, pest effects, disease, and drought all leave their marks on the ring sequence. The principal factor controlling tree ring width is water availability, as related to periods of high and low rainfall. Trees that are defoliated by insects also show a period of decrease in annual ring width starting the year following the damage. The extent of decrease is related to the degree of defoliation. Climatological records extending back more than a thousand years that are unequaled in length of time and accuracy of the dated record are now available. The question whether dendrochronologies can provide useful information related to changes in greenhouse gases is still a matter of debate. Because trace elements are absorbed by

**FIGURE 14.1** Transverse section of wood of *Abies fraseri* (Pinaceae) showing growth rings of varying widths, ×27. Courtesy of the Bailey-Wetmore Wood Collection, Harvard University.

trees and incorporated into the wood, there is growing interest in using the dendroanalytical method to monitor changes in environmental pollution levels.

## DENDROCHRONOLOGY

Growth rings typically vary in thickness from year to year across the radius of a tree cylinder because of particular effects of the environment. Tree growth is influenced by many interrelated variables, such as stored food, competition, disease, and water availability. The rings within a tree are thus a chronograph, a recording clock, reflecting alternating periods of slow and rapid growth over long periods of time. In some arid regions, where all conditions except water supply appear favorable for plant growth, ring width can be directly correlated with alternating periods of high and low rainfall for different tree species. Generally speaking, with abundant precipitation and soil moisture, trees grow fast and the rings are wide. Under drought conditions or less than average rainfall, tree growth slows and the rings are narrow or absent entirely. Rings are judged to be wide or narrow relative to all other rings within an individual tree. There is also developmental variation in ring width that is related to the age and maturity of the tree. Trees commonly grow rather rapidly in diameter until maturity, at which time growth is slowed and growth ring width is narrowed. A tree does not have to be cut down to examine its annual rings. Dendrochronologists use a tool called an **increment borer** to extract a sample of wood. An increment borer is a long metal tube that screws into a tree and cuts out a thin, pencillike cylinder of wood called a **core**. The core remains in the tube when the tube is extracted from the tree. The tree rings in the core sample appear as a series of parallel lines.

The discipline that studies the chronological sequence of annual growth rings in trees is termed **dendrochronology** and is used to date prehistoric and other undated wood materials and, by extension, early human history. Growth ring studies can also be used to reconstruct past weather patterns in order to predict future weather trends. The study of climate using growth ring data is termed **dendroclimatology.** Certain trees, such as bald cypress (*Taxodium distichum*), are especially useful in studying long-term environmental trends because they are long lived and particularly sensitive to climate, especially drought. Dicotyledonous wood also can be analyzed with respect to the occurrence of ring porosity, vessel diameter and density, and vessel size distribution to yield additional information of use in interpreting past climates. **Dendroecological** reconstruction also has demonstrated the value of ring width studies in establishing the precise dates of historical forest insect outbreaks and for assessing the duration, intensity, and frequency of insect attacks.

It is now clear that many tropical tree species growing in areas with seasonality in rainfall or flooding also produce distinct growth rings in their wood. For example, trees that grow in the inundation forests in the Amazon Basin form growth rings because the cambium becomes dormant during floods. However, not all trees are suitable for tree ring analysis. Trees growing under favorable conditions of temperature and rainfall are likely to show little ring variation. Comparative studies have further shown that tree species vary in their sensitivity to various environmental conditions, including water supply. Some species form rings in a very regular manner that deviate little in width over long periods of time. Such rings are termed **complacent ring series,** and the woods containing them are less informative as environmental calendars. Other species, notably certain forest border conifers from the southwestern United States such as pinyon pine (*Pinus edulis*), ponderosa pine (*Pinus ponderosa*), and Douglas fir (*Pseudotsuga menziesii*), produce a **sensitive growth ring series.** These trees typically grow in dry habitats where other vegetation is scarce and individual trees within the forest are growing under stress conditions. In this habitat, rainfall is sporadic and water is the principal limiting growth factor. The width of individual growth rings within a sensitive ring series has been shown to be directly proportional to the amount of water that was available to the tree. The correlation is so good that the growth rings of long-preserved timbers can be statistically analyzed to deduce prehistoric rainfall patterns.

The analysis of tree rings supplies accurate quantitative information that finds wide usage in many fields of inquiry. There is no better example of this application than in understanding the early history of the Native Americans in the southwestern United States. By studying tree rings within the coniferous timbers prepared by pre-Columbian Indians to build the floors and roofs of their dwellings (pueblos), it is possible to learn the exact time of construc-

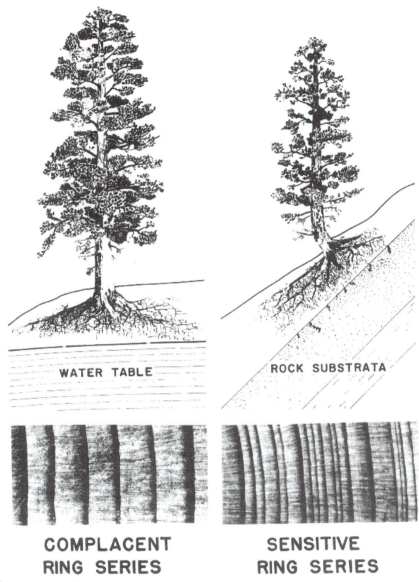

WATER TABLE

ROCK SUBSTRATA

## COMPLACENT RING SERIES

## SENSITIVE RING SERIES

**FIGURE 14.2**  Comparison of a uniformly spaced complacent growth ring series and a variable sensitive ring series. Reprinted with permission from Stokes and Smiley (1968), "An Introduction to Tree-Ring Dating," University of Chicago Press.

tion of those great settlements that remain. It also is possible to interpret the drought periods that cause settlements to move or subdivide.

## Methods of Dendrochronology

The study and interpretation of growth rings in trees requires experience. Difficulties arise when double rings ("false rings") form if the cambium initiates latewood and then resumes producing earlywood prior to the termination of the growing season. Other difficulties arise in analyzing rings due to eccentric growth patterns, diffuse boundaries of rings, occurrence of reaction wood, and the effects of fire. Rings may be very narrow or even appear to be missing, or they may be locally absent around the circumference of an axis (partial, locally absent, or missing rings). If these conditions are present, corrective procedures must be employed in order to obtain an accurate ring sequence. There are two major sources of error in tree ring dating. The first is the accuracy of the annual ring record itself. The second type of error concerns the sensitivity of the measurements. For most dating purposes, precision in measuring ring width is critical. Improved techniques, such as the use of better optical instruments, computers, and X-ray analysis, have improved the process of collecting and handling these data.

The idea of utilizing tree rings to interpret past climatic events was conceived by an American, Andrew Ellicott Douglass, at the University of Arizona. Douglas was an astronomer who developed an interest in solar changes, especially the sunspot cycle. In about 1901, he speculated that such cycles might influence the amount trees grow during a year. And if a causal relationship did exist, then it would be possible to observe sunspot cycles in the width of tree rings and the record of sunspots then could be extended backward several centuries. Douglas's musings ultimately led to the scientific study of tree rings and opened up new applications for the study of plant anatomy.

The basis for all tree ring studies is the construction of a **master chronology,** or tree ring calendar, for a given region that extends unbroken back in time. This requirement is satisfied by utilizing a technique known as **cross dating** and is the most significant step in tree ring studies. Ring identification and interpretation gains accuracy by establishing the year-to-year chronology of varying ring widths, which reflect the conditions under which the trees grew. To successfully construct a master ring sequence for the region under study, there must exist mature living trees with a similar ring pattern, as well as a wide choice of preserved wood in the form of critical timber specimens or artifacts made from the forests of the past. In this classic method, the long inside set of rings of the youngest trees, both living and dead, are matched against the pattern of wide and narrow outside rings of the next older specimen. For example, one area on the core might show a wide ring followed by three narrow rings, or two wide rings occurring on either side of a very narrow ring. In this manner, each specimen not only matches the next older but at the same time extends the ring record back in time. Because the inside ring of the most recently cut tree can be dated with certainty, a precise date can be given to each ring in the sequence. Provided that a complete series of specimen timbers is located, the tree ring calendar can be unbroken for thousands of years and can provide a standard sequence of rings with every ring definitely dated at a particular year. Gaps in the calendar can be bridged only by locating the critical missing specimens. A complete cross-dated ring sequence is acceptable only if it can be replicated in many wood samples of the same age. Thus each new location requires a long period of preparatory study before dates can be established for archaeological sites.

If a piece of wood is obtained from a region for which there exists a master chronology, a **skeleton plot** of the sample's ring history can be prepared. If the specimen was contemporaneous with any part of the known and dated

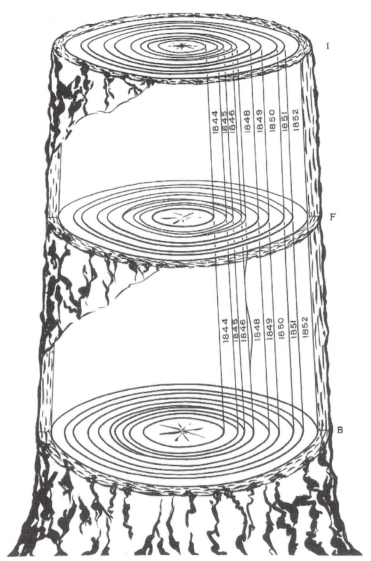

**FIGURE 14.3**  Schematic illustration of the base of a tree with the growth rings at different levels connected by a vertical line. Note that the ring representing 1847 is missing on the lowest section, appears as a lens between B and F, and shows as a smaller ring in sections F, and I. Reprinted with permission from Stokes and Smiley (1968), "An Introduction to Tree-Ring Dating," University of Chicago Press.

master ring sequence, the skeleton plot can be matched to the ring sequence of the master chronology and a precise date can be determined for the falling of the tree and the construction of the dwelling or artifact. For a piece of wood to serve as a basis for dating, it must meet the following requirements: (1) the wood must possess growth rings of variable width, with each one representing a 1-year growth increment, (2) tree growth must have been controlled by only one critical and variable limiting factor, preferably rainfall, and (3) the

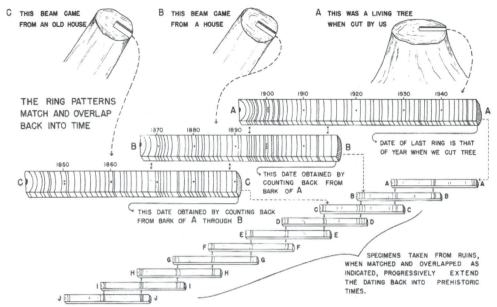

**FIGURE 14.4**  Construction of a tree ring calendar by cross dating. The inside ring sequence of the youngest specimen at the right (A) matches the outside ring sequence of the next older tree (B), the inside of this one matches the outside of the next (C) and so on. Reprinted with permission from Stokes and Smiley (1968), "An Introduction to Tree-Ring Dating," University of Chicago Press.

controlling factor must have been effective over a large geographic area. A continuous chronology for the American Southwest has been extended back to 273 B.C. by means of this technique. The record belongs not to giants like the *Sequoia*, but to the unimposing but long-lived bristlecone (*Pinus longaeva*) in the eastern California mountains. Its chronograph is over 8000 years in length. The oldest known bristlecone pine tree is over 4600 years in age and examination of fallen timber has permitted scientists to extend the chronology even further. The rate of growth of these plants is so reduced that 1100 annual rings have been counted in a piece of wood only 12.7 cm thick.

Tree ring chronology also has assumed considerable importance as a tool for testing the accuracy of conventional radiocarbon dates. Radiocarbon researchers have determined that the initial $^{14}C$ content of different samples was not uniform over the centuries, as they had previously assumed. Radiocarbon dates have been checked by comparing them to tree ring samples from oak and giant sequoia of known age (Table 14.1). In all cases the measured age agrees with the known age to within acceptable margins of error. A detailed calibration curve has been prepared using dependable tree ring methods. This has resulted in more precise dates for archaeological materials and has reinforced the reliability of the tree ring method.

## Dendrochronology Case Studies

One of the most famous of the Indian ruins of the southwestern United States is Pueblo Bonito in Chaco Canyon, New Mexico. Pueblo Bonito is the oldest

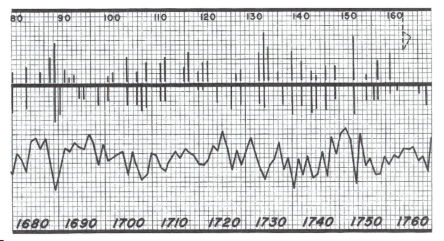

**FIGURE 14.5** Illustration of the method of matching a skeleton plot (top) along the dated master plot ring sequence (bottom). Reprinted with permission from Stokes and Smiley (1968), "An Introduction to Tree-Ring Dating," University of Chicago Press.

and largest of the prehistoric settlements in Chaco Canyon, and because the original inhabitants had neither a written language nor a calendar, the age of the ruins had long seemed an unsolvable riddle. However, by dating cores from beam timbers at the Pueblo Bonito ruins, it was determined that the oldest beam had been cut in A.D. 919 and the site had reached its golden age in 1067. The famous cliff dwellings at Mesa Verde, Colorado, date from 1073 and those in the Canyon de Chelly, Arizona, from 1060. To quote A. E. Douglass, "Some of these trees were cut a thousand years ago. From them we have learned the exact building dates of major ruins of the southwestern

**TABLE 14.1    Test of Dating Procedure With Tree Ring Samples of Known Age**

| TREE RING | NET RADIOCARBON ACTIVITY[a] (CARBON DAY$^{-1}$ MG$^{-1}$) | | RADIOCARBON DATE (YEARS) | EQUIVALENT DENDRO-CHRONOLOGICAL RANGE[b] | $\delta^{13}$C RELATIVE TO PDB[c] | COUTER NUMBER |
|---|---|---|---|---|---|---|
| | Sample | NBS oxalic acid | | | | |
| 31–37 B.C., sequoia | 13.06 ± 0.24 | 17.20 ± 0.22 | A.D. 32 ± 190 | 130 B.C. to A.D. 305 | −21.1 | C-4 |
| 31–37 B.C., sequoia | 13.09 ± 0.22 | 17.20 ± 0.19 | A.D. 44 ± 160 | 100 B.C. to A.D. 276 | −21.2 | C-6 |
| A.D., 36–42, sequoia | 12.74 ± 0.24 | 16.96 ± 0.19 | B.C. 69 ± 190 | 380 B.C. to A.D. 200 | −20.6 | C-5 |
| A.D., 1844–1850, oak | 15.80 ± 0.13 | 16.75 ± 0.19 | A.D. 1870 ± 120 | Not available | −24.1 | C-12 |

[a]Corrected for background; error includes errors in background.
[b]Maximum ranges reported.
[c]Values relative to the Pee Dee belemnite (PDB) standard were measured by A. P. Irsa.
Reprinted with permission from Harbottle, Sayre, and Stoenner (1979).

United States as definitely as we have been able to fix dates of Old World monuments of the ancients whose records are inscribed on stone."

The annual variations of the width of tree rings are much less pronounced in Central Europe's moderate climate than in the dry conditions of the southwestern United States. In Europe, oaks and other hardwoods have been used more extensively in developing tree ring chronologies. Missing or double rings are only minor problems in European woods because growth rings are usually not absent in ring-porous oaks, because the foliage appears only after formation of the spring wood. For some European sites, an oak chronology for over 1000 years has been established. This long chronology has supplied historians with important insights regarding the architectural history of some medieval buildings and towns.

Interpreting the ages of New Stone Age and Bronze Age lakeshore and marsh settlements in Switzerland and southern Germany by the pile dweller cultures is another outstanding achievement of dendrochronology. These central European settlements were built one on top of another by constructing dwellings supported by numerous timber piles anchored in lake bottoms or lakeshores. This low-oxygen microenvironment of cold water enabled the wood to be preserved in excellent condition and later used for ring analysis.

## DENDROCLIMATOLOGY

On August 18, 1587, on what is now the North Carolina coastal island called Roanoke Island, a daughter was born to Ananias and Eleanor Dare, a couple belonging to the small group of English settlers to the New World. They named the child Virginia, in honor of both the colony where they lived and of Elizabeth I, queen of England. Virginia Dare, as history tells us, was the first English child born in America. Four days later, on August 22, 1587, a few individuals left Roanoke Island for England, with the thought of returning with supplies. However, England was preparing for war with Spain, and no ships could be spared for the return to America. When ships did return to Roanoke from England 3 years later, no evidence of the stranded colonists could be found. Virginia Dare had vanished without a trace. Ever since, the question of what happened to the now famous Lost Colony has been one of the long-standing mysteries in American history.

Recent analyses of tree rings of 800–year-old bald cypress trees (*Taxodium distichum*) in southeastern Virginia and northeastern North Carolina not far from Roanoke Island, have provided new insight into the probable fate of the Lost Colony. Bald cypress trees, which commonly live 600 to 800 years (and which sometimes live up to 1700 years), are the longest lived trees in eastern North America. The reconstructed climatological history for the southeastern United States, derived from tree ring data, indicates that an extraordinary drought attended the first English attempts to colonize the New World at Roanoke Island and Jamestown, Virginia. Tree ring data tells us that the most extreme drought in 800 years occurred over a 3-year period from 1587 to 1589, just when the Roanoke settlers arrived. The Jamestown Colony arrived in 1607, in the middle of the driest 7-year period in 770 years. It is now hypothesized that food shortages and other hardships brought about by the drought were a major cause of the Lost Colony's failure.

## TREE RINGS AND OTHER ENVIRONMENTAL FACTORS

Historically, most dendrochronological analysis has been carried out in conjunction with archaeology or studies of past climate. As we have noted, analyses of wood anatomical features have been used to interpret the seasonality of climate by the presence or absence of growth rings and the conditions of growth by the width of the rings. In addition to water availability, studies have shown that some sequences of tree rings show variations in ring width that correlate with temperature fluctuations. Despite intensive study, however, correlations between climate and plant growth appear to be very complex and much less clear than previously assumed. Recent interest has turned to the response of trees to environmental factors other than climate, including atmospheric $CO_2$ concentration, UV-B radiation, and nitrogen deposition. For example, a considerable reduction in ring width over the last few decades can be observed in trees subjected to smog. This suggests that tree ring analysis can act as a tool to diagnose the effect of pollution before a tree dies. During the past decade the topic of global warming and the potential dangers of the "greenhouse effect" has provided controversy. Growth rings from the rare, long-lived (longer than 3600 years) South American conifer *Fitzroya cupressoides* have recently shown no evidence that the climate there is being warmed by human activity. The annual rings of a 3613-year-old *Fitzroya,* the second longest living tree species known after bristlecone pine, reveal that temperatures near South America's west coast have warmed and cooled many times but have not increased during the industrial age. This analysis, of course, does not disprove indications of warming in other areas of the world and perhaps suggests that South America may be "protected" by its last tracts of forest.

Attention also has been directed at using fire scars in wood to reconstruct the occurrences of surface fires, such as those that have burned periodically in the giant sequoia (*Sequoiadendron giganteum*) groves during the past 2000 years. Dates of past fires can be determined by observing the location of fire-caused lesions (fire scars) within annual rings and giant sequoia tree rings contain long and well-preserved fire records. At one site, the oldest dated fire burned in 1125 B.C. and an average of about 63 fire dates have been recorded per sampled tree. Comparisons with independently derived dendroclimatic reconstructions document that regional fires are inversely correlated with rainfall, and directly related to changing temperature regimes. Fire scar wood surfaces often contain other fire-associated ring structures, such as growth releases resulting in an increased ring width lasting 2 or 3 years up to several decades after a fire scar event, double latewood in which a narrow zone of thinner walled earlywood cells is present within the latewood band, ring separations in which tracheary cells are separated at the boundary between two rings, resin ducts formed parallel to ring boundaries, and narrow fire rings representing the next year's growth after a fire.

Partially burnt or oxidized wood in the form of charcoal is the most frequently recovered material at archaeological sites. The examination of thin sections of charcoal samples using both light and scanning electron microscopy often can identify the fragments at the genus or family level. Transverse, radial, and tangential sections of the charcoal specimen are prepared and studied to understand the organization and structure of the wood cells, including growth ring features, vessel distribution, vessel element type, and ray histology. Charcoal fragments provide evidence of the types of woods selected by ancient peoples for fuelwood and for construction of their artifacts and other structures. Paleoethnobotanists describe charcoal assemblages by quantifying the amount of each taxon present at a site. Growth rings can be analyzed to determine when a tree or shrub was cut. The charcoal identified from firepits can be compared with modern plant associations to reconstruct past vegetation. Assuming that the charcoal recovered at a site resulted from the burning of locally available woody plants, the identified taxa are indicators of former plant communities and thus of the prehistoric environment near the site. This method of environmental reconstruction may be particularly useful in projects to revegetate now barren areas or in areas where nonnative plants now flourish. Identification of charcoal taxa from multiple samples of different ages also can be used to develop a chronology of vegetation change or alteration in a given region.

## DIET AND ARTIFACTS OF PREHISTORIC PEOPLES

Anthropologists are interested in accumulating information on all aspects of the lives of ancient peoples. Microscopic analysis of human coprolites (fossilized excrement) can supply important clues of the food habits of certain prehistoric human populations. Specimens obtained from cave sites in Nevada contain plant remains that are anatomically identifiable as root fragments of the semiaquatic herb *Sagittaria* (Alismataceae). When observed with the scanning electron microscope, transverse sections of these fragments show three distinct tissue regions. A central vascular bundle cylinder contains a single prominent vessel element with simple perforation plates. The vascular tissue is surrounded by a layer of thick-walled cells that may represent an endodermis. The stele is enclosed by a lacuna that is devoid of recognizable cellular structure. The outermost region is several layers thick and is composed of thick-walled parenchyma. The morphology and anatomy of these and other organ fragments are remarkably similar to what is found in *Sagittaria*, suggesting that the early peoples of this region relied strongly on emergent vegetation for subsistence and, by extension, on foraging to obtain their food.

Among the artifacts of prehistoric peoples are also found textiles and cordage produced from plant materials. Braided cordage that reveals a composition of plant fibers has recently been described from southern Louisiana. Scanning electron microscopy shows that the cordage strands were made of material derived from roots of a monocotyledonous plant, possibly a grass or sedge. Transverse sections of the cordage strand reveal an outer layer, an endodermis of thick-walled cells, a layered pericycle, and an inner area composed of a mixture of metaxylem elements and smaller fibrous cells. The cor-

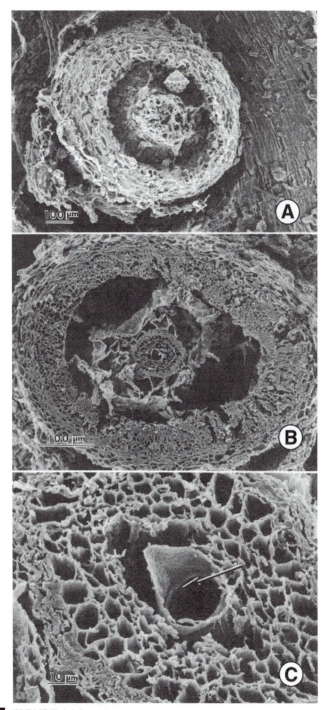

**FIGURE 14.6**   SEMs of plant remains from human coprolite. (A) Whole mount of remnant with protrusion. (B) Transverse section through protrusion. The protostele is evident. Note the region of degraded wall materials surrounding stele. (C) Central region of stele in transverse view showing a single vessel element. The arrow points to the simple perforation plate of the end wall. Reprinted with permission of Neumann *et al.,* Determination of prehistoric use of Arrowhead (*Sagittaria,* Alismataceae), *Econ. Bot.* **43,** 287–296. Copyright ©1989, New York Botanical Garden.

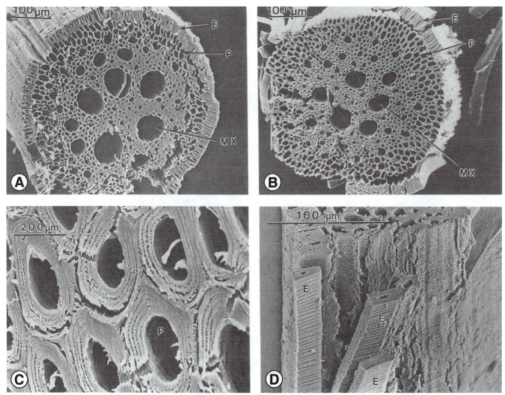

**FIGURE 14.7**   Scanning electron micrographs of early cordage from Louisiana. (A) Cross section through the source plant root, with pattern of cells similar to that seen in (B) (letters indicate points of similarity). (B) Cross section through a cordage strand. (C) High magnification of one of the plant roots showing the thick-walled fiber cells that give strength to the cordage strands (five layers are evident in the wall of the cell at P). (D) Longitudinal view along the edge of a root. Thick-walled endodermal cells shown at E. Reprinted with permission of Kuttruff *et al.* (1995), *Southeastern Archaeol.* **14,** 69–83.

tex and epidermis are absent in the cortex. The thick-walled fibers and endodermal cells would have given the cordage considerable strength.

## SUMMARY

One form of evidence of past human activity is the plant materials that humans have utilized as food or clothing, as housing materials for dwelling construction, in rituals, or as fuel. The anatomical record of these materials can be used to reconstruct a picture of the life of prehistoric people and to extend our knowledge of the climatic conditions under which they lived. Preserved wood and charcoal samples are especially valuable in dating historical buildings and archaeological sites and for interpreting past climates. This can be accomplished by studying the successive growth rings within a tree trunk.

Tree ring widths vary with the climate. Abundant precipitation and soil moisture result in wide rings, whereas under drought conditions or less than average rainfall the rings are narrow or absent. Fire, pests, and disease also leave their marks within the ring sequence. The discipline that studies the chronological sequence of annual growth rings in trees is dendrochronology, whereas the study of climate using growth rings data is termed dendroclimatology. The basis for all tree ring studies is the construction of a regional master chronology that extends back in time without interruption. This requirement of identifying and dating annual growth rings is achieved by utilizing a technique known as cross dating. In this method the long inside set of rings of the youngest tree, both living and dead, are matched against the pattern of wide and narrow outside rings of the next older specimen. Because the inside ring of the most recently cut tree can be dated with certainty, a precise data can be given to each ring in the sequence. If a piece of wood is obtained from a region for which a master chronology is known, a skeleton plot of the sample's ring history can be prepared. These techniques have provided exact building dates of many major archaeological ruins in the southwestern United States.

Charcoal specimens are examined microscopically to understand the organization and structure of the wood cells. Charcoal fragments provide evidence of the tree species selected by ancient peoples for fuelwood, building construction, and many other artifacts. Identification of charcoal taxa from multiple samples of different ages also can be used to develop a chronology of vegetation change or alteration in a given region. Microscopic analysis of human coprolites can also furnish important clues about food habits of ancient peoples.

## ADDITIONAL READING

1. Bailey, I. W., and Barghoorn, Jr., E. S. (1942). Identification and physical condition of the stakes and wattles from the fishweir. *In* "The Boylston Street Fishweir," Vol. 2, pp. 82–89. Robert S. Peabody Foundation for Archaeology, Andover, Massachusetts.
2. Brown, P. M., and Swetnam, T. W. (1994). A cross-dated fire history from coast redwood near Redwood National Park, California. *Can. J. For. Res.* **24**, 21–31.
3. Chaloner, W. G., and Creber, G. T. (1973). Growth rings in fossil woods as evidence of past climate. *In* "Implications of Continental Drift to the Earth Sciences"(T. H. Tarling and S. K. Runcorn, Eds.), pp. 425–437. Academic Press, New York.
4. Cook, E. R., and Jacoby, G. C., Jr. (1977). Tree-ring drought relationships in the Hudson Valley, New York. *Science* **198**, 399–401.
5. Cook, E. R., and Kairiukstis, L. A., Eds. (1990). "Methods of Dendrochronology." Kluwer Academic Publishers, Dordrecht, The Netherlands.
6. Cook, E., Bird, T., Peterson, M., et al.. (1992). Climatic change over the last millennium in Tasmania reconstructed from tree-rings. *Holocene* **2/3**, 205–217.
7. Curtis, B. A., Tyson, P. D., and Dyer, T. G. J. (1978). Dendrochronological age determination of *Podocarpus falcatus*. *S. Afr. J. Sci.* **74**, 92–95.
8. Douglass, A. E. (1929). The secret of the Southwest solved by talkative tree rings. *Nat. Geographic Mag.* **56**, 736–770.
9. Douglass, A. E. (1937). Tree rings and climate. *Univ. Arizona Phys. Sci. Bull.* **1**, 1–36.
10. Dunwiddie, P. W., and La Marche, V. C., Jr. (1980). A climatically responsive tree-ring record from *Widdringtonia cedarbergensis*, Cape Province, South Africa. *Nature* **286**, 796–797.
11. Dyer, T. G. J. (1982). South Africa. *In* "Climate from Tree Rings" (M. K. Hughes, P. M. Kelly, J. R. Pilcher, and V. C., LaMarche, Jr. Eds.), pp. 18–21. Cambridge University Press, Cambridge.

12. February, E. C., and Stock, W. D. (1998). The relationship between ring width measures and precipitation for *Widdringtonia cedarbergensis*. *S. Afr. J. Bot.* **64**, 213–21?

13. Ferguson, C. W., and Graybill, D. A. (1983). Dendrochronology of bristlecone pine: A progress report. *Radiocarpon* **25**, 27–288.

14. Fritts, H. C. (1971). Dendroclimatology and dendroecology. *Quaternary Res.* 419–449.

15. Fritts, H. C. (1972). Tree rings and climate. *Sci. Am.* **226**, 92–100.

16. Fritts, H. C. (1976). "Tree Rings and Climate." Academic Press, New York.

17. Harbottle, G., Sayre, E. V., and Stoenner, R. W. (1979). Carbon-14 dating of small samples by proportional counting. *Science* **206**, 684–685.

18. Hitch, C. J. (1982). Dendrochronology and serendipity. *Am. Sci.* **70**, 300–305.

19. Jacoby, G. C., and D'Arrigo, R. D. (1997). Tree rings, carbon dioxide, and climate change. *Proc. Natl. Acad. Sci.* USA **94**, 8350–8353.

20. Jacoby, G. C., Williams, P. L., and Buckley, B. M. (1992). Tree ring correlation between prehistoric landslides and abrupt tectonic events in Seattle, Washington. *Science* **258**, 1621–1623.

21. Kubiak-Martens, L. (1996). Evidence for possible use of plant foods in Palaeolithic and Mesolithic diet from the site of Calowanie in the central part of the Polish Plain. *Veg Hist. Archaeobot.* **5**, 33–38.

22. Kuttruff, J. T., Standifer, M. S., Kuttruff, C., and Tucker, S. C. (1995). Investigations of early cordage from Bayou Jasmine, Louisiana. *Southeastern Archaeol.* **14**, 69–83.

23. Lara, A., and Villalba, R. (1993). A 3620–year temperature record from *Fitzroya cupressoides* tree rings in southern South America. *Science* **260**, 1104–1106.

24. Neumann, A., Holloway, R., and Busby, C. (1989). Determination of prehistoric use of arrowhead (*Sagittaria*, Alismataceae) in the Great Basin of North America by scanning electron microscopy. *Econ. Bot.* **43**, 287–296.

25. Norton, D. A., Briffa, K. R., and Salinger, M. J. (1989). Reconstruction of New Zealand summer temperatures to 1730 A.D. using dendroclimatic techniques. *Intern. J. Climatol.* **9**, 633–644.

26. Schmid, R., and Schmid, M. J. (1975). Living links with the past. *Nat. Hist.* **84**, 38–45.

27. Schulman, E. (1958). Bristlecone pine, oldest known living thing. *Nat. Geo. Mag.* **113**, 354–372.

28. Schweingruber, F. H. (1993). "Trees and Wood in Dendrochronology. Morphological, Anatomical, and Tree-Ring Analytical Characteristics of Trees Frequently Used in Dendrochronology," Springer Series in Wood Science. Springer-Verlag, Berlin, Heidelberg, New York.

29. Schweingruber, F. H. (1996). "Tree Rings and Environment Dendroecology." Paul Haupt, Berne, Stuttgart.

30. Stahle, D. W., Cleaveland, M. K., Blanton, D. B., Therrell, M. D., and Gay, D. A. (1988). The Lost Colony and Jamestown droughts. *Science* **280**, 564–567.

31. Stallings, W. S., Jr. (1949). Dating prehistoric ruins by tree-rings. *Lab. Anthropol., Santa Fe New Mexico, Gen. Ser.* **8**, 1–18.

32. Stockton, C. W., and Meko, D. M. (1975). A long-term history of drought occurrence in western United States as inferred from tree rings. *Weatherwise* **28**, 244–249.

33. Stokes, M. A., and Smiley, T. L. (1968). "An Introduction to Tree-Ring Dating." University of Chicago Press, Chicago.

34. Swetnam, T. W. (1993). Fire history and climate change in giant *Sequoia* groves. *Science* **262**, 885–889.

35. Tippo, O., and Stern, W. L. (1977). "Humanistic Botany." W. W. Norton and Co., New York.

36. Weber, U. M., and Schweingruber, F. H. (1995). A dendroecological reconstruction of western spruce budworm outbreaks (*Choristoneura occidentalis*) in the Front Range, Colorado, from 1720–1986. *Trees* **9**, 204–213.

# 15

# PROPERTIES AND UTILIZATION OF WOOD

We are all familiar with the secondary plant tissue that is technically known as secondary xylem but is more commonly called **wood.** Everyone has sawed wood, driven nails into wood, split wood for the fireplace, sat on wooden furniture, or used a wooden product. Wood touches our lives in many ways, as raw material for construction in industry, as fuel, and by offering aesthetic value to our surrounding. Hardwood and softwood tree species furnish valuable raw material for the manufacture of furniture, furniture components, and other domestic articles. After millenia of human habitation, wood still remains our most widely used building material, and the global demands for wood as a construction material continue to increase. Wood equals or surpasses most other construction materials in strength, performance, toughness, and rigidity. In fact, straight-grained timber possesses the rare and valuable combination of good strength characteristics, moderate stiffness, and excellent toughness, properties matched by only a few other materials. All the uses of wood in building, engineering, and manufacturing rely directly or indirectly on a knowledge of wood structure and of its physical and mechanical properties.

Because wood is of biological origin, it is to be expected that variations in structure and consequently in properties are present. Woods of different species possess individual characteristics and differ in a number of properties, including grain pattern, durability, strength, density, and color. Differences in wood quality and behavior also exist among samples taken from the same species in different geographic locations and even different parts of the same tree. A reliable knowledge of wood anatomical properties and the behavior of wood under stress is essential for engineers, architects, carpenters, and woodworkers in order to use timber more efficiently. Builders have incurred substantial losses by using timber unsuited for a particular job or by employing more or less lumber than was necessary for a project. It is important for the architect or builder to have knowledge of the relationships between physical appearance, anatomical structure, mechanical and chemical properties, and tree growth characteristics. In some older literature, this branch of inquiry was called timber physics, but it is now more commonly referred to as **wood technology.** The uses and therefore the desired physical and structural features of wood change as technology changes. At one time, for example, the selection of wood for the manufacture of airplanes represented a critical problem that required wood having a particular combination of mechanical and physical properties. The design and building of wooden bridge trestles also required a fundamental understanding of the relationship between wood structure and its mechanical properties.

This chapter considers some of the basic concepts of wood property relationships at a nonmathematical level. Emphasis is placed on the general areas of wood figure, strength, shrinkage patterns, and deformation, and how these properties combine to produce woods suitable for specific applications.

## FEATURES AND APPEARANCE OF SAWED OR SPLIT LUMBER

Broadly speaking, **figure** refers to any feature or combination of features that makes a particular wood distinctive and lends beauty to processed lumber. For this reason figure is useful for distinguishing between different kinds of wood and assumes added importance when selecting wood for certain construction needs. In commerce the word *figure* is often used in a more restricted sense to describe an especially decorative surface pattern. Figure can be due mainly to a single unique feature but is frequently produced by the interaction of two or more factors. These include **color,** resulting from the deposition of pigments, gums, resins, tannins, and other substances within the cell lumina of the heartwood; **luster,** a quality related to the reflection of light; and **surface patterns** caused by growth rings, ray structure, and in a few cases wood axial parenchyma. Some woods have an especially attractive figure, and these are selected for interior finish work such as decorative paneling. Although not strictly components of figure, **weight** and **scent** (resulting from the infiltration of natural oils and aromatic compounds) can also contribute distinguishing qualities to wood.

The concepts of grain and texture in wood represent additional aspects of figure that are widely employed in the lumber trade and by woodworkers. Although the two terms are sometimes loosely used as synonyms, in a strict sense they have separate meanings. **Grain** is a general term that can describe different structural conditions as well as how a wood reacts when worked with tools. In common usage, grain describes the arrangement, size, appearance, and direction of alignment of the principal wood cells and, if growth rings are present, the width and construction of the rings. Variability in wood grain has led to such descriptive terms as fine or coarse, even or uneven, rough or smooth, as well as straight, cross, spiral, twisted, wavy, and curly. To have technical meaning, the term grain must be qualified. The influential American wood anatomist Samuel J. Record recommended that a wood be called **coarse grained** if its growth rings are wide and **fine grained** if growth rings are narrow following slow growth. **Even** and **uneven grain** refers to the regularity or irregularity of the growth rings. **Straight-grained** wood has its fibrous elements oriented in a plane parallel to the long axis of the stem in which they occur. If the elements are arranged spirally, or if the elements interweave or are twisted in various patterns, sometimes in alternating bands, a so-called **spiral, interlocked, cross,** or **twisted** grain pattern is formed.

From a mechanical standpoint, grain pattern is closely correlated with wood strength. Straight-grained woods consistently show greater strength than those with spiral grains. As a result, straight-grained logs or boards that are split along the grain are stronger and undergo less warping than those that are sawed. Early settlers invariably preferred wooden tool and farm implement handles that were produced by splitting lumber rather than sawing it. The terms **rough grained** and **smooth grained** apply to the manner in which a wood responds to tooling. The grain pattern in trunk wood typically becomes altered at the base of a limb so that fibers extend from the stem into the limb on the lower side. This arrangement forms a localized wavy or curly grain and reduces the cleavability of the wood at the region of the knot. Irregular cambial activity can sometimes form distinctive grain patterns, such as a "bird's eye" grain and others on the tangential section. A swirling distorted grain pattern is often called **burl**. Burl can be very dramatic and attractive and is used for decorative pieces.

In contrast to grain, various qualifying adjectives of the word **texture** are properly used in reference to the relative size (i.e., diameter) and proportional number of wood elements. A **coarse-textured** wood contains numerous large tracheary elements, either tracheids or vessel elements, whereas a **fine-textured** wood has conducting cells with uniformly small diameters and cell lumina. If all the wood elements are more or less uniform in diameter, the wood has an **even** or **uniform texture,** such as we find in maple, birch, and white pine. Wood that contains elements of markedly different sizes, such as a ring-porous species like oak, has an **uneven texture.** A common concern among individuals involved in wood finishing is whether the grain is open or closed. Woods such as walnut and oak have large pores that create what is described as **open grain.** This condition normally requires filling in order to create a smooth surface during finishing. Small pores in wood create what is commonly called **closed grain** that does not need special attention before finishing. We find this in birch, maple, beech, pine, and fir. The terms open grain and closed grain more accurately describe wood texture.

The planes, or surfaces, in which roundwood is sawed or cut have a significant effect upon the quality, behavior, and figure of the resulting lumber. For centuries woodworkers have recognized that timber should be split or sawed specifically for the purpose for which it will be used and the wood properties that are desired. For the detailed microscopic study of structural features, small blocks of wood must be trimmed to expose all three principal planes of section. These are the **transverse section** (cross section) and two longitudinal sections, the **radial** and **tangential.** By comparing the transverse and longitudinal sections, we can determine that certain wood cells are oriented with their long axes parallel with the long axis of the stem. The ray system is the most useful tissue for distinguishing between radial and tangential sections. In transverse section the growth rings, if present, are visible as concentric or partial rings, and the wood rays extend in the radii of the woody cylinder. In this section only the width of the rays is visible. The radial section is formed by cutting a log in the plane of the stem radius, that is, through the center of the log. Because the cut is made along the radius of a ray, the rays appear as parallel bands that are exposed in lateral or side view. The tangential section, on the other hand, is a longitudinal cut at right angles to the radial plane, so that in a large piece of wood the cut is approximately parallel to the growth rings. In this section, the rays appear in end view so that their vertical height becomes evident.

In the commercial lumber industry the transverse cut is essentially used only to create board length. It occasionally can be seen in such items as wood chopping blocks and mallets because the exposed cell ends (end grain) are simply mashed down during pounding and the wood does not readily chip away in slivers. Logs are most commonly sawed tangentially to their perimeter. Board lumber cut in this standard manner is said to be **plain sawn,** also called **flat sawn** or **slab cut.** Most plain sawed lumber that has been derived from temperate trees is easily recognized because the latewood component of the growth rings forms conspicuous patterned markings on the wide, longitudinal surface of the boards. Because growth increments within a tree occur as a series of superimposed cones that are inclined obliquely from the base of the trunk to the top and are therefore at different angles relative to the stem surface and at various depths along its length, boards cut parallel to the tangential longitudinal plane of the log will have variable but characteristic surface patterns or grain. Wood that has been sawed or planed in this fashion, such as that used in making most furniture and interior wall finishings of buildings, will show dark, curved growth rings resembling irregular parabolas or ellipses (U- and V-shaped patterns). The width, degree of irregularity, and position of growth rings in the log are correlated with a bold or delicate pattern or design on the boards. As wood dries or seasons, the greatest stresses occur in the direction of the circumference of the growth rings. Consequently, plain-sawn

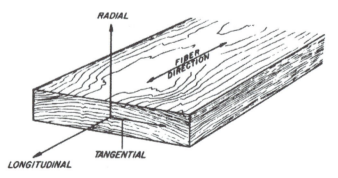

**FIGURE 15.1**   The three principal axes of wood with respect to grain direction and growth rings. Reproduced from Agricultural Handbook 72.

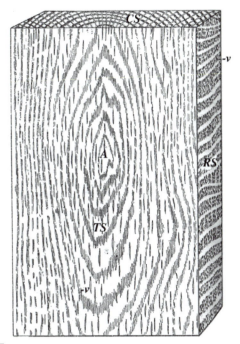

**FIGURE 15.2**   Block of oak wood. Abbreviations: CS, cross section; RS, radial section; TS, tangential section; v, vessels or pores; A, slight curve in log which appears in section as an islet. Reprinted from Roth (1895), Bull. No. 10, U.S. Department of Agriculture, Division of Forestry.

lumber with its curved growth rings tends to expand and twist when a change in moisture content of the wood occurs. Nevertheless, plain-sawn lumber results in minimum waste at the sawmill because the log can be rotated and cut continuously at right angles to the rays.

An infrequent practice is to saw logs along the radial longitudinal plane. In commercial practice wood surfaces cut in this manner are reported to be **quarter sawn** or **rift sawn,** although the boards are often not cut strictly parallel to the rays. In quarter-sawn lumber the rays extend as close as possible to right angles with the width of the board. To achieve this, each log is initially cut into quarters, lengthwise, and then each quarter is sawed into boards. Because the log is cut along a radial plane in harmony with the grain, the rays not only appear in lateral view, but the growth rings also form successive parallel bands. In rift sawing, long and short boards are cut alternately from the log along the radial plane. If the rays are large, as they are in oak, they appear as prominent broad bands or ribbons on the exposed surface. Because quarter-sawn boards have been cut along the grain and the growth rings have only slight curvature, they are less likely to warp than plain-sawn lumber. Lumber of this type is desirable as table tops, chests of drawers, and fine wood flooring and planking. Although quarter sawing was originally done to reduce or prevent warping, lumber sawed in this way is highly prized because of the unusual figure produced by the rays. The practice of using long-wearing quarter-sawn boards as flooring also can be attributable to the closely spaced and vertically exposed

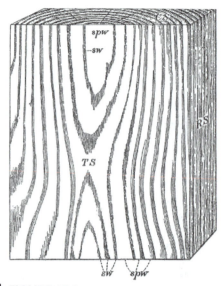

**FIGURE 15.3** Block of pine wood. Abbreviations: CS, cross section; RS, radial section; TS; tangential section; sw, summer wood; spw, spring wood. Reprinted from Roth (1895), Bull. No. 10, U.S. Department of Agriculture, Division of Forestry.

bands of highly lignified latewood elements that are considered to provide greater strength and durability. Although quarter-sawn lumber possesses many highly desirable qualities, logs are rarely cut this way in commercial mills because of the expense of wasted timber and the fact that the sawing operation is more difficult. Some characteristics and advantages of plain-sawn and quarter-sawn lumber are presented in Table 15.1.

Grain, texture, and planes of cut are basic to an understanding of wood cleavability. When an ax strikes a piece of wood, the wood will most

**TABLE 15.1   Some Advantages of Plain-Sawn and Quarter-Sawn Lumber**

| PLAIN SAWN | QUARTER SAWN |
|---|---|
| Figure patterns resulting from the annual rings and some other types of figures are brought out more conspicuously by plain sawing. | Types of figure due to pronounced rays, interlocked grain, and wavy grain are brought out more conspicuously. |
| | It shrinks and swells less in width and thickness. |
| Round or oval knots, which may occur in plain-sawn boards, affect the surface appearance less than spike knots that may occur in quarter-sawn boards. Also, a board with a round or oval knot is not as weak as a board with a spike knot. | Raised grain caused by separation in the annual rings does not become so pronounced. It twists and cups less. It surface checks and splits less in seasoning and in use. |
| Shakes and pitch pockets, when present, extend through fewer boards. | It wears more evenly. |
| It is less susceptible to collapse in drying. It may cost less because it is easier to obtain. | It does not allow liquids to pass into or through it so readily in some species. |

**FIGURE 15.4**  Wood cleavability. Wood readily splits along the radius of the grain, a to b, when struck by an ax. Reproduced from Roth (1895), Bull. No. 10, U. S. Department of Agriculture, Division of Forestry.

naturally split along two planes. Cleavage along the radial plane is the easiest to achieve because the fibrous cells, tracheary elements, and parenchymatous rays are aligned in the radial direction. This is especially evident in woods of conifers, but the wide rays in oak also allow for easy cleavage along the radial axis. If the wood has prominent growth rings, or if the wood is cross grained, tangential cleavage along a plane parallel to an annual ring may be readily attained. This is particularly true in woods that are ring porous. Irregularities in the alignment of fibers so as to form grain patterns that deviate from a straight-grained condition will result in a condition that resists splitting. Wood also divides more easily when fully wet than when dry because water softens the wood and reduces the lateral adhesion of elements.

## PHYSICAL PROPERTIES OF WOOD

Wood can be characterized according to its cellular organization and chemical properties. The **physical properties** of wood are those roughly based upon molecular constitution, cell wall organization, and interaction of structural units. These properties can be loosely divided into features of exterior appearance or physical properties that are determined by measurement, such as weight and density. The physical properties of wood are affected by influences such as water content, and include shrinkage and swelling, which may cause warping and checking, and affect the wood's thermal and electrical properties. Thermal properties are a measure of heat flow and how quickly the wood absorbs heat. Electrical properties include differences in electrical conductivity.

### Specific Gravity and Density

**Weight** is controlled by two major variables: the amount of solid cell wall material present and the quantity of water contained within the sample. Both factors vary with location within a tree, position within a growth ring, and degree to which a wood sample has dried. Weight is usually expressed in terms of specific gravity and density. **Wood specific gravity** refers to the ratio of the weight of oven-dried wood to the weight of an equal volume of displaced water at 4°C (39.2°F). One cubic foot of pure water at this temperature weighs 62.43 pounds. Dividing the weight (in pounds) of a cubic foot of thoroughly dried wood by 62.43 will provide the specific gravity of the sample. Because the value of the weight of the displaced volume of water depends on the volume of the wood sample, which in turn is related to the moisture content of the wood, it is necessary to indicate the moisture content of the wood sample at the time its volume was determined. Specific gravity increases as the moisture content of wood decreases. As a result, specific gravity must be computed for different moisture contents.

Relative **density,** on the other hand, expresses the weight (mass) of wood per cubic volume. Density is expressed as pounds per cubic foot (kilograms per cubic meter or grams per cubic centimeter) at a specific moisture content. Because both specific gravity and density are largely a direct reflection of the size of fibrous cells and the amount of cell wall substance that is present, these values increase as cell wall thickness increases and cell lumina size decreases. In the case of wood, relative density is related to the ratio of secondary wall to primary wall. Consequently, latewood or wood composed of very thick-walled fibers is denser and also has a higher specific gravity than earlywood or wood containing thin-walled, wide-lumened elements. In the United States, woods with specific gravities of 0.36 or less are classified as light; 0.36 to 0.50, as moderately light to moderately heavy; and above 0.50 as heavy. Indian sola wood (*Aeschynomene aspera* and *A. indica*) has a specific gravity of 0.04 and is probably the lightest known wood. It is characterized by the large-scale replacement of thick-walled fibers with thin-walled, air-filled, fusiform wood cells and a very low frequency of vessels.

## Wood and Water

Water is located in wood as **bound water** in the cell wall and as **free water** in the numerous spaces that compose the cell lumina and intercellular cavities. Free water does not affect cell wall and wood volume, whereas bound water does. As water enters the cell wall, the cellulosic microfibrils expand and the framework of the wall is extended and distorted until a balance is reached. The point at which the cell walls are saturated and wood ceases to swell is called the **fiber saturation point** of the wood. Wood at the fiber saturation point also is at maximum volume. The addition of free water beyond this level does not affect the volume or strength of wood. The amount of water in a tree can typically vary across the radial diameter from the heartwood to sapwood. When using wood for furniture manufacture or other purposes, logs should be cut with the wood at the moisture content it will have when in use. If wood is cut wet and undergoes subsequent shrinkage, cracking will develop. **Seasoning** is the process by which wood dries, although the state of absolute dryness is rarely attained because all the water is never entirely lost. As "ordinary" drying occurs, water is lost most rapidly from the transverse surface. The properties of wood change appreciably after drying. As wood continues to experience a uniform and thorough water loss, either by natural drying or under artificially controlled conditions, cell wall strength, stiffness, hardness, and durability (susceptibility to deterioration) increase significantly. By way of example, the strength of a sample of freshly cut undried spruce lumber increases approximately fourfold when completely dry. But if wood is resoaked, it will be weaker than it was originally. Although water is lost throughout the wood tissue, it is only the dehydration of cell walls that results in increased wood strength. Although dehydration in some tissues increases density values, dehydration of thick-walled tissues such as wood decreases relative density because there occurs an increase in the ratio of cell wall to protoplasm per unit volume but no accompanying decrease in absolute tissue volume. The lignin component of wood cell walls is hydrophobic and acts to prevent rapid changes in the moisture content of walls as well as rapid changes in the mechanical properties of the xylem. Because the water content of wood plays a major role in determining the overall properties of wood, engineering tests carried out to determine wood strength properties must be done under controlled conditions and the samples must have a uniform moisture content.

As the walls of individual cells lose water, they become thinner and the wood shrinks. The changes in cell wall dimension result from internal and unequal stresses that are created within the molecular system of the wall by interactions between the strong microfibrils and the amorphous matrix of lignin and hemicellulosic materials. These changes in cell wall infrastructure have profound effects on wood behavior and represent important considerations in the use of wood for manufacturing purposes. In normal wood, cells do not contract to any appreciable extent along the longitudinal axis. This means that wood cell length does not change significantly during the drying process but that cell diameter and wall thickness decrease in proportion to the original wall thickness. When placed in water or in a humid environment, the cell walls of dry wood will absorb water and the wood will regain some of its volume. This

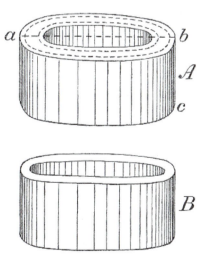

**FIGURE 15.5**  Shrinkage of wood. Short section of wood fibers, one thick walled, the other thin walled. When a fiber dries, its wall grows thinner as indicated by dotted lines (A). Cell width, a to b, becomes smaller but cell length, b to c, remains the same. In the thinner-walled fiber in B the effect is the same but the total change is much less. Reprinted from Roth (1895), Bull. No. 10, U. S. Department of Agriculture, Division of Forestry.

ability to gain or lose water is called **hygroscopicity.** It causes the moisture content of wood to fluctuate with changing climatic conditions, as anyone who has had to deal with stuck windows and doors or wood joints that come apart knows all too well. Whether the cell lumina are empty or filled with free water is of no consequence in determining the cell's dimensions.

Because wood is a heterogeneous tissue composed of cells of markedly different sizes, shapes, wall thicknesses, and arrangements, dehydration produces a nonuniform pattern of shrinkage. This is illustrated by the fact that thick-walled latewood elements shrink more than thin-walled earlywood cells. Radially aligned ray cells also undergo shrinkage, but they shrink and swell relatively little along their own long axes. Because wood shrinks to a much greater degree tangentially than it does radially, permanent splits may form as timber dries because this is the only way for stresses to be relieved and for the circumference to contract. Generalized shrinkage curves showing the magnitude of change in each direction are available to determine dimensional changes. Woods with irregular grain patterns or exceptionally large rays undergo extreme shrinkage and thus large pieces of wood are not dimensionally stable. Because wood expands and contracts more across the grain than along it, cross-banded veneer construction made by gluing successive layers of thin sheets with longitudinal cells oriented at right angles to one another provides both increased strength and also minimizes changes in wood size and warpage. The fabrication of plywood takes advantage of these facts.

Due to the stresses incurred in different directions by individual wood cells during the drying process and the complicated patterns of shrinkage that result, drying lumber tends to warp. The degree of warping or "curling" is directly related to the grain pattern, the degree of uniformity in cell size and

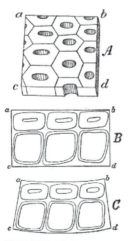

**FIGURE 15.6** Warping in wood. (A) Thin, transverse section showing several fibers. As drying occurs each fiber wall becomes thinner and the dimensions a to b and c to d become shorter. If a piece of wood is composed of an assortment of thick-walled and thin-walled fibers as in (B), then the thick-walled cells shrink more than the thin-walled elements and the wood becomes unevenly shrunk or warped as shown in (C). Reproduced from Roth (1895), Bull. No. 10, U. S. Department of Agriculture, Division of Forestry.

wall thickness, the uneven distribution of water within a timber, and the manner in which the timber has been cut and dried. Irregular or rapid drying increases the strains within a board. As internal stresses accumulate during unequal shrinkage, progressively more prominent longitudinal cracks or **checks** become visible on the ends surface of lumber pieces. Often the cracks are most evident along the rays where the stresses are most severe. Improperly seasoned wood, such as rapidly dried lumber, will form numerous checks. In all cases, splits or checks are regarded as strength-reducing defects and often cause problems in both finishing the wood and gluing it. To counterbalance the damaging consequences of checks and splits in construction timbers, S-shaped and C-shaped metal wedges are frequently inserted across the cracks.

## Permeability

**Permeability** is a wood characteristic that describes the ease with which fluids flow through or penetrate wood in response to an absolute pressure gradient. Species such as white oak and post oak produce especially impermeable wood. The presence of cell deposits and incrustations on the membranes of

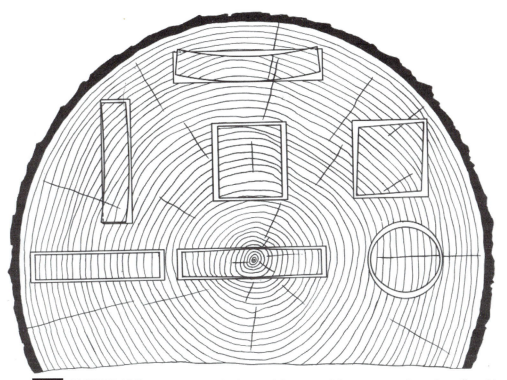

**FIGURE 15.7** Characteristic shrinkage and distortion of flats, squares, and rounds as affected by the direction of the growth rings. Tangential shrinkage is about twice as great as radial. Reproduced from Agricultural Handbook 72.

**FIGURE 15.8**  Effects of tangential shrinkage on logs. Reproduced from Roth (1895), Bull. No. 10, U. S. Department of Agriculture, Division of Forestry.

pits connecting wood cells is a major factor in determining resistance to liquids. From an applied standpoint, the basic principles underlying permeability and the effect of the minute structure of wood upon permeability are of considerable importance in the drying process, as well as in wood preservation or fire-retardant treatments where creosote and other toxic fluids must penetrate the wood in all directions. Permeability studies also are useful to determine the treatability or penetrability of coatings such as paint and water repellents.

Dry wood is considerably more permeable than green wood. When wood decays, its penetrability increases further. At the microscopic level, permeability is related to such structural properties as the pore size distribution, the conducting cell structure, and the extent to which cells become clogged with tyloses and gums. The possible penetration of the cell wall by liquids also must be taken into account. Lateral and horizontal movement of liquids and gases is much slower than movement in the longitudinal direction. In the case of coniferous wood, movement of liquids primarily takes place through conducting cell cavities and is regulated by the distribution and orientation of the specialized openings in the secondary wall known as bordered pits. Wall pits allow for communication between two adjacent cells. The bordered pits of tracheids are provided with a delicate pit membrane that contains a perforated outer region, the margo, and a circular thickened central portion, the torus. The torus can be displaced against the rim of the pit aperture, to one side or the other, in a valvelike action, effectively sealing one of the pit apertures of the pit pair. When this occurs, the pit is said to be fully "aspirated." Because this occurs in numerous pits throughout the wood, it prevents the penetration of air or liquids even under pressure. The structure and function of bordered pits are thus of fundamental importance in understanding wood permeability. Various methods of improving permeability have been described, including biological and chemical treatments to extract material from the pit membrane or to degrade the pit membrane itself.

Surprisingly, the dense latewood portion of the growth ring of many conifers is less resistant to penetration than the more porous earlywood. The structure and behavior of bordered pits appear to play a central role in this phenomenon. In pine, the pit membranes of latewood elements appear thickened and rigid and are not as flexible as the pit membranes in earlywood. Thus there is no tendency for pits to become aspirated. As pressures increase, however, most pits in earlywood coniferous tracheids are readily deflected to one side and are therefore very efficient at preventing the movement of liquids and gases. The physiological significance of this observation is unclear. The greater volume of air in springwood also appears to reduce the penetration of creosote. The role of rays and of resin and gum canals in the permeability of softwoods and hardwoods is not well understood.

## Sapwood and Heartwood

As a tree matures, the older xylem ceases to contain any living ray or axial parenchyma cells so any accumulating materials are converted into various extractive substances, including tannins, oils, gums, and resins. This centrally located wood is generally darker in color and is called **heartwood**. The heartwood is surrounded by lighter-colored sapwood containing living cells and starch. The wood of young trees is generally composed entirely of sapwood. As the tree ages, all the living cells in the central core of the trunk and large branches die, darken, and lose their physiological function. The most significant change in the transformation of sapwood into heartwood is death of ray and axial parenchyma cells. This can be observed in the gradual disappearance of nuclei in the parenchyma. Heartwood formation is also accompanied by anatomical changes such as an increased aspiration of pits in gymnosperm tracheids and the formation of tyloses in dicot vessel elements.

As a general rule, heartwood is more resistant to penetration by preservatives than is sapwood. It follows, therefore, that heartwood is not suitable for timbers that will be impregnated with these fluids. Heartwood also is more difficult to season without checking. The primary reason for the drastic loss of permeability when going from sapwood to heartwood appears to be the irreversible aspiration of bordered pits and the encrustation and occlusion of the conducting elements by secondary metabolites and deposits. Heartwood cells and pit cavities become filled with hardened deposits, and the pit membrane tori often cover the bordered openings into the pit and cell cavities.

## MECHANICAL PROPERTIES OF WOOD

The mechanical properties of wood are measurable expressions that result from applied forces. **Strength** is the characteristic of wood that enables it to sustain different primary stresses without changing shape or breaking. Strength is a complex wood characteristic that is determined by several variables, some of which are related to the rate at which the tree grew. Exposing

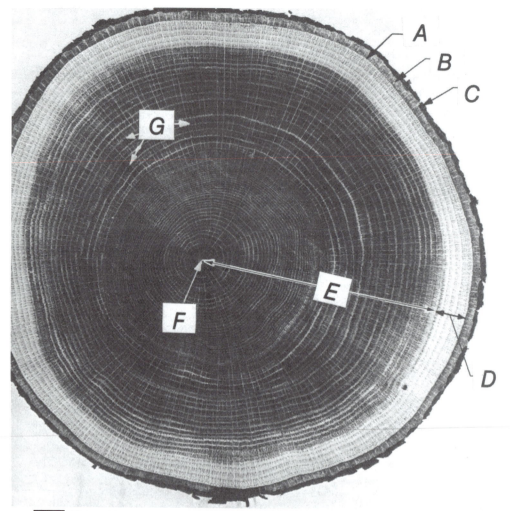

**FIGURE 15.9**   Cross section of white oak tree trunk showing tissue regions. (A) Cambium layer; (B) inner bark; (C) outer bark; (D) sapwood; (E) heartwood; (F) pith; (G) wood rays. Reproduced from Agricultural Handbook 72.

wood to various salts (such as those in sea water) can significantly increase certain strength properties.

Because different categories of force can act upon wood, woods exhibit different types of strength characteristics. Forces acting upon wood may result in changes to wood tissue without destroying the cohesion of cells; this resiliency is measured as indicated by a wood's elasticity, flexibility, and toughness. However, external forces can produce changes that are accompanied by the destruction of cell cohesion. When this occurs, a fracture is the likely result. When stresses in wood reach sufficiently high levels, **deformation** or strain occurs. Two types of deformation are significant: (1) load or external stress, such as those encountered by a beam during bending or by a column under axial load, and (2) internal stress, such as that caused by the effects of moisture and temperature gradients. Stresses may develop while the tree is growing or after the lumber has been cut. Deformation can further be divided into changes that are instantaneous or delayed and those that are recoverable or irrecoverable. To fully understand the properties of wood and its behavior under the influence of applied forces, we must investigate its mechanical behavior across the entire scale of units down to the cell wall and the molecular level.

Resistance to applied forces is a function of the total amount and proportion of cell wall material (cellulose, lignin) in a wood sample and the amount of extractives in the cell lumen. Both weight and strength can vary in different parts of the same tree trunk from center to periphery, and also at different ages of growth. The ability of wood to resist stress also is closely related to the duration of the stress and the presence of strength-reducing defects such as knots, checks, and splits. Reaction wood also differs significantly from normal wood in its deformation characteristics under both short and long term stresses. A principal objective of wood deformation studies is to predict the mechanical characteristics of different timbers from their structural and molecular properties, and therefore to be able to predict both the normal and the catastrophic failures of structures composed of wood. It is possible to make certain predictions about wood properties from careful anatomical examination.

## Stiffness and Strength

The degree to which a piece of wood will resist bending and other distortions of form, or being bent by a weight or load, is a measure of its **stiffness.** If the force is removed and the board straightens, it is said to show **elasticity.** The stiffness of wood is usually characterized as the "modulus (measure) of elasticity" and is related to the breadth, depth, and length of a piece of wood in addition to its cellular structure. Stiffness decreases geometrically as length increases so that doubling the length of a board reduces its stiffness eightfold. Doubling the width, on the other hand, doubles the stiffness. As builders are well aware, if one doubles the length of a supporting beam but requires the same stiffness, the thickness of the beam also must be doubled. Stiffness will further increase if the growth rings extend vertically to the load. Cross-grained, wet, or knotty lumber will show decreased stiffness because the orientation of its fibers differs. Because specific gravity is largely a measure of the amount of cell wall material present, it also is the best parameter to use when predicting wood strength properties. Accordingly, woods with many thick-walled cells will be stronger, heavier, and stiffer than wood with thin-walled elements. Furthermore, the growth characteristics of forest trees are directly related to their wood properties. Slow-growing conifers form dense, strong wood because the growth rings are close together and the number of latewood elements per unit area is high. In ring-porous hardwoods, however, slow growth results in weaker, less dense wood because of the numerous, closely spaced springwood vessels.

The force or load that causes a piece of wood to break is a reflection of the wood's maximum load-carrying capacity or **absolute strength,** also known as the modulus of rupture in bending. As one might suspect, timber strength properties can be measured in different ways and vary with respect to wood type and board thickness and length, but not in the same proportion as stiffness. There appears to be no obvious relationship between the bending and physical and mechanical properties of wood. Because of induced stresses and compressive failures in bending, however, bent wood is weaker than straight specimens. Bending strength is further influenced by moisture content and grain pattern. As a rule, coniferous woods exceed hardwoods in stiffness but show reduced bending strength. When a piece of wood is gradually subjected to pulling or stretching forces, **tension** is created. When tension is high, wood fibers are pulled apart from adjoining cells or broken, and the wood breaks or fails. Resistance to the forces of tension is related to the size, arrangement, and lateral adhesion of wood elements. Grain also affects resistance to tension. Straight-grained lumber is much stronger than cross-grained wood and clear wood is stronger than wood containing knots.

In contrast to tension, **compression** occurs when wood is pushed inward along the grain at the ends to shorten the wood or reduce its volume. Failure under excessive compressive forces is a complex phenomenon related to the magnitude of the stresses and to cellular construction. Wood will exhibit different properties when measured along different axes. Stress in compression can be measured parallel to the grain or perpendicular to the grain. Compressive strength is correlated with wood's specific gravity; as the specific gravity increases, the strength increases. As long as individual fibrous elements hold together and offer collective resistance, comparatively intense forces are necessary to cause mechanical instability and localized buckling. Failure occurs when individual cells begin to separate and behave as independent units that undergo their own deformation. When this occurs, wood can no longer resist the deformation caused by bending and the wood breaks. Because compressive strength along the grain is normally about one half to one fourth of the tensile strength, structural lumber is more likely to fail in compression than in tension. In living trees, the situation is different. The periphery of the trunk is in tension, whereas the inner wood is in longitudinal compression developed in growth; therefore, breaks most often result from tensile failures. A straight-grained wood with a regular arrangement of vertically oriented fibers offers the highest resistance. Cross-grained wood with fibers in obliquely oriented patterns show reduced strength characteristics. The behavior of wood cells to stress is based to large degree on the structure of their secondary walls, specifically the helical arrangement of the cellulose microfibrils in the $S_2$ wall layer, the largest layer in the wall and its major load-bearing component. The helical arrangement of the $S_2$ layer provides a mechanism for absorbing irreversibly great amounts of energy, thereby preventing structural failures.

## Hardness and Toughness

In addition to the mechanical properties just described, woods vary with respect to **hardness** and **toughness.** Hardness is a measure of the degree of indentation to the surface of a wood following the application of localized pressure. The harder the wood, the better it will resist wear, scratches, and dents. Hardness, like specific gravity and strength, is an indication of the amount of cell wall material that is present. Heavy wood is harder than light wood; latewood is harder than earlywood. Moisture softens wood, whereas seasoning hardens wood. Wood toughness refers to the ability of dry wood to absorb energy under impact of a sudden blow without breaking or cracking, as is required in an ax handle or a baseball bat. Hickory and ash are the woods of choice for handles of striking tools such as axes, hammers, hatchets, picks, and sledges as well as sporting equipment because of their impact resistance, toughness, resilience, stiffness, and hardness. Hickory handles are tough and absorb impact forces that would break most other woods. The selection of trees for maximum toughness and strength is in part a function of tree age and the conditions of growth. Toughness is a complex property that relates to several strength characteristics and again has its foundation in the ultrastructure of the cell wall. Straight-grained lumber is very tough in a plane perpendicular to the grain.

When impact forces are sufficiently high to result in breakage, the break occurs in a progressive manner, fiber by fiber, rather than suddenly and abruptly. Woods in which the natural toughness is greatly reduced and that consequently break abruptly following the absorption of energy and with little or no bending are **brash**. Brash wood is abnormal, and its strength properties are significantly reduced. When stressed, it breaks suddenly and completely. Brash wood shows a low ratio of tensile to compression strength along the grain, resulting in failure at low levels of stress. The most common causes of brashness are reduction in density, ultrastructural alterations in cell walls, changes in chemical composition, and presence of compression damage. As we noted, reduced density is attributable to a below-average proportion of fibers in hardwoods, reduced thickness of cell walls, inclusion of "juvenile" wood, or presence of fungal decay. Because brashness can be caused by the degradation of different combinations of structural, physical, chemical, and mechanical properties, it is often difficult to determine its exact cause. Nevertheless, the impact of wood structural failure due to this condition can be enormous.

## APPLICATION OF WOOD ANATOMY TO THE FIELD OF BIOMEDICAL RESEARCH

Experimentation has shown that carbonized wood can be successfully used as a matrix for bone regeneration in some animals. The wood used for such purposes must meet the following wood anatomical considerations: (1) the structure must be sufficiently homogeneous to ensure reproducible results, (2) it must retain its porous structure following carbonization, (3) it must possess vessels of suffcient size to permit the free invasion of bone tissue, (4) the frequency of vessels must fall within a range that ensures a suitable wood strength, and (5) the vessels must be devoid of tyloses and other obstructions. The fact that porous woods having vessel diameters of at least 100 μm are required rules out the use of softwoods and many hardwood species. The temperate climber *Clematis vitalba* has proven an ideal species for wood

**TABLE 15.2**   **Summary of the Time of Implantation and the Amount of Bone in Growth**

| RABBIT NO. | IMPLANTATION TIME (DAYS) | BONE IN GROWTH | | |
|---|---|---|---|---|
| | | *Parthenocissus tricuspidata* | *Terminalia superba* | *Clematis vitalba* |
| 16[a] | — | — | — | — |
| 17 | 47 | — | — | — |
| 18 | 55 | — | + | ++ |
| 19 | 55 | + | + | ++ |
| 20 | 65 | — | — | ++ |
| 21 | 56 | — | + | ++ |
| 22 | 56 | — | + | ++ |
| 23 | 56 | + | — | ++ |
| 24 | 56 | + | + | ++ |
| 25 | 56 | + | + | ++ |

[a]Died 10th postoperative day.
+, present; ++, significant; —, absent or limited marginal in growth only.
Reprinted with permission from Colville, *et al.* (1979), *IAWA Bull.*1.

implants, facilitating significant bone in growth. The results of experimentation on rabbits to assess the amount of bone in growth using carbonized wood implants from three species are shown in Table 15.2. Excellent biocompatibility has been demonstrated for carbonized wood. In *Clematis vitalba* implants, bone tissue can be observed growing throughout the large vessel elements of the surgically implanted wood.

## SUMMARY

Wood technology deals with the physical appearance, anatomical structure, and mechanical and chemical properties of this material. All these features affect wood figure, strength, shrinkage patterns, and deformation. Figure refers to any feature or combination of features that makes a particular wood distinctive and lends beauty to processed lumber. Wood figure is produced by color, luster, surface patterns, weight, grain, and texture. If wood elements are arranged spirally or if the elements interweave or are twisted in various patterns, a spiral or twisted grain pattern is formed. Straight-grained wood consistently shows greater strength than spiral grained. Logs are most commonly sawed tangentially to their perimeter. Board lumber cut in this manner is said to be plain sawed. In quarter-sawn lumber the boards are cut parallel to the rays. Because quarter-sawn boards have been cut along the grain and the growth rings have only slight curvature, they are less likely to warp than plain-sawed lumber.

The physical properties of wood are ultimately derived from its particular molecular constitution, its cell wall organization, and the interaction of its structural units. They are affected by external influence such as water content, shrinkage and swelling, which may cause warping and checking. Weight is controlled by two major variables: the amount of solid cell wall material present and the quantity of water in the sample. Wood specific gravity refers to the ratio of the weight of oven-dried wood to the weight of an equal volume of displaced water at 4°C. Water is located in wood as bound water in the cell wall and as free water in the numerous spaces of the cell lumina and intercellular cavities. As the walls of individual cells lose water, they become thinner and shrinkage of the wood occurs. Due to the stresses that are incurred in different directions by individual wood cells during the drying process and the complicated patterns of shrinkage that result, dried lumber tends to warp. As internal stresses accumulate during unequal shrinkage, progressively more prominent longitudinal cracks or checks occur in the boards. Permeability describes the ease with which fluids flow through or penetrate wood. The presence of cell deposits and pit membrane structure are major factors affecting resistance to liquids.

The mechanical properties of wood are the multitude of measurable expressions that result from applied forces. These include strength properties, stiffness, elasticity, hardness, and toughness. Woods in which the natural toughness is greatly reduced and that break abruptly with little or no bending are described as brash.

## ADDITIONAL READING

1. Adkins, J. (1980). "The Wood Book." Little and Brown, Boston.
2. Bailey, I. W. (1913). The preservation treatment of wood. II. The structure of the pit membranes in the tracheids of conifers and their relation to the penetration of gases, liquids and finely divided solids into green and seasoned wood. *For. Quart.* **11,** 12–20.
3. Bailey, I. W. (1915). The effect of the structure of wood upon its permeability. No. 1. The tracheids of coniferous timbers. *Bull. Am. Railway Engineering. Assoc.* **174,** 835–853.
4. Colville, J., Baas, P., Hoikka, V., and Vainio, K. (1979). Wood anatomy and the use of carbonised wood as a matrix for bone regeneration in animals. *IAWA Bull.* 1, 3–6.
5. Côté, W. A. (1963). Structural factors affecting the permeability of wood. *J. Polym. Sci. Part C* **72,** 231–242.
6. Côté, W. A. (1967). "Wood Ultrastructure." Univ. of Washington Press, Seattle.
7. Erickson, H. D., and Rees, L. W. (1940). The effect of several chemicals on the swelling and crushing strength of wood. *J. Agr. Res.* **60,** 593–603.
8. Goldstein, I. S. (1991). "Wood Structure and Composition." Dekker, New York.
9. Harris, J. M. (1954). Heartwood formation in *Pinus radiata* D. Don. *New Phytol.* 53, 517–524.
10. Hoadley, R. B. (1980). "Understanding Wood." Taunton Press, Newton, Connecticut.
11. Keating, W. G. (1982). "Characteristics, Properties and Uses of Timbers." Texas A & M University Press, College Station.
12. Koch, P. undated. "Utilization of Hardwoods Growing on Southern Pine Sites," Agriculture Handbook No. 605, Vol. II. U.S. Department of Agriculture Forest Service, Washington, D.C.
13. Kovach, E. G. (Ed.) (1975). "Properties of Wood in Relation to its Structure." The Report of a NATO Science Committee Conference Held at Les Arcs, France 17th–21st November, 1975. Scientific Affairs Division, NATO, Brussels, Belgium.

14. Larson, P. R. (1969). Wood formation and the concept of wood quality. *Bull. Yale Univ., Sch. For.* **74**, 1–54.

15. Lutz, J. F. (1978). Wood veneer: Log selection, cutting, and drying. U.S Department of Agriculture Forest Service Technical Bulletin 1577. U.S. Government Printing Office, Washington, D.C.

16. Record, S. J. (1919). "Identification of the Economic Woods of the United States, Including a Discussion of the Structural and Physical Properties of Wood." John Wiley & Sons, New York.

17. Ross, J. D. (1956). Chemical resistance of western woods, *For. Prod.* J. **6**, 34–37.

18. Roth, F. (1895). "Timber: An Elementary Discussion of the Characteristics and Properties of Wood," Bull. No. 10. U.S. Department of Agriculture, Division of Forestry, Washington, D.C.

19. Panshin, A. J., and de Zeeuw, C. (1980). "Textbook of Wood Technology: Structure, Identification, Properties, and Uses of the Commercial Wood of the United States and Canada." 4th ed. McGraw-Hill, New York.

20. "Wood Handbook: Wood as an Engineering Material." United States Department of Agriculture, Agriculture Handbook 72. U.S. Department of Agriculture. Rev. 1987, Washington, D.C.

# 16
## ■ THE ARTS AND ANTIQUES

The arts comprise the sum total of human creative work that has resulted in things that have form and beauty. The arts are generally defined to include the disciplines of painting, drawing, sculpture, architecture, music, drama and dance. In the widest application of this term, art also encompasses aesthetic artifacts such as antique furniture, clothing, textiles, manuscripts and other antiquities. The world of art is a diverse enterprise and the tangible **objets d'art** that receive our attention are varied in structure and composition. Each category has unique requirements for analysis, care, and conservation.

At first glance, the role that plant anatomy plays in the advancement, appreciation, understanding, conservation, and restoration of art may not be obvious. However, objects made from plant materials form a significant portion of the art housed in collections around the world. In many instances, art objects, antiques, and musical instruments owe their unique qualities directly to their botanical origin. As a consequence, it is sometimes possible to apply the special techniques of plant anatomical research to such materials and resolve fundamental questions relating to history, preservation, and conservation, as well as performance in the case of musical instruments. In some cases, the questions and problems are so complex and so diverse that they can be analyzed effectively only through the sustained effort of groups of physical, chemical and biological experts working in close collaboration. All major museums have conservation departments, which employ individuals technically trained in the conservation of artifacts made from plant materials. In fact, in the larger institutions conservators often work in groups that specialize in either wood, paper, or textiles.

To better understand how a knowledge of plant structure plays an important role in the arts, it is necessary to point out some of the basic questions and issues that art dealers, museum directors, and students of art encounter. Is the work authentic? How was the piece created? What materials were used in its construction? When was it created? Where did it originate? Why does it have a particular quality for musical performance? Does it show signs of deterioration and can the decay be repaired and prevented? Anatomical investigation can sometimes provide the most economical and rapid means of answering these and other questions. Anatomical evidence can be useful in several ways in art historical research:

1. as criteria for better verifying and documenting the attribution of a work of art.
2. as an important tool for dating.
3. to reevaluate the technique and style of execution of a work of art and to determine how it might fit into the larger area of art historical development.
4. as a tool to investigate physical deterioration and to suggest how a piece of art might best be conserved.
5. as a method to identify the materials used in construction of the piece and to suggest their historical significance.

## IDENTIFICATION AND DETERIORATION OF PLANT TISSUES AND CELLS USED IN OBJECT CONSTRUCTION

Materials of plant origin form a sizable component of all the materials that have been worked by human beings to create objects of beauty and worth. The ability to identify different plant tissues and cell types correctly is basic to understanding more about a work of art and its possible deterioration patterns. Often only fragments of materials are available for examination, and thus the ability to provide correct identification based on one or a few cells is essential. In order to make these judgments, the samples often must be meticulously prepared and the individual must use proper methods of observation and characterization.

A number of museums throughout the world possess wooden Egyptian antiquities. During the last several years, researchers at the Royal Botanical Gardens in Kew, England, have been working in conjunction with the British Museum to identify ancient Egyptian woods and thereby unlock a potentially important source of information about the life of the ancient Egyptians. Although few large trees are indigenous to Egypt, many wooden cultural artifacts have been recovered from tombs throughout the Nile Valley. It is also clear that the ancient Egyptians valued imported exotic woods and used them for making ceremonial or ritual items. Investigators have suggested that the comparative analysis of wood use from dynasty to dynasty may indicate a preferential or functional selection of wood for different items. It would be of interest to know, for example, whether specific species were selected for particular ritual purposes or were credited with special powers. At our present state of knowledge, it is only possible to identify most of these woods to the level of genus.

Although most tissues and plant parts have been utilized at some time, fiber and wood are the two most common materials that are the most ubiquitous materials used across cultures and over time. Because of their length, strength, and flexibility, fibers are the major cell type in such items as baskets, textiles, and paper. The trained art conservator has the ability to distinguish between wood fibers, phloem fibers, and monocot leaf fibers. As a prerequisite to undertaking techniques of conservation, the cellular composition of paper can be studied by carefully teasing apart small pieces of soaked paper and identifying the individual cells microscopically. Such examination can be used to verify the types of cells in the paper and the source of a particular sheet used in the formation of prints, drawings, and manuscripts. Small fragments of wood removed from antique furniture and other wooden objects also can be studied to determine the type of wood used in construction and to answer other questions. Variations in wood anatomy also can affect the application of surface coatings (shellac, varnish, paint) on furniture and other restoration work.

The leaves of some palms (*Corypha umbraculifera* and *C. talieri*) were once used as writing material in Ceylon. The remarkable preservation of these old palm leaf manuscripts has been shown to be related to leaf structure. Writing on the leaves of these plants caused the writing stylus to puncture the cuticle and epidermis of the lamina, thereby opening a direct channel for modern preserving fluids to enter the mesophyll. As a result, leaf areas without writing respond differently to the application of liquids than regions with writing. As drying oil is sprayed onto the leaf surface, it is absorbed into the inner mesophyll through the openings, where it hardens, and strengthens the leaf from the inside.

## DETERIORATION

One of the special problems that curators face is the continuous deterioration of antiques and other cultural property. After removal from their original environments, art objects and musical instruments often are subjected to numerous hostile environmental influences that cause decay and may ultimately lead to their total destruction. Anatomical examination sometimes can enable the conservator to detect the degradation and take preventive action to arrest its spread. The form of deterioration varies with the type of plant material and will require different methods of treatment. A knowledge of the anatomical and cellular basis of the decay is essential to arrest it, and the histological qualities of specific plant materials will affect the choice of conservation treatments.

Plant tissues undergo deterioration as a result of four general categories of decay processes: physical, chemical, mechanical, and biological. For example, the chemical composition of plant cell walls changes over time following thermal degradation and chemical reaction. Microbiological agents such as bacteria and fungi can further break down the cellulosic component of cell walls, resulting in changes such as loss in tissue toughness and strength. Exposure to ultraviolet radiation can lead to the degradation of cell wall pits, followed by a gradual breakdown of wood fibers and other cells. Extensive anatomical damage can occur from insect attack. Wooden art objects and other antiques are particularly vulnerable to dimensional changes brought about by variations in the moisture content of the tissue. Such changes can lead to major physical alterations such as checking, bowing, cupping, twisting, and crooking. The degree of alteration is related to the amount of internal stress that is created in different directions by individual wood cells upon drying. This type of physically induced defect is further related to the thickness, pattern of grain, fiber type, and orientation within the wood and can have a major effect on both the appearance and usefulness of the art object. Wooden altarpieces are particularly vulnerable to physically induced internal stresses caused by improper drying and storage. To correct these harmful conditions, an extensive knowledge of wood deformation patterns is required. By attaching a dimension-stabilizing back to a panel, the original appearance of the piece can be restored and the expansion and contraction of the boards is accommodated. The conservation of water-logged wood objects and timbers presents special problems that require sophisticated techniques of stabilization and restoration. Developments in the treatment of water-logged wood and experiments on the diffusion and exchange of liquids inside wood are based upon knowledge of the detailed structure of different wood types.

Although wooden objects such as statues, coffins, and furniture may appear to be well preserved, the ultrastructural examination of wooden artifacts has shown that most specimens possess some degree of deterioration. Despite being buried in a dry tomb chamber for thousands of years, ancient Egyptian woods typically show soft rot and brown rot fungal decay as well as a nonbiological form of deterioration that produces cracks and fissures within the secondary wall and delamination of the middle lamella. Decay by soft rot fungi is characterized by the presence of longitudinal cavities within the secondary cell wall layers. In an advanced stage of decay, the wall loses essentially all its original strength. On the other hand, brown rot fungi produces secondary walls that are swollen, porous, or distorted. The recognition of deterioration in archaeological woods such as this provides valuable information that enables the museum curators to assess the general condition of the wood and to develop appropriate conservation and restoration procedures, as well as to enable them to plan for long-term preservation.

## DENDROCHRONOLOGICAL DATING

We saw earlier that dendrochronology was a multidisciplinary study, contributing important information to meteorology, archaeology, anthropology, architecture, and other fields. The study of growth rings to determine the age of wooden artifacts also plays an important role in the arts. Research in the field of visual art and musical history often lacks objective methods of dating. Not only is it not known exactly when certain paintings were completed, but, in the absence of such information, it is not possible to interpret style changes

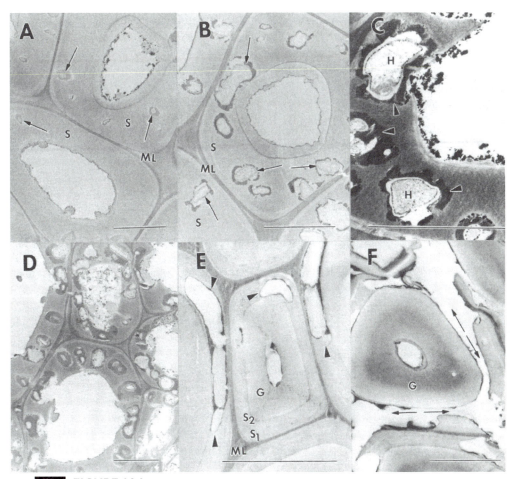

**FIGURE 16.1**   Transmission electron micrographs showing soft rot cavities within transverse sections of cell walls from wooden Egyptian objects in the Museum of Fine Arts, Boston. Reprinted with permission of Blanchette *et al.* (1994), *J. Am. Inst. Conserv.* **33,** 55–70. (A) MFA 21.899; (B) MFA 01.7431; (C) and (D) MFA 21.897; and in the Metropolitan Museum of Art: (E) and (F) MMA 27.9.5. (A) Incipient stages of decay showing small cavities (arrows) within the secondary wall layers (S). Middle lamellae (ML) are indicated. (B) Large cavities (arrows) within the secondary wall (S), with remnants of fungal hyphae located within the cavities. (C) Soft rot cavities within the cell wall have a large central void with fungal hyphae (H) present. The area of the secondary wall around the degraded hole is extremely electron dense (arrowheads). (D) Advanced stages of decay, with numerous cavities throughout secondary wall layers. Cavities coalesce, forming large holes within the cell wall that reduce wood strength and integrity. Middle lamellae regions remain relatively free of attack but often fragment due to the lack of intact secondary wall material. (E) An unusual form of soft rot with cavities in the interior portions of the woody cells. The thick wall layer within cells resembles the gelatinous layer (G) found in tension wood. Cavities (arrowheads) are restricted to the reduced $S_1$ and $S_2$ regions of the cell wall and appear elongated. (F) Cells with advanced decay have eroded $S_1$ and $S_2$ layers (arrows) and fragmented middle lamellae. The thick gelatinous layer (G) is not degraded except for small diameter cavities that resemble hyphal penetration sites. Bar = 5 μm.

by an artist over time. Dendrochronology offers a practical method for precise dating and thus is one of the most important tools available to the art historian. The evidence it provides can serve as positive evidence of art historical attribution or suggest that an attribution should be reevaluated.

Researchers at the University of Hamburg have shown that dendrochronological analysis is especially valuable for anonymous oil paintings done on wooden panels, especially those on oak boards from the Middle Ages to the period of the 17th century. Paintings of this type were often neither signed nor dated and are sometimes ascribed to an artist by indirect evidence. In some instances, dendrochronology can determine if an original wood panel painting is the correct age to have been done by a particular artist, or if it is a copy. The great value of this approach is that it establishes a felling date for the tree from which the painted panel wood was cut, thereby suggesting reference points in time for the painting's completion. By this method, we can establish whether different paintings were done on boards cut from the same tree or from different trees. For example, the central panel and separate side panels of an altarpiece may have come from the same or different trees. Individual boards were sometimes also sawed in sections and used in different panels. It also is important to know whether the frame is contemporary with the panels. If it can be shown that different boards came from the same tree, it is sometimes possible to make further attribution to a particular workshop. By studying wood structure, it often is possible to provide evidence for the reconstruction of an altarpiece and to show that pieces were added at a later date.

Dendrochronological analysis uses the fact that many trees in northern temperate regions produce secondary xylem with recognizable climate-determined growth rings of varying width, which indicate periods varying in favorability for plant growth. By carefully measuring the width of each ring across the wood cross section and recording ring width and age, a characteristic tree ring calendar or curve can be established for a specific tree species from a particular geographic area. By using the technique of cross-dating discussed in Chapter 12, a chronology can be constructed that extends back hundreds or thousands of years. The dated master calendar can then be matched against the measured growth ring pattern of an undated wood to resolve questions of age. Long-term calendars for oak are now available from parts of Germany and elsewhere in Europe. Unfortunately, softwood panels were utilized in some parts of Europe. Only woods with at least 50 growth rings can be used in tree ring analysis. The youngest rings of a tree, those present in the sapwood, provide the essential information for determining the felling date of a tree on whose sawn panels oil paintings were created. Because dendrochronological dating is of the wood rather than the painting, only estimates can be given of the length of time between tree felling and the creation of the painting. The elapsed time during which the wood was stored and dried could have varied widely, making this a controversial step in the dating process. To determine the exact felling date of the tree, a panel must contain at least 50 to 100 growth rings and sapwood must be present in order to have evidence of the latest growth rings. Unhappily for the art historian, early craftsmen routinely cut the sapwood off many panels, so a somewhat controversial method of estimating the number of sapwood rings has been used. Some old panels

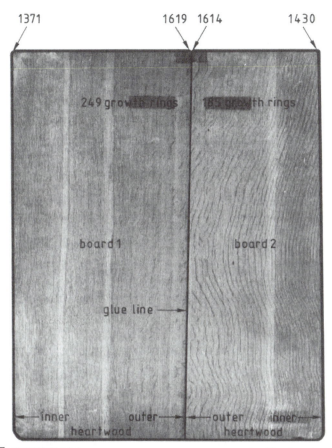

1371        1619   1614        1430

249 growth rings    185 growth rings

board 1        board 2

glue line ⟶

inner     outer ⟶ ⟵ outer   inner

heartwood        heartwood

**FIGURE 16.2**   Woodbiological investigation on panels of Rembrandt paintings. A 17th century oak panel of two boards, size 62.2 cm × 50 cm, with thickness 1 cm. Reprinted with permission of Bauch and Eckstein (1981), *Wood Sci. Technol.* **15**, 251–263.

were reused as overpainted boards many years after the original tree felling, so the art historian must be aware of this possibility and use x-ray analysis to determine if there was an earlier painting.

In spite of these caveats, dendrochronological research has proven valuable in judging the authenticity of many paintings, given that a painting could not have been executed prior to the year of felling of the tree from which the panel wood came. If it can be determined that a tree was felled after an artist is known to have died, then the painted panel in question could only have been produced by another individual. By applying dates to panels, it also may be possible to observe subtle style changes in a painter's work over time that may not have been obvious otherwise. The value of this approach can be illustrated with a typical example from the Detroit Institute of the Arts. The Flemish wood panel painting entitled *St. Jerome in His Study* has been attributed to the artist Jan Van Eyck, although for stylistic reasons questions have arisen as to whether it perhaps represents a skillful forgery. Dendrochronological examination of the oak panel indicated that the board was obtained from a tree

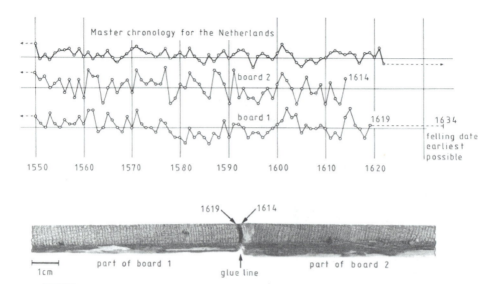

**FIGURE 16.3** Example of the dendrochronological dating of the oak wood of the 17th century panel pictured in Figure 16.2. Reprinted with permission of Bauch and Eckstein (1981), *Wood Sci. Technol.* **15,** 251–263.

felled in 1415 ± 5 years. Because it is estimated that boards in 15th century Flanders were usually dried in storage for about 20 years prior to use, the wooden panel would date fromVan Eyck's lifetime.

It is sometimes also possible to trace the trade routes of oak timbers from their provenance to the site of their use. This is evident in the westward movement of oak timber from Prussia and Poland to the the Netherlands and England, where they were used in the creation of late medieval paintings. Unfortunately, because complete standard growth ring curves exist for only a few hardwood species, it is not possible to apply the technique of dendrochronology to all the woods used in panel painting.

## GREEN-STAINED WOOD

Recent examination of the green-stained wood from the panels of the Italian Renaissance Gubbio studiolo in possession of the Metropolitan Museum of Art in New York City represents an outstanding example of the close relationship between plant anatomical research and the arts. The Gubbio **studiolo** is a small, intricately paneled room that was built for the statesman Federico da Montefeltro for his ducal palace in Urbino, Italy, around 1480. The room, which is considered one of the most significant creations in Renaissance art, consists of panels that show highly refined and delicately composed works of art using the technique of **intarsia** or wood inlay. This technique uses numerous small, carefully cut wood inlays about 5 mm thick placed into a wooden matrix to form detailed images. Intarsia craftsmen were extremely skilled and had to be very knowledgeable of wood structure. Small

**FIGURE 16.4**   *Saint Jerome in His Study,* c. 1435, Jan van Eyck. Oil on linen paper on oak.
Reproduced with permission of The Detroit Institute of Arts.

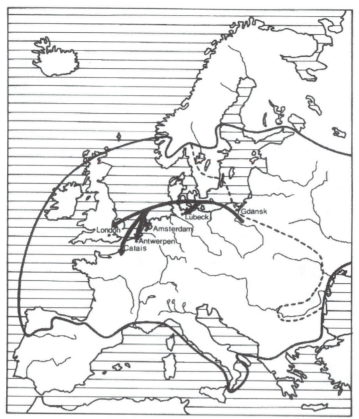

**FIGURE 16.5** The regions of the natural distribution of oak. Distribution of *Quercus robur* L. (European oak) is shown as a heavy line; distribution of *Q. petraea* Liebl. (sessile oak) is shown as an interrupted line. European oak originates farther northeast than does sessile oak. The source of oak timbers of a particular tree ring pattern and the places of their utilization as panels are indicated by arrows. Reproduced by permission of The Metropolitan Museum of Art from *Petrus Christus, 1994.*

pieces of wood representing different planes of wood and grain patterns were expertly used to create extraordinary three-dimensional designs and paintings. Attempts to fully recreate these images today have generally been unsuccessful. As noted by researchers, subtle variations in wood grain and color were utilized in the production of these masterpieces.

Locally available woods ranging from very light in color to extremely dark were included in the panels. In some instances, an unusual green wood of brilliant color was employed. In the intarsia panels of the Gubbio studiolo, this wood created the effect of natural scenery, trees, and foliage. Until recently it had been widely assumed that the green wood was produced using artificial stains. During restoration work an effort was made to clarify the origin of the green wood pieces. Small segments of green wood were subjected to careful anatomical observation at the light and electron microscope levels. As a result of this research, it is now known that this color is not produced by the application of inorganic stains but is a natural condition, caused by the fungal genus *Chlorociboria*. If inorganic dyes had been used, the stain would be uniformly distributed throughout the xylem tissue and would be especially evident in the

**FIGURE 16.6**   Green-stained wood in intarsia masterpieces. (A) Photograph of a hand-colored print showing a variety of naturally colored wood blocks from a German book published in 1773. A block of green-stained *Populus* wood is located in the center. This figure is used courtesy of Smithsonian Institution Libraries, Cooper-Hewitt Branch, New York. (B, C) Greenish blue fruiting bodies of *Chlorociboria* on the surface of wood from *Populus* sp. collected from the forest. (D, E) Green stain within wood from dead *Populus* sp. trees recently collected from the forest. (F, G) Light micrographs of transverse sections made from green-stained wood in the Gubbio *studiolo*. A combination of dark green and yellowish-orange pigments are located in ray parenchyma cells and in some fibers and vessels. Extensive coloration is found within the ray parenchyma cells. Reprinted with permission of Blanchette *et al.* (1992), *Holzforschung* **46**, 225–232. (See Color Plate.)

large vessel elements. Thin sections of green-stained wood from the intarsia panels, however, show colored deposits concentrated only in cells where fungal hyphae are present. Green and yellow-orange deposits usually were observed within the ray parenchyma cells adjacent to the rays. Vessels contained only sparse deposits. Cells with hyphae also showed some evidence of erosion of the secondary wall. The biological research demonstrated that fungal-induced green-stained wood was used by intarsia workers as early as the 15th century. Similar colored wood can still be collected naturally today.

## IDENTIFICATION OF ANTIQUE UPHOLSTERY FILLING AND WOODEN FURNITURE

Today's art historians and conservators use a diverse blend of disciplines to study objects and answer many fundamental questions. Plant anatomy is one course of study that has enabled researchers to reinterpret historic upholstery in order to identify the original filling materials that they contain. Documenting the filling of early padded seating has potential value for finding possible patterns in the use of plant materials, discovering regional preferences for particular materials, and even in clarifying the place of origin of a specific piece of furniture. Although it is known that plant materials were used as filling during colonial times, little detailed information on species identification is available.

By the 17th century the use of plant materials as seat stuffing in upholstered furniture was well established, although by the second quarter of the 18th century curled horsehair began to be more commonly used. Grasses and grasslike taxa were the plants of choice in both England and America. A microscopic examination of the original upholstered furniture in the collection at Colonial Williamsburg has revealed 15 species of true grasses, in addition to rushes, sedges, and species from assorted other families. Anatomical characters are important in grass systematics. Microscopic characters of the lamina, such as epidermal cell patterns in surface view, the distribution and frequencies of long cells and short cells, the condition of dermal appendages, and stomatal patterns, are all diagnostic but then require analysis in order to maximize the amount of information obtainable. In some instances the identification of plant species found only in America can be used to certify a chair's American origin.

An important concern of furniture curators and collectors is the correct identification of the woods used by early cabinetmakers. Harry A. Alden has applied an understanding of wood anatomy in an effort to document woods from antique furniture at the Henry Francis du Pont Winterthur Museum. He has discovered that not all furniture pieces described as mahogany are made of true mahogany (*Swietenia* sp.). Following an analysis of 150 pieces of so-called mahogany, furniture researchers found that the primary wood in six pieces was from a plant called sabicu, belonging to the tropical genus *Lysiloma*. *Lysiloma latisiliquum* is classified in the family Leguminosae and grows naturally in the West Indies and adjacent areas. The wood of sabicu was imported into the United States from the Bahamas during the late 1700s as a commercially important wood. It was used as a primary wood in various pieces of colonial furniture, either by mistake or purposefully, because of its unusual grain pattern. In either case, it was generally sold as mahogany under the names blonde or black mahogany.

## ANTIQUE MUSICAL INSTRUMENTS

The role of anatomy in historical analysis can also be illustrated by the identification of woods in musical instruments. Very little reliable information is available concerning the materials used in antique instruments, despite the fact that such information can be of critical importance in establishing the geographic origins of historical instruments. For modern instrument makers who replicate the acoustic qualities of historical violins, harpsichords, and pianos, information about the original woods is invaluable. The identification of certain woods in particular musical instruments and pieces of furniture also can clarify patterns of trade during the colonial period, and tell us whether an individual piece was of American or European origin.

Professor John Koster of the Shrine to Music Museum and the University of South Dakota has provided the following interesting case study as part of an extensive anatomical study of early (pre-1840) keyboard instruments. J. C. Schleip was an active piano maker in Berlin between 1816 and 1850. Schleip used woods of as many as 10 species in the construction of a single piano. These included both traditionally used native woods (*Abies, Acer, Fagus, Picea*, and *Quercus*) as well as nonnative woods with desirable physical and aesthetic properties. Some tropical woods also were incorporated into his instruments. According to Professor Koster, the use of foreign woods can be used to document the desire for exotic timbers in Europe in the early 19th century and also to trace the path of global trade in these materials. Especially interesting is Schleip's use of rattan palm (*Calamus*) "wood" for his hammer shanks. This is the earliest known instance of a Southeast Asia/Pacific Rim forest product being used in a Western musical instrument and is an almost unique use of monocotyledonous material. Previous reports of bamboo being used in musical instruments during the 18th century are probably erroneous.

## MUSICAL INSTRUMENTS AND WOOD TECHNICAL PROPERTIES

### Stringed Instruments

Historically, wood has been the most important class of material used in the manufacture of musical instruments. Each plant species possesses a unique combination of properties that interact to produce the overall technical quality of musical resonance. Because wood properties are variable, resonance properties also vary with the condition of the wood. The art of instrument making lies in an understanding of wood organization and properties, along with the ability to work those woods having the desired qualities. Instruments of the violin family are constructed as an almost totally enclosed wooden sounding box, with a set of tightened strings connected to an upper plate called the belly. When the strings are vibrated by the bow, energy from the motion of the strings is directed to the box and the closed air space. The dimensions, anatomy, and treatment of the wood of which the box is made are responsible for the sound quality produced by the instrument. Although the value of wood in the manufacture of musical instruments has long been appreciated, much remains to be learned about the precise relationships between structural, physical, and mechanical properties that govern its overall response and acoustic behavior. These properties change with time and with different drying procedures and chemical treatments.

Numerous attempts have been made to clarify and quantify these complex relationships by carrying out detailed tests and measurements followed by complex theoretical calculations. This experimentation has only reinforced the view that a full understanding of vibration theory and wood resonance properties remains an elusive goal. This is due in part to the complex nature of wood, its irregular properties, and the fact that the woods chosen by instrument makers vary with location and with time. A discussion of the physics of acoustic transmission theory is beyond the scope of this book; however, a few important features of wood that determine the behavior and properties of wooden instruments can be mentioned. The sound or tone produced when a thin piece of wood is struck or set to vibrating is directly related to the shape, size, type, and condition of the wood. In the case of stringed instruments and pianos, the vibration of the strings is communicated to the wood sounding board that is exposed to the vibration, and the wood itself undergoes reinforced vibration at about the same frequency as the strings, thereby changing the string's vibrating energy into the amplified sound waves that we hear. Such reinforcement and prolongation of a sound by reflection or by the vibration of other bodies is called **resonance**. The tone that a particular wood produces is variously influenced by its density, deformation, and the variable and measurable characteristics of elasticity along and across the grain. Wood stiffness, shear, growth ring width, moisture content, texture, and uniformity of grain throughout the wood are also relevant features. The thickness and degree of arching of the wood further influence its vibration properties.

Internal friction (also called internal damping) is another important acoustic feature. Damping is a measure of the ratio of energy dissipated to the energy stored elastically. This property results in a loss of acoustic energy and therefore reduces the ability of an instrument to produce sound. Damping is generally expressed as the quality factor or Q and is directly affected by the water absorbed by wood and by fluctuations in temperature. The higher the Q value, the lower the damping. A violin is composed of a top and back plate. Traditionally these plates are carefully crafted from solid or glued pieces of wood. The top plate is made from two pieces of straight-grained, quarter-cut spruce (*Picea abies*) joined down the middle. The back plate is usually maple (*Acer platanoides*) that has a curly grain pattern. Because the vibrational characteristics of both top and back plates are determined by the structure of their woods, they also reflect the life history of the particular tree from which they were derived and the conditions under which it grew. Violin makers must arrive at the correct proportions among such variables as wood size, thickness, and stiffness in order to produce the desired resonance of a violin.

The most famous stringed instruments ever made were crafted at the close of the Renaissance by a group of Italian artists working in the city of Cremona. These Cremonese instruments were produced by such masters as Nicolo Amati, Antonio Stradivari, and Giuseppi Guarneri and are generally regarded to be the most nearly perfect European instruments. Their richness and evenness of tone has never been matched by contemporary Western craftsmen. Researchers have tried to explain why the Cremona sound has not been duplicated. Some experts have suggested that the varnishes and other chemical wood treatments were critical factors, whereas others have focused on the construction aspects and wood acoustic properties.

One recent hypothesis has directed attention to the anatomical features of violin woods. When a violin is played, not only does the wood vibrate but the air trapped within the wood-conducting cells also vibrates. Air-dried, untreated softwood tracheids normally have most of their pit apertures covered by pectin-rich tori. One examination of tracheids from small pieces of Guarneri violin wood appeared to reveal a very high percentage of open, nonaspirated bordered pits. Some have theorized that this open microstructure of the wood cells, coupled with particular varnishes, created optimal qualities. It also has been hypothesized that this structural condition is directly related to the procedure of soaking wood in sea water prior to construction. The longer the period of soaking, the greater the number of open pits. The practice of soaking wood allows minerals to diffuse into the wood. Furthermore, stiffness is greatly reduced across the grain compared to along the grain. Because wood was normally delivered and stored in mineral-rich Italian bays, it also may have incidentally become mineralized. Other workers have found no evidence for this theory. Careful scanning electron microscope observations of a variety of early instruments have revealed no undamaged bordered pits with membranes absent or destroyed by soaking in salt water.

Absolutely straight-grained wood is essential for quality instrument construction. The growth rings must have a smooth transition between earlywood and latewood and the densities of earlywood and latewood must fall within a restricted range. Wood containing large pores, knots, cross grain, resinous ducts, rays of different sizes, and irregular growth rings is rejected as unsuitable. For critical parts, violin and piano makers knowingly or unknowingly selected woods according to anatomical criteria, that is, clear, perfectly straight-grained, well-seasoned wood of uniform structure and proper texture and density. The wood is quarter sawn or split, thus producing a grain orientation that ensures maximum dimensional stability and strength.

Pianos are made by using several species of wood. Each wood is selected on the basis of its strength, weight, stiffness, and hardness characteristics and the stresses that will be applied. High-quality spruce has been the wood of choice for making sounding boards because it is straight grained and strong and has outstanding acoustical properties. One of the most desirable properties of spruce is its high ratio of stiffness to density, which translates into a high sound velocity. All solid spruce sound boards are installed with the grain running diagonally in order to capture the best tone. To achieve the best resonance, wood is not excessively stressed during the construction process. Individual fibers must be free to vibrate. An example of how a small

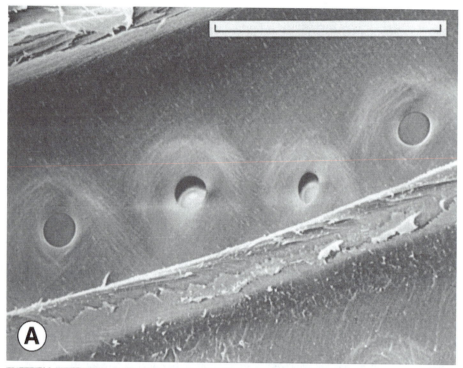

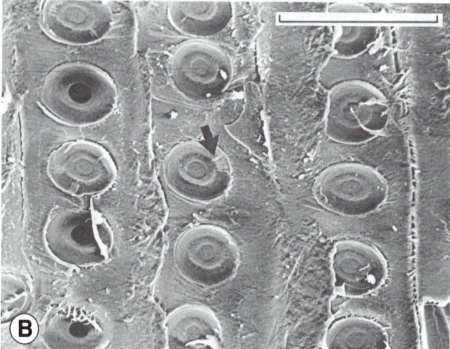

**FIGURE 16.7**   Anatomy of old violins. (A) Sample from a 1704 Stradivari violin. All the pits are aspirated, but the membranes in the central pair of pits are attached to the lower halves of the pits and are bright due to charging, scale bar = 20 μm. (B) Sample from a violin by Maggini, © 1610. Part of the structure of the margo in some of the pits is visible as radial filaments (arrow), scale bar = 50 μm. Reprinted with permission of Barlow and Woodhouse (1990), *J. Microscopy* **160**, pt. 2, 203–211, copyright © The Royal Microscopical Society.

anatomical variation can affect acoustical properties is seen among woods with the infrequent feature of indented growth rings. Since the 17th century, many famous Italian violin makers have preferred spruce wood with indented growth rings for the construction of soundboards. Recent data have shown that the occurrence of such growth ring indentations clearly affects the elastic and acoustical properties of the wood. Of course, musical instruments possess their own individual characteristics, and it remains a mystery why only some instrument makers were able to construct instruments with exceptional tonal qualities.

## Wind Instruments

Woodwind instruments, such as the clarinet, flute, oboe, and bassoon, produce a musical tone when a thin strip of flexible material vibrates against the opening of the mouthpiece when a current of air is blown into the narrow opening. Woodwind musicians use a thin strip of plant stem tissue called a reed to produce sound, and the most commonly used reed is derived from the stem of the giant grass *Arundo donox,* a plant native to the Mediterranean region.

We now know that there is a relationship between the anatomical characteristics of these reeds and their musical performance. Reed cane growers try to understand this relationship. Detailed comparisons of reeds grown under cultivated plantation conditions and those originating from agricultural windbreak plants have revealed a number of anatomical characteristics that show statistically significant differences in musical performance categories. Superior reeds possess vascular bundles in the inner cortex with a higher proportion and area of fibrous cells ensheathing each bundle in a continuous ring. These bundles also contain a lower proportion of vascular tissue. No correlation is evident between musical quality and the numerical density of vascular bundles in the stem cortex. The correlation between fiber abundance and the acoustical properties of the *Arundo donax* reed can be related to reed stiffness. Stiff reeds result in good performance quality and the extraordinary stiffness of good *Arundo* reeds is due to the high numbers of fibrous cells in the inner cortex.

## SUMMARY

In many instances, art objects, antiques, and musical instruments owe their unique qualities directly to their botanical origins. As a consequence, it is possible to apply the special techniques of plant anatomy to such materials in an effort to help solve a number of fundamental questions relating to their history, preservation, and conservation. Plant anatomical evidence can be used to document attribution, date, and degree of deterioration and to identify the materials used in object construction. Often only fragments of materials are available for examination and the ability to provide correct identification based on one or a few cells is essential. The trained art conservator has the ability to distinguish plant cell and tissue types and to relate them to particular groups of plants. Wooden objects often possess some cellular deterioration, and anatom-

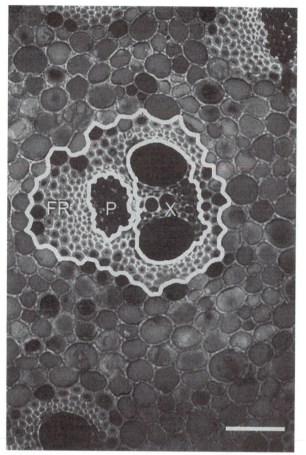

**FIGURE 16.8**   Confocal image of vascular bundle in inner cortex of clarinet reed manufactured from *Arundo donax.* Abbreviations: FR, fiber ring; P, phloem; X, xylem. Bar = 100 μm. Reprinted with permission of Kolesik *et al.* (1998), *Ann. Bot.* **81,** 151–155.

ical examination at the light and electron micrograph levels is essential in order to repair the damage. Deterioration occurs as a result of four general categories of decay processes: physical, chemical, mechanical, and biological. The recognition of decay provides valuable information that enables the museum curator to develop appropriate conservation and restoration procedures.

Dendrochronology, the study of growth rings in wood, offers an important and practical method for definitive dating and is one of the most important tools available to the art historian. Dendrochronological analysis is especially valuable for dating oil paintings on wooden panels, such as those produced on oak boards from the Middle Ages until the 17th century. Dendrochronological research has proven valuable in judging the authenticity of paintings given that a painting could not have been executed prior to the year of felling of the tree from which the panel board was constructed. It also is possible to trace the trade routes of timbers from their place of origin to the site of their utilization. Plant anatomy has also provided answers to questions relating to materials used in the construction of antique objects, furniture, and musical instruments.

**TABLE 16.1** Comparison of the Anatomical Characteristics and Musical Performance of Clarinet Reeds Manufactured from *Arundo Donax* and Derived from Cultivated Plantations or Agricultural Windbreaks

| ANATOMICAL CHARACTERISTIC OF VASCULAR BUNDLES IN INNER CORTEX | LOCALITY | | | | | | | | | | | | |
| --- | --- | --- | --- | --- | --- | --- | --- | --- | --- | --- | --- | --- | --- |
| | Plantation | | | | | | Windbreak | | | | | | |
| | MUSICAL QUALITY | | | | | | | | | | | | |
| | Good | | Fair | | Poor | | Good | | Fair | | Poor | | n |
| | Mean | s.e.m. | Mean | s.e.m. | Mean | s.e.m. | Mean | s.e.m. | Mean | s.e.m. | Mean | s.e.m. | |
| % fiber | 63.2[a] | 0.76 | 59.5[a,b] | 0.60 | 56.0[b] | 0.77 | 60.6[a] | 0.82 | 56.2[b] | 0.83 | 50.2[c] | 0.79 | 50 |
| Area of fiber (1000 $\mu m^2$) | 44.0[a] | 1.28 | 38.1[a,b] | 1.33 | 32.3[b] | 1.00 | 36.8[a,b] | 0.99 | 37.8[a] | 1.23 | 32.3[b] | 0.95 | 50 |
| % xylem | 27.2[a] | 0.63 | 28.8[a,b] | 0.46 | 32.4[b] | 0.69 | 29.3[a] | 0.64 | 33.1[b] | 0.69 | 37.1[c] | 0.69 | 50 |
| 50% phloem | 9.6[a] | 0.26 | 11.7[b] | 0.25 | 11.6[b] | 0.35 | 10.0[a] | 0.31 | 10.8[a] | 0.26 | 12.8[b] | 0.24 | 50 |
| % with continuous fiber ring | 88.2[a] | 2.05 | 82.0[a] | 2.23 | 50.3[b] | 4.64 | 65.6[a] | 3.35 | 46.6[b] | 4.67 | 57.7[a,b] | 5.99 | 10 |

Different superscripts within characteristics at both localities indicate a significant difference ($P<.05$); s.e.m., standard error of means; *n*, number of replicates.
Reprinted with permission from Kolesik *et al.* (1998). *Ann. Bot.* **81**, 151–155.

All woods and stems possess a certain combination of interacting properties that produce the overall technical quality of resonance. Resonance is important when selecting wood for the construction of musical instruments. The construction and wood acoustic properties of old violins have received considerable scientific attention. Recent investigations have directed our attention to fine details of wood anatomy (e.g., pit structure) in relationship to musical properties. An important connection between the anatomical characteristics of reeds and their musical performance has been found. Reeds showing superior musical performance possess vascular bundles in the inner cortex with a high proportion and area of fibrous cells ensheathing each bundle in a continuous ring.

## ADDITIONAL READING

1. Barlow, C. Y., and Woodhouse, J. (1990). Bordered pits in spruce from old Italian violins. *J. Microscopy* **160**, (2), 203–211.
2. Bauch, J. (1978). Dendrochronology applied to the dating of Dutch, Flemish and German paintings. *In* "Dendrochronology in Europe"(J. Fletcher, Ed.), pp. 307–314. BAR International, Ser. 51.
3. Bauch, J., and Eckstein, D. (1970). Dendrochronological dating of oak panels of Dutch seventeenth-century paintings. *Studies in Conservation* **15**, 45–50.
4. Bauch, J., and Eckstein, D. (1981). Woodbiological investigations on panels of Rembrandt paintings. *Wood Sci. Technol.* **15**, 251–263.
5. Bauch, J., Eckstein, D., and Meier-Siem, M. (1972). Dating the wood of panels by a dendrochronological analysis of the tree-rings. *Nederlands Kunsthistorisch Jaarboek* **23**, 485–496.
6. Blanchette, R. A., and Simpson, E. (1992). Soft rot and wood pseudomorphs in an ancient coffin (700 B.C.) from Tumulus MM at Gordion, Turkey. *IAWA Bull N. S.* **13**, 201–213.
7. Blanchette, R. A., Cease, K. R., Abad, A. R., Koestler, R. J., Simpson, E., and Sams, G. K. (1991). An evaluation of different forms of deterioration found in archaeological wood. *Intern. Biodeterioration* **28**, 3–22.
8. Blanchette, R. A., Haight, J. E., Koestler, R. J., Hatchfield, P. B., and Arnold, D. (1994). Assessment of deterioration in archaeological wood from Ancient Egypt. *J. Am. Inst. Conserv.* **33**, 55–70.
9. Blanchette, R. A., Nilsson, T., Daniel, G., and Abad, A. (1990). Biological degradation of wood. *In* "Archaeological Wood: Properties, Chemistry, and Preservation"(R. M. Rowell and R. J. Barbour, Eds.), pp. 141–174. Advances Chem. Ser. No. 225.
10. Blanchette, R. A., Wilmering, A. M., and Baumeister, M. (1992). The use of green-stained wood caused by the fungus *Chlorociboria* in Intarsia masterpieces from the 15th century. *Holzforschung* **46**, 225–232.
11. Bucar, V. (1995). "Acoustics of Wood."CRC Press, Boca Raton, Florida.
12. Dunlop, J., and Shaw, M. (1991). Acoustical properties of some Australian woods. *Catgut Acoust. Soc. J.* **1**, 17–20.
13. Eckstein, D., Wazny, T., Bauch, J., and Klein, P. (1986). New evidence for the dendrochronological dating of Netherlandish paintings. *Nature* **320**, 465–466.
14. Florian, M.L E., Kronkright, D. P., and Norton, R. E. (1990). "The Conservation of Artifacts Made From Plant Materials." The J. Paul Getty Trust.
15. Goodway, M. (1987). Fiber identification in practice. *J. Am. Inst. Conserv.* **26**, 27–44.
16. Harris, M. (Ed.) (1954). "Handbook of Textile Fibers." Harris Research Lab, Washington, D.C.
17. Hinckley, F. L. (1960). "Directory of the Historic Cabinet Woods." Bonanza Books, New York.
18. Hutchins, C. M. (1962). The physics of violins. *Sci. Am.* **207**, 79–93.

19. Hutchins, C. M. (1981). The acoustics of violin plates. *Sci. Am.* **245**, 170–186.

20. Klein, P. (1991). The differentiation of originals and copies of netherlandish panel paintings by dendrochronology. *In* "Le Dessin Sous-Jacent dans la Peinture"(Verougstraete-Marq and R van Schoute, Eds.), pp. 29–42. College Erasme, Louvain-la-Neuve.

21. Klein, P. (1994). Dendrochronological analysis of panels attributed to Petrus Christus. *In* "Petrus Christus, Renaissance Master of Bruges" (M. W. Ainsworth), pp. 213–215. The Metropolitan Museum of Art, New York.

22. Kolesik, P., Mills, A., and Sedgley, M. (1998). Anatomical characteristics affecting the musical performance of clarinet reeds made from *Arundo donax* L. (Gramineae). *Ann. Bot.* **81**, 151–155.

23. McCrone, W. C. (1994). Polarized light microscopy in conservation: A personal perspective. *J. Am. Inst. Conserv.* **33**, 101–114.

24. Montegut, D., Indictor, N., and Koestler, R. J. (1991). Fungal deterioration of cellulosic textiles: A review. *Intern. Biodeterioration* **28**, 209–226.

25. Nagyvary, J. (1988). The chemistry of a Stradivarius. *Chem. Eng. News*, 24–31.

26. Ritman, K. T., and Milburn, J. A. (1991). Monitoring of ultrasonic and audible emissions from plants with or without vessels. *J Exp. Bot.* **42**, 123–130.

27. Rocaboy, F., and Bucur, V. (1990). About the physical properties of wood of twentieth century violins. *Catgut Acoust. Soc. J.* **1**, 21–28.

28. Schaffer, E. (1981). Fiber identification in ethnological textile artifacts. *Stud. Conserv.* **26**, 571–585.

29. Williams, M. A. (Ed.) (1990). Upholstery conservation. *Am. Conserv. Consort.*, East Kingston, New Hampshire.

# GENERAL ANATOMICAL REFERENCES

1. Ayensu, E. S. (1972). "Anatomy of the Monocotyledons. VI. Dioscoreales" (C. R. Metcalfe, Ed.). Clarendon Press, Oxford.
2. Bailey, I. W. (1954). "Contributions to Plant Anatomy." Chronica Botanica, Waltham, Massachusetts.
3. Biebl, R., and Germ, H. (1950). "Praktikum der Pflanzenanatomie." Springer Verlag, Wien.
4. Boureau, E. (1954–1956, 1957). "Anatomie végétale," 3 vols. Presses Universitaire de France, Paris.
5. Côté, W. A., Jr. (Ed.) (1965). "Cellular Ultrastructure of Woody Plants." Syracuse University Press, Syracuse, New York.
6. Carlquist, S. (1961). "Comparative Plant Anatomy." Holt, Rinehart & Winston, New York.
7. Cutler, D. F. (1969). "Anatomy of the Monocotyledons. IV. Juncales" (C. R. Metcalfe, Ed.). Clarendon Press, Oxford.
8. Cutler, D. F. (1978). "Applied Plant Anatomy." Longman Group Limited, London.
9. Cutler, D. F., and Gregory, M. (Eds.) (1998). "Anatomy of the Dicotyledons. IV. Saxifragales (sensu Takhtajan 1983)," 2nd ed., Clarendon Press, Oxford.
10. Cutter, E. G. (1969). "Plant Anatomy: Experiment and Interpretation. I. Cells and Tissues." Addison-Wesley Publishing. Co., London.
11. Cutter, E. G. (1971). "Plant Anatomy: Experiment and Interpretation. II. Organs." Addison-Wesley Publishing Co., London.
12. DeBary, A. (1884). "Comparative Anatomy of the Vegetative Organs of the Phanerogams and Ferns" (F. O. Bower and D. H. Scott, Engl. Trans). Clarendon Press, Oxford.
13. Eames, A. J., and MacDaniels, L. H. (1947). "An Introduction to Plant Anatomy," 2nd. ed., McGraw-Hill, New York.
14. Esau, K. (1953). "Plant Anatomy." John Wiley & Sons, New York (2d. ed., 1965).
15. Esau, K. (1960). "Anatomy of Seed Plants." John Wiley & Sons, New York.
16. Esau, K. (1965). "Vascular Differentiation in Plants." Holt, Rinehart & Winston, New York.
17. Esau, K. (1969). The Phloem. *In* "Encyclopedia of Plant Anatomy," Band V, Teil 2. Gebrüder Borntraeger, Berlin.

18. Fahn, A. (1967). "Plant Anatomy." Pergamon Press, Oxford.
19. Fahn, A. (1990). "Plant Anatomy," 4th ed. Pergamon Press, Oxford, New York.
20. Foster, A. S. (1949). "Practical Plant Anatomy," 2d ed. D. Van Nostrand Co., New York.
21. Goebel, K. (1900–1905). "Organography of Plants," Parts 1 and 2. Clarendon Press, Oxford.
22. Guttenberg, H. von (1926). "Die Bewegungsgewebe." In "Handbuch der Pflanzenanatomie" (K. Linsbauer, Ed.), Vol. 5(2). Gebrüder Borntraeger, Berlin.
23. Guttenberg, H. von (1940). Die primäre Bau der Angiospermenwurzel. In "Handbuch der Pflanzenanatomie" (K. Linsbauer, Ed.), Vol. 8(3). Gebrüder Borntraeger, Berlin.
24. Guttenberg, H. von (1943). "Die physiologischen Scheiden. In "Handbuch der Pflanzenanatomie" (K. Linsbauer, Ed.), 5(4). Gebrüder Borntraeger, Berlin.
25. Guttenberg, H. von (1960). "Grundzuge der Histogenese Hoherer Pflanzen. I. Die Angiospermen." Gebrüder Borntraeger, Berlin.
26. Guttenberg, H. von (1961). "Grundzuge der Histogenese Hoherer Pflanzen II. Die Gymnospermen." Gebrüder Borntraeger, Berlin.
27. Haberlandt, G. (1914). "Physiological Plant Anatomy," (M. Drummond, Engl. Trans.), 6th ed. Macmillan & Co., London.
28. Hayward, H. E. (1938). "The Structure of Economic Plants." Macmillan & Co., New York.
29. Hooke, R. (1665). "Micrographia or some physiological descriptions of minute bodies made by magnifying glasses with observations and inquiries thereupon." The Royal Society (reprinted 1961, Dover Publ.,).
30. Huber, B. (1961). "Grundzüge der Pflanzenanatomie." Springer-Verlag, Berlin.
31. Jane, F. W. (1965). "The Structure of Wood." Adam & Charles Black, London.
32. Jeffrey, E. C. (1917). "The Anatomy of Woody Plants." University of Chicago Press, Chicago.
33. Kaussmann, B. (1963). "Pflanzenanatomie." Gustav Fischer Verlag, Jena.
34. Korsmo, E. (1954). "Anatomy of Weeds." Grondahl, Oslo.
35. Küster, E. (1925). "Pathologische Pflanzenanatomie," 3rd ed. Gustav Fischer, Jena.
36. Linsbauer, K. (1930). Die Epidermis. In "Handbuch der Pflanzenanatomie" (K. Linsbauer, Ed.), Vol. 4. Gebrüder Borntraeger, Berlin.
37. Mahlberg, P. G. (1972). "Laboratory Program in Plant Anatomy." William C. Brown Co. Publ., Dubuque.
38. Mansfield, W. (1916). "Histology of Medicinal Plants." John Wiley & Sons, New York.
39. Mauseth, J. D. (1988). "Plant Anatomy." Benjamin/Cummings, Menlo Park, California.
40. Metcalfe, C. R. (1960). "Anatomy of the Monocotyledons. I. Gramineae." Clarendon Press, Oxford.
41. Metcalfe, C. R. (1971). "Anatomy of the Monocotyledons. V. Cyperaceae." (C. R. Metcalfe, Ed.). Clarendon Press, Oxford.
42. Metcalfe, C. R. (1987). "Anatomy of the Dicotyledons. III. Magnoliales, Illiciales, and Laurales (sensu Takhtajan)," 2nd ed., Clarendon Press, Oxford.
43. Metcalfe, C. R., and Chalk, L. (1950). "Anatomy of the Dicotyledons," 2 vols. Clarendon Press, Oxford.
44. Metcalfe, C. R., and Chalk, L. (1979). "Anatomy of the Dicotyledons." I. Systematic Anatomy of Leaf and Stem, With a Brief History of the Subject," 2nd ed. Clarendon Press, Oxford.
45. Metcalfe, C. R., and Chalk, L. (1983). "Anatomy of the Dicotyledons. Wood Structure and Conclusion of the General Introduction," 2nd ed. Clarendon Press, Oxford.
46. Meyer, F. J. (1962). Das trophische Parenchyma A. Assimilationsgewebe. In "Handbuch der Pflanzenanatomie" (K. Linsbauer, Ed.) Vol. 4(7A). Gebrüder Borntraeger, Berlin.
47. Meylan, B. A., and Butterfield, B. G. (1972). "Three-Dimensional Structure of Wood: A Scanning Electron Microscope Study." Syracuse University Press, Syracuse, New York.
48. Miller, R. H. (1960). "Morphology and Anatomy of Roots." Scholar's Library, New York.
49. Möbius, M. (1927). Die Farbstoffe der Pflanzen. In "Handbuch der Pflanzenanatomie" (K. Linsbauer, Ed.), Vol. 3(1/1). Gebrüder Borntraeger, Berlin.
50. Molisch, H. (1947). "Anatomie der Pflanze," 5th ed. Gustav Fischer, Jena.
51. Napp-Zinn, K. (1966). Anatomie des Blattes. I. Blattanatomie der Gymnospermen. In "Encyclopedia of Plant Anatomy," Band VIII, Teil 1. Gebrüder Borntraeger, Berlin.
52. Napp-Zinn, K. (1973). Anatomie des Blattes. II. Blattanatomie der Angiospermen. In "Encyclopedia of Plant Anatomy," Band VIII, Teil 2A, Vol. 1. Gebrüder Borntraeger, Berlin.
53. Napp-Zinn, K. (1974). Anatomie des Blattes. II. Blattanatomie der Angiospermen. In "Encyclopedia of Plant Anatomy." Band VIII, Teil 2A, Vol. 2. Gebrüder Borntraeger, Berlin.

54. Napp-Zinn, K. (1984). Anatomie des Blattes. II. Blattanatomie der Angiospermen. *In* "Encyclopedia of Plant Anatomy," Band VIII, Teil 2B. Gebrüder Borntraeger, Berlin.

55. Netolitzky, F. (1929). Die Kieselkorper als Zellinhaltskorper. *In* "Handbuch der Pflanzenanatomie" (K. Linsbauer, Ed.), Vol. 3(la). Gebrüder Borntraeger, Berlin.

56. Netolitzky, F. (1932). Die Pflanzenhaare. *In* "Handbuch der Pflanzenanatomie" (K. Linsbauer, Ed.), Vol. 4(4). Gebrüder Borntraeger, Berlin.

57. O'Brien, T. P., and McCully, M. E. (1969). "Plant Structure and Development. A Pictorial and Physiological Approach." Macmillan Co., London.

58. Pfeiffer, H. (1928). Die pflanzlichen Trennungsgewebe. *In* "Handbuch der Pflanzenanatomie" (K. Linsbauer, Ed.), Vol. 5(3). Gebrüder Borntraeger, Berlin.

59. Popham, R. A. (1952). "Developmental Plant Anatomy." Long's College Book Co., Columbus, Ohio.

60. Popham, R. A. (1966). "Laboratory Manual for Plant Anatomy." C. V. Mosby Co., St. Louis.

61. Rauh, W. (1950). "Morphologie der Nutzpflanzen." Quelle & Meyer, Heidelberg.

62. Record, S. J. (1934). "Identification of the Timbers of Temperate North America." John Wiley, New York.

63. Roelofsen, P. (1959). The plant cell wall. *In* "Handbuch der Pflanzenanatomie" (K. Linsbauer, Ed.), III (4). Gebrüder Borntraeger, Berlin.

64. Romberger, J. A. (1963). "Meristems, Growth and Development in Woody Plants." U.S. Department of Agriculture Technical Bulletin. No. 1293.

65. Roth, I. (1981). Structural patterns of tropical barks. *In* "Handbuch der Pflanzenanatomie" (K. Linsbauer, Ed.) Vol. IX (3). Gebrüder Borntraeger, Berlin.

66. Rudall, P. (1995). "Anatomy of the Monocotyledons. VIII. Iridaceae" (D. F. Cutler and M. Gregory, Eds.). Clarendon Press, Oxford.

67. Sachs, J. von (1875). "Textbook of Botany." Clarendon Press, Oxford.

68. Schnee, L. (1939). Ranken and Dornen. *In* "Handbuch der Pflanzenanatomie" (K. Linsbauer, Ed.), Vol. 9, pp. 1–24. Gebrüder Borntraeger, Berlin.

69. Schürhoff, P. (1924). Die Plastiden. *In* "Handbuch der Pflanzenanatomie" (K. Linsbauer, Ed.), Vol. 1. Gebrüder Borntraeger, Berlin.

70. Solereder, H. (1908). "Systematic Anatomy of the Dicotyledons" (Boodle and Fritsch, Engl. trans.) 2 vols. Clarendon Press, Oxford.

71. Solereder, H., and Meyer, F. J. (1928–33). "Systematische Anatomie der Monokotyledonen." Gebrüder Borntraeger, Berlin.

72. Stover, E. L. (1951). "An Introduction to the Anatomy of Seed Plants." Heath & Co., Boston.

73. Tobler, F. (1957). Die mechanischen Elemente und das mechanische System. *In* "Handbuch der Pflanzenanatomie" (K. Linsbauer, Ed.), Vol. IV. Gebrüder Borntraeger, Berlin.

74. Tomlinson, P. B. (1961). "Anatomy of the Monocotyledons II. Palmae" (C. R. Metcalfe, Ed.). Clarendon Press, Oxford.

75. Tomlinson, P. B. (1969). "Anatomy of the Monocotyledons III. Commelinales-Zingiberales" (C. R. Metcalfe, Ed.). Clarendon Press, Oxford.

76. Tomlinson, P. B. (1982). "Anatomy of the Monocotyledons. VII. Helobiae (Alismatidae)" (C. R. Metcalfe, Ed.). Clarendon Press, Oxford.

77. Torrey, J. G., and Clarkson, D. (Eds.) (1975). "The Development and Function of Roots." Academic Press, San Diego.

78. Troll, W. (1937–1942). "Vergleichende Morphologie der Höheren Pflanzen," Band 1–3. Gebrüder Borntraeger, Berlin.

79. Tschirch, A. (1889). "Angewandte Pflanzenanatomie," 2 vols. Urban & Schwarzenberg, Wien.

80. Uphof, J. C. Th. (1962). Plant hairs. *In* "Handbuch der Pflanzenanatomie." (K. Linsbauer, Ed.), Vol. IV. Gebrüder Borntraeger, Berlin.

81. Zach, O. (1954). Die Anatomie der Blutenpflanzen." Franckhsche Verlagshandlung, Stuttgart.

# PLANT ANATOMY AND THE WORLD WIDE WEB

The following Web sites provide information dealing with plant anatomy. Many features may be viewed with explanatory text.

1. http://www4.ncsu.edu/unity/lockers/class/wps202002/Hardwoods/
   http://www2.ncsu.edu/unity/lockers/class/wps202001/VEL/vel.html

These programs were developed by Dr. Elisabeth Wheeler and deal with wood anatomy and wood properties. She may be contacted at the following address:

Dr. Elisabeth A. Wheeler
Department of Wood and Paper Science
North Carolina State University
P.O. Box 8005
Raleigh, North Carolina 27695-8005 USA
Phone: 919-515-5728
Fax: 919-515-3496
E-mail: xylem@unity.ncsu.edu

2. Atlas de Anatomia Vegetal—http://felix.ib.usp.br/atlasveg

This program deals with the stem, leaf, and root anatomy of dicotyledons and monocotyledons. This program was developed by Juliana Pisaneschi and Jane E. Kraus, who can be contacted at the following address:

Dr. Juliana Pisaneschi and Dr. Jane E. Kraus
Instituto de Biociencias da Universidade de Sao Paulo
Rua do Matao-trav. 14 No. 321. Caixa Postal 11461
CEP 05422-970
Sao Paulo, Brasil

3.  Computer-Assisted Wood Identification Systems

The following programs were compiled by Mr. J. Ilic and are on three high-density 3.5-inch disks: GUESS, IDENT8, CSIROID, WOOD, INTKEY, XID, GUYANA. Each of these systems is compressed using PKZIP. To install each of the systems do the following.

- Create a separate directory for each.
- Copy the compressed file (file with the extension ZIP) and PKUNZIP.EXE to the directory.
- At the DOS prompt of the newly created directory, type PKUNZIP *<file name>* (e.g., C:/GUESS>PKUNZIP GUESS.ZIP).
- Press ENTER.

The compressed file will decompress and create many new files.

## GUESS

GUESS program was developed by Dr. Elisabeth Wheeler et al. She may be contacted at the following address:

> Dr. Elisabeth A. Wheeler
> Department of Wood and Paper Science
> North Carolina State University
> P.O. Box 8005
> Raleigh, North Carolina 27695-8005 USA
> Phone: 919-515-5728
> Fax: 919-515-3496
> E-mail: xylem@unity.ncsu.edu

A reference manual entitled "Computer-Aided Wood Identification" is available from Dr. Wheeler for a small fee.

Start the program by typing GUESS at the DOS prompt. Read the introduction and follow the instructions as they appear on the screen. This operates in a batch mode. Select all or most of your characters and then search. All the other programs are interactive.

## IDENT8

Read the information contained in the READ.ME file. It will explain how to get started. The character list is not part of the program. A character list and publication explains how IDENT8 works.

## CSIROID

To start the program, type CSIROID. Follow the instructions on the screen. There are also help screens available (F1 key).

For more information about the program contact:

Mr. J. Ilic
Division of Forest Products
CSIRO
Private Bag 10
Clayton, Victoria 3168
Australia
Phone: +61 3 9545 2127
Fax: +61 3 9545 2133
E-mail: jugoi@forprod.csiro.au
*or* jugo.Ilic@ffp.csiro.au

There are also several books which complement the program, and these can be purchased from Mr. Ilic.

## WOOD

This program is based on the program INTKEY and the DELTA system. There are German and English versions. To start the English version, type WOOD.

There are many help screens to get you started. INTKEY is the most flexible and powerful of all the programs, but it is also initially the most complicated. There is one image in the data set. From the menu, select ILLUSTRATE and then select CHARACTERS.

For more information about DELTA, contact:

Dr. Mike Dallwitz
CSIRO, Division of Entomology
GPO Box 1700
Canberra, ACT 2601
Australia
Fax: +61 2 6246 4000
E-mail: delta@ento.csiro.au

## INTKEY

This package uses the same program as WOOD (see earlier discussion). To start, type INTKEY. The main difference between WOOD and INTKEY is the character list and data. Again, for more information about DELTA and INTKEY contact Mike Dallwitz.

## XID

To start this program, type XIDAUTH. Use the help screens or the tutorial. This is only a demo package, and there is no wood data with it.

## GUYANA

To start this program, type IDENT2. Help screens are available, but to maximize the use of the program, it should be used with its accompanying book. The book is entitled "Major Timber Trees of Guyana: A Lens Key" by Brunner, Kucera, and Zurcher. It was published by the TROPENBOS Foundation in 1994 and can be obtained from either TROPENBOS in Wageningen, The Netherlands, or the Swiss Federal Institute of Technology in Zurich, Switzerland.

# GLOSSARY

**Abscission zone,** region of structural weakness at base of petiole, peduncle, and pedicel that results in the separation of leaves, flowers, and fruits from the plant.

**Acicular crystal,** slender, needle-shaped crystal; not occurring in a multicrystal bundle.

**Acropetal,** applied to tissues and structures that develop in succession toward the apex.

**Actinocytic stoma,** mature stoma surrounded by subsidiary cells that are more or less radially elongated.

**Adventitious,** roots that arise from any organ other than a root or, in the case of buds, from other than terminal or axillary structures.

**Aerenchyma,** parenchymatous tissue containing large intercellular air spaces (lacunate.)

**Albuminous cell,** parenchymatous cell intimately associated with a sieve cell in gymnospermous phloem but not typically derived from the same initial as the sieve cell.

**Aliform axial parenchyma,** paratracheal wood parenchyma with winglike lateral extensions as viewed in transverse section.

**Alternate pitting,** multiseriate intervascular pitting in which the pits are distributed in diagonal rows; when pits are crowded, the outlines of the borders tend to become hexagonal in frontal view.

**Amphicribral vascular bundle,** concentrically structured vascular bundle with phloem completely surrounding the xylem.

**Amphivasal vascular bundle,** concentrically structured vascular bundle with xylem completely surrounding phloem.

**Amyloplast,** a colorless plastid that stores starch.

**Anastomosis,** union of one vascular bundle with another.

**Angular collenchyma,** living cell with the primary wall thickened at the corners; regarded as the commonest type of collenchyma.

**Anisocytic stoma,** mature stoma surrounded by three subsidiary cells of which one is distinctly smaller. (*Syn.* Cruciferous or Unequal-celled type.)

**Anisotropic,** having different optical properties along different axes. Optically, anisotropic materials polarize light resulting in the property of birefringence.

**Annular collenchyma,** living cell in which the primary wall is more or less uniformly thickened.

**Anomocytic stoma,** mature stoma surrounded by a limited number of cells that are indistinguishable in size, shape, or form from those of the remainder of the epidermis. (*Syn.* Ranunculaceous or Irregular-celled type.)

**Anticlinal division,** cell division at right angles, or perpendiclar, to the surface. In the cambium, a division of a fusiform initial in the radial longitudinal plane; also called pseudotransverse division.

**Apical,** at the point or summit of a structure or organ.

**Apical meristem,** collection of cells at the growing apices of shoot and root that divide to ultimately produce the primary plant body and maintain itself as a region of cell division.

**Apical zonation,** apex structure characterized by definable areas composed of cells that differ morphologically and physiologically from cells in other zones.

**Apoplast,** free space in tissue composed of cell walls and intercellular spaces.

**Apotracheal parenchyma,** axial wood parenchyma not associated with the vessels.

**Areole,** smallest area of leaf tissue surrounded by veins, which, taken together, form a contiguous field over most of the area of the lamina.

**Articulated laticifer,** a latex-containing system consisting of cells or a series of fused cells, which form a vessellike system.

**Aspirated pit,** bordered pit in which the pit membrane is laterally displaced so as to block the pit aperture.

**Astrosclereid,** branched sclerenchymatous element, typically having a stellate appearance.

**Atactostele,** stelar pattern characteristic of many monocotyledonous stems in which the vascular bundles appear "scattered" within the ground tissue.

**Axial cell,** a term of convenience in wood anatomy for all cells derived from the fusiform cambial initials.

**Axial wood parenchyma,** wood parenchyma derived from fusiform cambial initials and composing a part of the axial system.

**Banded wood parenchyma,** axial wood parenchyma distributed as narrow or wide tangential lines or bands; either independent of vessels or associated with vessels.

**Bark,** a nontechnical term used for all the tissues outside the vascular cambium. In older trees, usually divisible into inner living tissue (phloem) and outer dead cork.

**Basipetal,** applied to tissues and structures that develop in succession toward the base.

**Bast,** a term referring to the phloem, the inner fibrous portion of the bark.

**Bast fiber,** a fiber of the phloem or bark.

**Bicollateral vascular bundle,** vascular bundle construction with phloem on two sides (outer and inner) of the xylem.

**Bordered pit,** pit in which the pit membrane is overarched by the secondary wall.

**Bordered pit pair,** an intercellular pairing of two bordered pits.

**Brachysclereid,** a more or less isodiametric, massively thick-walled and lignified sclerenchymatous element. The wall is often conspicuously laminated and may contain ramiform pits. (*Syn.* Stone cell.)

**Bulliform cell,** thin-walled cell of the grass epidermis that is larger and more inflated than neighboring cells and together form a band parallel with the long axis of the leaf.

**Bundle sheath,** the single or multiple layer(s) of parenchymatous or sclerenchymatous cells surrounding a vascular bundle.

**Callose,** glucose polymer that is deposited during ontogeny in the sieve areas of sieve elements; usually lining each sieve pore. Callose is especially abundant following tissue injury.

**Cambial initial,** an individual cell of the cambium.

**Cambial zone,** term of convenience for the region of varying width composed of cambial initials and their immediate undifferentiated derivatives.

**Cambium,** an actively dividing layer of cells that lies between, and gives rise to, the secondary plant body; a lateral meristem. (*Syn.* Vascular cambium.)

**Casparian strip,** waxy band encircling some endodermal cells along their radial and transverse walls.

**Cauline,** pertaining to the stem; vascular bundles of the stem.

**Cavitation,** a break in the water column within a tracheary cell resulting from the formation of a gas-filled void that forms in a water column under tension; ultimately resulting in xylem dysfunction.

**Cell,** basic structural and functional unit within the plant body. A chamber or compartment that at some time contains a protoplast.

**Cell plate,** structure formed during telophase of mitosis between the two new nuclei and ultimately separating the protoplasts of contiguous cells; composed of microtubules.

**Cell wall,** the structurally compound, enclosing outer envelope of a plant cell; in mature cells it consists ontogenetically of different superimposed layers.

**Cellulose,** a polysaccharide composed of hundreds of glucose molecules linked in a chain and found in plant cell walls.

**Central cylinder,** the vascular system of an axis; the stele.

**Chambered crystalliferous cell,** a crystalliferous cell that is divided into compartments by septa.

**Check,** fissure in the secondary cell wall.

**Chimera,** cellular zones of different genetic or structural constitution in the same region of the plant.

**Chloroplast,** organelle in green plant cells that functions in photosynthesis.

**Chromoplast,** cellular organelle containing an abundance of yellow or orange carotenoid pigments.

**Circular bordered pit,** bordered pit having a circular outline to the pit border as seen in face view.

**Cisterna,** a flattened or saclike space between membranes of the endoplasmic reticulum and Golgi body.

**Closed vascular bundle,** vascular bundle lacking a vascular cambium and thus lacking the potentiality for secondary growth.

**Coalescent pit apertures,** slitlike inner pit apertures united into spiral grooves.

**Coarse-grained wood,** wood with wide growth rings.

**Collateral vascular bundle,** type of vascular bundle construction with phloem on one side of the xylem only.

**Collenchyma,** living supporting cell of the primary body; more or less elongated with an unevenly thickened primary wall.

**Colleter,** multicellular glandular structure, usually composed of an elongated axis bearing oval to very elongate glandular epidermal cells; secretes mucilages, gums, or resins.

**Columella cell,** cell in the central region of a root cap; characterized by occurring in orderly rows and containing many amyloplasts.

**Commissural vein,** small transverse vein interconnecting large parallel bundles in a monocot leaf and usually lying at some angle to them.

**Companion cell,** parenchymatous cell positioned next to a sieve tube element as a sister cell and arising by division of a sieve tube element mother cell; intimately connected with the sieve tube member and retains a nucleus and dense cytoplasm at maturity.

**Complementary tissue,** loosely organized cells composing a lenticel; suberized or unsuberized. (*Syn.* Filling tissue.)

**Compound middle lamella,** term of convenience for the compound layer between adjacent plant cells consisting of two primary walls and an intercellular layer (middle lamella); of varying thickness.

**Compound sieve plate,** sieve plate composed of two or more well-defined sieve areas.

**Compound stem,** an "anomalous" structural condition in which a central vascular cylinder is surrounded by three, five, or more cylinders (peripheral steles) that are separated by ground tissue. (*Syn.* Multistelar stem.)

**Compression wood,** reaction wood formed typically on the lower sides of branches and leaning or crooked stems of coniferous trees; characterized anatomically by heavily lignified tracheids that are rounded in transverse section and bear spiral wall checks.

**Concentric vascular bundle,** vascular bundle type with one vascular tissue surrounding the other.

**Confluent axial parenchyma,** aliform wood parenchyma forming irregular tangential or diagonal bands as seen in transverse section.

**Conjunctive tissue,** a special type of parenchyma associated with included phloem.

**Cork,** nontechnical term for phellem. An external, secondary tissue impermeable to water and gases; often formed in response to wounding or infection.

**Cork cambium,** meristematic layer that produces the periderm. (*Syn.* Phellogen.)

**Corpus,** mass of cells below the surface tunica in an angiosperm shoot apex that divide in various planes.

**Cortex,** region of primary tissue in the plant axis located between the vascular tissue and epidermis; composed predominantly of parenchyma.

**Cotyledon,** an embryonic leaf.

**Crassula(e),** the thicker, generally arching portion of the compound middle lamella above and below a bordered pit. (*Syn.* Bar(s) of Sanio.)

**Crista(e),** folded inner membrane of a mitochondrion.

**Cristarque cell,** sclereid with eccentrically thickened wall and centrally positioned crystal.

**Crystal,** solitary or compound and variously shaped mineral deposition; most commonly composed of calcium oxalate.

**Crystal sand,** a granular mass composed of very fine individual calcium oxalate crystals.

**Crystalliferous cell,** cell containing one or more crystals.

**Crystalliferous sclereid,** sclereid containing one or more crystals.

**Cross-dating,** the procedure of comparing wood annual ring sequences in ancient timbers with those in living trees to produce backward extending datable chronology.

**Cross field,** a term of convenience for the pitting formed by the walls of a ray cell and a neighboring axial tracheid, as seen in radial section. Used mainly for conifers.

**Cuticle,** a thin, waxy sheet of noncellular material that covers the outer surface of all parts of the primary shoot and reproductive structures; consists primarily of wax and cutin.

**Cutin,** biopolymer composed of fatty and hydroxy fatty acids that is a major chemical component of the cuticle.

**Cyclocytic stoma,** mature stoma surrounded by subsidiary cells forming one or two narrow rings around the guard cells; number of cells usually four or more.

**Cystolith,** internal stalked concretion that projects into the cell lumen and is composed of a peg of cellulose covered with calcium carbonate; can be variously shaped and sometimes completely fill a cell.

**Cytoplasm,** the nonnuclear contents of the protoplasm; enclosed within the plasma membrane.

**Dendrochronology,** discipline of investigating tree growth rings in order to interpret tree age and past climatic conditions.

**Derivative,** undifferentiated cell formed by the division of a meristematic initial.

**Dermal tissue,** outer primary or secondary covering tissue of the plant body.

**Dermatogen,** the cell or histogen in the root apex from which the epidermis is derived.

**Desmotubule,** proteinaceous tubular structure that extends through a plasmodesmata and is attached to the endoplasmic reticulum on either side.

**Determination,** the commitment of a cell to a single developmental pathway, from among several alternatives.

**Diacytic stoma,** mature stoma enclosed by a pair of subsidiary cells whose common wall is at right angles to the guard cells. (*Syn.* Caryophyllaceous or Cross-celled type.)

**Diaphragm,** transverse layer in the pith composed of firm-walled cells and alternating with regions of soft tissue; may collapse with age.

**Diarch,** structural condition in a root having primary xylem with two protoxylem poles.

**Dichotomous,** forking regularly by pairs.

**Dictyosome,** cellular organelle composed of a stack of flattened membranous vescicles; involved in cellular transport of materials and secretion.

**Differentiation,** the chemical and morphological changes associated with the developmental process in a cell or organism.

**Diffuse axial parenchyma,** single wood parenchyma strand or pairs of strands distributed irregularly among the fibrous elements.

**Diffuse in aggregates parenchyma,** axial wood parenchyma strands grouped into short discontinuous tangential or oblique lines among the fibrous elements.

**Diffuse porous wood,** wood in which the pores (vessels) are of fairly uniform size or only gradually changing in size and distribution throughout a growth ring.

**Diffuse sclereid,** sclereid dispersed in the leaf mesophyll without any relationship with a vein ending; occurring as solitary or grouped elements.

**Divided stem,** structural condition in which a vascular cylinder is radially divided into four or five major segments by parenchymatous tissue.

**Double (or mulitiple) annual ring,** an annual ring composed of two (or more) growth rings.

**Druse,** compound calcium oxalate crystal of more or less spherical shape, in which the many component crystals protrude from the surface giving the cluster a star-shaped appearance; either attached to the wall by a peg or lying free in the cell.

**Earlywood,** less dense, larger-celled, first-formed part of a growth ring. (*Syn.* Springwood.)

**Element,** general term used for an individual cell. Used in wood anatomy, particularly to distinguish between vessels and the individual cells of which they are composed.

**Embolism,** a partial vacuum in a tracheary element caused by an air bubble and producing an obstruction in the water column.

**End wall,** term of convenience for (a) a wall at right angles to the longitudinal axis of a parenchyma cell (i.e., for the tangential walls of ray cells or the transverse walls of an axial parenchyma cell) and (b) the oblique or transverse wall between two vessel elements.

**Endarch,** primary xylem strand in which the radial differentiation of the cells progresses centrifugally; that is, the oldest elements (protoxylem) are closest to the center of the axis; typical of stems in seed plants.

**Endodermis,** a layer of cells without intercellular spaces and forming the innermost zone of the cortex separating the cortex from vascular tissue; distinguished by a waxy Casparian strip; found in roots and only some stems and leaves.

**Endogenous,** produced within from deep-seated cells as contrasted with a superficial origin.

**Endoplasmic reticulum,** irregularly joined and flattened double membranes that permeate the cytoplasm and connect to the nuclear membrane; involved in cellular packaging and transport.

**Epicuticular wax,** wax deposition of variable morphology on the surface of aerial plant organs.

**Epidermis,** outermost layer of cells on the primary plant body; often with strongly thickened and cutinized outer walls; usually consisting of one cell layer, but sometimes of multiple layers.

**Epithelial cell,** cell of the epithelium.

**Epithelium,** layer of secretory parenchymatous cells that surround an intercellular canal or cavity.

**Eustele,** dissected stele in which individual vascular bundles are collateral or bicollateral in structure and arranged in a ring.

**Exarch,** primary xylem strand in which the radial differentiation of cells progresses centripetally; that is, the oldest elements (protoxylem) are farthest from the center of the axis. Typical of roots in seed plants.

**Exodermis,** outermost layer or layers of the cortex in some roots; sometimes characterized by Casparian bands.

**Extended pit aperture,** an inner pit aperture whose outline, in surface view, extends beyond the outline of the border.

**Extraxylary fiber,** fiber located in tissues other than the xylem.

**False Annual Ring,** one of the growth rings of a double, or multiple, annual ring.

**Fascicular cambium,** vascular cambium that develops between the primary xylem and phloem within a vascular bundle.

**Fascicular tissue,** vascular tissue.

**Fiber,** term of convenience for any long, narrow, usually thick-walled and lignified cell of the primary and secondary body. Often qualified as wood fibers or bast fibers. Used loosely for wood elements in general.

**Fiber tracheid,** a fiberlike tracheid; commonly thick-walled with a small lumen, pointed ends, and bordered pits having lenticular to slitlike apertures.

**Fibrous root,** root system composed of many fine, threadlike, or slender roots.

**Filiform sclereid,** much elongated, slender sclereid resembling a fiber.

**Fine-grained wood,** slow-growing wood with narrow growth rings.

**Fundamental tissue,** tissue of the plant body other than the epidermis and vascular tissues. (*Syn.* Ground tissue.)

**Fusiform initial,** cambial initial giving rise to an axial, or vertical, element of the secondary xylem or phloem; it is typically spindle shaped (fusiform) as seen in tangential section.

**Gelatinous fiber,** fiber having a more or less unlignified inner wall with a gelatinous appearance. (*Syn.* Mucilaginous fiber.)

**Girder,** sclerenchymatous connection on a leaf between a vascular bundle and the epidermis; characteristic of some monocots.

**Gland,** a secreting structure or surface.

**Glandular trichome,** trichome bearing a knob-like secretory swelling (head) at its tip.

**Granum(a),** series of stacked membranes within a chloroplast that contain photosynthetic pigments. Composed of individual membranes termed thylakoids.

**Ground meristem,** primary meristematic tissue that will give rise to the primary cortex and pith; composed of vacuolated cells.

**Growth layer,** a layer of wood or bark produced apparently during one growing period; frequently, especially in woods of the temperate zones, divisible into early and latewood or bark. (*Syn.* Growth ring, annual ring.)

**Growth ring,** see Growth layer.

**Growth Ring Boundary,** the outer limit of a growth ring.

**Gum,** complex polysaccharidal substances formed by cells in reaction to wounding or infection.

**Gum duct,** an intercellular canal containing gum.

**Half-bordered pit pair,** an intercellular pairing of a simple and bordered pit.

**Halophyte,** plant living on basic soil abundantly supplied with chloride ions.

**Heartwood,** inner layer of wood that, in the growing tree, has ceased to contain living cells and from which the reserve materials have been removed or converted to heartwood substances. It is generally darker in color than sapwood, though not always clearly differentiated.

**Herbaceous,** applies to a higher plant that does not develop woody tissues.

**Heterocellular ray,** wood ray composed of cells of different morphological type (i.e., in dicotyledons both procumbent and square or upright cells).

**Heterochrony,** displacement in time of ontogenetic appearance and development of one organ with respect to another.

**Heterogenous ray tissue,** wood ray tissue in which the individual rays are composed wholly or in part of square or upright cells.

**Heterophyllous,** with leaves of different sizes and/or shapes.

**Hexacytic stoma,** mature stoma resembling the tetracytic pattern but having an additional pair of lateral subsidiary cells.

**Histogen,** one of a group of vertically superimposed meristematic cells in the root apex from which primary tissues are derived; individual histogens are envisioned to give rise to the epidermis (dermatogen), cortex (periblem), stele (plerome), and root cap (calyptrogen.)

**Homocellular ray,** wood ray composed of cells of the same morphological type (i.e., all procumbent or all square or upright).

**Homogenous ray tissue,** wood ray tissue in which the individual rays are composed wholly of procumbent cells.

**Hydathode,** specialized structures at vein endings through which water is discharged under certain atmospheric conditions to the leaf surface.

**Hydrophyte,** plant growing in water, as free floating, submerged, floating and anchored, or emergent and anchored.

**Hypodermis,** subsurface or outermost layer or layers of cells that is not always distinguishable from the epidermis but is derived from the ground meristem or mesophyll; may be supportive or protective in function; when Casparian band is present, it is termed exodermis.

**Idioblast,** any cell that differs markedly in form and contents from other neighboring cells of the same tissue.

**Imperforate tracheary element,** water-conducting cell with bordered pits having intact pit membranes to congeneric elements; a tracheid.

**Included pit aperture,** an inner pit aperture whose outline, in surface view, is included within the outline of the border.

**Included phloem,** scattered or isolated phloem strands within the secondary xylem of certain dicotyledons. (*Syn.* Interxylary phloem.)

**Inclusion,** body occurring in the cytoplasm; nonprotoplasmic substance in the form of granules, droplets, crystals.

**Initial,** actively dividing cell of a meristem.

**Inner pit aperture,** the opening of the pit canal into the cell lumen.

**Intercalary meristem,** meristematic tissue separated from the apical meristem in the primary body by more or less mature tissues.

**Intercellular canal,** tubular intercellular duct of indeterminate length and surrounded by an epithelium; generally containing secondary plant products such as resins and gums.

**Intercellular layer,** layer between adjacent cells; it is isotropic and lacks cellulose. (*Syn.* Middle lamella.)

**Intercellular space,** space between adjacent cells.

**Interfascicular cambium,** vascular cambium that arises in the parenchymatous region between the primary vascular bundles.

**Internode,** section or region of stem between nodes.

**Internal phloem,** primary phloem positioned internal to the primary xylem. (*Syn.* Intraxylary phloem.)

**Intraxylary phloem,** phloem positioned next to the pith. (*Syn.* Internal phloem.)

**Isobilateral,** leaf construction in which palisade mesophyll occurs on both the upper (adaxial) and lower (abaxial) sides of the lamina.

**Kranz structure,** structure condition of foliar veins surrounded by conspicuously enlarged parenchymatous bundle sheath cells containing chloroplasts.

**Lacuna,** an air space in a tissue.

**Lateral meristem,** meristem in a lateral position on the axis that produces secondary tissues; includes vascular and cork cambia.

**Lateral sieve area,** differentiated sieve area on the lateral wall of a sieve element.

**Latewood,** denser, smaller-celled, later-formed part of a growth ring. (*Syn.* Summerwood, Autumnwood.)

**Latex,** an often milky fluid formed in latex tubes and containing various substances, including starch, proteinaceous enzymes, lipids, alkaloids, crystals, and rubber.

**Latex tube,** tube of modified cells or a series of cells containing latex. (*Syn.* Latex canal.)

**Laticifer,** general term for a cell containing latex; may be a single cell or a series of tubular cells.

**Leaf gap,** parenchymatous interruption in the cylinder of stem conducting tissues at the level where a leaf trace connects with the stele.

**Leaf trace,** vascular connecting strand from leaf base to stem bundle; inclusively, the complex of all the bundles that supply a leaf.

**Lenticel,** specialized portion of the periderm, variously shaped but often lenticular, consisting of loosely arranged cells that are never more than slightly suberized; serving for the exchange of gases through the otherwise impermeable periderm.

**Libriform fiber,** an elongated, commonly thick-walled cell of the secondary xylem with simple pits.

**Lignin,** complex polymer that provides strength and rigidity to the secondary cell wall; a principal component of wood cell walls.

**Lithocyst,** an enlarged cell containing a cystolith.

**Lumen,** the cell cavity or space resulting from breakdown of cytoplasm and bounded by the cell wall.

**Lysigenous space,** space in a tissue formed by the disintegration, dissolution, or death of cells.

**Macrosclereid,** columnar sclerenchymatous element. (*Syn.* Rod cell.)

**Meristem,** collection of cells capable of active cell division, thereby adding new cells to the plant body and maintaining itself as a dividing cell system; embryonic tissue.

**Mesarch,** primary xylem strand in which the radial differentiation of cells begins in the center and then proceeds both centripetally and centrifugally; that is, the oldest elements (protoxylem) are in the center of the strand.

**Mesogenous stoma,** stoma in which guard cells and subsidiary cells arise from a common initial.

**Mesophyll,** all of the internal, usually parenchymatous tissue (excluding the vascular bundles) located between the epidermal layers of a leaf.

**Mesophyte,** plant growing under medium moisture conditions.

**Mestome,** the inner bundle sheath of a two-layered leaf vascular bundle sheath.

**Metaxylem,** later-formed primary xylem; typically composed of pitted tracheary elements.

**Microfibril,** smallest, threadlike structural component of cell walls; composed of cellulose polymer chains and associated polysaccharides of other types; visible with the electron microscope.

**Microfilament,** a fine proteinaceous filament composed of actin and myosin; capable of contraction when supplied with ATP.

**Microtubule,** thin, hollow proteinaceous threadlike filament in the cytoplasm of eukaryotic cells.

**Middle lamella,** thin, usually pectic wall layer between adjacent cells; it is isotropic and generally lacks cellulose. In woody tissues, pectin is replaced by lignin. Special techniques may be required to distinguish it. (*Syn.* Intercellular cementing layer.)

**Mitochondrion,** double-membraned cellular organelle in which aerobic respiration occurs in eukaryotic cells.

**Mucilage cell,** specialized parenchymatous cell containing mucilage; typically rounded in outline and enlarged; not always possible to distinguish oil cells and mucilage cells by their appearance.

**Multilacunar node,** structural condition having five or more leaf gaps at the node.

**Multiple epidermis,** horizontally divided multiple epidermal layers that are derived from the division of protodermal or leaf epidermal cells. See also *Velamen*.

**Multiseriate ray,** wood ray two or more cells wide as viewed in tangential section.

**Myrosin cell,** internal secretory cell containing myrosin.

**Nectary,** secretory structure associated with the secretion of sugar-containing solutions; positioned on the flower (floral nectary) or vegetative organs (extrafloral nectary).

**Node,** region of stem from which a leaf, leaves, or branches arise.

**Nonarticulated laticifer,** intrusively growing, very elongated, and often branched coenocytic cell containing latex.

**Nuclear membrane,** membrane surrounding the nucleus in eukaryotic cells. (*Syn.* Nuclear envelope.)

**Nucleolus,** specialized, more or less spherical body within the nucleus; contains RNA.

**Nucleus,** spherical protoplasmic organelle surrounded by a double nuclear membrane and containing the chromosomes of a eukaryotic cell; governs cell metabolism and hereditary features.

**Oil cell,** specialized cell of the parenchyma containing oil; typically, but not always, rounded in outline and enlarged.

**Open vascular bundle,** vascular bundle possessing a vascular cambium and thus having the potentiality for secondary growth.

**Opposite pitting,** multiseriate intervascular pitting in which the pits are arranged in horizontal pairs or in short horizontal rows; when the pits are crowded, the outlines of the borders tend to become rectangular in outline.

**Organelle,** subcellular body that performs a particular physiological function (or functions).

**Osteosclereid,** columnar sclerenchymatous cell with dilated, lobed, or ramified ends.

**Outer pit aperture,** the opening of the pit canal into the pit chamber.

**Palisade,** region of leaf mesophyll in which the elongate and parallel chlorenchymatous cells are vertically oriented below the upper, or less commonly both upper and lower, epidermal layer.

**Paracytic stoma,** mature stoma accompanied on either side by one or more subsidiary cells parallel to the long axis of the pore and guard cells. (*Syn.* Rubiaceous or parallel-celled type.)

**Paradermal section,** a section made parallel with the expanded surface of an organ, such as a leaf blade.

**Paratracheal axial parenchyma,** axial wood parenchyma associated with or in contact with the vessels or vascular tracheids.

**Paraveinal mesophyll,** region of leaf mesophyll between the palisade and spongy regions composed of cells with conspicuously extended arms and that extend horizontally between the veins.

**Parenchyma,** living, nucleated, usually thin-walled cell that is isodiametric, multisided, or brick shaped, with simple pits or pit fields; primarily concerned with metabolism, storage, and distribution of food materials. Termed wood parenchyma or xylem parenchyma if occuring in the xylem, and phloem parenchyma if in the phloem. (*Syn.* Soft tissue, storage tissue.)

**Passage cell,** thin-walled endodermal cell possessing a Casparian strip and associated with endodermal cells having thick secondary walls; commonly lies on the same radius as a protoxylem pole.

**Pectin,** a methylated polymer of galacturonic acid found in the middle lamella and the primary wall of plants.

**Perforate tracheary element,** cell involved in water conduction with conspicuous openings on the end walls from one element to another; a vessel element.

**Perforation plate,** term of convenience for the area of the wall (originally imperforate) involved in the coalescence of two vessel elements.

**Periblem,** a root apical histogen from which the cortex is derived.

**Periclinal division,** cell division parallel to the organ surface; division of a cambial initial in the tangential longitudinal plane.

**Pericycle,** parenchymatous layer lying adjacent to the endodermis and forming the outermost region of the stele; may become multilayered.

**Periderm,** outer protective layers produced by the phellogen that replace the epidermis as the impermeable covering of older stems and roots; consisting of phellem (cork), phellogen, and phelloderm.

**Perigenous stoma,** stoma in which the guard cells arise from a single mother cell and the surrounding subsidiary cells arise independently.

**Peripheral cell,** surface cell of a root cap; secretes mucilage that assists the root's movement through the soil; sloughed off during growth.

**Perivascular fiber,** fiber located on the outer periphery of the vascular region in stems; originating outside the primary phloem.

**Phellem,** tissue produced externally by the phellogen in a stem or root; cell walls are generally suberized, and, in thick-walled kinds, there may be additional lignified layers toward the cell lumen; cork.

**Phelloderm,** living tissue produced internally by the phellogen and that generally resembles cortical parenchyma in appearance.

**Phellogen,** a lateral meristematic layer that produces the periderm. (*Syn.* Cork cambium.)

**Phi layer,** a layer of cells in the root immediately outside the endodermis and characterized by elements with lignified wall deposits on the radial walls.

**Phloem,** complex food-conducting tissue of vascular plants. It occurs both as primary and secondary tissue and is usually, but not invariably, associated with xylem. The basic cell types of which it is composed are sieve elements, parenchyma cells, fibers, and sclereids.

**Phloem protein,** a striated, fibrous protein cytoplasmic component of a sieve tube element.

**Phloem ray,** ribbonlike aggregate of cells extending radially in the secondary phloem; derived from ray initials of the vascular cambium.

**Phyllotaxis,** arrangement of leaves on the stem.

**Pit,** recess in the secondary wall of a cell, together with its external closing membrane; opens internally to the lumen.

**Pit aperture,** opening or mouth of a pit.

**Pit border,** the overarching part of the secondary cell wall of a bordered pit.

**Pit canal,** passage from the cell lumen to the chamber of any bordered pit or simple pit with thick walls.

**Pit cavity,** entire space within a pit from the membrane to the lumen.

**Pit chamber,** space between the pit membrane and the overarching pit border of a bordered pit.

**Pit membrane,** that part of the compound middle lamella that limits a pit cavity externally.

**Pit pair,** two complementary pits of adjacent cells.

**Pith,** central core in the stem and occasionally in the root; consisting chiefly of parenchymatous or soft tissue.

**Pith fleck,** aggregation of parenchymatous wound tissue within the wood.

**Plasma membrane,** outermost cytoplasmic membrane surrounding the entire protoplasm and lying adjacent to the cell wall. (*Syn.* Plasmalemma.)

**Plasmodesmata,** pore and associated delicate strand of cytoplasm that traverses the primary cell wall and connects the protoplasm of two adjacent cells.

**Polyarch,** structural condition in a root having xylem with many protoxylem groups.

**Polysaccharide,** a large organic molecule composed of many units of simple sugar.

**Pore,** a term of convenience for the cross section of a vessel. Used primarily in wood anatomy.

**Pore chain,** series or line of adjacent solitary pores in wood.

**Pore cluster,** cluster of two or more pores crowded together in an irregular grouping in wood.

**Primary body,** that part of the plant body derived from the activity of the shoot and root apical meristems and their derivative meristematic tissues.

**Primary endodermis,** endodermis composed of cells having a narrow to broad Casparian band of lignin and suberin encircling the radial and transverse cell walls.

**Primary meristematic tissue,** meristematic tissue derived from the shoot or root apical meristem; protoderm, ground meristem, and procambium.

**Primary phloem,** phloem of the primary plant body; in stem and root it is differentiated below or behind the apical meristem.

**Primary pit field,** thinner area of the compound middle lamella within the limits of which one or more pit pairs usually develop.

**Primary thickening meristem,** meristem located near the shoot apex of virtually all monocotyledons and responsible for an increase in diameter of the shoot axis.

**Primary tissue,** tissue differentiating from the derivatives of the shoot and root apical meristems and their derivative meristematic tissues.

**Primary wall,** wall of the meristematic cell modified during differentiation to form the first anisotropic layer of the cell.

**Primary xylem,** xylem of the primary plant body; it is differentiated behind the apical meristem.

**Primordium,** an organ, a cell, or an organized series of cells in their earliest stage of development.

**Prismatic crystal,** solitary rhombohedral or octahedral crystal composed of calcium oxalate; birefringent under polarized light.

**Procambium,** undifferentiated primary vascular tissue; a primary meristematic tissue composed of densely protoplasmic cells. (*Syn.* Procambial strand.)

**Proembryo,** early stage of embryo development prior to formation of the main embryo body and suspensor.

**Prosenchyma,** a general descriptive term for elongated cells with tapering ends; used as opposed to parenchymatous in shape.

**Protective layer,** layer of cells composing the abscission zone; divide to form a suberized covering and protecting zone over the leaf, branch, or fruit scar.

**Protoderm,** undifferentiated surface cell layer of the primary plant body; a primary meristematic tissue; precursor to epidermis.

**Protoplast,** the mass of protoplasm enclosed by a cell wall and plasma membrane.

**Protoxylem,** first-formed primary xylem, composed of tracheary elements with annular or spiral secondary wall thickenings.

**Provascular strand,** see *Procambium.*

**Quiescent center,** a mitotically inactive region, or region with low mitotic activity, in the root or stem apical meristem.

**Ramiform pit,** simple pit with coalescent, canal-like cavities, as in some sclereids.

**Radial pole multiple,** group of two or more wood pores crowded together in radial files with flattened tangential walls between them.

**Radial section,** longitudinal section of wood cut in a plane of a radius of the stem. (*Syn.* Quarter sawn.)

**Raphide(s),** needle-shaped crystal occurring as one of a closely packed, sheathlike bundle.

**Ray initial,** cambial initial giving rise to a ray cell; usually one of a group and often more or less isodiametric as seen in tangential section.

**Ray tracheid,** tracheid forming part of a ray; occurring at the margins (tops and bottoms) of some softwood rays; characterized by the presence of bordered pits.

**Ray-vessel pitting,** pitting between a ray cell and a vessel member.

**Reaction wood,** wood with more or less distinctive anatomical characters, formed typically in parts of leaning or crooked stems and in branches; tends to restore the original position. In dicotyledons this consists of tension wood and in conifers of compression wood.

**Rhytidome,** phellem and tissues isolated by it; often enclosing pockets of cortical or phloem tissues. A technical term for the outer bark. The rhytidome may be shed or leave a smooth trunk, or be retained as a thick, fibrous or corky layer.

**Ribosome,** subcellular, nonmembranous organelle that contains RNA and protein and is involved in protein synthesis.

**Ring porous wood,** wood in which the pores of the earlywood are distinctly larger than those of the latewood and form a well-defined zone or ring.

**Rootcap,** collection of cells that covers and protects the root apical meristem and that assists the root's movement through the soil.

**Sapwood,** portion of wood in the living tree that contains living cells and reserve materials.

**Scalariform intervessel pitting,** elongated or linear intervessel pits arranged in a ladderlike series.

**Scalariform perforation plate,** perforation plate with multiple, ladderlike openings that are elongated and parallel; the remnants of the plate between the openings are called bars.

**Scanty axial parenchyma,** wood parenchyma distributed as incomplete sheaths or occasional parenchyma cells around the vessels.

**Schizogenous cavity,** space or lacuna formed by the separation of cells caused by the splitting of the common wall between adjacent cells or dissolution of pectic substances and plasmodesmata.

**Sclereid,** a cell that is usually not markedly elongate but that has a thick, typically lignified secondary wall and that typically lacks a protoplast when mature. (*Syn.* Sclerotic cell.)

**Sclerenchyma,** cell of variable shape and provided with a secondarily thickened, typically lignified wall; generally, but not invariably, devoid of protoplast at maturity; often supportive or protective in function.

**Sclerophyllous,** leaf that is leathery and tough; often resulting in desiccation resistance.

**Secondary body,** part of the plant body derived from the activity of the lateral meristems; composed of secondary tissues.

**Secondary endodermis,** endodermis composed of cells in which a thin suberin layer has been deposited inside the entire primary wall with Casparian band.

**Secondary growth,** growth resulting from the activity of the vascular and cork cambia; results in an increase in axis diameter.

**Secondary phloem,** phloem tissue formed by the activity of the vascular cambium; part of the bark.

**Secondary thickening meristem,** lateral meristem in some monocotyledons that is removed from the shoot apex and produces chains of secondary vascular bundles in a parenchymatous ground tissue.

**Secondary tissue,** tissue derived from activity of a lateral meristem or cambium.

**Secondary wall,** a strongly anisotropic wall formed inside the primary wall of some cells.

**Secondary xylem,** xylem or wood produced by a vascular cambium.

**Secretory cell,** parenchymatous idioblast filled with oil or mucilage.

**Secretory duct,** tubular intercellular canal surrounded by an epithelium.

**Secretory structure,** an external or internal structure that produces a secretion.

**Septate fiber,** fiber with thin transverse walls across the lumen.

**Sheath cell,** upright wood ray cell located along the side of a broad ray, as viewed in tangential section, that is larger than cells of the central part of the ray one of a series of cells forming a sheath around a multiseriate ray.

**Sieve area,** depressed area in the wall of a sieve element, perforated by a sievelike cluster of minute pores through which the protoplast is connected with that of a contiguous sieve element.

**Sieve cell,** long, slender, primitive conducting cell of the phloem that does not form a constituent element of a sieve tube but that is provided with relatively unspecialized sieve areas, especially in the tapering ends of the cell that overlap those of other sieve cells.

**Sieve element,** cell in the phloem tissue that functions principally in the longitudinal conduction of organic solutes.

**Sieve plate,** specialized part of the end wall of a sieve tube element that has a solitary sieve area (simple sieve plate) or several closely spaced sieve areas often arranged in a scalariform or reticulate manner (compound sieve plate).

**Sieve tube,** food-conducting tube of the phloem made up of an axial series of sieve tube elements.

**Sieve tube element,** long conducting cell of the phloem of flowering plants that forms one of an axial series of such cells arranged end to end to form a sieve tube, the common walls, which may be inclined or transverse, being sieve plates; sometimes with additional, less specialized sieve areas elsewhere in the walls. (*Syn.* Sieve tube member.)

**Silica body,** spheroidal or irregularly shaped particles composed of silicon dioxide. (*Syn.* Silica grain, Silica inclusion.)

**Simple perforation plate,** single and usually large and more or less rounded opening in the perforation plate of a vessel element.

**Simple pit,** pit in which the cavity becomes wider or remains of constant width or only gradually narrows during the growth in thickness of the secondary wall.

**Simple sieve plate,** sieve plate composed of a single sieve area.

**Simple stem,** structural condition of a single stele in the stem. (*Syn.* Undivided stem.)

**Solitary vessel,** wood vessel completely surrounded by xylem ground cells.

**Spiral grain,** wood in which axial elements are interwoven or are not constantly aligned in one general direction. (*Syn.* Interlocked grain, Cross grain.)

**Spiral thickenings,** helical ridges on the inner face of, and part of, the secondary cell wall of secondary xylem tracheary elements or fibers.

**Spongy mesophyll,** region of leaf mesophyll composed of chlorenchyma cells that are irregular in shape and separated from one another by an extensive system of intercellular air spaces; usually in an abaxial position within the leaf lamina.

**Springwood,** see Earlywood.

**Starch sheath,** innermost layer(s) of the stem cortex when it contains a conspicuous accumulation of starch.

**Stegmata,** small, isodiametric silica cell that contains a single small silica body; typically occurs in longitudinal files adjacent to vascular and fibrous bundles of the stem and leaf of some monocots.

**Stele,** central column of primary vascular tissues and associated ground tissues of the stem and root; located inside the cortex.

**Stoma,** a pair of guard cells together with the aperture between them.

**Storied cambium,** cambium structure in which the initials are arranged in horizontal series or seriation on tangential surfaces. (*Syn.* Stratified cambium.)

**Straight grain,** wood in which the axial elements extend parallel to the long axis of the tree.

**Strand,** sclerenchymatous extension in the leaf lamina free from a vascular bundle; may be subepidermal or occur anywhere in the mesophyll. Characteristic of some monocots.

**Stratified cambium,** cambium structure characterized by the arrangement of initials in horizontal series as viewed on the tangential surface. (*Syn.* Storied cambium.)

**Styloid,** large, elongated crystal, typically about four times as long as broad, with pointed or square ends.

**Suberin,** waxy substance that impregnates cell walls and is especially characteristic of cork cells and the suberin lamella of the endodermis.

**Subsidiary cell,** an epidermal cell lying next to a guard cell and differing structurally from other epidermal cells in size, shape, or contents.

**Successive cambia,** type of anomalous growth in which a series of supernumerary cambia arise and are arranged approximately concentrically and produce several increments of secondary tissues.

**Summerwood,** see *Latewood*.

**Suspensor,** superimposed group of cells that are derived from the zygote and located at the base of an embryo; not part of the embryo proper and function to push the embryo into nutritive tissue of the seed.

**Symplast,** sum total of the interconnected protoplasm of a plant.

**Sympodium(a),** axial stem vascular bundle and its associated leaf and branch traces that compose an open primary vascular system.

**Tangential collenchyma,** living cell of the stem and leaf in which the wall is thickened mainly on the inner and outer tangential walls.

**Tangential section,** longitudinal cut or section of wood that is at right angles to the rays. (*Syn.* Slab cut.)

**Tap root,** persistent, well-developed primary root.

**Tension wood,** reaction wood formed typically on the upper sides of branches and leaning or crooked stems of dicotyledonous trees and characterized anatomically by a lack of cell wall lignification and often by the presence of an internal gelatinous wall layer.

**Terminal sclereid,** sclereid positioned at a foliar veinlet ending.

**Tertiary endodermis,** endodermis composed of cells in which a secondary wall has been deposited over the suberin wall layer; the added wall material may be uniformly or asymmetrically thickened and be lignified or unlignified.

**Tetracytic stoma,** mature stoma surrounded by four subsidiary cells, two of them being parallel to the guard cells and the other pair being polar and often smaller.

**Tetrarch,** structural condition in a root having xylem with four protoxylem poles.

**Thylakoid,** internal chloroplast membrane with chlorophyll pigments; typically occurring as an individual membranous component of a granum.

**Tile cell,** special type of apparently empty upright ray cell of approximately the same height as the procumbent ray cells and occurring in indeterminate horizontal series usually interspersed among the procumbent cells.

**Tissue,** a collection of cells of common origin having essentially the same structure and performing the same functions.

**Torus,** a central, thicker part of a bordered pit membrane.

**Tracheid,** imperforate water-conducting cell with bordered pits to cogeneric elements.

**Tracheary element,** principal water-conducting cell of the xylem; includes tracheids and vessel elements.

**Transfer cell,** cell with irregular ingrowths from the cell wall, providing an extensive interface between the wall and the protoplast to facilitate a more efficient apoplastic movement of solutes.

**Translocation,** movement of water and nutrients through the plant.

**Transverse section,** section cut at right angles or across the long axis of an organ.

**Traumatic parenchyma,** see *Wound parenchyma.*

**Traumatic ring,** a zone of traumatic tissue produced by a cambium that has been injured. Common causes are frost, drought and fire. (*Syn.* Frost ring, Drought ring.)

**Triarch,** structural condition in a root having xylem with three protoxylem poles.

**Trichoblast,** root epidermal cell that forms a root hair.

**Trichome,** epidermal outgrowth of diverse form, structure, and function; often integrading with surface emergences of subdermal origin.

**Trichosclereid,** type of branched sclereid with thin hairlike branches extending into intercellular spaces.

**Trilacunar node,** structural condition having three leaf gaps at a node.

**Tunica,** zone composed of one to several peripheral layers of the cells at the angiosperm shoot apex, within which divisions are predominantly anticlinal.

**Tylosis(ses),** outgrowth from an adjacent ray or axial parenchyma cell that extends through a pit cavity in a vessel wall and enters a vessel element, partially or completely blocking the vessel lumen; may be thin- or thick-walled.

**Unilacunar node,** structural condition having one leaf gap at a node.

**Uniseriate ray,** ray one cell wide as seen on tangential section.

**Unstratified cambium,** vascular cambium structure in which the initials are not arranged in horizontal series as viewed on the tangential surface. (*Syn.* Nonstoried cambium.)

**Upright ray cell,** ray cell with its longest axis vertical.

**Vacuole,** a membrane-bound, fluid-filled cavity in the cytoplasm.

**Vascular bundle,** strandlike association of primary xylem and phloem that extends throughout the plant body.

**Vascular cambium,** lateral meristem responsible for the production of secondary xylem and secondary phloem.

**Vascular tissue,** specialized conducting tissue composed of xylem and phloem.

**Vasicentric axial parenchyma,** paratracheal wood parenchyma forming a complete sheath around a vessel; of variable width and circular or slightly oval in transverse outline.

**Vasicentric tracheid,** short, irregularly formed tracheid in the immediate proximity of a wood vessel and not forming part of a definite axial row.

**Velamen,** multiple epidermis covering the aerial roots of some tropical epiphytic orchids and aroids.

**Vessel,** an axial series of cells (vessel elements) that have coalesced to form a tubelike structure of indeterminate length; the lateral wall pits of cogeneric elements are bordered.

**Vessel element,** one of the cellular, perforate components of a vessel. (*Syn.* Vessel member.)

**Vestured pit,** bordered pit with the pit cavity wholly or partially lined with small projections from the cell wall.

**Wavy grain,** wood in which the elements undulate in a regular manner without successive layers crossing.

**Wood,** secondarily formed, strengthening, and principal water conducting tissue of stems and roots. (*Syn.* Xylem.)

**Wood density,** mass of wood per unit of volume.

**Wood figure,** general term referring to any design or distinctive markings on the longitudinal surface of wood.

**Wood grain,** predominant arrangement and direction of alignment of wood cells.

**Wood ray,** ribbonlike aggregate of cells extending radially in the secondary xylem; derived from the ray initials of the vascular cambium.

**Wood specific gravity,** ratio of the oven-dry weight of a piece of wood to the weight of the water displaced by the wood at a given moisture content.

**Wood texture,** term referring to the size and proportional amount of the elements in wood.

**Wound parenchyma,** parenchyma cells of irregular size, shape, and distribution resulting from injury to the cambium. (*Syn.* Traumatic parenchyma.)

**Xerophyte,** plant growing under dry conditions.

**Xylary fiber,** fiber forming the ground tissue of wood.

**Xylem,** principal water-conducting tissue of the primary and secondary plant body; a complex tissue composed of tracheids, vessel elements, fibers, and parenchyma.

# INDEX